Biological Oceanography

Biological Oceanography

Edited by **Suzy Bullock**

SYRAWOOD
PUBLISHING HOUSE
New York

Published by Syrawood Publishing House,
750 Third Avenue, 9th Floor,
New York, NY 10017, USA
www.syrawoodpublishinghouse.com

Biological Oceanography
Edited by Suzy Bullock

International Standard Book Number: 978-1-68286-358-9 (Hardback)

Printed in China.

Contents

	Preface	**VII**
Chapter 1	**Modelling of underwater light fields in turbid and eutrophic waters: application and validation with experimental data** B. Sundarabalan and P. Shanmugam	**1**
Chapter 2	**Phytoplankton blooms on the western shelf of Tasmania: evidence of a highly productive ecosystem** J. Kämpf	**21**
Chapter 3	**Thermohaline properties in the Eastern Mediterranean in the last three decades: is the basin returning to the pre-EMT situation?** V. Cardin, G. Civitarese, D. Hainbucher, M. Bensi and A. Rubino	**32**
Chapter 4	**A parameter model of gas exchange for the seasonal sea ice zone** B. Loose, W. R. McGillis, D. Perovich, C. J. Zappa and P. Schlosser	**46**
Chapter 5	**Improved sea level record over the satellite altimetry era (1993–2010) from the Climate Change Initiative project** M. Ablain, A. Cazenave, G. Larnicol, M. Balmaseda, P. Cipollini, Y. Faugère, M. J. Fernandes, O. Henry, J. A. Johannessen, P. Knudsen, O. Andersen, J. Legeais, B. Meyssignac, N. Picot, M. Roca, S. Rudenko, M. G. Scharffenberg, D. Stammer, G. Timms and J. Benveniste	**58**
Chapter 6	**An alternative method for correcting fluorescence quenching** L. Biermann, C. Guinet, M. Bester, A. Brierley and L. Boehme	**71**
Chapter 7	**The CO_2 system in the Mediterranean Sea: a basin wide perspective** M. Álvarez, H. Sanleón-Bartolomé, T. Tanhua, L. Mintrop, A. Luchetta, C. Cantoni, K. Schroeder and G. Civitarese	**83**
Chapter 8	**Temporal variations of zooplankton biomass in the Ligurian Sea inferred from long time series of ADCP data** R. Bozzano, E. Fanelli, S. Pensieri, P. Picco and M. E. Schiano	**107**
Chapter 9	**Is coccolithophore distribution in the Mediterranean Sea related to seawater carbonate chemistry?** A. Oviedo, P. Ziveri, M. Álvarez and T. Tanhua	**120**

Chapter 10 **Using empirical orthogonal functions derived from remote-sensing reflectance for the prediction of phytoplankton pigment concentrations** 140
A. Bracher, M. H. Taylor, B. Taylor, T. Dinter, R. Röttgers and F. Steinmetz

Chapter 11 **Fate of colloids during estuarine mixing in the arctic** 160
O. S. Pokrovsky, L. S. Shirokova, J. Viers, V. V. Gordeev, V. P. Shevchenko, A. V. Chupakov, T. Y. Vorobieva, F. Candaudap, C. Causserand, A. Lanzanova and C. Zouiten

Chapter 12 **Turbulent dispersion properties from a model simulation of the western Mediterranean Sea** 179
H. Nefzi, D. Elhmaidi and X. Carton

Chapter 13 **Secchi depth in the Oslofjord–Skagerrak area: theory, experiments and relationships to other quantities** 188
E. Aas, J. Høkedal and K. Sørensen

Chapter 14 **Adapting to life: ocean biogeochemical modelling and adaptive remeshing** 211
J. Hill, E. E. Popova, D. A. Ham, M. D. Piggott and M. Srokosz

Chapter 15 **The role of subpolar deep water formation and Nordic Seas overflows in simulated multidecadal variability of the Atlantic meridional overturning circulation** 232
K. Lohmann, J. H. Jungclaus, D. Matei, J. Mignot, M. Menary, H. R. Langehaug, J. Ba, Y. Gao, O. H. Otterå, W. Park and S. Lorenz

Chapter 16 **Interannual correlations between sea surface temperature and concentration of chlorophyll pigment off Punta Eugenia, Baja California, during different remote forcing conditions** 247
H. Herrera-Cervantes, S. E. Lluch-Cota, D. B. Lluch-Cota and G. Gutiérrez-de-Velasco

Chapter 17 **Sources of 21st century regional sea-level rise along the coast of northwest Europe** 258
T. Howard, A. K. Pardaens, J. L. Bamber, J. Ridley, G. Spada, R. T. W. L. Hurkmans, J. A. Lowe and D. Vaughan

Permissions

List of Contributors

Preface

This book was inspired by the evolution of our times; to answer the curiosity of inquisitive minds. Many developments have occurred across the globe in the recent past which has transformed the progress in the field.

Oceanography is a broad discipline comprising the study of the varied aspects of the ocean such as marine organisms, fluid dynamics, ocean currents, geology of the ocean bed, etc. This book aims to study these wide categories that fall under this discipline. It brings forth researches from across the globe on ocean physics and chemistry, air-sea interactions, various types of ocean models, etc. This book is a ripe text for students and professionals engaged in the field of oceanography and associated disciplines.

This book was developed from a mere concept to drafts to chapters and finally compiled together as a complete text to benefit the readers across all nations. To ensure the quality of the content we instilled two significant steps in our procedure. The first was to appoint an editorial team that would verify the data and statistics provided in the book and also select the most appropriate and valuable contributions from the plentiful contributions we received from authors worldwide. The next step was to appoint an expert of the topic as the Editor-in-Chief, who would head the project and finally make the necessary amendments and modifications to make the text reader-friendly. I was then commissioned to examine all the material to present the topics in the most comprehensible and productive format.

I would like to take this opportunity to thank all the contributing authors who were supportive enough to contribute their time and knowledge to this project. I also wish to convey my regards to my family who have been extremely supportive during the entire project.

Editor

Modelling of underwater light fields in turbid and eutrophic waters: application and validation with experimental data

B. Sundarabalan and P. Shanmugam

Ocean Optics and Imaging Laboratory, Department of Ocean Engineering, Indian Institute of Technology Madras, Chennai-600036, India

Correspondence to: P. Shanmugam (pshanmugam@iitm.ac.in)

Abstract. A reliable radiative transfer (RT) model is an essential and indispensable tool for understanding the radiative transfer processes in homogenous and layered waters, analyzing measurements made by radiance sensors and developing remote-sensing algorithms to derive meaningful physical quantities and biogeochemical variables in turbid and productive coastal waters. Existing radiative transfer models have been designed to be applicable to either homogenous waters or inhomogeneous waters. To overcome such constraints associated with these models, this study presents a radiative transfer model that treats a homogenous layer as a diffuse part and an inhomogeneous layer as a direct part in the water column and combines these two parts appropriately in order to generate more reliable underwater light-field data such as upwelling radiance (L_u), downwelling irradiance (E_d) and upwelling irradiance (E_u). The diffuse model assumes the inherent optical properties (IOPs) to be vertically continuous and the light fields to exponentially decrease with depth, whereas the direct part considers the water column to be vertically inhomogeneous (layer-by-layer phenomena) with the vertically varying phase function. The surface and bottom boundary conditions, source function due to chlorophyll and solar incident geometry are also included in the present RT model. The performance of this model is assessed in a variety of waters (clear, turbid and eutrophic) using the measured radiometric data. The present model shows an advantage in terms of producing accurate L_u, E_d and E_u profiles (in spatial domain) in different waters determined by both homogenous and inhomogeneous conditions. The feasibility of predicting these underwater light fields based on the remotely estimated IOP data is also examined using the present RT model. For this application, vertical profiles of the water constituents and IOPs are estimated by empirical models based on our in situ data. The present RT model generates L_u, E_d and E_u spectra closely consistent with the measured data. These results lead to a conclusion that the present RT model is a viable alternative to existing RT models and has an important implication for remote sensing of optically complex waters.

1 Introduction

Knowledge of the transmission and distribution of light fields within the water body is essential for the solution of many problems in optical remote sensing, underwater visibility, underwater imaging, underwater communication and naval operations. In the past decades, several radiative transfer (RT) models have been developed to compute the reflectance and transmittance of direct and diffuse solar fluxes at the ocean surface and in the water column. For instance, Kirk (1981) presented the Monte Carlo simulation scheme for studying the radiative transfer processes in the ocean and other natural waters. Stamnes et al. (1988) summarised an advanced and thoroughly documented discrete ordinate method (DISORT) for time-independent radiative transfer calculations in vertically inhomogeneous, non-isothermal, plane-parallel media. Mobley (1994) developed the HydroLight software, which is a radiative transfer numerical model based on the invariant imbedding technique that computes spectral radiance distributions within and leaving the natural water bodies. Haltrin developed a method for estimating the underwater light-field parameters in the homogeneous water column illuminated by the direct sun light and sky light (Haltrin and Kattawar,

1993; Haltrin, 1998a, b). Lee and Liou (2007) developed a radiative transfer model for a coupled atmosphere–ocean system using the analytic four-stream approximation. Hollstein and Fischer (2012) provided radiative transfer solutions for coupled atmosphere–ocean systems using the matrix operator technique. These RT models developed based on numerical as well as analytical solutions perform well in clear oceanic waters but have limitations in turbid coastal and productive waters. The key problems associated with some of the above models include the assumption of flat or randomly chosen slope of the sea surface, the treatment of material reflectance instead of the effective bottom reflectance (taking into account the material reflectance and configuration of the seabed), the constant phase function along the depth and the inadequate source function (especially for turbid and productive waters often optically shallow, vertically stratified, or vertically mixed) which introduce significant errors in the simulated underwater light fields (Sundarabalan et al., 2013). Conversely, the radiative transfer models developed for a inhomogeneous medium do not account for diffuse radiance in the water column, where the influence of inherent optical properties (IOPs) from a particular (adjacent) layer is not the only factor affecting underwater light fields in that layer but the subsequent layers (with non-uniform IOPs) would have potential contributions to modifying the underwater light-field environment. Moreover, the assumption of the homogenous water column in some of the RT models is not valid in many coastal waters where the water constituents would vary with depth (e.g. an increasing trend of turbidity with depth in many coastal regions). Thus, a reliable RT model is needed accounting for the vertically varying IOPs and treating the surface and bottom boundary conditions adequately in order to provide accurate underwater light-field data in turbid coastal waters.

Ocean colour data provided by modern day sensors (e.g. NASA's SeaWiFS on board its SeaStar satellite and MODIS on board its Aqua satellite, ESA's MERIS on board its Envisat satellite, Ocean Colour Monitor (OCM) from the Indian Space Research Organisation (ISRO) on board its Indian Remote Sensing (IRS) satellite and more recently Geostationary Ocean Colour Imager (GOCI) from the Korea Aerospace Research Institute (KARI) onboard its Communication, Ocean and Meteorological Satellite (COMS) satellite) are a vital resource for a wide variety of operational forecasting and oceanographic research, and related applications. With the advent of these new sensors, the prospects of better algorithms to enable the interpretation of ocean colour in coastal oceanic waters have improved vastly. Some of the potential applications of these sensors include monitoring and assessment of the spatial and temporal variability of algal blooms (instrumental in characterising variability of marine ecosystems and is a key tool for research into how marine ecosystems respond to climate change and anthropogenic perturbations), coastal marine pollution, river plumes, global carbon budgets, ocean radiant heat budgets and climate change impacts. Many of these applications can be achieved by estimating IOPs from the remote-sensing data, since light transmission in the water column is determined by these properties that depend mainly on the contents of chlorophyll (Chl), suspended sediments (SS) and coloured dissolved organic matter (CDOM). In most oceanic waters IOPs are determined primarily by phytoplankton and its associated detrital matter, which in turn determine the distribution and spectral quality of the underwater light fields (Morel, 1988; Hoepffner and Sathyendranath, 1992). Though several inversion models have been developed to estimate IOPs from remote-sensing data, they are often reported to yield large uncertainties in turbid coastal waters. Similar problems also exist with the retrieval of the water constituents' concentrations from satellite observations in these waters (O'Reilly et al., 1998; O'Reilly et al., 2000; Shanmugam, 2011a). The errors of more than 10 % in retrieval of IOPs (Stramski, 2001) and even much higher (20 times higher than measurements) in retrieval of chlorophyll are reported (Wozniak and Stramski, 2004).

The surface chlorophyll concentration estimated from satellite ocean colour data is used as an important parameter for reconstruction of its vertical profile in the water column (Morel, 1988; Platt and Sathyendranath, 1988; Sathyendranath and Platt, 1988; Morel and Berthon, 1989; Antoine et al., 1996; Uitz et al., 2006). The generalised Gaussian profile (Lewis et al., 1983) is used to predict the average dimensionless chlorophyll profile, superimposed onto a constant background concentration. The shape of the chlorophyll profiles directly depends on the subsurface chlorophyll maximum (SCM) $[Chl]_{max}$ and depth of chlorophyll maximum (DCM) Z_{max} which are parameterised using the surface chlorophyll data. While the background is generally considered as the surface chlorophyll concentration, some of the existing models assume it to decrease progressively with depth (Martin et al., 2010; Stramska and Stramski, 2005; Arrigo et al., 2011; Cherkasheva et al., 2013). Previously, Morel and Berthon (1989) modelled Chl profile shapes for nine tropic categories and developed a global algorithm for SCM, DCM and other parameters, regardless of region and season. The estimation of these profile parameters from the existing algorithms is applicable for certain seasons and regions. After the comprehensive study of DCM (Martin et al., 2010), it is confirmed that the published global, statistical relationships between the surface Chl and profile parameters lead to a severe underestimation when the SCM is sharp and intense in clear oceanic waters. This investigation motivates us to determine the new relation between the surface Chl versus profile parameters for predicting the column integrated chlorophyll profiles. In coastal waters, suspended sediments also play a major role in the determination of underwater light fields. Like the chlorophyll profiles, the surface suspended sediment concentration is used to extrapolate the SS along the depth. Previously, Ramakrishnan et al. (2013) used the power-law function to predict the vertical SS profiles from OCM data.

It is well known that the underwater radiometric parameters directly depend on IOPs of the water body (Shanmugam et al., 2010, 2011). The IOPs are mainly absorption (a), attenuation (c), scattering (b) and backscattering (b_b) coefficients which are generally derived as a function of the chlorophyll concentration in oceanic waters. Over the past decades, several models have been developed to estimate IOPs in case 1 waters (Prieur and Sathyendranath, 1981; Ahn, 1990; Bricaud et al., 1995; Babin et al., 2003; Matsuoka et al., 2011). However, the overly simplified parameterisations do not account for much of the optical variability observed in natural waters, leading to large uncertainties in case 2 waters (Babin et al., 2003; Dmitriev et al., 2009). The variations in these parameters can be attributed to three water constituents, such as phytoplankton, yellow substances (CDOM) and non-algal particles (both organic and inorganic). The particulate absorption (a_p) is mainly dominated by the non-algal particles, but phytoplankton becomes the dominant contributor in algal bloom waters (Wang et al., 2011). Stramski et al. (2001) explain that the mineral particles could be important for scattering and backscattering. Recently, Gokul et al. (2014) have developed models to predict IOPs and their vertical profiles using the remote-sensing reflectance data.

This work intends to derive a generalised radiative transfer model for predicting the underwater light fields in a variety of waters (including turbid coastal waters and eutrophic waters). The model is run with the in situ IOP data and predicted IOP data from remote-sensing data and its results are compared with the measured radiometric data. The results of the present RT model are further discussed for a variety of waters around southern India.

2 Data and methods

In situ measurements of the various optical and physical properties together with the water sampling were conducted in relatively clear and turbid coastal waters off Point Calimere, Chennai and eutrophic (lagoon) waters around Chennai during August 2012 and August, November and December 2013. The nature and characteristics of these waters have been investigated in a recent study by Pravin and Shanmugam (2014). For each station, water samples were collected from discrete depths and filtered and analysed for the determination of Chl, SS and CDOM contents (Gokul et al., 2014). Table 1 summarises some important symbols and notations used in this paper. The data used for this study and the sampling stations are described in Table 2. The vertical profiles of IOPs and other properties were measured with WET-Labs AC-S, BB9 and FLNTU sensors. Necessary corrections (for temperature, salinity and scattering effects) were applied to the AC-S data to obtain more reliable absorption and attenuation data (Pegau et al., 1997). Physical properties of the seawater such as conductivity, temperature and depth were measured by a SBE-CTD sensor to support the above data processing and analysis.

Field radiometric measurements were carried out using RAMSES (TriOS) hyperspectral radiometers; RAMSES ARC and ACC were used to measure the upwelling radiance (L_u), upwelling irradiance (E_u) and downwelling irradiance (E_d) in the water column. The irradiance sensor has an inbuilt pressure sensor which provides the corresponding depth in the water column. Both these sensors measure the radiance signal in the visible and near-infrared (0.350–0.950 μm) with a field of view of 7° and spectral accuracy of 0.0033 μm. Since the radiance sensor was immersed in water, the immersion factors (wavelength-dependent correction factors) were used to correct the measured radiance signal (Pravin and Shanmugam, 2014). Similarly, the above-surface measurements were made with another set of TriOS sensors that provided the sky radiance L_{sky}, downwelling irradiance $E_d(0^+)$ and total radiance L_t. The desired water-leaving radiances were determined after eliminating the surface-reflected light contributions to the total radiance signal (Pravin and Shanmugam, 2014).

3 Modelling

Figure 1 shows the schematic flow diagram of the present RT model. It can simulate the underwater light fields based on the measured IOPs (right part) or predicted IOPs from the remote-sensing data (left part) for the same solar incident geometry and surface and bottom boundary conditions. For the second part, surface Chl and SS are estimated from remote-sensing reflectance data and their vertical profiles are subsequently predicted using the known functions. These vertical profiles are used to derive the IOPs (hereafter referred to as Pred IOP). Finally, the Pred IOPs are used along with the other input parameters to simulate the underwater light fields. The step-by-step procedure is detailed in what follows.

3.1 Radiative transfer model

Radiative transfer is the physical phenomenon of energy transfer in the form of electromagnetic radiation. The propagation of radiation through a medium is affected by absorption, emission and scattering processes. The equation of radiative transfer describes these interactions mathematically. The basic RT equation that connects the radiance and IOPs is expressed as follows:

$$\cos\theta \frac{\mathrm{d}L(z,\theta,\phi,\lambda)}{\mathrm{d}z} = -c(z,\lambda)L(z,\theta,\phi,\lambda) + \int_{4\pi} L(z,\theta',\phi',\lambda) \times \beta(z;\theta',\phi' \rightarrow \theta,\phi;\lambda)\mathrm{d}\Omega' + S(z,\theta,\phi,\lambda). \quad (1)$$

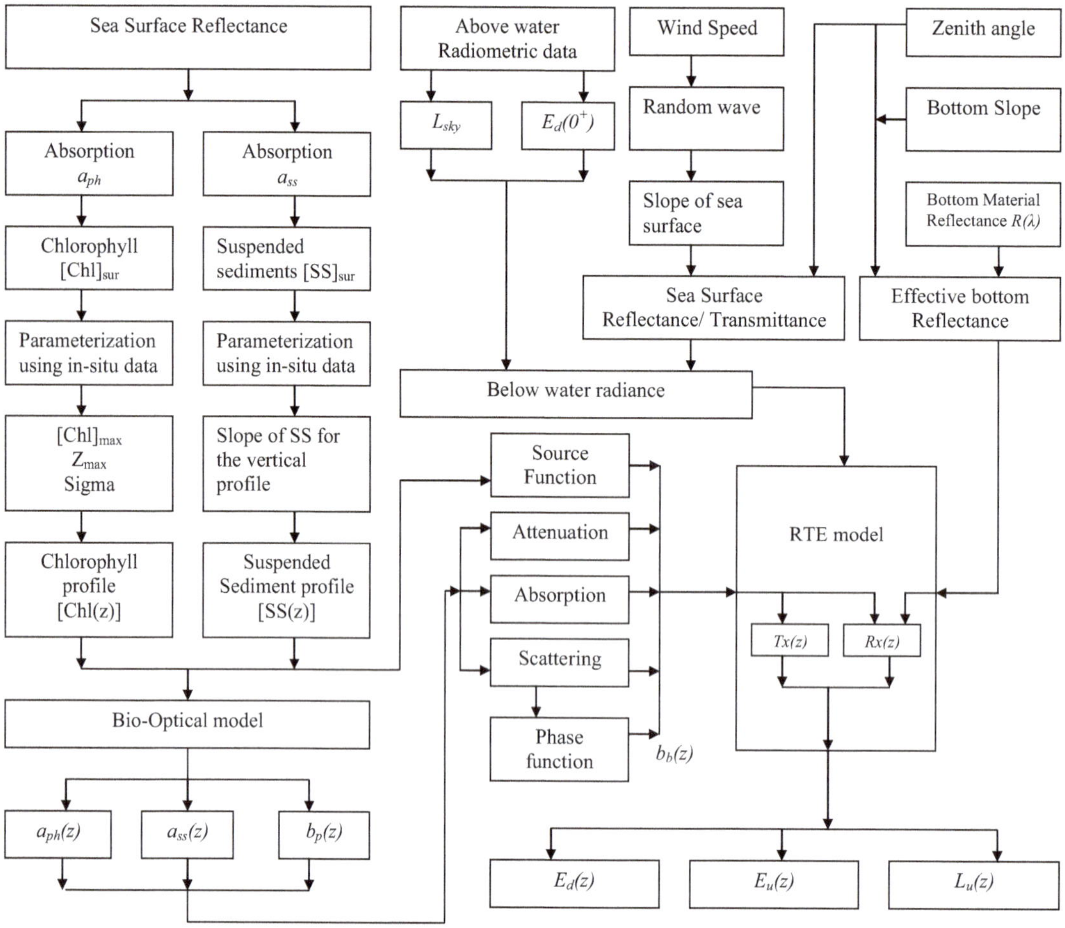

Figure 1. The schematic block diagram indicates the overall system for the simulation of underwater light fields using the present RT model with the in situ IOP (right) or predicted IOP from remote-sensing data (left) along with other input parameters.

The scattering angle ψ in the volume scattering function (VSF) is the angle between the incident direction $(\theta\prime, \varphi\prime)$ and the scattered direction (θ, φ). The source term $S\,(z, \theta, \varphi, \lambda)$ describes either an internal light source such as bioluminescence or inelastically scattered light from other wavelengths.

Figure 2 shows the schematic representation of the in-water radiative transfer technique for homogenous and inhomogeneous waters with the direct and diffuse terms. For practical applications, it is important to consider the influences of homogenous (diffuse term) and inhomogeneous (direct term) layers of the water column on the underwater light-field parameters. The homogenous and inhomogeneous effects are included in the present RT model (by taking the average of these two terms) to simulate the underwater light fields in a wide variety of waters (including relatively clear, turbid and eutrophic waters). The downwelling irradiances along the depth for both homogenous (HE_d) and inhomogeneous waters (E_d) can be calculated using

$$E_d(z) = \left(\frac{1}{2}\right)(HE_d(z) + E_d(z))\,. \tag{2}$$

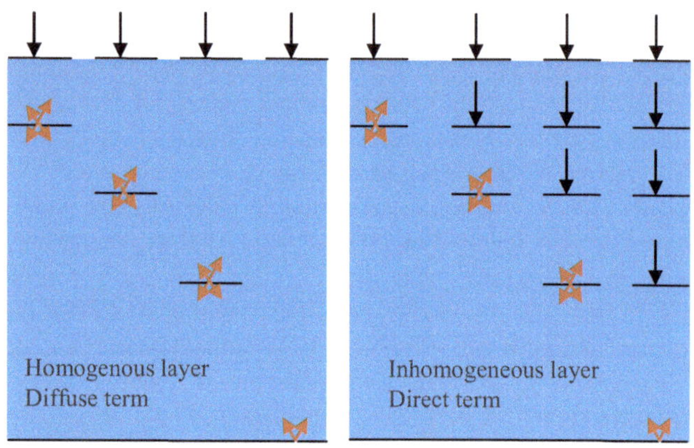

Figure 2. Schematic diagram showing the homogenous (left) and inhomogeneous radiative transfer concepts (right) with the diffuse and direct terms for the layers in the water column from the sea surface to sea bed.

Table 1. Symbols and notations.

Symbol	Description	Unit
U	Wind speed	$m\,s^{-1}$
θ_z	Solar zenith angle	degree
z	Depth	m
n_W	Refractive index	
μ	Average cosine	
q	Transmittance at air–water interface	
Tx_F	Fresnel transmittance at air–water interface	
Rf_F	Fresnel reflectance at air–water interface	
$T(z)$	Transmittance at depth z	
$R(z)$	Reflectance at depth z	
R_s	Diffuse reflectance (direct solar light)	
R_∞	Diffuse reflectance (diffuse light)	
c	Attenuation	m^{-1}
a	Absorption	m^{-1}
b	Scattering	m^{-1}
$E_d(0^+)$	Downwelling irradiance just above the water	$mW\,cm^2\,\mu m^{-1}$
$E_d(0^-)$	Downwelling irradiance just below the water	$mW\,cm^2\,\mu m^{-1}$
$E_d(z)$	Downwelling irradiance	$mW\,cm^2\,\mu m^{-1}$
$E_u(z)$	Upwelling irradiance	$mW\,cm^2\,\mu m^{-1}$
$L_u(z)$	Upwelling radiance	$mW\,cm^2\,\mu m^{-1}\,sr^{-1}$
$[Chl]_{sur}$	Surface chlorophyll concentration	$mg\,m^{-3}$
$[SS]_{sur}$	Surface suspended sediments concentration	$mg\,m^{-3}$
Z_{max}	Depth of chlorophyll maximum	m
$[Chl]_{max}$	Subsurface chlorophyll maximum	$mg\,m^{-3}$
σ	Standard deviation	m

From the downwelling irradiance and reflectance at different depths, the upwelling irradiance (for both homogenous, HE_u, and inhomogeneous, E_u) can be calculated from

$$E_u(z) = \left(\frac{1}{2}\right)[HE_u(z) + E_u(z)] + \left[\frac{S(z)}{4\pi}\right]. \tag{3}$$

These equations provide more accurate upwelling irradiances along the depth. The upwelling radiance along the depth (for homogenous and inhomogeneous) can be calculated from

$$L_u(z) = \left[\frac{E_u(z)}{2\pi\mu}\right]. \tag{4}$$

Using the above equations one can generate the underwater light fields and study their fluctuations in both clear and turbid waters. The inhomogeneous or direct term includes the phase function, source term and surface and bottom boundary conditions which are solved in equal interval along the water column. The homogeneous term or diffuse term is directly calculated from the IOPs for any arbitrary depth (Haltrin, 1998b).

3.1.1 Boundary conditions

Surface transmittance

The propagation of light through the sea surface is calculated from the reflected and transmitted angles using the Fresnel function (Gjerstad et al., 2003). Since the resulting underwater light fields strongly depend on the exact shape of the wave, the slope of the sea surface plays an important role in determining the transmitted and reflected angles. In this study, the shape of the irregular sea surface is generated based on the wind speed using the Pierson and Moskowitz (PM) spectrum (Pierson and Moskowitz, 1964) which is obtained from the field data. The slope is calculated based on the generated irregular sea surface. From this slope, the tilt angle is calculated based on Snell's law. The equation used to determine the tilt angle (γ) of the sea surface from the slope is expressed as follows:

$$\gamma = \tan^{-1}\left[\frac{\Delta z}{\Delta x}\right], \tag{5}$$

where Δx and Δz are the differential space of sea surface waves in the horizontal and vertical directions, respectively. Snell's law for the flat surface is given by $n_1 \sin(\theta_i) = n_2 \sin(\theta_t)$, where θ_i and θ_t are the incident and transmitted angles and n is the refractive index. After including the slope of the titled angle in the above equation, the transmitted angle for the sea surface is calculated from

$$\theta_t = \sin^{-1}\left[\frac{n_1}{n_2}\sin(\theta_i + \gamma)\right] - \gamma. \tag{6}$$

The modified transmitted and incident angles are applied in the Fresnel equation as follows:

$$\begin{aligned} \mathrm{Rf_F} = & \frac{1}{2}\left[\left(\frac{\sin(\theta_i - \theta_t)}{\sin(\theta_i + \theta_t)}\right)^2 + \left(\frac{\tan(\theta_i - \theta_t)}{\tan(\theta_i + \theta_t)}\right)^2\right], \quad \theta_i \neq \theta_t \\ = & \left(\frac{n_w - 1}{n_w + 1}\right)^2, \quad \theta_i = \theta_t \\ \mathrm{Tx_F} = & 1 - \mathrm{Rf_F}. \end{aligned} \tag{7}$$

The transmittance calculated from the above equation is used as the interface between the air and water for the downwelling irradiance (notations given in Table 1).

Bottom reflectance

The effective reflectance R_{eff} of the bottom (considering the bottom material and morphology) is calculated according to Zaneveld and Boss (2003):

$$\frac{R_{\mathrm{eff}}(\lambda)}{R_b(\lambda)} = 0.5 \times \cos[\theta_z + \theta_b] + 0.5 \times \cos[\theta_z - \theta_b], \tag{8}$$

where $\theta_b = a\tan(4A_b/L_b)$ the angle of the bottom slope (due to ripples on the sea bed), and θ_z is the zenith angle of the irradiance. A_b and L_b are the amplitude and wavelength, respectively, for the triangular shaped bottom. The effective reflectance spectra of the sea bottom are not same for different materials since the reflectance is about to vary for different materials.

3.1.2 Optical properties in the water column

Phase function

The angular distributions of scattered radiance are mainly explained in terms of the phase function, which plays an important role in coastal waters. The characteristic of phase functions in natural volumetric media is sharply peaked in the forward scattering direction, with only a few percent of backscatter in the total angular redistribution of a single scattering event (Sundarabalan et al., 2013). Of several phase function models developed in the past, the Fournier–Forand (FF) model is an analytic form of the phase function giving better results when compared to other models (Mobley et al., 2002). The FF phase function is given by

$$\begin{aligned} \beta(\theta) = & \frac{1}{4\pi(1-\delta)^2\delta^v}\left\{\left[v(1-\delta) - (1-\delta^v)\right]\right. \\ & \left. + \left[\delta(1-\delta^v) - v(1-\delta)\sin^{-2}\left(\frac{\theta}{2}\right)\right]\right\} \\ & + \frac{1-\delta_{180}^v}{16\pi(\delta_{180}-1)\delta_{180}^v}\left(3\cos^2\theta - 1\right) \\ v = & \frac{3-\mu}{2}, \quad \delta = \frac{4}{3(n-1)^2}\sin^2\left(\frac{\theta}{2}\right). \end{aligned} \tag{9}$$

Here μ is the slope of the Junge particle distribution, n is the real index of refraction and δ_{180} is calculated by considering $\theta = 180°$. Based on previous studies (Twardowski et al., 2001; Freda and Piskozub, 2007; Sundarabalan et al., 2013), the parameters μ and n are modelled using the IOPs, attenuation $c(520)$ and scattering $b(520)$. Finally, the backscattering b_b coefficients are computed from the phase function for scattering angles between 90 and 180°,

$$b_b(520) = \int_{\frac{\pi}{2}}^{\pi} \beta(\theta)d\theta. \tag{10}$$

The spectral variation of the backscattering b_b coefficients can be expressed as (Haltrin, 2002)

$$b_b(\lambda) = b_b(520) \times \left(\frac{520}{\lambda}\right)^{1.1}. \tag{11}$$

This phase function is mainly used to determine the b_b coefficients along the depth, which are more compatible for turbid coastal waters.

Transmittance along the depth

Haltrin (1998b) derived the transmittance as a function of depth $T(z)$ based on the self-consistent method, which depends on the IOPs of seawater. The transmittance function $T(z)$ is expressed as

$$T(z) = \frac{1 + q\{\mu_s\varepsilon(z) + hR_sF_s(z)[(2+\overline{\mu})\mu_s + 1]\}}{1 + q\mu_s}. \tag{12}$$

Several important parameters that depend on the IOPs are used to calculate $T(z)$. μ_s is the cosine function related to the solar angle which is calculated based on the refractive index of seawater (n) and the solar elevation h_s. $\overline{\mu}$ is the average cosine that connects with Gordon's parameter g, which depends on the absorption and backscattering coefficients. The result obtained by solving the RT equation is α_∞ which is found by dividing absorption (a) by the average cosine ($\overline{\mu}$). μ_0 is another cosine function which depends on the average cosine $\overline{\mu}$. The reflectance parameters involved in the calculation of $T(z)$ are the diffuse reflectance (R_s) of a deep sea layer optically illuminated by direct solar light and the diffuse reflectance (R_∞) of the optically deep sea illuminated by the diffuse light. Both are calculated as a function of the cosine functions as follows:

$$R_s = \frac{(1-\overline{\mu})^2}{1+\overline{\mu}\mu_s(4-\overline{\mu}^2)}, \quad R_\infty = \left(\frac{1-\overline{\mu}}{1+\overline{\mu}}\right)^2. \tag{13}$$

The average cosine function used in the above equation is defined as

$$\overline{\mu} = \left\{ \frac{1-g}{1+2g+[g(4+5g)]^{1/2}} \right\}^{1/2},$$
$$\mu_0 = \frac{1+\overline{\mu}^2}{\overline{\mu}(3-\overline{\mu}^2)},$$
$$\mu_s = \left[1 - \left(\frac{\cosh_s}{n_w} \right)^2 \right]^{1/2}, \quad (14)$$

where $c = a + b$ is the beam attenuation coefficient, $q = \frac{E_s^{\perp}}{E_d^0}$ is the transmittance at air–sea interface, $\alpha = a + 2b_b$ is the re-normalised attenuation coefficient, $\alpha_\infty = \frac{a}{\overline{\mu}}$ is the intermediate parameter which depends on IOPs and $g = \frac{b_b}{a+b_b}$ is the Gordon's parameter (Gordon et al., 1975; Haltrin, 2003) which depends on absorption and scattering.

The IOP-dependent intermediate parameters are defined as

$$\varepsilon(z) = \exp\left[-\alpha z \left(\frac{1}{\mu_s} - \frac{1}{\mu_0} \right) \right], \quad h = \frac{(1+\overline{\mu})^2}{2(1+\overline{\mu}^2)},$$
$$F_s(z) = \left(1 - \exp\left[-\alpha z \left(\frac{1}{\mu_s} - \frac{1}{\mu_0} \right) \right] \right) \Big/ \left(\frac{1}{\mu_s} - \frac{1}{\mu_0} \right),$$
$$\mu_s \neq \mu_0,$$
$$F_s(z) = \alpha z, \quad \mu_s = \mu_0. \quad (15)$$

The above parameters, $\varepsilon(z)$, h, and $F_s(z)$, are the functions calculated based on the IOPs and solar elevation (Haltrin, 1998b).

Reflectance along the depth

The reflectance along the depth $R(z)$ is generally calculated based on the IOPs ($b_b/(a+b_b)$), but it is also highly influenced by the bottom material and solar zenith angle (Lee et al., 1998, 1999). Thus, the reflectance along the depth is calculated from Lee et al. (1998), which takes into account the bottom material effect and the IOPs of the water column. The model parameters are defined as follows:

$$R(z) = R_{iop}(z) + R_{btm}(z). \quad (16)$$

The reflectance influenced by the IOPs and bottom effect are calculated from the following equations:

$$R_{iop}(z) = r_{rs}(z) \times \left\{ 1 - \exp\left[-\kappa(z) H \left(D_u^c + \left(\frac{1}{\cos\theta_w} \right) \right) \right] \right\},$$
$$R_{btm}(z) = R_{effbtm} \times \left\{ -\exp\left[-\kappa(z) H \left(D_u^b(z) + \left(\frac{1}{\cos\theta_w} \right) \right) \right] \right\}, \quad (17)$$

where D_u^c is the path-elongation factor for scattered photons from the water column which varies with the IOPs. The optical path-elongation factor D_u^b for the bottom mainly depends on the bottom reflectance. These factors are defined as

$$D_u^c = 1.03\sqrt{(1+(2.4 \times u))}, \quad (18)$$
$$D_u^b = 1.04\sqrt{(1+(5.4 \times u))}. \quad (19)$$

r_{rs} is the subsurface remote-sensing reflectance which is a function of the IOPs at a given depth and expressed as

$$r_{rs} = u \times [(u \times 0.170) + 0.084]. \quad (20)$$

Also, θ_w is the subsurface solar zenith angle, H is the bottom depth, R_{effbtm} is the effective bottom reflectance and κ and u are the inherent optical parameters which can be obtained from

$$u = \frac{b_b(z)}{a(z) + b_b(z)}, \quad (21)$$
$$\kappa = a(z) + b_b(z). \quad (22)$$

Based on the IOPs along the depth and effective reflectance of the bottom, the reflectance functions along the depth $R(z)$ can be calculated from these equations.

Source function

Since the source function affects the underwater light fields (by way of re-emitting photons by phytoplankton at longer wavelengths after absorption at shorter wavelengths), it is also included in the present RT model. The source function can be computed as follows (Gower et al., 2004):

$$\mathrm{Fl}(\lambda) = \frac{0.15 \times \mathrm{Chl}}{1 + 0.2 \times \mathrm{Chl}}. \quad (23)$$

The source function is calculated as a function of the chlorophyll fluorescence as follows:

$$S(\lambda) = \mathrm{Fl}(\lambda) \times h(\lambda), \quad (24)$$

where $h(\lambda)$ is the fluorescence emission function per unit wavelength calculated based on the Gaussian distribution,

$$h(\lambda) = \frac{1}{\sqrt{2\pi\sigma^2}} \exp\left(-\frac{(\lambda - \lambda_0)^2}{2\sigma^2} \right). \quad (25)$$

The wavelength of maximum emission is $\lambda_0 = 0.685\,\mu$m and the standard deviation $\sigma = 0.011\,\mu$m.

3.1.3 Inhomogeneous term: underwater light-field parameters

The downwelling irradiance $E_d(0^-)$ just below the water for the inhomogeneous (or layer by layer) condition can be calculated from the downwelling irradiance $E_d(0^+)$ just above the water along with the transmittance derived from the Fresnel equation (Eq. 7),

$$E_d(0^-) = E_d(0^+) \times \mathrm{Tx_F}. \quad (26)$$

Once the downwelling irradiance is transmitted through the water surface, the intensity of the downwelling irradiance based on the transmittance is purely dependent on the IOPs. The downwelling irradiances for the first and subsequent layers of the depth are calculated from

$$E_d(z_1) = E_d(0^-);\ E_d(z_2) = E_d(z_1) \times \mathrm{Tx}(z_2). \tag{27}$$

The downwelling irradiance along the depth can be calculated from an explicit method with the corresponding depth transmittance $\mathrm{Tx}(z)$. The common equation for calculating $E_d(z)$ along the depth is given as

$$E_d(z) = E_d(z-1) \times \mathrm{Tx}(z). \tag{28}$$

The upwelling irradiances below the water along the depth are calculated from

$$E_u(z) = (E_d(z) \times R(z)), \tag{29}$$

where $E_d(z)$ is downwelling irradiance and $R(z)$ is the reflectance for the corresponding depth which includes the effect of the bottom reflectance and IOPs from the bottom boundary condition.

3.1.4 Homogeneous term: underwater light-field parameters

The downwelling irradiance equation developed by Haltrin (1998b) for the homogenous water column takes the following expression,

$$\mathrm{HE}_d(z) = E_d^0 \exp(-\alpha_\infty z) + E_s^\perp \exp(-\alpha z/\mu_s) \tag{30}$$

$$+E_s^\perp h R_s \{1+\mu_s(2+\overline{\mu})\} F_s(z) \exp(-\alpha_\infty z). \tag{31}$$

The upwelling irradiance for the homogenous water column can be calculated with the equation of Haltrin (1998b):

$$\mathrm{HE}_u(z) =$$

$$E_d^0 R_\infty \exp(-\alpha_\infty z) + \mu_s E_s^\perp R_\infty \exp(-\alpha z/\mu_s) \tag{32}$$

$$+ E_s^\perp h R_s \{\mu_s(2-\overline{\mu}) - 1\} F_s(z) \exp(-\alpha_\infty z). \tag{33}$$

The input parameters used in the calculation of HE_d and HE_u are explained in the transmittance section. Here, E_s is the diffuse term of the above-water irradiance which is obtained from the sky radiance L_{sky}. The total underwater light-field parameters (E_d, E_u and L_u) can be obtained by applying Eqs. (26)–(29) in Eqs. (2)–(4). The model presented in this study is much easier to implement when compared to the existing RT models.

3.2 Prediction of remotely sensed IOPs along the water column

3.2.1 Bio-optical model

This section presents methods to estimate the surface chlorophyll $[\mathrm{Chl}]_{sur}$ and suspended sediments $[\mathrm{SS}]_{sur}$ (concentrations) from the normalised water-leaving radiance ($\mathrm{nL_w}$). The slope values (S_{nLw}) required for fixing the threshold limit (i.e., scenarios in Table 3) for different waters are calculated as follows:

$$S_{\mathrm{nLw}} = 100 \times \left[\frac{\mathrm{nL_w}(0.443) - \mathrm{nL_w}(0.547)}{0.443 - 0.547}\right]. \tag{34}$$

Since the normalised water-leaving radiance values are larger than the remote-sensing reflectance, the former quantity is used to better determine the used scenarios (Ahn and Shanmugam, 2006; Shanmugam, 2011b). Based on the S_{nLw} and $\mathrm{nL_w}$ data, three scenarios are found adequate for the various water types. The first scenario represents the open-ocean waters where S_{nLw} is less than 0.5 and [Chl] is based on the ratio of R_{rs} (0.488) and R_{rs} (0.547) (Table 3). The second scenario indicates turbid coastal waters where the same band ratio is used for the [Chl] parameterisation. The [SS] parameterisations are different for both these scenarios. The third scenario is developed for inland and eutrophic waters based on the exponential function that uses the R_{rs} values at three different bands (0.690, 0.700 and 0.760 μm) (Zhang et al., 2009). The coefficients of the exponential equation are obtained based on the IR_{rs} value which is defined as follows:

$$\mathrm{IR}_{rs} = \left(\frac{1}{R_{rs}(0.690)} - \frac{1}{R_{rs}(0.700)}\right) \times R_{rs}(0.760). \tag{35}$$

For the [SS] parameterisation, there is a shift of peak between 0.547 and 0.488 μm in clear waters (first scenario) and the ratio of R_{rs} (0.620) to the maximum value of R_{rs} (0.488) and R_{rs} (0.547) is found to be suitable for these waters. In turbid coastal waters (second scenario), the reflectance peak at 0.547 μm dominates the R_{rs} values at 0.488 μm and the relative change of these values is used in terms of the ratio to estimate [SS] in turbid waters. Considering inland and eutrophic waters (third scenario), the ratio of R_{rs} at 0.620 and 0.720 μm is used for the estimation of [SS] in these waters.

3.2.2 Vertical profiles of chlorophyll and suspended sediments

The chlorophyll and suspended sediments along the vertical column are determined from the surface [Chl] and [SS] data. For the chlorophyll profile, Lewis et el. (1983) found the generalised Gaussian distribution model which captures the major features of the observed vertical profile. The major parameters used to determine the chlorophyll profile are the surface chlorophyll $[\mathrm{Chl}]_{sur}$, maximum chlorophyll $[\mathrm{Chl}]_{max}$, depth chlorophyll maximum Z_{max} and σ (standard deviation that controls the thickness of the $[\mathrm{Chl}]_{max}$ layer and determines the vertical spread). The $[\mathrm{Chl}]_{max}$ is the value of maximum chlorophyll in the water column and Z_{max} is the depth of $[\mathrm{Chl}]_{max}$. The schematic representation of these chlorophyll profile parameters is shown in Fig. 3. The determination of the chlorophyll profile based on the above parameters

Table 2. Station details and the observed environmental parameters used for this study.

Station	Date	Time (LT)	Latitude	Longitude	θ_z	Depth (m)	U (ms^{-1})	[Chl] ($mg\,m^{-3}$)	[SS] ($mg\,m^{-3}$)	$a_{CDOM}(490)$ (m^{-1})
St-1	31/08/2013	13:00	13°08.715 N	80°21.041 E	21.9	19.8	7.7	0.2–1.1	7.9–17.7	0.181
St-2	31/08/2013	15:00	13°08.715 N	80°21.041 E	43.6	19.8	7.7	0.3–0.5	8.2–11.1	0.175
St-3	26/08/2013	15:00	10°20.714 N	80°08.604 E	43.2	17.7	5	1.0–6.7	7.0–15.8	0.83
St-4	26/08/2013	11:45	10°22.103 N	79°57.720 E	4.55	6.3	7.7	0.8–2.3	33.8–41.9	1.15
St-5	17/08/2012	15:10	10°20.992 N	80°05.573 E	44.9	17.9	4	6.2–18.5	10.9–15.7	0.92
St-6	17/08/2012	15:55	10°20.714 N	80°08.604 E	55.9	22	3	5.3–20.4	8.4–16.6	0.85
St-7	08/11/2013	14:00	13°07.408 N	80°17.565 E	42.8	5	0.25	18.3–18.5	21.6–24.0	0.98
St-8	10/11/2013	14:50	12°48.474 N	80°14.204 E	53.2	1	0.25	52.5	70.4–71.6	1.48
St-9	10/11/2013	15:10	12°48.321 N	80°14.239 E	59.5	1	0.25	54.1	68.2–87.3	1.51
St-10	16/12/2013	13:15	12°48.474 N	80°14.204 E	42.5	1	0.25	72.1	63.6–65.5	1.34
St-11	16/12/2013	13:30	12°48.321 N	80°14.239 E	43.8	1	0.25	73.8	62.1–64.8	1.25

Table 3. Empirical relationships between the in situ R_{rs} and in situ surface concentration ($[Chl]_{sur}$ and $[SS]_{sur}$) in a variety of waters for three different scenarios. Note that S.No. 3 will have another expression $[Chl]_{sur} = 17.128 \times \exp(0.888 \times IR_{rs})$ if one intends to use $R_{rs}(0.680)$ instead of $R_{rs}(0.690)$ in Eq. (37). The empirical coefficients presented in this table are derived based on our in situ data.

$[CHL]_{sur}$	$[SS]_{sur}$	Scenario
$1.123 \times \left(\frac{R_{rs}(0.488)}{R_{rs}(0.547)}\right)^{-3.7144}$	$21.503 \times \left(\frac{R_{rs}(0.620)}{\max[R_{rs}(0.488), R_{rs}(0.547)]}\right)^{0.3998}$	$S_{nLw} < 0.5$
$0.409 \times \left(\frac{R_{rs}(0.488)}{R_{rs}(0.547)}\right)^{-7.567}$	$34.01 \times \left(\frac{R_{rs}(0.547)}{R_{rs}(0.488)}\right)^{-1.2615}$	$S_{nLw} > 0.5$
$18.267 \times \exp(1.9913 \times IR_{rs})$	$38.685 \times \left(\frac{R_{rs}(0.620)}{R_{rs}(0.720)}\right)^{-0.554}$	$R_{rs}(0.547) > R_{rs}(0.488)$& $R_{rs}(0.680) < R_{rs}(0.690)$

is given below:

$$\mathrm{Chl}(z) = \mathrm{Chl}_{sur} + \mathrm{Chl}_{max} \times \exp\left[-\frac{(z - z_{max})^2}{2\sigma^2}\right]. \quad (36)$$

The chlorophyll profile parameters calculated from the surface [Chl] are shown in Table 4 and the basic equations are taken from Gokul et al. (2014). The equations were developed based on a large number of in situ data. Conversely, the SS profile is estimated using the power-law function,

$$\mathrm{SS}(z) = \mathrm{SS}_{sur} \times \left(\frac{z}{z_{sur}}\right)^{0.0383}. \quad (37)$$

The slope is calculated based on the mean slope values from a number of measured SS profiles. Here, the SS_{sur} of the concentration of suspended sediments at the near surface and Z_{sur} is the depth at which the SS_{sur} is measured ($Z_{sur} = 1$ m).

3.2.3 IOP model

Clear and turbid coastal waters

A simplified model is used to estimate the vertical profiles of IOPs using chlorophyll and suspended sediment profile data. Though the IOPs may be determined by more than three sub-

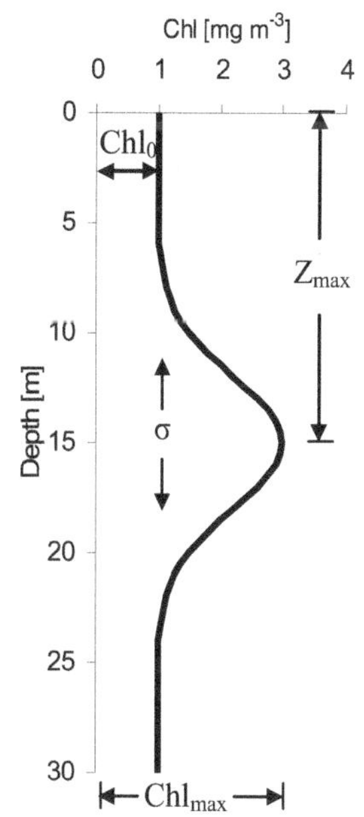

Figure 3. Schematic diagram represents the vertical profile of the chlorophyll along the depth including the distribution of profile parameters using the Gaussian distribution function.

Table 4. Empirical relationships of the chlorophyll profile parameters ($[\mathrm{Chl}]_{\mathrm{max}}$, Z_{max} and σ) for the various ranges of surface chlorophyll $[\mathrm{Chl}]_{\mathrm{sur}}$ from different stations.

$\mathrm{Chl}_{\mathrm{sur}}$	$\mathrm{Chl}_{\mathrm{max}}$	Z_{max}	Σ
$1.0 < \mathrm{Chl}_{\mathrm{sur}} < 10$	$2.019 \times \mathrm{Chl}_{\mathrm{sur}} - 0.2328$	$0.7595 \times \mathrm{Chl}_{\mathrm{sur}} + 4.082$	$0.3456 \times \mathrm{Chl}_{\mathrm{sur}} + 4.104$
$0.5 < \mathrm{Chl}_{\mathrm{sur}} < 1$	$-8.288 \times \mathrm{Chl}_{\mathrm{sur}} + 8.3628$	$-12.247 \times \mathrm{Chl}_{\mathrm{sur}} + 15.479$	$3.1606 \times \mathrm{Chl}_{\mathrm{sur}} - 0.199$
$0 < \mathrm{Chl}_{\mathrm{sur}} < 0.5$	$-8.288 \times \mathrm{Chl}_{\mathrm{sur}} + 2.6628$	$-12.247 \times \mathrm{Chl}_{\mathrm{sur}} + 15.479$	$3.1606 \times \mathrm{Chl}_{\mathrm{sur}} - 0.199$

stances, it is assumed that the absorption and scattering coefficients in clear and turbid coastal waters are mainly determined by water itself, suspended sediment particles and phytoplankton (both living and non-living). Thus, the total absorption coefficient of seawater, $a(\lambda, z)$, is the sum of the absorption of seawater $a_{\mathrm{w}}(\lambda, z)$, dissolved organic $a_{\mathrm{CDOM}}(\lambda, z)$, and particulate matter $a_{\mathrm{p}}(\lambda, z)$. The total absorption coefficient observed at any given wavelength can be expressed as

$$a(\lambda, z) = a_{\mathrm{w}}(\lambda, z) + a_{\mathrm{CDOM}}(\lambda, z) + a_{\mathrm{p}}(\lambda, z), \tag{38}$$

$$a_{\mathrm{p}}(\lambda, z) = a_{\mathrm{ph}}(\lambda, z) + a_{\mathrm{ss}}(\lambda, z). \tag{39}$$

Here the total absorption (pure water, chlorophyll, suspended sediments and coloured dissolved organic matter) is estimated from Morel (1991) and the absorption coefficient of suspended sediments is calculated from Gokul et al. (2014):

$$a(\lambda, z) = \left(a_{\mathrm{w}}(\lambda) + 0.06 a_C^*(\lambda)[\mathrm{Chl}(z)]^{0.65}\right) \times \left(1 + 0.2\exp(-0.014(\lambda - 0.400) \times 10^{-3})\right) + a_{\mathrm{ss}}(\lambda, z). \tag{40}$$

The seawater absorption coefficients were taken from Pope and Fry (1997). The absorption coefficient of suspended sediments is estimated as follows:

$$a_{\mathrm{ss}}(\lambda, z) = a_{\mathrm{ss}}(\lambda_r) \times \exp(-0.0104(\lambda - \lambda_r)), \tag{41}$$

where the absorption at a reference wavelength $0.443\,\mu\mathrm{m}$ is calculated from the power fit shown in Fig. 4a and the equation is given as

$$a_{\mathrm{ss}}(0.443) = 0.0007[\mathrm{SS}(z)]^{1.7653}, \quad [R^2 = 0.9]. \tag{42}$$

Similarly, the total scattering coefficient of sea water is the sum of the scattering due to pure seawater $b_{\mathrm{w}}(\lambda, z)$ and particulate matter $b_{\mathrm{p}}(\lambda, z)$ (due to phytoplankton and suspended sediments). The pure sea water scattering coefficients were taken from Smith and Baker (1981). Since the contribution of CDOM is negligible, it is omitted leading to the total scattering $b(\lambda, z)$,

$$b(\lambda, z) = b_{\mathrm{w}}(\lambda, z) + b_{\mathrm{ph}}(\lambda, z) + b_{\mathrm{ss}}(\lambda, z). \tag{43}$$

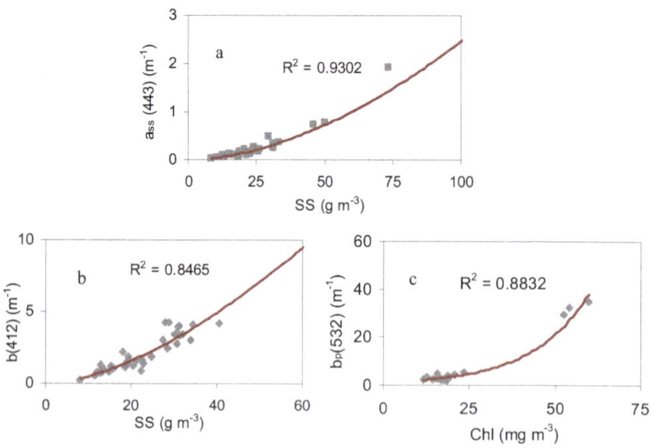

Figure 4. (a) Relationship between the suspended sediment concentration and absorption by suspended sediment $a_{\mathrm{ss}}(443)$, used to calculate the absorption coefficient of suspended sediments $a_{\mathrm{ss}}(443)$. **(b)** Relationship between the suspended sediment concentration and scattering $b(412)$, used to calculate the scattering (b_{ss}) coefficient of suspended sediments. **(c)** Relationship between the Chlorophyll and particulate scattering $b_{\mathrm{p}}(532)$ which is used to calculate the particulate scattering directly from the chlorophyll concentration. This is applicable only for surface Chl should it be greater than $15\,\mathrm{mg\,m^{-3}}$.

The scattering due to phytoplankton (b_{ph}) depends on the chlorophyll concentration which is derived from Gordon and Morel (1983):

$$b_{\mathrm{ph}}(\lambda, z) = \left(\frac{0.550}{\lambda}\right)^S \times \left(0.3[\mathrm{Chl}(z)]^{-0.3}\right), \tag{44}$$

where S is the slope which is calculated based on the chlorophyll concentration (Huot et al., 2008). When the Chl is greater than $2\,\mathrm{mg\,m^{-3}}$, S is considered to be zero. If Chl is less than $2\,\mathrm{mg\,m^{-3}}$, the slope will be calculated from the following equation:

$$S = 0.5 \times (\log([\mathrm{Chl}(z)]) - 0.3). \tag{45}$$

The scattering due to suspended sediments is estimated using the power-law model (Fig. 4b) which is expressed as

$$b_{\mathrm{ss}}(\lambda, z) = \frac{1}{2}\left(\frac{0.412}{\lambda}\right)^{0.88} \times \left(0.0043[\mathrm{SS}(z)]^{1.9657}\right). \tag{46}$$

Finally, the attenuation coefficients $\alpha(\lambda, z) (= c(\lambda, z))$ are calculated based on the total absorption $a(\lambda, z)$ and total scattering $b(\lambda, z)$ coefficients.

Table 5. New spectral absorption coefficients of water particles derived from the in situ data, which are used only when the concentration of $[\mathrm{Chl}]_{\mathrm{sur}}$ is greater than $15\,\mathrm{mg\,m^{-3}}$, for the calculation of the particulate absorption.

Wavelength (μm)	a_p^*	Wavelength (μm)	a_p^*
0.401	0.195342	0.575	0.038799
0.425	0.208779	0.601	0.044386
0.451	0.168103	0.625	0.056005
0.475	0.129094	0.651	0.053105
0.501	0.098007	0.675	0.080339
0.525	0.061625	0.701	0.019868
0.551	0.042149	0.725	0.002783

Eutrophic and phytoplankton-dominated waters

For eutrophic and phytoplankton-dominated waters, the variations in absorption and scattering coefficients are poorly documented as most of the previous studies on IOPs were conducted in relatively clear and open-ocean waters. In this study, absorption and scattering by particles are estimated using separate models for these waters. In the previous studies, phytoplankton absorption $a_{\mathrm{ph}}(\lambda)$ was generally calculated based on the specific phytoplankton absorption $a_{\mathrm{ph}}^*(\lambda)$. Here the specific particulate absorption coefficients $a_p^*(\lambda)$ are used to calculate the particulate absorption coefficients $a_p(\lambda)$. The values of new $a_p^*(\lambda)$ are given in Table 5. The particulate absorption coefficients are then derived as a function of chlorophyll as follows:

$$a_p(\lambda, z) = a_p^*(\lambda) \times [\mathrm{Chl}(z)] \times \left(\frac{[\mathrm{Chl}(z)]}{54}\right). \tag{47}$$

Similarly, the particulate scattering is calculated directly based on the exponential function of chlorophyll (Fig. 4c) as follows:

$$b_p(\lambda, z) = 1.4 \times \exp(0.0525 \times [\mathrm{Chl}(z)]) \times \left(\frac{\lambda}{0.532}\right)^{-0.3}. \tag{48}$$

This equation is derived from the relationship between in situ chlorophyll and particulate scattering (c–a from WetLabs AC-S). Both particulate absorption and scattering coefficients are added with the respective pure water coefficients to obtain the total absorption and scattering coefficients.

4 Results and discussion

Results are categorised into four parts: (1) comparison of the model versus measured IOP profile data, (2) effects of homogeneous and inhomogeneous water column conditions, (3) comparison of the underwater light fields predicted by the RT model, based on the measured and estimated IOP profiles from remote-sensing data, and (4) apparent optical properties from underwater light fields. Figure 5 shows the examples of measured remote-sensing reflectances from different waters (clear, turbid and eutrophic waters) used for construction of the vertical profiles of IOPs and simulation of the underwater light fields. The specific spectral features of IOPs for these waters have already been described in a recent study (Pravin and Shanmugam, 2014). It should be mentioned that the underwater light-field parameters are simulated from the RT model using MATLAB 2007 with a 4 gigabyte RAM computer. The run time for this model is 8 milliseconds for the entire wavelength at one depth for a given sample. In fact, this can be reduced if the model is implemented in FORTRAN with a high performance computer. The performance of the model in simulating the underwater light-field parameters is assessed using the mean relative error (MRE =(model − in situ)/in situ).

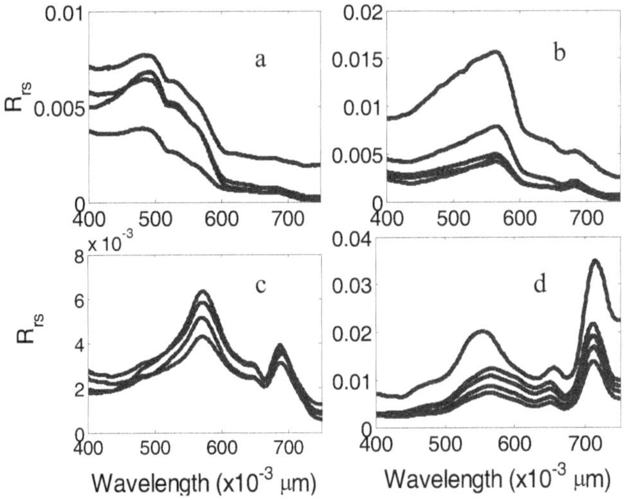

Figure 5. The typical remote-sensing reflectance spectra $R_{\mathrm{rs}}(\lambda)$ measured at four stations in the study area during 2012 and 2013. **(a)** Relatively clear waters off Chennai, **(b)** moderately turbid waters off Point Calimere, **(c)** phytoplankton-dominated harbour waters of Chennai and **(d)** eutrophic waters.

4.1 Prediction of the IOP profiles from remote-sensing data

This section is focused on the determination of the vertical profiles of IOPs based on the seawater constituents (chlorophyll and suspended sediments at the surface level) that are estimated from the above-water remote-sensing reflectance (R_{rs}) data.

4.1.1 Retrieval of the seawater constituents

The spectral information of the remote-sensing reflectance (R_{rs}) was used to estimate [Chl] and [SS] in surface wa-

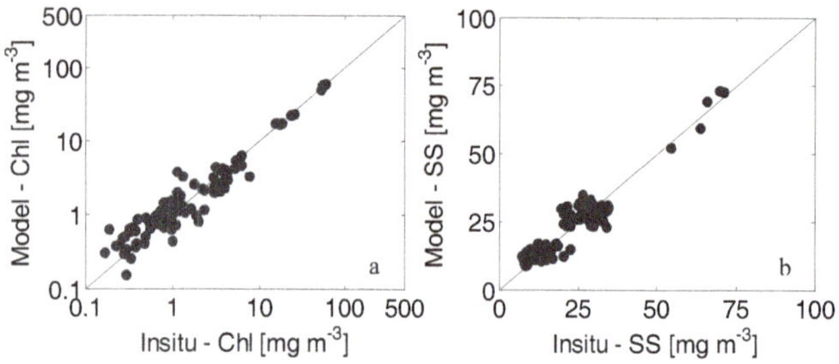

Figure 6. **(a)** Scatter plot showing the comparison of the estimated chlorophyll concentration from the remote-sensing reflectance R_{rs} with the in situ chlorophyll from different waters. **(b)** Scatter plot showing the comparison of the estimated suspended sediment concentration from the remote-sensing reflectance R_{rs} with the in situ suspended sediment concentration from different waters.

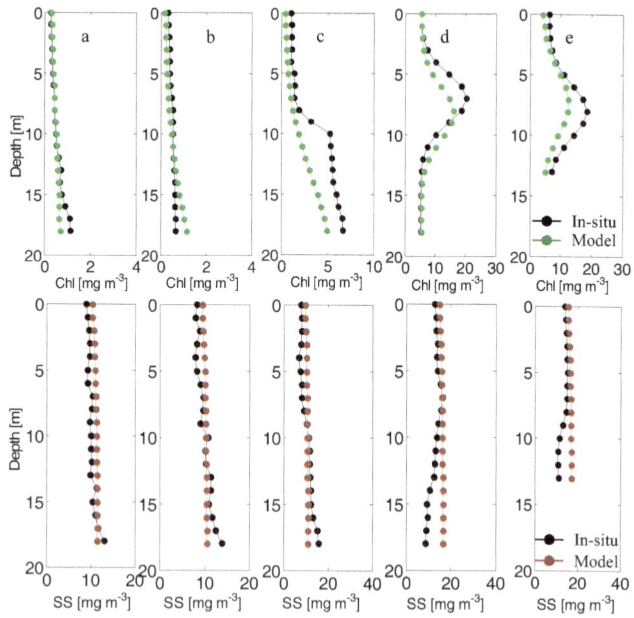

Figure 7. The examples of the modelled and measured vertical profiles of [Chl(z)] for five different cases (top row). **(a and b)** St-1 and St-2 from relatively clear waters off Chennai, **(c)** St-4 from moderately turbid water with chlorophyll settled at the bottom of seabed. **(d and e)** St-6 and St-5 from the chlorophyll-dominated regions. The second row shows the modelled and measured vertical profiles of [SS(z)] for the same locations. The [SS(z)] modelled profiles are almost uniform along the depth.

ters off Point Calimere and Chennai. Since the models are based on the different spectral slopes of R_{rs}, different scenarios were used to estimate these water constituents accurately. The estimated [Chl] and [SS] show good agreement with measured data from different waters (Fig. 6a), where [Chl] ranged from 0.1 to 100 mg m^{-3}. The statistics analyses indicate low errors and high slopes and correlation coefficients (MRE = 0.17, RMSE = 0.18, slope = 1.0, bias = −0.031, R^2 = 0.91, N = 98). Similarly, the estimated [SS] agree closely with the in situ [SS] (Fig. 6b), with good statistics (MRE = 0.01, RMSE = 0.09, slope = 0.98, bias = −0.01, R^2 = 0.84, N = 98). These results clearly demonstrate consistency between the estimated and measured data for a wide range of waters.

4.1.2 Vertical profiles of chlorophyll and suspended sediments

On the basis of surface chlorophyll $[\text{Chl}]_{\text{surf}}$ and [SS] estimated from remote-sensing data, the vertical profiles of [Chl(z)] and [SS(z)] were constructed in relatively clear and turbid coastal waters. It is observed that the modelled and measured chlorophyll profiles agree well in relatively clear waters off Chennai (31 August 2013 at 13:00 and 15:00 LT) (Fig. 7a and b). The Chl concentration is low in surface waters (0.3 mg m^{-3}) and gradually increases along the depth. For relatively clear waters off Point Calimere in August 2013 (Fig. 7c), the surface chlorophyll is very low (0.8 mg m^{-3}) and the depth of chlorophyll maximum (Z_{max}) shifts to the seabed exponentially. This profile is the indication of more light attenuation towards the sea bed. The subsurface chlorophyll maximum $[\text{Chl}]_{\text{max}}$ might occur due to the influences of benthic resuspension caused by tides and currents. Though the modelled chlorophyll profile typically follows the measured chlorophyll profile at this station, there is a slight deviation of the modelled chlorophyll profile observed at the intermediate depth. The $[\text{Chl}]_{\text{surf}}$ in surface waters off Point Calimere is relatively high (5 mg m^{-3}) during August 2012 (Fig. 7d). As the depth increases, [Chl] increases with a maximum value ($[\text{Chl}]_{\text{max}}$) around 7 m depth (20 mg m^{-3}) and then decreases following the surface $[\text{Chl}]_{\text{surf}}$. This trend typically follows the Gaussian distribution function, and thus there is better consistency between the modelled and measured Chl profiles (Fig. 7e). Another station towards the coast of Point Calimere during the same period was found to have a similar Gaussian profile indicating that the euphotic zone lies horizontally at a depth of 7 m. The corresponding measured and modelled SS profiles [SS(z)] are shown in Fig. 7a–c for these stations. Generally, the measured [SS] profiles are uniform along the depth and the power-law function captures their depth variations adequately.

4.1.3 Modelling of IOPs based on the Chl and SS profiles

The [Chl(z)] and [SS(z)] profiles constructed from the models were used to estimate the IOP profiles. Figure 8 shows the comparison of estimated (black colour) and predicted (grey colour) IOPs (plotted for three wavelengths 0.440, 0.555 and 0.676 μm) with the in situ IOP data, where the three clusters correspond to different waters (bottom – clear waters, middle – turbid coastal waters, top – eutrophic waters). Since a wide variety of waters is considered in this study, separate models were developed to treat the different water types. In Fig. 8

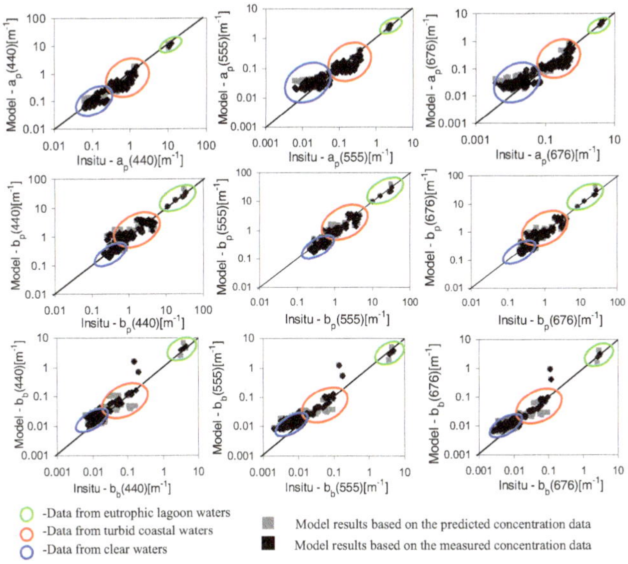

Figure 8. Comparisons of the modelled and in situ IOP data (particulate absorption a_p, particulate scattering b_p, and backscattering b_b) at three wavelengths: 0.440, 0.555 and 0.676 μm. Black colour represents model results based on the in situ concentrations ([Chl(z)] and [SS(z)]) and grey colour represents model results from the predicated profile data (from the estimated Chl and SS concentrations using remote-sensing reflectance data).

(top row), the model provides good estimates of particulate absorption a_p across the entire visible wavelengths (MRE -0.06–0.0774), although there is a slight overestimation (at higher wavelengths) especially in low-Chl waters. The predicted a_p values are also well consistent with in situ a_p data. Figure 8 (second row) presents the comparison of modelled and measured particulate scattering b_p values (at 0.440, 0.555 and 0.676 μm). Interestingly, the model performs well in different waters with a wide range of chlorophyll concentrations. The MRE values associated with this model are very low (0.1169 at 0.440 μm, 0.5816 at 0.555 μm, and 0.2316 at 0.676 μm). Similarly, b_p predicted by the model has low MRE values (-0.1473 at 0.440 μm, 0.3152 at 0.555 μm and 0.0726 at 0.676 μm). Comparison of the modelled backscattering (b_b) with in situ b_b data (at 0.440, 0.555 and 0.676 μm as shown in third row of Fig. 8) shows that the model works slightly better in relatively clear, turbid and eutrophic waters but slightly overestimates b_b values in clear waters. This problem could be attributed to measurement errors in very clear waters. The MRE values are -0.0735 at 0.440 μm, -0.1309 at 0.555 μm and -0.0876 at 0.676 μm for the estimated b_b using in situ Chl and SS data and -0.0604 at 0.440 μm, -0.1194 at 0.555 μm and 0.0763 at 0.676 μm for the predicted b_b using estimated Chl and SS data.

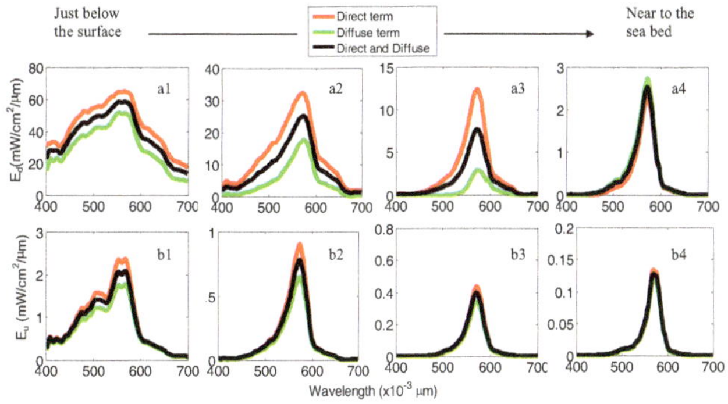

Figure 9. Spectral representation of the direct and diffuse terms of the underwater light fields for four discrete depths (**a1–a4** for E_d and **b1–b4** for E_u) in moderately turbid waters off Point Calimere. The red colour represents the direct term, the green colour indicates the diffuse term and the black colour represents the contribution of both direct and diffuse terms in the underwater light fields.

4.2 Spectral and depth variations of diffuse and direct terms

This section examines the spectral and depth variations of diffuse and direct terms of the underwater light fields for moderately turbid waters off Point Calimere. As shown in Fig. 9, the magnitude of the direct term (red) is high when compared to the diffuse term (green), especially in the upper and middle layers of the water column. However, as the light propagates through the air–water interface, the contribution of direct term makes a significant change in the downwelling irradiance at consequent depths. The contribution of the diffuse component is less since it is mainly influenced by the water column. Note that the direct and diffuse terms have a large variation in the blue-green region. Thus, considering either one of these terms in RT models is not appropriate in shallow/coastal waters and may lead to large errors along the depth (see Fig. 9a1–a4). To arrive at an approximate solution, the present study considers that both the terms weigh equally. Results of the combined effect of these terms are shown in Fig. 9a1–a4. The diffuse and direct terms are not significantly affected in upwelling irradiance (Fig. 9b1–b4) as compared to those observed in downwelling irradiance. Slight variation is noticed in the direct and diffuse terms of upwelling irradiance at near-surface depths due to the effect of downwelling irradiance. As the depth increases, both these terms generate the spectrum with almost the same magnitude of upwelling irradiance. Results of the combined effect of these two terms on underwater light fields are validated with in situ data and further discussed in the subsequent section.

4.3 Underwater light-field parameters

The results of the present RT model (with the consideration of solar zenith angle, sea surface slope, IOP-dependent phase function, bottom slope/material) based on the in situ

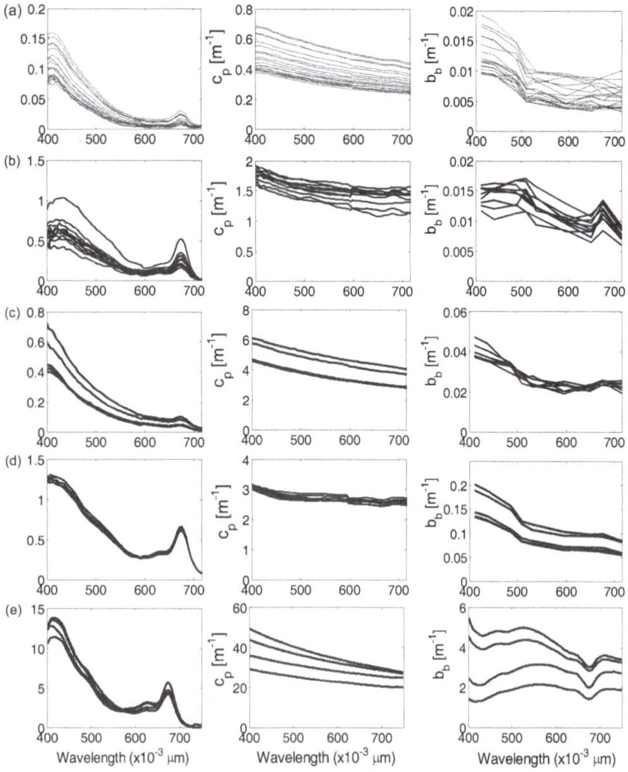

Figure 10. Spectral plots of the in situ IOPs (particulate absorption a_p, particulate attenuation c_p, and particulate backscattering b_b) from different waters. **(a)** Relatively clear waters off Chennai (St-1), **(b)** moderately turbid waters off Point Calimere (St-5), **(c)** turbid waters off Point Calimere (St-4), **(d)** phytoplankton-dominated harbour waters off Chennai (St-7) and **(e)** Eutrophic waters off Chennai (St-8–11).

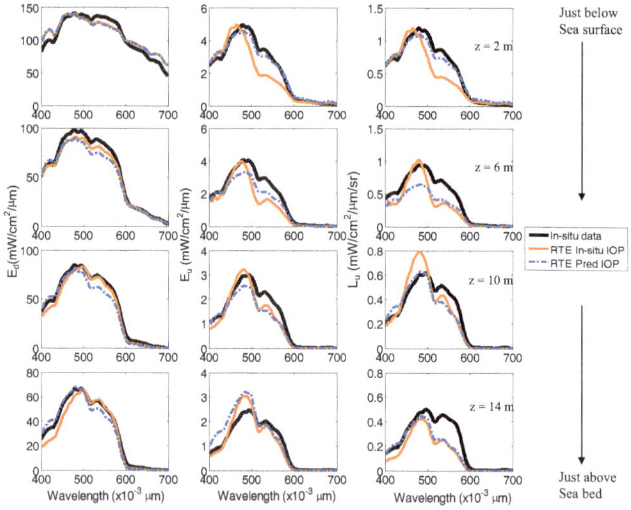

Figure 11. Spectral comparisons of the simulated and measured underwater light fields (E_d, E_u and L_u) for four discrete depths in clear waters off Chennai (St-1). The black colour represents the measured data, orange colour represents results from the present RT model using in situ IOP data and the blue colour represents results from the present RT model using the predicted IOPs from the remote-sensing reflectance data.

and predicted IOP profile data are evaluated by comparison with measured underwater radiometric data such as downwelling irradiance $E_d(\lambda, z)$, upwelling irradiance $E_u(\lambda, z)$ and upwelling radiance $L_u(\lambda, z)$. The predicted (from remote-sensing reflectance data) in situ IOP profile data were used as inputs for the present RT model. The IOP data measured from different waters (Fig. 10) include the particulate absorption (a_p), particulate attenuation (c_p) and backscattering (b_b).

4.3.1 Clear ocean waters

Figure 11 shows the comparisons of simulated and measured underwater light fields (E_d, E_u and L_u) for four discrete water depths in relatively clear waters off Chennai. Looking at the E_d spectra (orange colour – present RT model using in situ IOP data and blue colour – present RT model using Pred IOP data from remote-sensing reflectance), the maximum value is seen at $0.490\,\mu$m throughout the water column. As the depth increases, E_d becomes attenuated in the blue region (0.400–$0.500\,\mu$m) but progressively more attenuated in the red region ($> 0.600\,\mu$m). The E_d values from the present RT model using the in situ and predicted IOP data are closely consistent with measured E_d across the entire visible wavelengths. The simulated E_d yields very low MRE (at $0.555\,\mu$m), 0.03 and -0.01 for the present RT model using in situ IOP data and predicted IOP data, respectively. Conversely, the spectral pattern of E_u is different from that of E_d (second column) displaying two peaks – one at $0.490\,\mu$m (primary) and another at $0.520\,\mu$m (secondary). The E_u from the present RT model using the in situ IOP data closely matches with measured E_u at most wavelengths, except in the region of secondary peak at $0.520\,\mu$m. The MRE (at $0.555\,\mu$m) is very low (-0.3) for the present model using the predicted IOP data. Similar peaks – primary peak around $0.500\,\mu$m and secondary peak around $0.550\,\mu$m – are also seen in the L_u spectrum. As the depth increases the secondary peak becomes more pronounced due to an increase in turbidity level. The deviation in L_u is very small for the present RT model as its predicted L_u spectra are well consistent with the measured L_u spectra, because of the inclusion of the effective bottom reflectance instead of the material reflectance.

4.3.2 Moderately turbid waters

Figure 12 shows the spectral variations of simulated and measured underwater light-field data for four discrete depths in moderately clear waters (with elevated chlorophyll in relatively deep waters as compared to the previous case) off Point Calimere (August 2012). At this location, the vertical profile of chlorophyll typically followed the Gaussian pattern, with

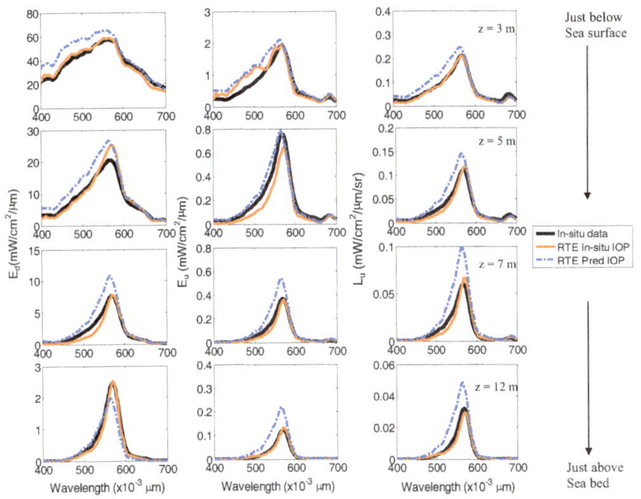

Figure 12. Spectral comparisons of the simulated and measured underwater light fields (E_d, E_u and L_u) for four discrete depths in moderately turbid waters off Point Calimere (St-5). The black colour represents the measured data, orange colour represents results from the present RT model using in situ IOP data, and the blue colour represents results from the present RT model using the predicted IOPs from the remote-sensing reflectance data.

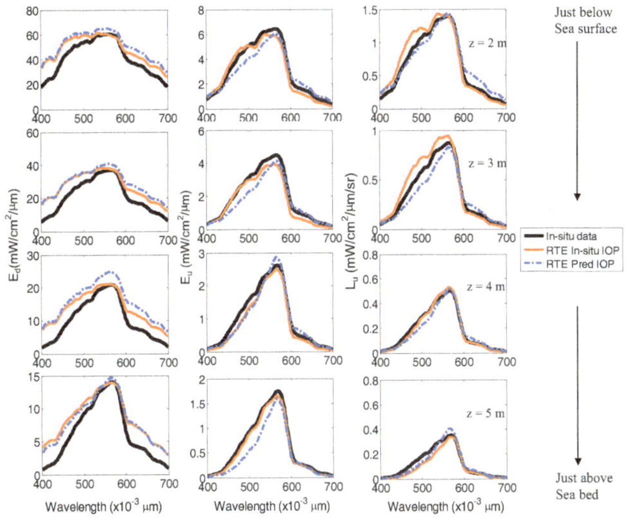

Figure 13. Spectral comparisons of the simulated and measured underwater light fields (E_d, E_u and L_u) for four discrete depths in turbid coastal waters off Point Calimere (St-4). The black colour represents the measured data, orange colour represents results from the present RT model using in situ IOP data, and the blue colour represents results from the present RT model using the predicted IOPs from the remote-sensing reflectance data.

the surface chlorophyll 6.2 mg m^{-3} and maximum chlorophyll 18.5 mg m^{-3} at 7 m. The range of suspended sediments varied from 10.9 to 15.3 g m^{-3}. The solar zenith angle measured was 44.9°. Thus, the spectral pattern of E_d (first column) just below the surface resembles that of clear waters. As the depth increases, the pattern changes following the turbid water case with a major peak at 0.555 μm. The E_d spectra simulated from the present RT model using in situ IOP data have fairly good agreement with the measured E_d. Since the predicted IOPs are low at the surface, the magnitude of the predicted E_d is relatively high when compared to the measured and simulated E_d. The spectral pattern of E_u (second column) is slightly different for these waters as the primarily peak is located around 0.555 μm and a secondary peak around 0.685 μm due to the chlorophyll fluorescence. Similar features are also observed in the L_u spectra. The shape and magnitude of the E_u and L_u from the present RT model based on the in situ IOP data agrees well with measured data (with MRE values 0.1 and 0.04, respectively). By contrast, the E_u and L_u spectra generated from the present RT model using the predicted IOP data are improved although showing a slight overestimation in the green domain.

4.3.3 Turbid coastal waters

Figure 13 shows the spectral comparisons of simulated and measured underwater light fields (E_d, E_u and L_u) for four discrete depths in turbid coastal waters off Point Calimere during August 2013. This station is in the vicinity of the coast with a depth of 6.3 m, where the in situ measurements of apparent optical properties (AOPs), IOPs and other parameters were made when the solar zenith angle was 4.55°. At this station, the benthic resuspension and sediment transport noticeably increased the magnitude of the IOPs and turbidity. The measured E_d spectra (first column) show the maximum values at green wavelengths (0.555 μm) and minimum values at blue and red wavelengths throughout the water column, which are the characteristic features of turbid waters. Note that the magnitude of E_d spectra from the present model using the in situ and predicted IOP data is closer to the measured E_d spectra, with low MRE values of 0.006 and 0.03, respectively. The E_u and L_u spectra produced by the present RT model based on the in situ and predicted IOP data also match well with the measured E_u and L_u spectra (with relatively low MRE values of −0.06 to 0.1 and −0.1 to −0.09, respectively). Better results associated with the present RT model are due to the modified boundary conditions and new phase function which varies along the depth.

4.3.4 Phytoplankton-dominated harbour waters

Figure 14 depicts the differences between simulated and measured underwater light fields in phytoplankton-dominated harbour waters of Chennai. Since these waters were well mixed vertically, the vertical profiles of IOPs predicted by the respective models were considered continuous along the depth. The spectra of E_d, E_u and L_u obtained from these waters appear slightly different from the previous cases because of a well-pronounced peak around 0.555 μm and a fluorescence peak shifted from 0.685 μm to 0.700 μm. Inter-

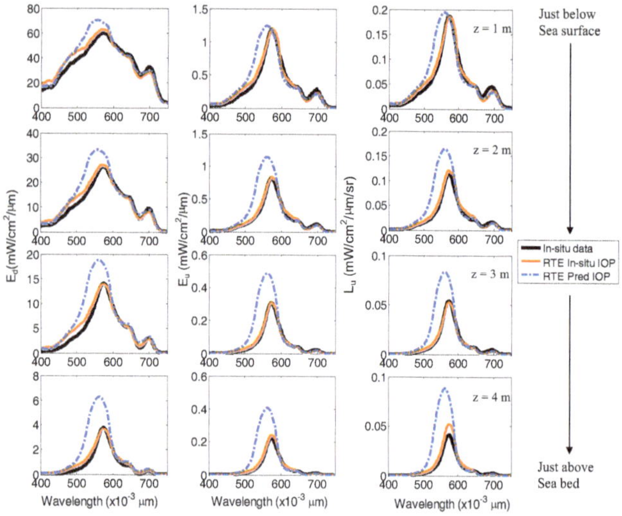

Figure 14. Spectral comparisons of the simulated and measured underwater light fields (E_d, E_u and L_u) for four discrete depths in phytoplankton-dominated harbour waters off Chennai (St-7). The black colour represents the measured data, orange colour represents results from the present RT model using in situ IOP data, and the blue colour represents results from the present RT model using the predicted IOPs from the remote-sensing reflectance data.

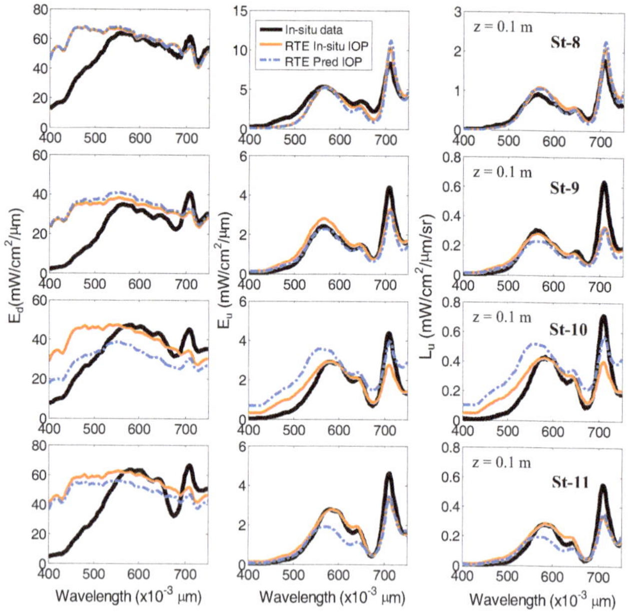

Figure 15. Spectral comparisons of the simulated and measured underwater light fields (E_d, E_u and L_u) just below the surface in eutrophic waters off Chennai (St-8–11). The black colour represents the measured data, orange colour represents results from the present RT model using in situ IOP data, and the blue colour represents results from the present RT model using the predicted IOPs from the remote-sensing reflectance data.

estingly, the E_d, E_u and L_u spectra of the present RT model from the in situ IOP data are well consistent with measured data across the entire visible wavelengths. As a result, its MRE values are reduced to 0.04 for E_d, 0.029 for E_u and L_u. Slight deviations are observed in the E_d, E_u and L_u spectra generated by the present RT model, which could be attributed to errors associated with the predicted IOPs and the assumption of the vertically homogenous water column (i.e., constant IOPs along the depth). This would eventually increase the MRE values to 0.3 for E_d, 0.2 for E_u and 0.26 for L_u.

4.3.5 Eutrophic waters

The performance of the RT models is also examined in highly complex eutrophic waters, which exhibit much higher magnitudes in IOP spectra compared to other waters (Fig. 10). Figure 15 provides the spectral comparisons of simulated and measured underwater light fields from four different discrete stations (profile data not collected due to shallow water body). At these stations, both Chl and SS reached beyond 73 and $71\,\mathrm{g\,m^{-3}}$, respectively (in addition to high CDOM) ultimately reducing the magnitude of E_d (more at station St-11) in the blue wavelength region and shifting the position of the red peak towards $0.715\,\mu$m (by the combined effect of fluorescence and backscattering by the phytoplankton cells). Thus, the RT models failed to generate the measured E_d in the blue wavelength region although their results are reasonably good in the green and red regions. The mean MRE values (for E_d at $0.555\,\mu$m) are 0.003 and 0.016 for the present RT model using the predicted IOPs and in situ IOPs, respectively. Interestingly, the present RT model based on the predicted and in situ IOP data gives accurate E_u and L_u in terms of magnitude and shape at all four stations. The shifted red peaks are also better captured by the present RT model due to the inclusion of the appropriate source function. The MRE values (at $0.555\,\mu$m) are 0.07 and -0.06 for the E_u and L_u (from the predicted IOP data) and 0.08 and -0.05 for the E_u and L_u (from the in situ IOP data). These results suggest that the present RT model can be used combined with remote-sensing data to simulate the underwater light fields in a wide variety of waters.

4.4 Depth variations in apparent optical properties

This study is further extended for the calculation of the AOPs such as irradiance reflectance $R = E_\mathrm{u}/E_\mathrm{d}$, and subsurface remote-sensing reflectance $r_\mathrm{rs} = L_\mathrm{u}/E_\mathrm{d}$ from the underwater light-field parameters. Figure 16a shows the comparison of the depth profiles of the model and in situ irradiance reflectances (R) at the wavelength of $0.555\,\mu$m for moderately turbid water. The R profiles increase (varying) with depth mainly due to the inhomogeneous water column conditions. Figure 16b shows the model and in situ subsurface remote-sensing reflectance r_rs profiles for moderately turbid water at the wavelength of $0.555\,\mu$m. Note that the r_rs profiles are not constant along the depth. Since the AOPs mainly depend on the concentration of the constituents in the water column

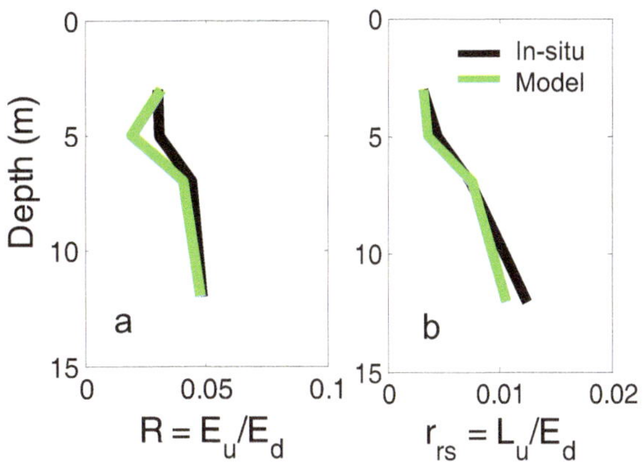

Figure 16. **(a)** Comparison of the measured and modelled reflectance ($R = E_u/E_d$) at 0.555 μm along the depth for moderately turbid water. **(b)** Comparison of the measured and modelled subsurface remote-sensing reflectance ($r_{rs} = L_u/E_d$) at 0.555 μm along the depth for moderately turbid water. There is good agreement between the modelled and measured reflectance profiles at this station.

as well as light-field conditions, variations in the profiles of AOPs are expected in coastal waters (Pravin and Shanmugam, 2014). These results show that there is only a minimum likelihood for neither uniform nor constant AOPs (R, r_{rs}) along the depth in such waters.

5 Summary and conclusion

Existing radiative transfer models to simulate the underwater light fields often lead to large uncertainties in turbid coastal and eutrophic waters. This could be due to the fact that some models treat the water column as homogenous (not considering the direct term), while others consider the water column as inhomogeneous (not considering the diffuse term). Assuming a constant phase function along the water column, a flat or oversimplified sea surface in a random manner, and considering the bottom material reflectance instead of the effective bottom reflectance (based on bottom slope and material reflectance) have already been reported to introduce large errors in the simulated underwater light fields (Sundarabalan et al., 2013). Thus, it is necessary to develop a reliable RT model to generate the underwater light-field parameters in a wide variety of waters.

To overcome the above issues, the proposed RT model can now handle more complex interactions of light with particulate matters with different surface and bottom boundary conditions. The direct and diffuse terms are included in this model to deal with the homogenous and inhomogeneous effects in the water column. The new sea surface boundary condition and the estimated sea surface transmittance have significant effects on the downwelling irradiance (E_d). As the sea surface slope increases, transmission through the air–water interface to the water column increases, but effects of the sea surface are most prominent when the sun is away from the nadir (Jin et al., 2006). The significance of the modified phase function is especially noticed in the simulated underwater light fields (Twardowski et al., 2001; Mobley et al., 2002; Freda and Piskozub, 2007; Sundarabalan et al., 2013). Since the sea bed is not uniform or flat, the inclusion of bottom morphology along with the material reflectance (effective bottom reflectance) in the bottom boundary shows better upwelling radiance L_u (Zaneveld and Boss, 2003). The bottom reflectance affects the entire water column, and is treated properly with IOPs along the depth (Lee et al., 1998). The source function based on the chlorophyll is included in the model and its effect is clearly seen in the simulated E_u and L_u with a florescence peak at 0.685 μm (Gower et al., 2004). Finally, the calculated transmittance $T(z)$ is purely based on the IOPs (Haltrin and Kattawar, 1993; Haltrin, 1998a, b) and the reflectance $R(z)$ based on the IOPs and bottom effects (Lee et al., 1998). The results of the present RT model based on the in situ IOPs have good agreement with the measured underwater light-field parameters.

In the context of satellite remote sensing, the present RT model has been extended to predict the underwater light fields from the remotely sensed IOP profiles and its results have been validated using the measured underwater radiometric data and those simulated from the same RT model using the measured IOP profile data. The remotely sensed IOP profiles were derived from the new parameterisations developed by considering the different types of waters. These IOP profiles were determined based on the estimated Chl and SS. The chlorophyll profile was determined by the Gaussian distribution function (Lewis et al., 1983) and SS profile by the power-law function. The vertical structures of these estimated Chl and SS had good agreement with the measured profile data. It was found that the predicted underwater light-field parameters (E_d, E_u and L_u) from the present RT model using the Pred_IOPs are reasonably good when compared with the measured radiometric data as well as those obtained from the same RT model using the in situ IOP data in clear waters, turbid waters, phytoplankton-dominated waters and eutrophic waters. The differences between the predicted and measured E_d, E_u and L_u may arise from the bio-optical parameterisations used to estimate the IOPs. Perhaps, the non-uniform trend of the chlorophyll pattern along the water column is also expected to cause these differences in the predicted E_d, E_u and L_u by the present RT model. Nevertheless, this comprehensive study demonstrates that the present RT model is capable of dealing with homogenous and inhomogeneous water conditions and has the ability to generate more realistic underwater light-field parameters (E_d, E_u and L_u) using the measured IOPs as well as those estimated from remote-sensing data. The present RT model is a viable alternative to existing models and has an important implication for remote sensing as well.

Acknowledgements. This research was supported by INCOIS under the grant (OEC1314117INCOPSHA) of the SATCORE program. We thank D. Rajasekhar, The Head, Vessel Management Cell (VMC), and the Director of National Institute of Ocean Technology (NIOT) for providing Coastal Research Vessels to Indian Institute of Technology (IIT) Madras, Chennai, India for conducting various bio-optical measurements in coastal waters off Point Calimere and Chennai on the southeast coast of India. We are thankful to the scientists N. Ravi and K. Shashikumar of VMC for arranging the vessels to make the necessary measurements. We are thankful to the anonymous reviewers for their valuable comments to improve the quality of this manuscript.

Edited by: O. Zielinski

References

Ahn, Y. H.: Optical properties of biogenous and mineral particles present in the ocean. Application: inversion of reflectance, PhD Thesis, Paris-VI University, 1990.

Ahn, Y. H. and Shanmugam, P.: Detecting red tides from Satellite Ocean color observations in optically complex Northeast-Asia coastal waters, Remote Sens. Environ., 103, 419–437, 2006.

Antoine, D., André, J. M., and Morel, A.: Oceanic primary production: 2.Estimation at global scale from satellite (coastal zone colour scanner) chlorophyll, Global Biogeochem. Cy., 10, 57–69, 1996.

Arrigo, K. R., Matrai, P. A., and van Dijken, G. L.: Primary productivity in the Arctic Ocean: Impacts of complex optical properties and subsurface chlorophyll maxima on large-scale estimates, J. Geophys. Res., 116, C11022, doi:10.1029/2011JC007273, 2011.

Babin, M., Stramski, D., Ferrari, G. M., Claustre, H., Bricaud, A., Obolensky, G., and Hoepffner, N.: Variations in the light absorption coefficients of phytoplankton, nonalgal particles, and dissolved organic matter in coastal waters around Europe, J. Geophys. Res., 108, C73211, doi:10.1029/2001JC000882, 2003.

Bricaud, A., Babin, M., Morel, A., and Claustre, H.: Variability in the chlorophyll-specific absorption coefficients of natural phytoplankton: analysis and parameterization, J. Geophys. Res., 100, 13321–13332, 1995.

Cherkasheva, A., Nöthig, E.-M., Bauerfeind, E., Melsheimer, C., and Bracher, A.: From the chlorophyll a in the surface layer to its vertical profile: a Greenland Sea relationship for satellite applications, Ocean Sci., 9, 431–445, doi:10.5194/os-9-431-2013, 2013.

Dmitriev, E. V., Khomenko, G., Chami, M., Sokolov, A. A., Churilova, T. Y., and Korotaev, G. K.: Parameterization of light absorption by components of seawater in optically complex coastal waters of the Crimea Peninsula (Black Sea), Appl. Opt., 48, 1249–1261, 2009.

Freda, W. and Piskozub, J.: Improved method of Fournier–Forand marine phase function parameterization, Opt. Exp., 20, 12763–8, 2007.

Gjerstad, K. I., Stamnes, J. J., Hamre, B., Lotsberg, J. K., Yan, B., and Stamnes, K.: Monte Carlo and Discrete-Ordinate Simulations of Irradiances in the Coupled Atmosphere-Ocean System, Appl. Opt., 42, 2609–2621, 2003.

Gokul, R. E., Shanmugam, P., and Sundarabalan, V. B.: Inversion models for deriving inherent optical properties and their vertical profiles in coastal waters, Cont. Shelf Res., 84, 120–138, 2014.

Gordon, H. R. and Morel, A.: Remote assessment of ocean colour for interpretation of satellite visible imagery: A review, Lecture notes on coastal and estuarine studies, Springer Verlag Heidelberg, Germany, 1983.

Gordon, H. R., Brown, O. B., and Jacobs, M. M.: Computed relationships between the inherent and apparent optical properties of a flat homogeneous ocean, Appl. Opt., 14, 417–427, 1975.

Gower, J. F. R., Brown, L., and Borstad, G. A.: Observation of chlorophyll fluorescence in west coast waters of Canada using the MODIS satellite sensor, Can. J. Remote Sens., 30, 17–25, 2004.

Haltrin, V. I.: Self-consistent approach to the solution of the light transfer problem for irradiances in marine waters with arbitrary turbidity, depth, and surface illumination. I. Case of absorption and elastic scattering, Appl. Opt., 37, 3773–3784, 1998a.

Haltrin, V. I.: Apparent optical properties of the sea illuminated by Sun and sky: case of the optically deep sea, Appl. Opt., 37, 8336–8340, 1998b.

Haltrin, V. I.: One-parameter two-term Henyey–Greenstein phase function for light scattering in seawater, Opt. Exp., 41, 1022–8, 2002.

Haltrin, V. I.: About nonlinear dependence of Remote Sensing and Diffuse reflection coefficients on Gordon's Parameter, Proceedings of the II International Conference "Current Problems in Optics of Natural Waters," ONW'2003, edited by: Levin, I. and Gilbert, G., St. Petersburg, Russia, 363–369, 2003.

Haltrin, V. I. and Kattawar, G. W.: Self-consistent solutions to the equation of transfer with elastic and inelastic scattering in oceanic optics, I. Model, Appl. Opt., 32, 5356–5367, 1993.

Hoepffner, N. and Sathyendranath, S.: Bio-optical Characteristics of coastal waters: Absorption spectra of phytoplankton and pigment distribution in the western North Atlantic, Limnol. Oceanogr., 37, 1660–1679, 1992.

Hollstein, A. and Fischer, J.: Radiative transfer solutions for coupled atmosphere ocean systems using the matrix operator technique, J. Quant. Spectrosc. Radiat. Transfer., 113, 536–548, 2012.

Huot, Y., Morel, A., Twardowski, M. S., Stramski, D., and Reynolds, R. A.: Particle optical backscattering along a chlorophyll gradient in the upper layer of the eastern South Pacific Ocean, Biogeosciences, 5, 495–507, doi:10.5194/bg-5-495-2008, 2008.

Jin, Z., Charlock, T. P., Rutledge, K., Stamnes, K., and Wang, Y.: Analytical solution of radiative transfer in the coupled atmosphere–ocean system with a rough surface, Appl. Opt., 45, 7443–7455, 2006.

Kirk, J. T. O.: Monte Carlo study of the nature of the underwater light field in, and the relationships between optical properties of, turbid yellow waters, Aust. J. Mar. Freshwater Res., 32, 517–532, 1981.

Lee, W. L. and Liou, K. N.: A coupled atmosphere–ocean radiative transfer system using the analytic Four-Stream approximation, J. Atmos. Sci., 64, 3681–3694, 2007.

Lee, Z. P., Carder, K. L., Mobley, C. D., Steward, R. G., and Patch, J. S.: Hyperspectral remote sensing for shallow waters. I. A semi-analytical model, Appl. Opt., 37, 6329–6338, 1998.

Lee, Z. P., Carder, K. L., Mobley, C. D., Steward, R. G., and Patch, J. S.: Hyperspectral remote sensing for shallow waters: 2. Deriving bottom depths and water properties by optimization, Appl. Opt., 38, 3831–3843, 1999.
Lewis, M. R., Cullen, J. J., and Platt, T.: Phytoplankton and thermal structure in the upper ocean: Consequences of non-uniformity in chlorophyll profile, J. Geophys. Res., 88, 2565–2570, 1983.
Martin, J., Tremblay, J. E., Gagnon, J., Tremblay, G., Lapoussiere, A., Jose, C., Poulin, M., Gosselin, M., Gratton, Y., and Michel, C.: Prevalence, structure and properties of subsurface chlorophyll maxima in Canadian Arctic waters, Mar. Ecol.-Prog. Ser., 412, 69–84, 2010.
Matsuoka, A., Hill, V., Huot, Y., Babin, M., and Bricaud, A.: Seasonal variability in the light absorption properties of western Arctic waters: Parameterization of the individual components of absorption for ocean colour applications, J. Geophys. Res., 116, C02007, doi:10.1029/2009JC005594, 2011.
Mobley, C. D.: Light and water: radiative transfer in natural waters, San Diego: Academic Press, Inc., 1994.
Mobley, C. D., Sundman, L. K., and Boss, E.: Phase function effects on oceanic light fields, Appl. Opt., 41, 1035–50, 2002.
Morel, A.: Optical modelling of the upper ocean in relation to its biogenous matter content (Case 1 waters), J. Geophys. Res., 93, 10749–10768, 1988.
Morel, A.: Light and marine photosynthesis: a spectral model with geochemical and climatological implications, Prog. Oceanogr., 26, 263–306, 1991.
Morel, A. and Berthon, J. R.: Surface pigments, algal biomass profiles, and potential production of the euphotic layer: Relationships reinvestigated in view of remote-sensing application, Limnol. Oceanogr., 34, 1545–1562, 1989.
O'Reilly, J. E., Maritorena, S., Mitchell, B. G., Siegel, D. A., Carder, K. L., Garver, S. A., Kahru, M., and McClain, C.: Ocean colour algorithms for SeaWiFS, J. Geophys. Res., 103, 24937–24953, 1998.
O'Reilly, J. E., Maritorena, S., O'Brien, M. C., Siegel, D. A., Toole, D., Menzies, D., Smith, R. C., Mueller, J. L., Mitchell, B. G., Kahru, M., Chavez, F. P., Strutton, P., Cota, G. F., Hooker, S. B., McClain, C. R., Carder, K. L., Karger, F. M., Harding, L., Magnuson, A., Phinney, D., Moore, G. F., Aiken, J., Arrigo, K. R., Letelier, R., and Culver, M.: Ocean Colour Chlorophyll a Algorithms for SeaWiFS, OC2 and OC4: Version 4, NASA Technical Memorandum, 11, 9–23, 2000.
Pierson, W, J. and Moskowitz, L.: Proposed spectral form for fully developed wind seas based on similarity theory of S. A. Kitaigorodskii, J. Geophys. Res., 69, 5181–5190, 1964.
Pegau, W. S., Gray, D., and Zaneveld, J. R. V.: Absorption and attenuation of visible and near-infrared light in water: dependence on temperature and salinity, Appl. Opt., 36, 6035–6046, 1997.
Platt, T. and Sathyendranath, S.: Oceanic Primary Production: Estimation by Remote Sensing at Local and Regional Scales, Science, 241, 1613–1620, 1988.
Pope, R. M. and Fry, E. S.: Absorption spectrum (380–700 nm) of pure water. II. Integrating cavity measurements, Appl. Opt., 36, 8710–8723, 1997.
Pravin, J. D. and Shanmugam, P.: A new model for the subsurface irradiance reflectance in clear and turbid waters, Opt. Exp., 22, 9548–9566, 2014.
Prieur, L. and Sathyendranath, S.: An optical classification of coastal and oceanic waters based on the specific spectral absorption curves of phytoplankton pigments, dissolved organic matter, and other particulate materials, Limnol. Oceanogr., 26, 671–689, 1981.
Ramakrishnan, R., Rajawat, A. S., and Chauhan, O. S.: Suspended sediment concentration profiles from synoptic satellite observations, IEEE J. Sel. Topics Appl. Earth Observ. Remote Sens., 6, 2051–2057, 2013.
Sathyendranath, S. and Platt, T.: Remote sensing of ocean chlorophyll: consequence of non-uniform pigment profile, Appl. Opt., 28, 490–495, 1988.
Shanmugam, P.: A new bio-optical algorithm for the remote sensing of algal blooms in complex ocean waters, J. Geophys. Res., 116, C04016, doi:10.1029/2010JC006796, 2011a.
Shanmugam, P.: New models for retrieving and partitioning coloured dissolved organic matter in the global oceans. Implications for remote sensing, Remote Sens. Environ., 115, 1501–1521, 2011b.
Shanmugam, P., Ahn, Y. H., Ryu, J. H., and Sundarabalan, V. B.: An evaluation of inversion models for retrieval of inherent optical properties from ocean colour in coastal and open sea waters around Korea, J. Oceanogr., 66, 815–830, 2010.
Shanmugam, P., Sundarabalan, V. B., Ahn, Y. H., and Ryu, J. H. A.: New Inversion Model to Retrieve the Particulate Backscattering in Coastal/Ocean Waters, IEEE Trans. Geosci. Remote Sens., 49, 2463–74, 2011.
Smith, R. C.: Ocean colour for the estimation of Global marine primary productivity, Workshop on the Global Ocean Flux study, Woods Hole, Masschusetts, 103–110, 1984.
Smith, R. C. and Baker, K. S.: Optical properties of the clearest natural waters (200–800 nm), Appl. Opt., 20, 177–184, 1981.
Stamnes, K., Tsay, S., Wiscombe, W., and Jayaweera, K.: Numerically stable algorithm for discrete-ordinate-method radiative transfer in multiple scattering and emitting layered media, Appl. Opt., 27, 2502–2509, 1988.
Stramska, M. and Stramski, D.: Effects of a non-uniform vertical profile of chlorophyll concentration on remote-sensing reflectance of the ocean, Appl. Opt., 44, 1735–1747, 2005.
Stramski, D., Bricaud, A., and Morel, A.: Modelling the inherent optical properties of the ocean based on the detailed composition of the planktonic community, Appl. Opt., 40, 2929–2945, 2001.
Sundarabalan, V. B., Shanmugam, P., and Manjusha, S. S.: Radiative transfer modelling of upwelling lightfield in coastal waters, J. Quant. Spectrosc. Radiat. Transfer., 121, 30–44, 2013.
Twardowski, M. S., Boss, E., Macdonald, J. B., Pegau, W. S., Barnard, A. H., and Zaneveld, J. R. V.: A model for estimating bulk refractive index from the optical backscattering ratio and the implications for understanding particle composition in case I and case II waters, J. Geophys. Res., 106, 14129–14142, 2001.
Uitz, J., Claustre, H., Morel, A., and Hooker, S. B.: Vertical distribution of phytoplankton communities in open ocean: An assessment based on surface chlorophyll, J. Geophys. Res., 111, C08006, doi:10.1029/2005JC003207, 2006.
Wang, Y., Liu, D., Song, K., Du, J., Wang, Z., Zhang, B., Tang, X., Lei, X., and Wu, Y.: Characterization of Water Constituents Spectra Absorption in Chagan Lake of Jilin Province, Northeast China, Chin. Geogra. Sci., 21, 34–345, 2011.

Wozniak, S. B. and Stramski, D.: Modelling the optical properties of mineral particles suspended in seawater and their influence on ocean reflectance and chlorophyll estimation from remote sensing algorithms, Appl. Opt., 43, 3489–3503, 2004.

Zaneveld, J. R. V. and Boss, E.: The influence of bottom morphology on reflectance: theory and two-dimensional geometry model, Limnol. Oceanogr., 48, 374–379, 2003.

Zhang, Y., Liu, M., Qin, B., Woerd, H. J., Li, J., and Li, Y.: Modelling remote-sensing reflectance and retrieving Chlorophyll-a concentration in extremely turbid case-2 waters (lake Taihu, China), IEEE Trans. Geosci. Remote Sens., 47, 1937–1948, 2009.

Phytoplankton blooms on the western shelf of Tasmania: evidence of a highly productive ecosystem

J. Kämpf

School of the Environment, Flinders University, Adelaide, Australia

Correspondence to: J. Kämpf (jochen.kaempf@flinders.edu.au)

Abstract. Satellite-derived chlorophyll *a* data using the standard NASA-OC3 (ocean colour) algorithm are strongly biased by coloured dissolved organic matter and suspended sediment of river discharges, which is a particular problem for the western Tasmanian shelf. This work reconstructs phytoplankton blooms in the study region using a quadratic regression between OC3 data and chlorophyll fluorescence based on the fluorescence line height (FLH) data. This regression is derived from satellite data of the nearby Bonney upwelling region, which is devoid of river influences. To this end, analyses of 10 years of MODIS-aqua satellite data reveal the existence of a highly productive ecosystem on the western Tasmanian shelf. The region normally experiences two phytoplankton blooms per annum. The first bloom occurs during late austral summer months as a consequence of upwelling-favourable coastal winds. Hence, the western Tasmanian shelf forms a previously unknown upwelling centre of the regional upwelling system, known as Great South Australian Coastal Upwelling System. The second phytoplankton bloom is a classical spring bloom also developing in the adjacent Tasman Sea. The author postulates that this region forms another important biological hot spot for the regional marine ecosystem.

1 Introduction

Physical processes that enrich the euphotic zone with nutrients are principal agents of coastal phytoplankton blooms. Such processes include upwelling, storm-induced mixing, internal waves and river plumes. Despite elevated nutrient levels, high turbidity levels in river plumes can suppress diatom growth (e.g. Chen et al., 2003). Except for the situation of coastal upwelling, vertical density stratification generally supports phytoplankton production, in particular when the surface mixed layer is relatively shallow and coincides with the euphotic zone. This situation supports the development of the spring bloom, which is characteristic of temperate North Atlantic, sub-polar and coastal waters (Mann and Lazier, 2006).

The southern shelves of Australia host a large seasonal coastal upwelling system (Kämpf et al., 2004; Kämpf, 2010). In response to south-easterly coastal winds, upwelling events occur in austral summer months (December to April). This upwelling system, referred to as the Great South Australian Coastal Upwelling System, consists of three upwelling centres (Fig. 1): the long-known Bonney upwelling (Rochford, 1977; Lewis, 1981; Schahinger, 1987; Griffin et al., 1997) along the so-called Bonney coast, and an upwelling centre off the southern tip of the Eyre Peninsula (Kämpf et al., 2004). The latter region plays a vital role in the life cycles of sardine (*Sardinops sagax*), anchovy (*Engraulis australis*), and southern bluefin tuna (*Thunnus maccoyii*) (Ward et al., 2006). Another smaller upwelling centre develops occasionally off the southwest coast of Kangaroo Island (see Kämpf et al., 2004).

This study focusses on the western Tasmanian shelf. Throughout the year, the Zeehan Current runs south-eastward confined to the shelf break along the continental shelf edge of western Bass Strait and western Tasmania (Cresswell, 2000). This current is the extension of the South Australian Current, which itself is the continuation of the Leeuwin Current (Ridgway and Condie, 2004), but seasonally also entrains warm water formed during summer months on the shelves of the western Great Australian Bight (Herzfeld, 1997). Currents within Bass Strait are created by tides, winds, incident continental shelf waves and

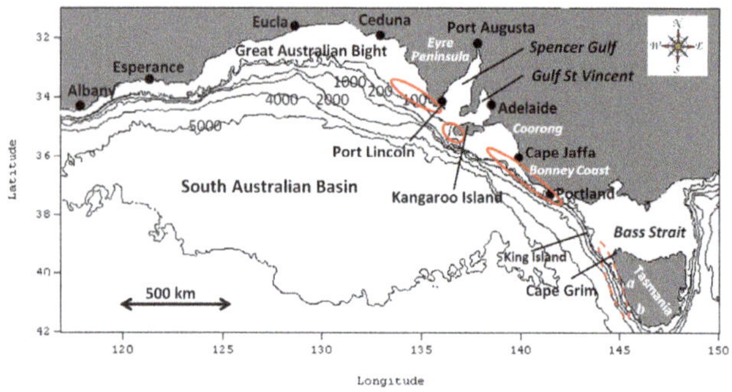

Figure 1. The geography of Australia's southern shelves. Isobath depths are in metres. Solid-line ellipses display known locations of coastal upwelling centres. The dashed-line ellipse highlights the region investigated in this paper. Letters a and b indicate the locations of the mouths of the Macquarie Harbour estuary and the Davey River.

density-driven flows (e.g. Sandery and Kämpf, 2005). Bass Strait is relatively shallow (~ 50–70 m) and the main pathway of currents is generally eastward (Sandery and Kämpf, 2007). Prominent oceanographic features of Bass Strait are the existence of tidal mixing fronts on both sides of the strait (Sandery and Kämpf, 2005), and the wintertime formation of a density-driven overflow on the eastern side of the strait in the vicinity of the Bass Canyon, known as the Bass Strait Cascade (Tomczak, 1985). Based on sparse field data, Gibbs et al. (1986) concluded that nutrient levels in Bass Strait are low overall ($< 1\,\mu$M in nitrate) except along the eastern edge where nutrient concentrations reach high levels (up to $7\,\mu$M in nitrate) in winter. According to these authors, chlorophyll *a* levels in the Bass Strait are also generally low (< 0.5 mg m^{-3}) but show the highest concentrations over the adjacent shelf, again in winter.

Earlier work (Connolly and Von der Borch, 1967) postulated that upwelling of cold sub-Antarctic waters is the main reason for the occurrence of extensive temperate carbonates on the southern Australian shelves. Isotopic studies (e.g. Wass et al., 1970) validated this upwelling model for the formation of cold-water carbonates. Interestingly, modern cold-water carbonate is also the predominant sediment on the western Tasmanian shelf (Rao and Green, 1983). Isotope-based findings of Rao and Adibi (1992) demonstrate upwelling seawater as the main process responsible for the formation of carbonates in western Tasmania. Motivated by the findings of Rao and Adibi (1992), this work explores phytoplankton blooms on the western Tasmanian shelf and their underlying physical processes.

2 Methodology

This work is based on satellite data, coastal wind data and time series of river discharges with a focus on the western Tasmanian shelf in comparison with the Bonney upwelling region. Averages of phytoplankton-related parameters were constructed from the satellite data of predefined target areas (see Fig. 1) as the average of 10 pixel data of the highest values for Sea-viewing Wide Field-of-view Sensors (SeaWiFS), and 50 pixels of the highest values for MODIS-aqua; this corresponds to combined areas of 810 and 800 km^2, respectively. Values were classified as invalid from a visual analysis of the level of cloud bias within the target areas.

Ocean colour data are based on standard blue / green ratio algorithms for chlorophyll *a* retrieval, which can substantially overestimate phytoplankton blooms in coastal waters in the presence of coloured dissolved organic matter (CDOM or *Gelbstoff*) and suspended sediment (e.g. Garcia et al., 2006; Shanmugam, 2011). SeaWiFS ocean colour data are based on the OC4 (ocean colour) algorithm, whereas MODIS-aqua data are often based on the OC3 algorithm (O'Reilly et al., 2000). Two different time periods are considered. The first period (1 January 1998 to 31 December 2000), also studied by Kämpf et al. (2004), is adopted to establish initial evidence of the phytoplankton dynamics on the western Tasmanian shelf. Data have been extracted from the South East Fishery (SEF) ocean movies (David Griffin, CSIRO, http://www.marine.csiro.au/~griffin/SEF/) based on the SeaWiFS database (9 km spatial resolution). It should be noted that of the 136 eight-day segments of OC4 data, 35 segments (26 %) are invalid for the western Tasmanian shelf due to cloud bias, mainly during austral winter–spring months. This first analysis indicates that ocean-colour-derived chlorophyll *a* data for the study region are substantially modified by river discharges.

The fluorescence line height (FLH) is a relative measure of the amount of radiance leaving the sea surface in the chlorophyll fluorescence emission band (e.g. Xing et al., 2007). The chlorophyll *a* fluorescence peak near 685 nm has been proposed as an alternative option of chlorophyll *a* retrieval (Neville and Gower, 1977; Gordon, 1979). Many researchers have studied the relationship between chlorophyll *a* concentration and FLH, and have found a good correlation between these properties (Maritorena et al., 2000; Gower et al., 2004; Gower and King, 2007) for relatively low chlorophyll *a* concentrations < 9 mg m^{-3}, which applies to this study. At higher chlorophyll *a* concentrations, the fluorescence peak shifts to longer wavelengths (e.g. Gitelson, 1993).

The second, extended study period (1 January 2005 to 31 March 2014) adopts MODIS-aqua satellite data (4 km spatial resolution), provided by NASA's Giovanni project (http://disc.sci.gsfc.nasa.gov/giovanni). Standard normalised FLH data were used to reconstruct phytoplankton blooms on the western Tasmanian shelf. In the absence of suitable field data, this reconstruction is based on a regression of normalised FLH and OC3 data for the nearby Bonney upwelling region, which is devoid of noticeable river influences. The reconstructed time series of chlorophyll *a* is then used in an event-based, nonparametric statistical analysis of phytoplankton blooms in response to possible nutrient-supply events, i.e. coastal upwelling events and/or river plumes. To this end, all relevant data are converted to 8-day segments in alignment with the satellite data, noting that SST data are not used in this event analysis. For the western Tasmanian shelf, 90 (21 %) of the total of 425 eight-day segments are unusable due to cloud bias.

Wind data from the Cape Grim weather station (see Fig. 1), provided by Australia's Bureau of Meteorology (http://www.bom.gov.au/), were used to calculate the classical upwelling index representative for the western Tasmanian shelf. This index is based on the theoretical offshore volume transport in the surface Ekman layer and it is calculated from

$$\mathrm{UI} = \frac{|\tau|}{\rho_o\,|f|}\cos\left(\alpha - \alpha'\right), \qquad (1)$$

where τ is wind stress, $\rho_o \approx 1026\,\mathrm{kg\,m^{-3}}$ is seawater density, $f \approx -0.9 \times 10^{-4}\,\mathrm{s^{-1}}$ is the value of the Coriolis parameter at 40° S, α is wind direction and α' is average coastline orientation, taken equivalent to 160° (based on the meteorological convention that 0° refers to northerly winds). Small variations of α' have little influence of the results (not shown). For initial comparison, we also calculated the upwelling index for the Bonney upwelling system for the period from 1 January 1998 to 31 December 2000.

River discharge data for the western coast of Tasmania are sparse. There are no stream-flow data for the Macquarie Harbour estuary, which is the largest freshwater source on the western Tasmanian shelf. The only continuous time series of relevant river discharges that the author could find is that of the Davey River, being located in the south of western Tasmania (see Fig. 1). This study assumes that the seasonal discharge pattern of the Macquarie Harbour estuary is similar to that of the Davey River. The source of river data is Tasmania's Department of Primary Industries, Parks, Water and Environment via the Water Information System of Tasmania (http://wrt.tas.gov.au/wist/ui).

The satellite data used in this study are subject to various errors including instrument and retrieval algorithm errors, errors arising from oceanic variability, and merging, gridding and analysis algorithm errors. While the accuracy of satellite SST data is generally $< 1\,°\mathrm{C}$, chlorophyll-a-related satellite products are less certain. Apart from biases due to CDOM and inorganic suspended particles, ocean colour data are influenced by seabed reflection in shallow (< 20 m) nearshore water (which is irrelevant in the context of this work) and a widely variable shape of the phytoplankton absorption spectrum depending on pigment packaging, physiological state of the cell and community composition (Sauer et al., 2012). Consequently, it has become a necessity to develop regional ocean colour algorithms rather than using the global algorithm (e.g. Volpe et al., 2007). Nevertheless, the global OC4 algorithm worked well in reproducing the range of in situ measured chlorophyll a levels during an upwelling event off the southern tip of the Eyre Peninsula (Kämpf et al., 2004).

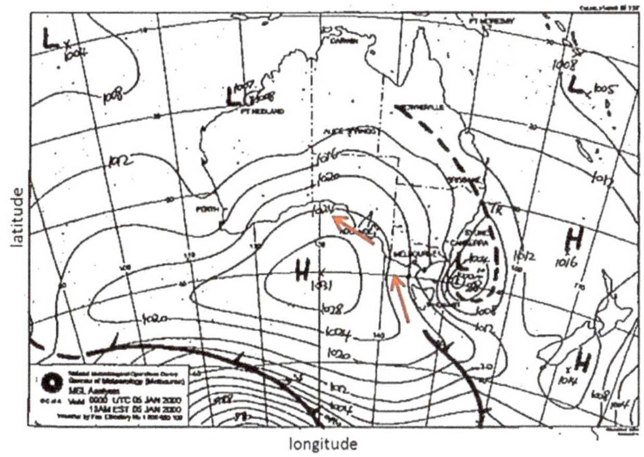

Figure 2. Mean sea level pressure for 6 January 2000, courtesy of the Bureau of Meteorology (Australia) (http://www.bom.gov.au/cgi-bin/charts/). Arrows indicate upwelling-favourable coastal winds, influenced by a blocking low-pressure cell over Tasmania.

On the other hand, the global OC3 and OC4 algorithms create similar ranges in chlorophyll a in the South Australian upwelling centres, which justifies their unmodified use for the South Australian region (but not the western Tasmanian shelf, as will be shown with the results).

Dominant factors influencing fluorescence distributions are chlorophyll concentration, pigment packaging effects on light absorption and light-dependent energy-quenching processes (Behrenfeld et al., 2009). As far as the author is aware, there is no in situ validation of satellite FLH data for the South Australian upwelling regions. Furthermore, Giovanni's normalised FLH data, used in this study, are based on a division by the cosine of the local solar zenith angle at the time of observation. This normalisation is based on the assumption that the fluorescence signal is proportional to local solar irradiance. Recently, Gower (2014) demonstrated that this assumption is invalid and can lead to significant errors. In this work, the use of normalised FLH values as a reasonably accurate measure of chlorophyll a levels (on the timescale of 8-day composites) is validated by a regression analysis between normalised FLH and OC3 data for the Bonney upwelling region (see Sect. 3.2).

3 Results and discussion

3.1 Initial evidence based on SeaWiFS data (1998–2000)

Coastal upwelling events on the southern shelves of Australia are associated with high-pressure weather systems that create south-easterly coastal winds (Kämpf et al., 2004). Due to their spatial scale and the geometry of Australia's coastline, such weather patterns can also initiate coastal upwelling on the western shelf of Tasmania. In early January 2000, for

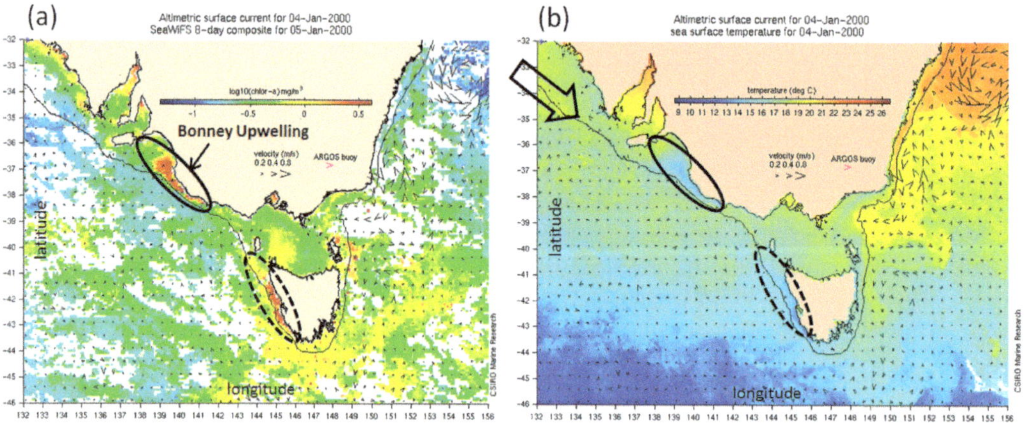

Figure 3. Occurrence of a pronounced coastal upwelling event in early January 2000, evident in satellite-derived distributions of **(a)** MODIS-OC3 chlorophyll *a* and **(b)** sea surface temperature. White regions in panel **(a)** are missing data due to clouds. The large arrow in panel **(b)** indicates the pathway of the South Australian Current. Data source: South East Fishery ocean movies, David Griffin, CSIRO, http://www.marine.csiro.au/~griffin/SEF/.

example, a high-pressure weather system developed centred over the South Australian Basin (Fig. 2). This high-pressure system became blocked by a low-pressure cell over Tasmania, triggering coast-parallel, upwelling-favourable winds along both Australia's southern shelves and the west coast of Tasmania. During this period, strong upwelling occurred in the upwelling centres on the southern shelves (Kämpf et al., 2004). During this event, chlorophyll *a* levels on the western Tasmanian shelf attained values of $\sim 3\,\mathrm{mg\,m^{-3}}$, the same order of magnitude as those observed in the upwelling centre along the Bonney coast (Fig. 3a). Upwelling-related negative SST anomalies can be identified in both regions (Fig. 3b). While the Bonney upwelling has pronounced temperature anomalies of 2–3 °C, temperature anomalies on the western Tasmanian shelf are relatively difficult to distinguish from those in the ambient ocean which are of a similar range.

The time series of the upwelling indices for both regions (Fig. 4) reveals that, similar to the upwelling centres of Australia's southern shelves, coastal winds along the west coast of Tasmania are, on average, upwelling favourable during austral summer months (December–April). This indicates that both regions share similar wind-forced upwelling characteristics. Earlier work by Kämpf et al. (2004) has overlooked this feature.

OC4 chlorophyll *a* levels off the Bonney coast develop clear peaks during the austral summer upwelling season; chlorophyll *a* levels on the western shelf of Tasmania attain a complex temporal structure (Fig. 5a). In particular, large discrepancies in OC4 chlorophyll *a* levels between the regions occur in austral winter months, which is the period of enhanced river discharges (Fig. 5c).

The time series of SST (Fig. 5b) reveals intermittent warming periods from May to July (i.e. during late austral autumn and early winter). These warming events, which are more pronounced along the Bonney coast than on the western Tasmanian shelf, where it appears as a shoulder on a declining head, are associated with incursions of the South Australian Current (Fig. 6 shows an example). Being of shelf origin, this current has low nutrient content (e.g. Herzfeld, 1997) and its appearance in the study regions can be identified by decreases in OC4 chlorophyll *a* levels in each year of the time series (see Fig. 5a).

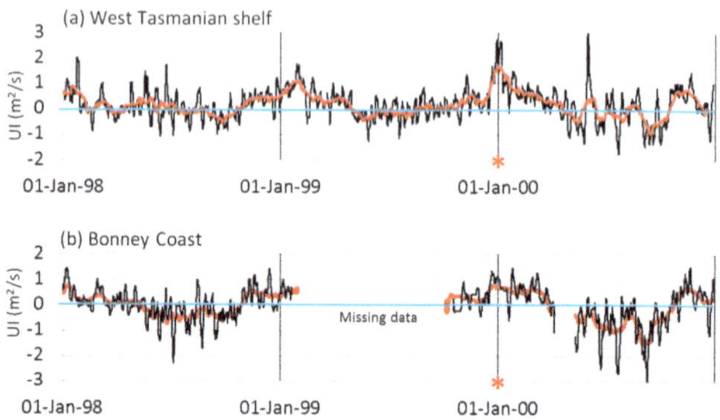

Figure 4. Time series (1 January 1998 to 31 December 2000) of the upwelling index ($\mathrm{m^2\,s^{-1}}$) for **(a)** the western Tasmanian shelf and **(b)** the Bonney coast. Thin, black (thick, red) curves are 4-day (20-day) moving averages. Stars highlight an upwelling event in early January 2000, corresponding to the spatial distributions shown in Fig. 3. Data source: Bureau of Meteorology, Australia.

The timing of some events of elevated OC4 chlorophyll *a* levels on the western Tasmanian shelf in October seem to coincide with the onset of spring blooms in the western Tasman Sea, where chlorophyll *a* levels seasonally peaked in October at a level of $0.8\,\mathrm{mg\,m^{-3}}$ in the years 1998–2000 (Tilburg et al., 2002). It should be noted that the Bonney upwelling region is devoid of such spring blooms.

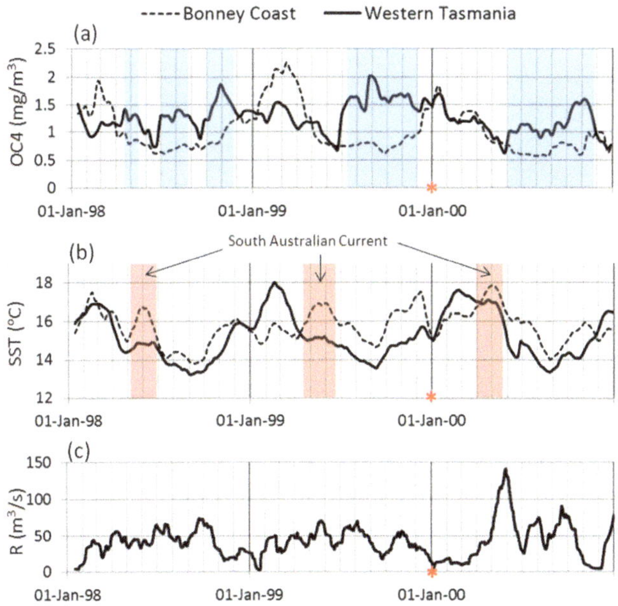

Figure 5. Time series (1 January 1998 to 31 December 2000) of satellite-derived data of **(a)** OC4 ($mg\,m^{-3}$), **(b)** sea surface temperature (SST, °C) for the Bonney coast and the western Tasmanian shelf, and **(c)** discharge from the Davey River ($m^3\,s^{-1}$). Running averages over an interval of 24 days have been used for both OC4 and river discharge. A running average over 20 days has been applied to the SST data. Shaded areas of panel **(a)** highlight periods of substantial differences between the time series. Shaded areas in panel **(b)** denote intermittent warming created by the South Australian Current. Stars highlight an upwelling event in early January 2000, corresponding to the spatial distributions shown in in Fig. 3. Data source: SEF ocean movies, David Griffin, CSIRO, Australia.

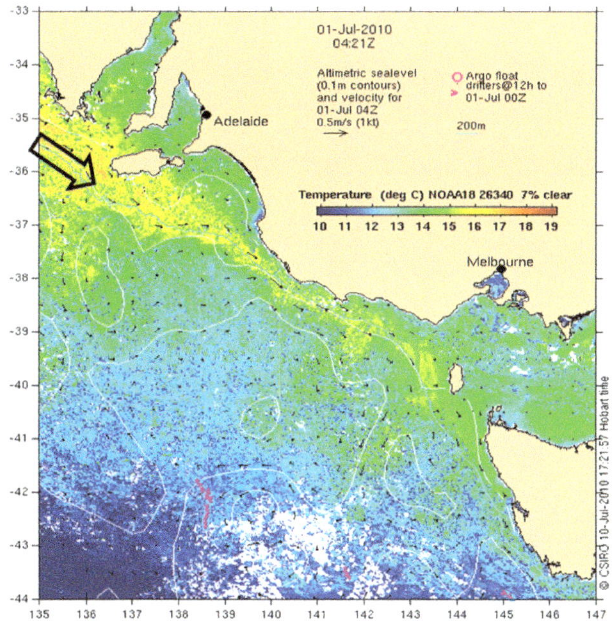

Figure 6. An example of the inflow of warm, nutrient-low South Australian Current that regularly appears along the Bonney coast and on the western Tasmanian shelf between May and July in every year. Data source: Integrated Marine Observing System (IMOS), http://oceancurrent.imos.org.au/Adelaide/2010/2010070104.html.

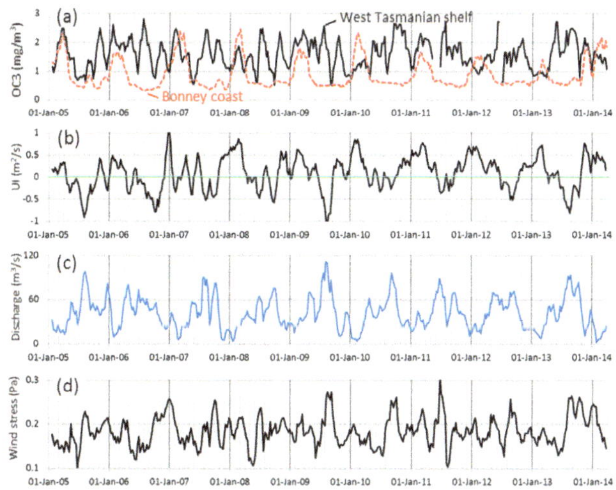

Figure 7. Time series (1 January 2005 to 31 March 2014) of **(a)** OC3, **(b)** upwelling index, **(c)** discharge from the Davey River and **(d)** wind stress. Data were first converted to 8-day segments and then smoothed with a running average over three segments. For comparison, panel **(a)** includes data for the Bonney coast. Data sources: NASA (Giovanni), Bureau of Meteorology (Australia) and Water Information System of Tasmania.

3.2 Detailed analysis based on MODIS-aqua data (2005–2014)

Given the absence of noticeable river influences, OC3 chlorophyll *a* data for the Bonney upwelling region can be taken as a proxy for phytoplankton levels. The time series for this region confirms the pronounced annual periodicity of phytoplankton blooms during January–May (Fig. 7a). Average peak chlorophyll *a* levels tend to slightly vary between the years, which reflects interannual variations of the frequency and intensity of individual upwelling events in this region. Middleton et al. (2007) speculated that the upwelling intensity is strongly modulated by ENSO events, but the satellite data shown here are devoid of any dramatic interannual variability that could be linked to ENSO variability. For instance, the moderate La Niña in 2007–2008 followed by a moderate El Niño in 2009–2010 coincided with phytoplankton blooms of similar magnitudes. Similarly, the strong El Niño in 1998–1999 followed by a strong La Niña in 1999–2000 did not lead to significantly different chlorophyll *a* levels during the upwelling period (see Fig. 5a).

In contrast to the Bonney upwelling, the OC3 data on the western Tasmanian shelf vary in a highly irregular fashion and all year round (see Fig. 7a), noting that both regions attain peak values between 2 and 3 $mg\,m^{-3}$. The upwelling in-

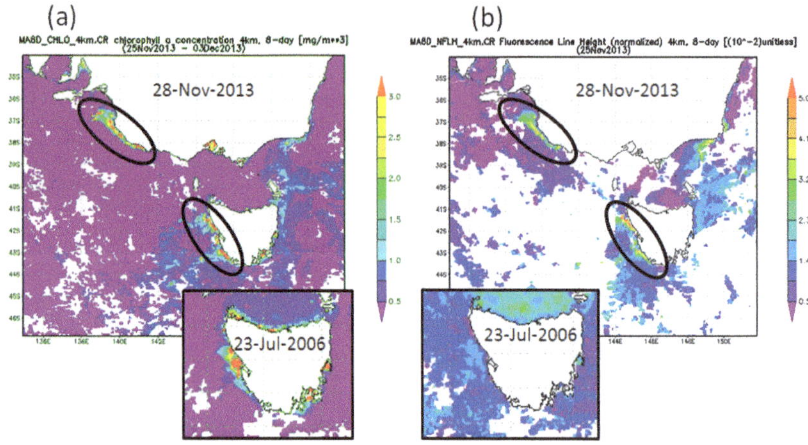

Figure 8. Snapshots of **(a)** OC3 ($mg\,m^{-3}$) and **(b)** normalised FLH (relative units) during a major upwelling event in late November 2013. The inserts show an example of a river discharge event in late July 2006. Data source: NASA (Giovanni).

dex for the western Tasmanian shelf generally attains positive values for austral summer months (Fig. 7b). In some years, this index indicates brief upwelling events outside the summer season. For instance, the year 2007 had such an event in June, and other years (e.g. 2005 and 2013) had early upwelling-favourable wind events in November, being mirrored by individual OC3 peaks (see Fig. 7a). On the other hand, the upwelling index also indicates events of strong downwelling-favourable winds, such as in August 2009. This particular example also coincided with high wind stress, strong river flow and elevated chlorophyll *a* levels.

Overall, the discharge from the Davey River tends to peak in austral winter–spring with markedly reduced flows during austral summer months (Fig. 7c). An exception is the year break of 2005–2006 which had a relatively strong riverine discharge occurring in December–January.

For completeness, the author also included a time series of wind stress magnitude (Fig. 7d), given that strong storms have the ability to modify phytoplankton blooms via changes in the mixed-layer depth and potential entrainment of nutrient-rich water from below. On the other hand, storms can also erode vertical structure of phytoplankton concentrations, thereby removing the surface appearance of a phytoplankton bloom. For instance, the existence of relatively strong winds (~ 2.5 Pa) in January 2007 might explain relatively low surface OC3 levels in western Tasmanian coastal water although winds were upwelling favourable. It should be noted that periods of 3–7 days of relaxed winds after a brief upwelling event are deemed optimal for phytoplankton accumulation (Wilkenson et al., 2006). This relaxation effect is not explored in the context of this work.

The NASA OC3 algorithm is unreliable for the western Tasmanian shelf due to strong river influences. While it gives reasonable results in rare occasions of coastal upwelling events in the absence of river flows, it often misinterprets low productive river plumes as phytoplankton blooms, which

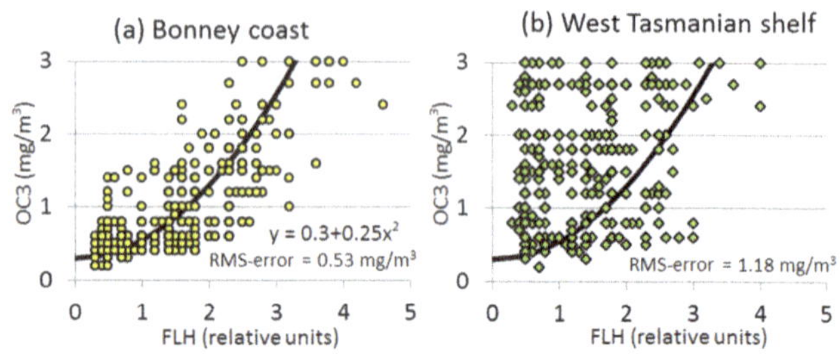

Figure 9. Correlation between OC3 values and normalised fluorescence line height (FLH), multiplied by a factor of 100, for **(a)** the Bonney coast and **(b)** the western Tasmanian shelf for the period from 1 January 2005 to 31 March 2014. Solid lines show a quadratic regression corresponding to a root mean square error (RMSE) of $0.53\,mg\,m^{-3}$ for the Bonney coast data. Data source: NASA (Giovanni).

are absent in the FLH data (Fig. 8). In order to reconstruct chlorophyll *a* levels for the western Tasmanian shelf, OC3 and FLH data can be related to each other for the Bonney upwelling region (Fig. 9a), where river influences are sparse. This comparison indicates a quadratic regression of the form of

$$y = 0.25 + 0.3x^2, \qquad (2)$$

where y is the reconstructed chlorophyll *a* value, and x is the FLH value (derived from the NASA Giovanni database) multiplied by 100. The associated root mean square error (RMSE) of $0.53\,mg\,m^{-3}$ is acceptable, noting that discrepancies can be attributed to absorption properties of the atmosphere and physiological variations in the phytoplankton (Xing et al., 2007). Nevertheless, the reconstructed chlorophyll *a* time series (based on Eq. 2) agrees well with the original OC3 data (Fig. 10a). As expected, there is a mismatch between OC3 and FLH data on the western Australian shelf (Fig. 9b) attributed to river discharges. In this region, the re-

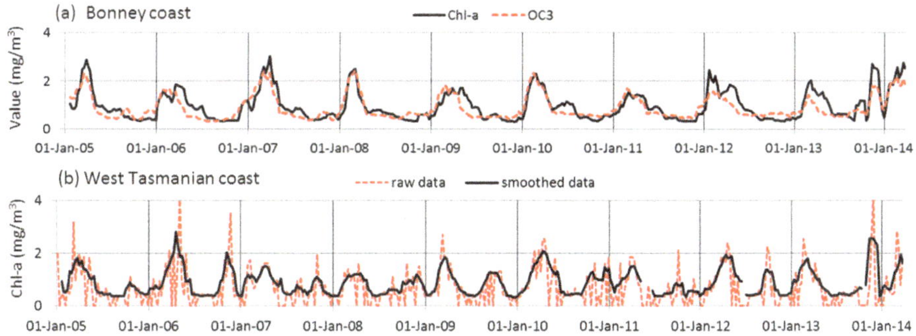

Figure 10. Reconstructed time series of chlorophyll *a* concentrations (solid lines) from normalised fluorescence line height (FLH) according to Eq. (2) for **(a)** the Bonney coast and **(b)** the western Tasmanian shelf for the period from 1 January 2005 to 31 March 2014. Shown are running mean averages over three successive data segments. The dashed line in panel **(a)** shows the corresponding OC3 data. The dashed line in panel **(b)** shows the raw 8-day composite data with zero values (ignored in the averaging procedure) being allocated to missing data. Data source: NASA (Giovanni).

constructed time series of chlorophyll *a* differs substantially from that derived from the OC3 algorithm (Fig. 10b, compared with Fig. 7a). To this the end, the reconstructed time series indicates the occurrence of two phytoplankton blooms every year. The first and larger bloom occurs in late austral summer (March–April), and a second smaller bloom occurs in October coinciding with the development of spring blooms in the adjacent Tasman Sea (Tilburg et al., 2002).

3.3 Event analysis (2005–2014)

When using window-averaged data, a standard cross-correlation analysis between upwelling index, OC3 or FLH data and river discharges does not give satisfactory results. Overall, the resultant correlation coefficients are insignificantly small and strongly biased by data smoothing and interpolation. Instead of this, statistically more relevant information can be derived from an event-based analysis, whereby all relevant data are averaged onto 8-day data segments. Each data segment is defined as an individual event. The underlying assumption is that phytoplankton blooms follow within ~ 3–7 days after a nutrient-supply event. This implies that there is a relatively high probability that both the physical event and the ecological response occur within the timescale (8 days) of a data segment. This assumption is consistent with observational evidence (Wilkerson et al., 2006). Data segments biased by cloud cover are referred to as invalid events in the following.

This approach, for instance, returns histograms of (reconstructed) chlorophyll *a* ranges for the study region in comparison with the Bonney upwelling region (Fig. 11). Higher levels $> 1.5\,\mathrm{mg\,m^{-3}}$ can be interpreted as phyto plankton blooms. Such levels are found off the Bonney coast $\sim 18.4\,\%$ of time, which is equivalent to roughly 67 days per year. In the study region, high chlorophyll *a* levels $> 1.5\,\mathrm{mg\,m^{-3}}$ occurred 14.6 % of time, which corresponds to around 53 days per year. Stronger phytoplankton blooms of

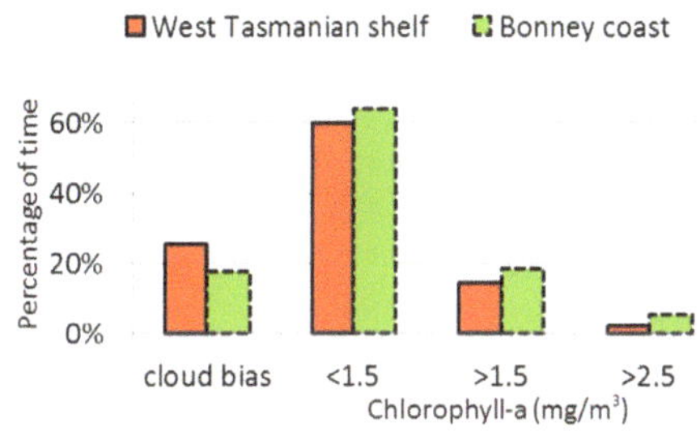

Figure 11. Event analysis. Histogram of 8-day segments for the period from 1 January 2005 to 31 March 2014 that fall within certain ranges of chlorophyll *a* concentrations for the western Tasmanian shelf and the Bonney coast. The result is expressed as time percentage with reference to the entire time period. Cloud bias refers to missing data.

chlorophyll *a* concentrations $> 2.5\,\mathrm{mg\,m^{-3}}$ occurred 5.4 % of time or 19.8 days per year off the Bonney coast and 2.4 % of time or 8.6 days per annum on the western Tasmanian shelf.

When grouping data segments into different intervals of upwelling index and river discharge, the largest number of records exists for large values of the upwelling index $> 0.25\,\mathrm{m^2\,s^{-1}}$ in the presence of relatively low river discharge $< 25\,\mathrm{m^3\,s^{-1}}$ (Table 1). This criterion returns 101 out of a total of 316 valid events ($\sim 32\,\%$) and reflects austral summer conditions. On the other hand, a relatively large number of valid events exist for stronger river discharges $> 50\,\mathrm{m^3\,s^{-1}}$ and negative values of the upwelling index $< -0.25\,\mathrm{m^2\,s^{-1}}$, returning 46 events out of 316 valid events ($\sim 15\,\%$). This criterion corresponds to periods of enhanced river discharges in austral winter and spring months when stronger northerly wind events occur (see Fig. 7). The other intervals considered have statistically satisfactory population sizes between 9 and 29 valid events.

Invalid events due to cloud bias occur for all parameter intervals considered (Table 2). The fraction of in-

Table 1. Number of valid events of the time series (1 January 2005 to 31 March 2014) grouped according to different intervals of upwelling index (UI) and river discharge (R).

↓ R ($m^3 s^{-1}$)/UI ($m^2 s^{-1}$) →	$< -1/4$	$-1/4$ to 0	0 to $+1/4$	$> +1/4$	all UI
< 25	9	18	21	101	149
25 to 50	16	13	22	29	80
> 50	46	11	15	15	87
all R	71	42	58	145	$\Sigma = 316$

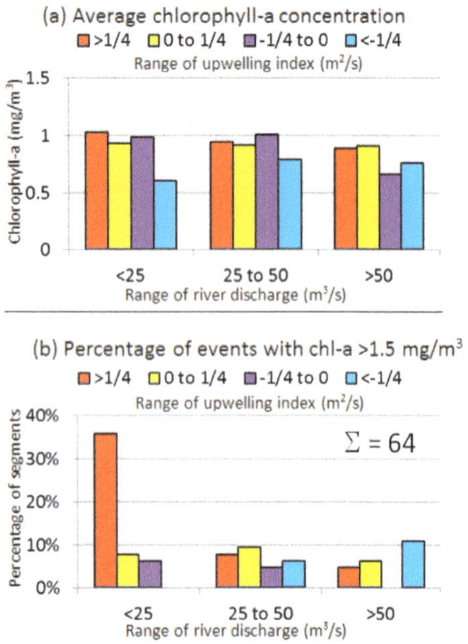

Figure 12. Event analysis of the time series (1 January 2005 to 31 March 2014). **(a)** Average chlorophyll a values based on data segments that fall into certain intervals of upwelling index and discharge from the Davey River. The total number of segments is 316. **(b)** Number of data segments of chlorophyll a concentration $> 1.5\,mg\,m^{-3}$, grouped as in panel **(a)**. In total, 64 of the 316 data segments satisfy this condition.

valid to total events is greater in times of enhanced river discharges $> 50\,m^3\,s^{-1}$ for which 52 out or the total of 139 events (or ~37.5 %) are invalid. In contrast, only 28 out of 177 events (~15.8 %) are invalid for events with river discharges $< 25\,m^3\,s^{-1}$.

Consistent with the aforementioned weak cross-correlations between parameters, average chlorophyll a concentrations are of the same order of magnitude in each parameter interval considered (Fig. 12a). This implies, for instance, that stronger upwelling-favourable winds do not always create phytoplankton blooms. The only exception is a noticeable reduction of chlorophyll a concentration in austral winter months (see Fig. 7b), which coincides with strong river discharges and downwelling-favourable winds.

With a focus on events of phytoplankton blooms of chlorophyll a concentrations $> 1.5\,mg\,m^{-3}$ the outcome is vastly different (Fig. 12b). Here, the by far largest number of blooms occurs for stronger upwelling-favourable winds (UI $> 0.25\,m^2\,s^{-1}$) (31 out of 64 events or 48.4 %). Half of these 31 events occur during low river discharges, which is characteristic of the austral summer season. Note that the proportion of such phytoplankton blooms increases from 48.4 to 69 % when accounting for upwelling-favourable wind events with UI $> 0.25\,m^2\,s^{-1}$ in preceding data segments. Hence, more than two thirds of the phytoplankton blooms on the western Tasmanian shelf can be attributed to the classical wind-driven upwelling mechanism. The remaining events are mainly associated with October spring blooms, which are regular features of the study region (see Fig. 10b).

While the outcome from the viewpoint of phytoplankton blooms gives conclusive results, there are a total of 145 events with UI $> 0.25\,m^2\,s^{-1}$ during the study period of which only 39 (22 %) triggered a phytoplankton bloom with chlorophyll a concentrations exceeding $1.5\,mg\,m^{-3}$. Another 13 events follow when accounting for UI values of the preceding data segment, which brings the total to 30 %. A number of possible processes could explain these missing phytoplankton blooms including (a) interference with river plumes that operate to suppress phytoplankton blooms, (b) lack of sufficiently long periods of relaxed wind after upwelling events (Wilkerson et al., 2006), (c) preceding downwelling periods that create a southward geostrophic coastal current and offshore transport in the bottom Ekman layer that, due to inertia effects, resist subsequent wind changes, (d) incursions of nutrient-poor water from the South Australian Current and (e) nutrient limitation. A more detailed analysis of possible causes of the missing blooms is beyond the scope of this study.

In order to compare phytoplankton productivity between the Bonney upwelling system and the western Tasmanian shelf, the author assumes that the spatial extent of their productivity zones is similar (see Figs. 3 and 8), and that 8-day composite values of reconstructed chlorophyll a concentration are proportional to the amount of phytoplankton formed during the 8-day period. This comparison is justifiable given (a) the relatively close proximity of the upwelling centres (which implies that the source properties of upwelled water are similar), (b) similar SST values during upwelling events (indicating similar upwelling intensities) and (c) similar values of FLH during upwelling events (indicating a sim-

Table 2. Number of invalid events (cloud bias) of the time series (1 January 2005 to 31 March 2014) grouped according to different intervals of upwelling index (UI) and river discharge (R).

$\downarrow R$ ($m^3 s^{-1}$)/UI ($m^2 s^{-1}$) $\rightarrow$	$< -1/4$	$-1/4$ to 0	0 to $+1/4$	$> +1/4$	all UI
< 25	3	5	8	12	28
25 to 50	4	5	6	14	29
> 50	35	3	5	9	52
all R	42	13	19	35	$\Sigma = 109$

Table 3. Relative production ($mg\,m^{-3}$) for the Bonney coast and the western Tasmanian shelf (1 January 2005 to 31 March 2014) for different chlorophyll a thresholds.

Criterion	Bonney coast	western Tasmanian shelf	% of Bonney coast
All events	355.4	289.3	81.4 %
Events with chl $a > 1.5\,mg\,m^{-3}$	180.0	126.4	70.2 %
Events with chl $a > 2.0\,mg\,m^{-3}$	108.0	48.8	45.2 %
Events with chl $a > 2.5\,mg\,m^{-3}$	79.1	31.1	39.3 %

ilar phytoplankton growth rate on the timescale of upwelling events).

To this end, relative production can be defined as the time integral of 8-day composite chlorophyll a values. This assumption gives a total of 289 mg (per unit volume) in relative production for the western Tasmanian shelf (Table 3), which is about 81.4 % of the relative production of the Bonney upwelling region. When only counting phytoplankton blooms with chlorophyll a concentrations $> 1.5\,mg\,m^{-3}$, the western Tasmanian shelf still accounts for 70 % of the relative production of the Bonney upwelling region. When considering exclusively rarer events of high chlorophyll a concentrations $> 2.5\,mg\,m^{-3}$, the relative productivity of the western Tasmanian shelf reduces to 40 % with reference to the Bonney upwelling region.

4 Conclusions

A detailed analysis of satellite-derived ocean colour data for the periods 1998–2000 and 2005–2014 suggest that the western Tasmanian shelf accommodates a highly productive ecosystem. This region forms another significant seasonal upwelling centre of the Great South Australian Coastal Upwelling System which develops in late austral summer (March–April). The addition of this newly discovered upwelling centre, which the author refers to as West Tasmanian Upwelling, makes the Great South Australian Coastal Upwelling System one of the largest (total spatial extension ~ 1500 km) seasonal coastal upwelling systems on Earth.

River plumes appear on the western Tasmanian shelf throughout the year, but continental runoff is markedly reduced during the upwelling season. Overall, there is no indication that river plumes contribute to primary production on the adjacent shelf, probably because of high concentrations of CDOM and suspended sediment (as indicated in OC-3 data) and the associated reduction in light levels. Interestingly, austral spring blooms develop regularly on the western Tasmanian shelf in October each year, but the reasons behind this development remains unclear.

The accuracy of satellite radar altimeter sea surface height measurement degrades in coastal region (Roesler et al., 2013) and cannot be used to identify upwelling jets here. Nevertheless, classical upwelling theory suggests the existence of such jets on the western Tasmanian shelf. The significance of these upwelling jets is that they operate to disperse nutrient-rich water northward along the shelf and possible into western Bass Strait. This advective process would explain elevated chlorophyll a levels in western Bass Strait – a typical feature of the region during austral summer months (see Fig. 3a). As such, upwelling on the western Tasmanian shelf presumably constitutes an important nutrient source for western Bass Strait waters.

The findings presented in this work were solely based on satellite data. Additional field work is required to explore the phytoplankton dynamics and potential differences between the upwelling centres off the Bonney coast and the western Tasmanian shelf in more detail.

Acknowledgements. The author thanks Paul Sandery (Bureau of Meteorology, Australia) for the provision of Cape Grim wind data and Emlyn Jones (CSIRO, Australia), who pointed out the unreliability of NASA-OC3 data for the western Tasmanian shelf. The author is also grateful to Piers Chapman and an anonymous reviewer for useful comments that improved the quality of the paper. This work has not received external funding. The author declares that there is no conflict of interests regarding the publication of this paper.

Edited by: M. Hoppema

References

Behrenfeld, M. J., Westberry, T. K., Boss, E. S., O'Malley, R. T., Siegel, D. A., Wiggert, J. D., Franz, B. A., McClain, C. R., Feldman, G. C., Doney, S. C., Moore, J. K., Dall'Olmo, G., Milligan, A. J., Lima, I., and Mahowald, N.: Satellite-detected fluorescence reveals global physiology of ocean phytoplankton, Biogeosciences, 6, 779–794, doi:10.5194/bg-6-779-2009, 2009.

Chen, C., Zhu, J., Beardsley, R. C., and Franks, P. J. S.: Physical-biological sources for dense algal blooms near the Changjiang River, Geophys. Res. Lett., 30, 1515, doi:10.1029/2002GL016391, 2003.

Conolly, J. R. and Von der Borch, C. C.: Sedimentation and physiography of the sea floor south of Australia, Sediment. Geol., 1, 181–220, doi:10.1016/0037-0738(67)90059-0, 1967.

Cresswell, G.: Currents of the continental shelf and upper slope of Tasmania, Papers and Proceedings of the Royal Society of Tasmania, 133, 21–30, 2000.

Garcia, V. M. T., Signorini, S., Garcia, C. A. E., and McClain, C. R.: Empirical and semianalytical chlorophyll a algorithms in the southwestern Atlantic coastal region (25–40° S and 60–45° W), Int. J. Remote Sens., 27, 1539–1562, 2006.

Gibbs, C. F., Tomczak, M., and Longmore, A. R.: The nutrient regime of Bass Strait, Aust. J. Mar. Fresh. Res., 37, 451–466, 1986.

Gitelson, A.: Algorithms for remote-sensing of phytoplankton pigments in inland waters, Adv. Space Res., 13, 197–201, 1993.

Gordon, H. R.: Diffuse reflectance of the ocean: the theory of its augmentation by chlorophyll a fluorescence at 685 nm, Appl. Optics, 18, 1161–1166, 1979.

Gower, J. and King, S.: Validation of chlorophyll a fluorescence derived from MERIS on the west coast of Canada, Int. J. Remote Sens., 28, 625–635, 2007.

Gower, J. F. R.: A simpler picture of satellite chlorophyll fluorescence, Remote Sens. Lett., 5, 583–589, doi:10.1080/2150704X.2014.940630, 2014.

Gower, J. F. R., Brown, L., and Borstad, G. A.: Observation of chlorophyll a fluorescence in west coast waters of Canada using the MODIS satellite sensor, Can. J. Remote Sens., 30, 17–25, 2004.

Griffin, D. A., Thompson, P. A., Bax, N. J., and Hallegraeff, G. M.: The 1995 mass mortality of pilchards: no role found for physical or biological oceanographic factors in Australia, Aust. J. Mar. Fresh. Res., 48, 27–58, 1997.

Herzfeld, M.: The annual cycle of sea surface temperature in the Great Australian Bight, Prog. Oceanogr., 39, 1–27, doi:10.1016/S0079-6611(97)00010-4, 1997.

Kämpf, J.: On the preconditioning of coastal upwelling in the eastern Great Australian Bight, J. Geophys. Res.-Oceans, 115, C12071, doi:10.1029/2010JC006294, 2010.

Kämpf, J., Doubell, M., Griffin, D., Matthews, R. L., and Ward, T. M.: Evidence of a large seasonal coastal upwelling system along the southern shelf of Australia, Geophys. Res. Lett., 31, L09310, doi:10.1029/2003GL019221, 2004.

Lewis, R. K.: Seasonal upwelling along the southeastern coastline of South Australia, Aust. J. Mar. Fresh. Res., 32, 843–854, 1981.

Mann, K. H. and Lazier, J. R. N.: Dynamics of Marine Ecosystems: Biological-Physical Interactions in the Oceans, Oxford, Blackwell Publishing Ltd., 2006.

Maritorena, S., Morel, A., and Gentili, B.: Determination of the fluorescence quantum yield by oceanic phytoplankton in their natural habitat, Appl. Optics, 39, 6725–6737, 2000.

Middleton, J. F., Arthur, C., van Ruth, P., Ward, T., McClean, J., Maltrud, M., Gill, P., and Middleton, S.: ENSO effects and upwelling along Australia's southern shelves, J. Phys. Oceanogr., 37, 2458–2477, 2007.

Neville, R. A. and Gower, J. F. R.: Passive remote sensing of phytoplankton via chlorophyll a fluorescence, J. Geophys. Res., 82, 3487–3493, 1977.

O'Reilly, J. E., Maritorena, S., O'Brien, M. C., Siegel, D. A., Toole, D., Menzies, D., Smith, R. C., Mueller, J. L., Mitchell, B. G., Kahru, M., Chavez, F. P., Strutton, P., Cota, G. F., Hooker, S. B., McClain, C. R., Carder, K. L., Muller-Karger, F., Harding, L., Magnuson, A., Phinney, D., Moore, G. F., Aiken, J., Arrigo, K. R., Letelier, R., and Culver, M.: Ocean color chlorophyll a algorithms for SeaWiFS, OC2, and OC4: Version 4, in: SeaWiFS postlaunch calibration and validation analyses: Part 3, edited by: Hooker, S. B. and Firestone, E. R., 11, 9–23, SeaWiFS Postlaunch Technical Report Series, NASA Goddard Space Flight Center, 2000.

Rao, C. P. and Adabi, M. H.: Carbonate minerals, major and minor elements and oxygen and carbon isotopes and their variation with water depth in cool, temperate carbonates, western Tasmania, Australia, Mar. Geol., 103, 249–272, doi:10.1016/0025-3227(92)90019-E, 1992.

Rao, C. P. and Green, D. C.: Oxygen- and carbon-isotope composition of cold shallow-marine carbonates of Tasmania, Australia, Mar. Geol., 53, 117–129, doi:10.1016/0025-3227(83)90037-3, 1983.

Ridgway, K. R. and Condie, S. A.: The 5500-km-long boundary flow off western and southern Australia, J. Geophys. Res.-Oceans, 109, c04017, doi:10.1029/2003JC001921, 2004.

Rochford, D. J.: A review of possible upwelling situation off Port MacDonald, South Australia, CSIRO Australian Division Oceanography Report No. 81, 4 pp., 1977.

Roesler, C. J., Emery, W. J., and Kim, S. Y.: Evaluating the use of high-frequency radar coastal currents to correct satellite altimetry, J. Geophys. Res.-Oceans, 118, 3240–3259, doi:10.1002/jgrc.20220, 2013.

Sandery, P. A. and Kämpf, J.: Winter-Spring Flushing of Bass Strait, South-Eastern Australia: A numerical modelling study, Est. Coast. Shelf Sci., 63, 23–31, doi:10.1016/j.ecss.2004.10.009, 2005.

Sandery, P. A. and Kämpf, J.: Transport timescales for identifying seasonal variation in Bass Strait, south-eastern Australia, Estuar. Coast. Shelf S., 74, 684–696, doi:10.1016/j.ecss.2007.05.011, 2007.

Sauer, M. J., Roesler, C. S., Werdell, P. J., and Barnard, A.: Under the hood of satellite empirical chlorophyll a algorithms: revealing the dependencies of maximum band ratio algorithms on inherent optical properties, Opt. Express, 20, 20920–20933, 2012.

Schahinger, R. B.: Structure of coastal upwelling events observed off the southern coast of South Australia during Feb 1983–April 1984, Aust. J. Mar. Fresh. Res., 38, 439–459, 1987.

Shanmugam, P.: A new bio-optical algorithm for the remote sensing of algal blooms in complex ocean waters, J. Geophys. Res., 116, C04016, doi:10.1029/2010JC006796, 2011.

Tilburg, C. E., Subrahmanyam, B., and O'Brien, J. J.: Ocean color variability in the Tasman Sea, Geophys. Res. Lett., 29, 125.1–125.4, doi:10.1029/2001GL014071, 2002.

Tomczak, M.: The Bass Strait water cascade during winter 1981, Cont. Shelf Res., 2, 55–87, 1985.

Volpe, G., Santoleri, R., Vellucci, V., Ribera d'Alcala, M., Marullo, S., and D'Ortenzio, F.: The colour of the Mediterranean Sea: Global versus regional bio-optical algorithms evaluation and implication for satellite chlorophyll estimates, Remote Sens. Environ., 107, 625–638, 2007.

Ward, T. M., McLeay, L. J., Dimmlich, W., Rogers, A., McClatchie, S., Matthews, R. L., Kämpf, J., and Van Ruth, P. D.: Pelagic ecology of a northern boundary current system: effects of upwelling on the production and distribution of sardine (Sardinops sagax), anchovy (Engraulis australis) and southern bluefin tuna (Thunnus maccoyii) in the Great Australian Bight, Fish. Oceanogr., 15, 191–207, doi:10.1111/j.1365-2419.2006.00353.x, 2006.

Wass, R. E., Connolly, R. J., and MacIntyre, R. J.: Bryozoan carbonates and sand continuous along southern Australia, Mar. Geol., 9, 63–73, doi:10.1016/0025-3227(70)90080-0, 1970.

Wilkerson, F. P., Lassiter, A. M., Dugdale, R. C., Marchi, A., and Hogue, V. E.: The phytoplankton bloom response to wind events and upwelled nutrients during the CoOP WEST study, Deep Sea Res.-Pt. II, 53, 3023–3048, 2006.

Xing, X.-G., Zhao, D.-Z., Liu, Y.-G., Yang, J.-H., Xiu, P., and Wang, L.: An overview of remote sensing of chlorophyll fluorescence, Ocean Sci. J., 42, 49–59, doi:10.1007/BF03020910, 2007.

3

Thermohaline properties in the Eastern Mediterranean in the last three decades: is the basin returning to the pre-EMT situation?

V. Cardin[1], G. Civitarese[1], D. Hainbucher[2], M. Bensi[1], and A. Rubino[3]

[1]OGS, Istituto Nazionale di Oceanografia e di Geofisica Sperimentale, Borgo Grotta Gigante 42/c, 34018 Sgonico, Trieste, Italy
[2]CEN, Institut für Meereskunde, University of Hamburg, Bundesstraße 53, 20146 Hamburg, Germany
[3]Università Ca' Foscari di Venezia, Dorsoduro 2137, 30121 Venice, Italy

Correspondence to: V. Cardin (vcardin@ogs.trieste.it)

Abstract. Temperature, salinity and oxygen data collected during April and June 2011 (M84/3 and P414 cruises respectively) are analysed to derive the oceanographic characteristics of the Eastern Mediterranean (EM) basin. These observed characteristics are compared with those from previous cruises over the period 1987–2011. As a result, the interannual and decadal variability of the EM thermohaline properties are discussed in the context of the evolution of the Eastern Mediterranean Transient (EMT) and of the general circulation of the basin. We found that the state of the EM is still far from the pre-EMT conditions, though the 2011 results possibly indicate a slow return to this status. In particular, a comparison between thermohaline property evolution deriving from interannual variability of the preconditioning and air–sea interaction (heat fluxes) in the South Adriatic and the Cretan Seas reveals aspects of the alternation of the two dense water sources (Adriatic and Aegean) during the last three decades, which have strong implications for the hydrographic characteristics of the intermediate and deep layers of the Ionian and Levantine basins.

1 Introduction

The circulation of the Eastern Mediterranean (EM) is deeply influenced by several driving forces, strong topographic constraints and internal dynamic processes (Robinson et al., 1991). The basin-scale thermohaline circulation of the EM is formed by two cells: the closed internal cell of the deep circulation encompassing the Adriatic–Ionian area and the Levantine basin, known as the EM conveyor belt, and the external thermohaline cell involving the exchange of water between the EM and the Western Mediterranean (Lascaratos et al., 1999). The first one, mainly driven by the formation of Adriatic Deep Water (AdDW), which eventually becomes the Eastern Mediterranean Deep Water (EMDW), is characterized by a renewal time of approximately 126 years (Roether and Schlitzer, 1991). Indeed, the major source of the deep waters of the EM can be considered the Adriatic Sea.

Recent decades have shown dramatic changes in the circulation of the EM. During the late 1980s and early 1990s, the EM conveyor belt underwent a major climate shift known as the Eastern Mediterranean Transient (EMT) (Roether et al., 1996, 2007; Theocharis et al., 2002), during which dense waters of Aegean origin replaced the AdDW in the deep layers of the EM. The structure of the South Aegean Sea (namely, the Cretan Sea; CS) water column changed as exceptionally dense and salty water started to fill the deep Cretan basin and to overflow the sills of the Cretan Arc straits. Observations conducted in 1987, 1991 and 1995 indicated a reduced transport of Atlantic Water (AW) into the Levantine basin, where the salinity increased, altering the circulation pattern of the Levantine Intermediate Water (LIW) from its formation site towards the Ionian Sea (Klein et al., 1999; Malanotte-Rizzoli et al., 1999). These drastic changes and the major contribution of dense water of Aegean origin continued for a period of 7–8 years. Due to its high density ($\sigma_\theta > 29.35\,\mathrm{kg\,m^{-3}}$), the Cretan Deep Water (CDW) continued to fill the deepest parts of the Levantine and Ionian basins, uplifting the resident and older deep and bottom waters of Adriatic origin (Schlitzer et al., 1991; Roether et al., 1996; Malanotte-Rizzoli et al., 1997). CDW appeared to be the most important source of the new type of EMDW, the AdDW not being dense enough to spread over the bottom layer of the EM. The Aegean source apparently disappeared by about 1995 (Theocharis et al., 2002).

Hydrographic measurements obtained in 2003 indicate a new ventilation of the deep layers in the Adriatic and Ionian Seas (see Hainbucher et al., 2006; Manca et al., 2006; Cardin et al., 2011). The necessary preconditioning for deep convection in the Southern Adriatic (SA) and the formation of water dense enough to spread below the bottom water of EM origin were only re-established after 2005, following a salinity increase linked with the influence of EM water at intermediate levels. However, bottom waters encountered in the deep layers of the Ionian basin were warmer and saltier between 2003 and 2007 (Rubino and Hainbucher, 2007). This implies the existence of a delayed link between characteristics of the intermediate layer of the SA and abyssal waters of the Ionian Sea (Bensi et al., 2013). Experimental observations show that the Northern Ionian Gyre (NIG) reverses on an almost decadal scale (Gačić et al., 2010). This reversal is driven by and responsible for internal mass redistribution within the EM. In particular, it seems mainly driven by variability in salinity (Gačić et al., 2011; Theocharis et al., 2014; Velaoras et al., 2014). Wind stress and other external forces play an important role as they interact with the above-mentioned processes to shape the variability (strengthening or weakening) of the surface circulation regime and of the sub-basin ocean structures (Molcard et al., 2002; D'Ortenzio and Prieur, 2012; Pinardi et al., 2013). The result is a highly complex nonlinear system acting at very different spatial and temporal scales.

In the present study, we have analysed temperature, salinity and oxygen data collected during April and June 2011 (M84/3 and P414 cruises, respectively) to examine the oceanographic characteristics of the EM basin. We focused our attention particularly on the properties and possible pathways of the AdDW, the EMDW and the LIW. Their evolutions are put in the context of the evolution of the EM large-scale oceanographic properties and of the circulation during the last three decades. Finally, we discuss the interannual variability of the preconditioning (salt redistribution) and air–sea interaction (heat fluxes) in the two main dense-water source areas for the EM, i.e. the South Adriatic and Cretan seas.

2 Data and methods

2.1 M84/3 and P414 cruise data sets

The hydrographic data analysed in this study were collected in the course of two German oceanographic cruises: *Meteor* M84/3 (April 2011) and *Poseidon* P414 (June 2011). The *Meteor* M84/3 cruise was set up to follow the demands and requirements of repeating hydrography as specified by the GO-SHIP group (http://www.go-ship.org/), i.e. to obtain a comprehensive set of physical and chemical parameters measured to the highest standards. The aim of the cruise was to cover the quasi-zonal section through the Mediterranean Sea, covering the main basins of the Mediterranean Sea, with dense sampling of physical and chemical parameters throughout the whole water column. The primary goal of the *Poseidon* P414 (June 2011) cruise was to identify the routes of the AdDW into the Ionian Sea, so as to study the spatial and temporal variability of the spreading and mixing of the EMDW (Fig. 1a).

The data set has two main components: the continuous profiles of conductivity, temperature and depth (CTD), and the discrete data from the water samples. It contains information from 61 hydrographic stations sampled during the April cruise (M84/3) (more information in Tanhua et al., 2013) and 33 full-depth standard hydrographic stations sampled in June 2011 during the P414 cruise. A SeaBird SBE911plus CTD-O_2 probe equipped with dual sensors of temperature, salinity and dissolved oxygen was employed, attached to a SeaBird carousel water sampler (24-bottle during the M84/3 and 12-bottle during the P414), together with an altimeter. Additionally, a fluorometer sensor was installed on the CTD during the M84/3 cruise and a digital reversing thermometer was attached to the probe for quality checking. At almost all stations, water samples for dissolved oxygen were taken at different depths throughout the whole water column. The oxygen samples were analysed onboard using a potentiometric Winkler titration method. From three depth levels, depending on the vertical profile of the stations, water samples were also taken for calibration of the salinity values, and they were analysed onboard using a Guildline Autosal Salinometer. Data were processed applying the Seabird software and a Matlab post-processing package. Spikes were removed from all data by applying the instrumental and climatological range criteria backed up by visual checks. Profiles were then averaged every 1 dbar. The overall accuracies are within 0.002 °C for temperature and 0.003 for salinity.

2.2 Long-term oceanographic variability

Vertical distributions of hydrographic data in the EM were examined to determine the long-term variability in the area since 1985. Three different approaches were applied:

1. Vertical distributions of temperature and salinity from the quasi-zonal section crossing the EM, from the Sicily Strait to the Levantine basin, were analysed by means of repeated surveys. We used data from all cruises throughout a trans-basin section: 1987 (Nellen et al., 1996), 1995 (Hemleben et al., 1996), 1999 (Pätzold et al., 2000), 2001 (Hemleben et al., 2003) and 2008 (Moutin et al., 2012). All these cruises took place as part

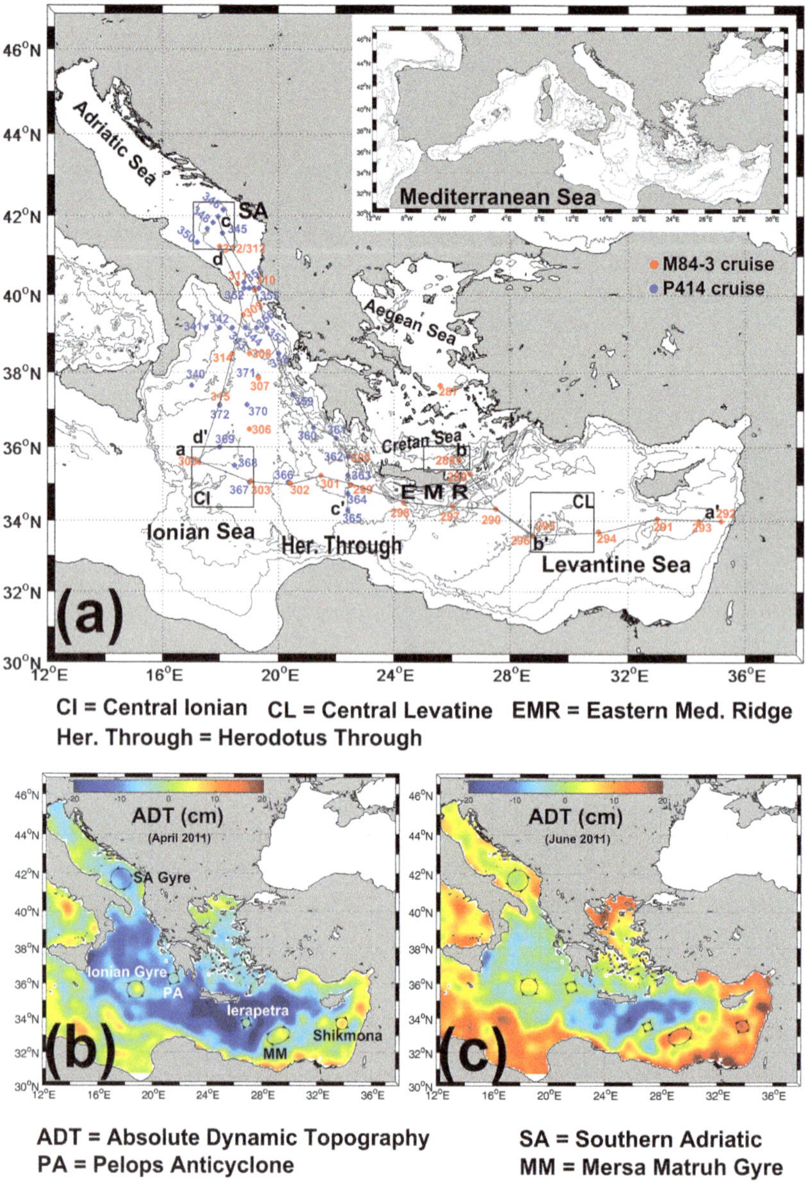

Figure 1. **(a)** CTD stations sampled in the Eastern Mediterranean Sea (EM) and Adriatic Sea during the M84/3 (red) and P414 (blue) cruises, and areas selected for the long-term analysis depicted. Hydrographic sections discussed in the text are identified as aa′ to dd′. **(b, c)** Monthly absolute dynamic topography (ADT) map as the rectangles of the Eastern Mediterranean for April and June 2011. Principal structures are highlighted in the maps.

of collaborative investigations of the EM: the POEM (Physical Oceanography of the Eastern Mediterranean) research programme, EMTEC (Eastern Mediterranean Transient and Ecosystem monitoring), and the BOUM (Biogeochemistry from the Oligotrophic to the Ultra-oligotrophic Mediterranean) experiment.

2. The long-term variability of the water mass structure in the EM was analysed on the basis of single profiles located in the central Ionian [17–19° E/34.5–36° N] and in the central Levantine [29–31.5° E/33–35° N]; see Fig. 1.

3. The temporal evolution of the water mass structure in the CS and SA was analysed on the basis of thermohaline parameters collected within the areas depicted in Fig. 1 between 1985 and 2011. Some of the data come from the MEDATLAS database (see Maillard et al., 2002; Manca et al., 2004), while some were collected within the framework of European and German National projects. The Hellenic Centre of Marine Research (HCMR) provided some of the profiles obtained between 2006 and 2011 for the CS sampled during the E1M3A observatory maintenance cruises. Further infor-

mation regarding the data set used for the SA can be found in Cardin et al. (2011).

For all data sets, derived parameters such as potential temperature, salinity and potential density were obtained from each original in situ temperature and conductivity profile. Hereinafter, temperature indicates potential temperature (θ), salinity is reported according to the practical salinity scale, density indicates the potential density excess (σ_θ), and the oxygen corresponds to the dissolved oxygen.

2.3 Atmospheric forcing

Air–sea heat fluxes at the air–sea interface were obtained from meteorological data such as mean sea level pressure, total cloud cover, wind speed at 10 m above the mean sea level, air temperature, dew point temperature at 2 m above the mean sea level, and skin temperature. These data were derived from the European Centre for Medium-Range Weather Forecasts (ECMWF) ERA-Interim data set with approximately 0.7° spatial resolution, and downloaded on a Gaussian grid with 0.25° × 0.25° resolution in latitude and longitude (interpolated by the ECMWF system from the original data). Total heat fluxes (Q_{net}) were obtained as the sum of the heat flux components: the solar radiation (Qs) was calculated using the modified Reed formula (Schiano, 1996), the net long-wave radiation (Q_{B}) was obtained by applying the Bignami et al. (1995) formula, while the sensible (Q_{H}) and latent (Q_{L}) heat fluxes were obtained from the bulk aerodynamic formulas (Kondo, 1975). More information on the formulas applied, procedures and coefficients can be found in Cardin and Gačić (2003). Total winter heat fluxes were calculated as a time-integral for the period spanning 1 December to 31 March for the years 1985–2011. Normalized heat fluxes for each area were calculated applying

$$Q_{\text{norm}} = \frac{X - \mu}{\sigma},$$

where Q_{norm} is normalized value; X is individual Q_{net} to normalize, μ is arithmetic long-term (1985–2011) winter average calculated from winter daily mean values and σ is long-term (1985–2011) winter standard deviation.

3 2011 Thermohaline conditions

The monthly average of the absolute dynamic topography (ADT) in the EM, referred to April and June 2011, are shown in Fig. 1b and c. Data are obtained from the Live Access Server of the AVISO web portal (http://las.aviso.oceanobs.com/las/getUI.do). The ADT is the sum of sea level anomaly and Synthetic Mean Dynamic Topography computed combining in situ drifter velocities and altimetry measurements. The error of the Synthetic Mean Dynamic Topography in the study area is in the range 1–3 cm (Rio et al., 2007). In this study, the ADT maps are used only to show the major circulation features in the upper thermocline, as some of the stations sampled during the M84/3 and P414 cruises were positioned at the edges or across dynamic structures that influence the water mass pathways. In the EM, several well-developed sub-basin-scale cyclonic or anticyclonic gyres, permanent or quasi-permanent, and interconnected by jets and meandering currents (Malanotte-Rizzoli et al., 1997), were observed in both April and June 2011.

Temperature and salinity along the quasi-zonal section crossing the EM, from the Sicily Strait to the Levantine basin, is shown in Fig. 2. In April 2011, the easternmost part of the transect was characterized by the presence of the Levantine Surface Water (LSW), which occupied the first 50 dbar and showed high temperature ($\theta > 18$ °C) and salinity ($S > 39.15$). Beneath this layer, a core of LIW with a salinity maximum of 39.24 on stations 293 (and 292, not shown) protruded westward (Fig. 2b), occupying the 150–350 dbar layer. Water characterized by $S > 39.20$ and $\theta > 17.50$ °C flowed out from the CS (St. 297) through the eastern Cretan Arc spreading into the Levantine Sea, and breaking the main LSW/LIW tongue with outcrops of the isohalines (see also Fig. 6). The westernmost part of the surface layer was mainly occupied by the AW within the upper 130 dbar and confined by the 38.80 isohaline, contained within the Ionian circulation. AW propagated eastward along the pathway observed from the dynamic heights (ADT, Fig. 1b, c). Hence, from the analysis of Fig. 1b, c and Fig. 2 we can deduce that the AW veered due to the action of an intense anticyclonic mesoscale structure centred over the deepest area of the Ionian (at ~ 19° E, 36° N), the edge of which we sampled at stations 305, 303 and 302. AW intrusion into the Levantine basin is very variable and may depend on the circulation present in the Ionian (Gačić et al., 2010). In 2011 the NIG circulation was just reversing to cyclonic, as reported in Gačić et al. (2014). Previously, the main effect of ~ 5 years of anticyclonic NIG (2006–2010), when the AW meandered in the northernmost part of the Ionian Sea, was a general increase in salinity in the Levantine basin (see the long-term analysis and discussion in Sect. 4) and a general decrease in salinity in the SA.

The deep layer below 2500 dbar (Fig. 2) is topographically divided into two by the EM Ridge (see its location in Fig. 1a, highlighted by the black dotted line). West of it, a large part of the Ionian basin was dominated by EMDW of Adriatic origin, still the coldest ($\theta < 13.48$ °C), freshest ($S < 38.73$) and densest ($\sigma_\theta \approx 29.20$ kg m^{-3}) water mass of the area, with the highest oxygen value (204 μmol kg^{-1}). The signature of the AdDW flowing out through the Otranto Strait was also found in profiles measured in the northernmost part of the Ionian, as will be shown later. East of it, the deepest part (below 2000 dbar) of the Levantine basin was filled by EMDW of Aegean origin, with similar σ_θ but slightly warmer and saltier ($\theta > 13.6$ °C, $S > 38.78$) than the EMDW found in

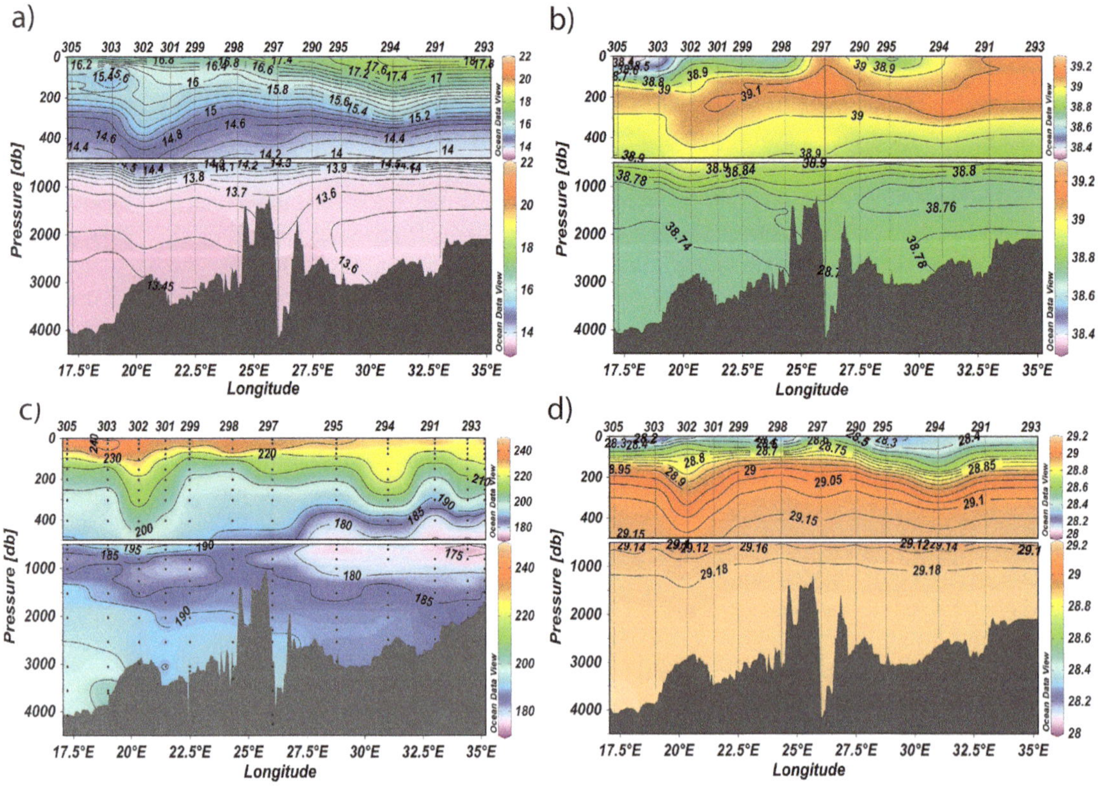

Figure 2. Water properties along a quasi-zonal east–west section in 2011 (marked as aa′ in Fig. 1a): **(a)** temperature (°C), **(b)** salinity, **(c)** oxygen (μmol kg^{-1}), **(d)** density (kg m^{-3}). The upper panel highlights the first 500 dbar. Ierapetra and Mersa Matruh anticyclones can be identified at ~26 and ~29° E respectively.

the Ionian, and with lower oxygen levels (between 185 and 190 μmol kg^{-1}) (see Fig. 2c).

The Cretan Sea water that flowed from the Kasos Strait showed salinity higher than ~39.15 occupying the first 250 dbar and protruding like a tongue into the Levantine (Fig. 3, transect B). However, its area of influence was mainly confined to the north part of the EM Ridge; south of this area it encountered the AW at the surface, while it mixed with the LIW in the intermediate layer. Very dense water occupying the deep layer of the CS remained "isolated" by virtue of the presence of the sill. Notably, the densest water was found in the CS at station 288 ($\theta \approx 13.95$ °C, $S \approx 39.06$ and $\sigma_0 > 29.30$ kg m^{-3}; Hainbucher et al., 2014). This water remained blocked in the CS (see St. 289), filling the layer below 1000 dbar. The T–S diagram (Fig. 3d) shows the characteristics of four stations along section bb′. Two of them (St. 290, green and St. 296, orange) are located in the Levantine basin and they revealed the typical long-lasting pronounced T–S inversions already described in Hainbucher et al. (2014) for this cruise, and explained by Roether et al. (2014) in detail. This inversion was caused by a noticeable increase in temperature and salinity throughout the EM deep waters due to an anomalously abundant and dense outflow of water of Aegean origin produced during the EMT. Station 289 (magenta) is located directly above the sill of the Kasos Strait where an inversion of higher temperature and salinity values at around 500 dbar is found. This might indicate that two kinds of Cretan deep water mass, an older and a newer, are present in this area. Station 288 (red) is located in the Aegean Sea and reaches its minimum salinity values at around 1150 m, indicating the presence of CDW at that level.

From its formation area westwards, the LIW pathway was directly influenced by the NIG and by the anticyclonic vortex present around 35.5° N/19.0° E that trapped low-salinity waters in its centre, as seen in Figs. 1b, c and 3. There, the LIW bifurcated: one branch flowed westward directly to the Sicily Channel while the other one flowed northwards along the eastern shelf-break of the Ionian. To describe the LIW pathway into the northern Ionian and into the Adriatic, temperature, salinity and density meridional sections along transect cc′ are presented in Fig. 4. The LIW seems to be influenced by the Pelops anticyclone (PA in Fig. 1b, at ~36.5° N), as revealed by the down-sloping of the isopycnals, and by the water exiting the CS through the Antikythera Strait (right part of the panels). The southernmost stations of the transect showed the presence of the AW in the first 100 dbar while the rest of the stations showed the LIW spreading northward into the Ionian basin as a warm ($\theta \approx 15.0$ °C) and

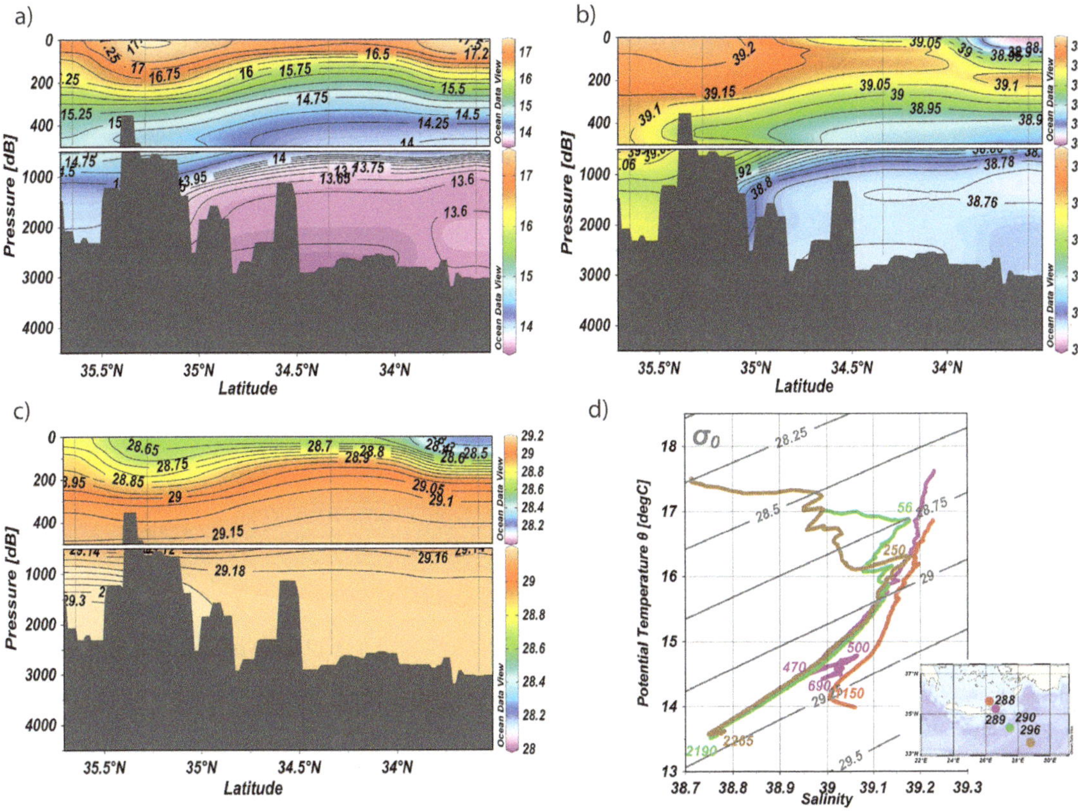

Figure 3. Water properties along the section crossing the Kasos Strait (marked as bb′ in Fig. 1a): **(a)** temperature (°C), **(b)** salinity, **(c)** density ($kg\,m^{-3}$), **(d)** T–S diagram of the stations considered on the transect. Positions of salinity inversions along the water column are indicated in dbar in the diagram and coloured similar to the station profile. The upper panels of **(a)**, **(b)** and **(c)** highlight the first 500 dbar.

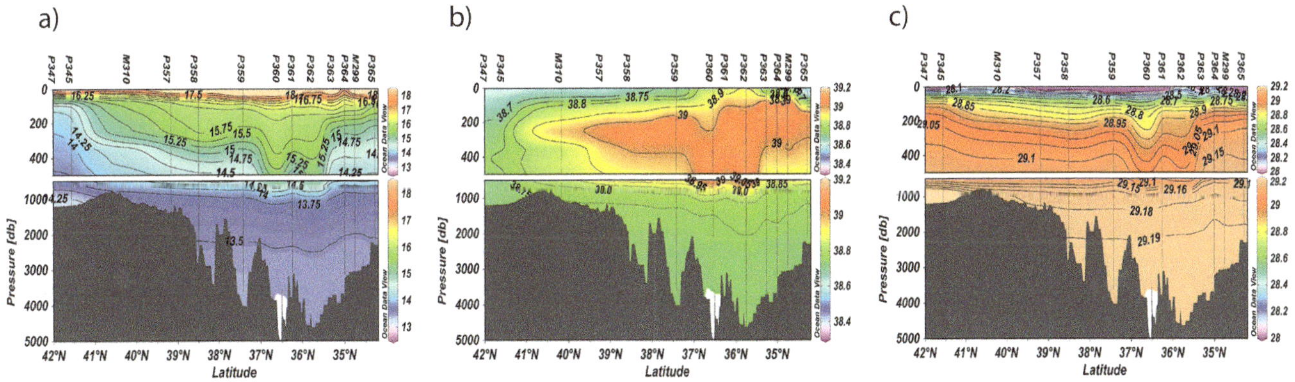

Figure 4. Water properties along the eastern longitudinal section (N–S) (marked as cc′ in Fig. 1a) following the traditional path of the saline core water towards the Adriatic Sea: **(a)** temperature (°C), **(b)** salinity, **(c)** density ($kg\,m^{-3}$). The upper panels highlight the first 500 dbar. *Poseidon* P414 stations are identified by the letter P and *Meteor* M84/3 by the letter M. Pelops anticyclone can be identified at ~36.5° N.

saltier (salinity maximum ranging from 38.8 to 39) tongue of water. At the time of the cruises (April and June) the LIW core ($S>=39.0$) did not enter the Adriatic through the Strait of Otranto, and only water with $S<38.80$ was advected over the Otranto sill, occupying the intermediate horizon in the Adriatic. The deep layer (> 800 dbar) of the SA (Fig. 5) was filled with AdDW ($\theta \approx 13.07\,°C$, $S \approx 38.73$, $\sigma_\theta \approx 29.275\,kg\,m^{-3}$). However, only a part of the AdDW overflowed the Otranto sill (depth ~ 800 dbar), cascading along the Ionian continental margin towards the deep layer of the central Ionian. The intense mixing affecting the AdDW branches sinking toward the abyssal plain of the Ionian Sea leads eventually to a warming and salinification (Manca et al., 2003; Hainbucher et al., 2006; Bensi et al., 2013) as well

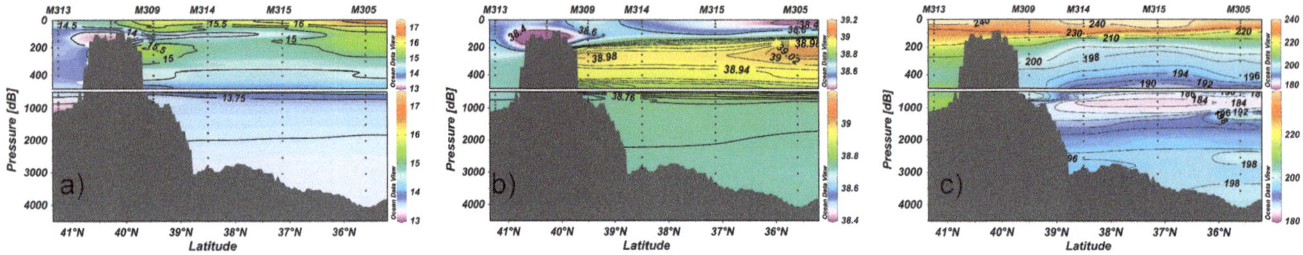

Figure 5. Water properties along the western longitudinal section (N–S) in the Ionian (marked as dd′ in Fig. 1a): **(a)** temperature (°C), **(b)** salinity, **(c)** oxygen (μmol kg^{-1}). The upper panels highlight the first 500 dbar. *Meteor* M84/3 stations are identified by the letter M.

as to a reduction in the relatively high dissolved oxygen content associated with the AdDW.

The western meridional (N–S) section dd′ (Fig. 5) links the SA and Ionian Seas along the Italian coast, ending at the western edge of the quasi-zonal transect crossing the EM. This field distribution complements the information inferred from Fig. 4, indicating that the LIW did not enter the Adriatic, while the cyclonic circulation observed is coherent with the ADT distribution shown in Fig. 1b and c. High-salinity waters characterized the intermediate layer spreading from east to west in a cyclonic circulation with the core of the LIW located between 200 and 300 dbar, whereas low-salinity waters were confined to the surface layer along almost the whole section shown from the northern part of transects (b) and (c), depicted in Figs. 4 and 5. The latter also shows two cores of high salinity water, the first located at St. M309 and the second at St. M305, both catching a less diluted signal of the LIW spreading westward from the Levantine. Beneath the LIW, low-salinity waters (38.72) with relatively high oxygen content (198 μmol kg^{-1}) filled the deepest part of the central Ionian. They can be attributed to relatively fresh and recently ventilated water of Adriatic origin. The oxygen minimum was located in the layer 1000–1300 dbar, with values around 180 μmol kg^{-1}, showing that this layer was scarcely renewed.

Maps of salinity distributions (Fig. 6a) at 100 dbar and at the isopycnal surface of $\sigma_\theta = 29.05$ kg m^{-3} (Fig. 6b), i.e. the typical horizon of the LIW (Malanotte-Rizzoli et al., 1997), provide evidence of the major water masses occupying the basin from the Sicily Channel to the Cretan Passage and their circulation patterns. The AW entered the Ionian Sea from the Sicily Channel and protruded towards east and north, occupying an extended portion of the west half of the Ionian basin. A front running from 36 to 40° N separated low-salinity waters of Atlantic origin (< 38.45) from Ionian Surface Waters (38.60). This pattern confirms that the anticyclone positioned at 19° E (Ionian gyre in Fig. 1b and c) acts as a sorting point from which the AW divides into two branches, one flowing to the east and the other proceeding northward. Bifurcated pathways of LIW were also recognizable from the salinity distribution at 29.05 kg m^{-3} (Fig. 6b): (i) the tongue defined by $S \sim 39.0$–39.05 spread directly from the Cretan Passage towards the Sicily Channel; (ii) the other branch of LIW was embedded in the NIG, and extended north of the Peloponnesus peninsula along the eastern border of the Ionian basin. On its way towards west and northwest the LIW mixed with the surrounding waters, decreasing its salinity.

4 Long-term variability in the Eastern Mediterranean

The long-term variability along the quasi-zonal section, from the Sicily Strait to the Levantine basin, is discussed here through the analysis of the vertical distributions of temperature and salinity sampled during six repeated surveys carried out in different oceanographic stages of the EMT: 1987, which represents the pre-EMT phase; 1995 when the EMT was in full swing – i.e. a large amount of Aegean-derived deep water was still deposited in the near-bottom waters around the Cretan Arcs and the Adriatic salinity maximum decreased; 1999 and 2001 showing the evolution of the EMT; and, finally, data from cruises carried out in 2008 and 2011 which reflect the more recent oceanographic situation. Figures 7 and 8 show the vertical distributions of temperature and salinity respectively. For a detailed discussion on the major findings from the cruises carried out in 1987, 1995 and 1999, the reader is referred to the works by Klein et al. (1999, 2000), Malanotte-Rizzoli et al. (1999), Manca et al. (2003) and Roether et al. (2007). A similar approach was used by Touratier and Goyet (2011) to determine the distributions of properties (θ, S, oxygen) along a west–east transect in the EM, but considering all stations (from the Medar/Medatlas II database) within a band of 200 km and for years 2000–2001.

Data from the M5/6 cruise in 1987 show the hydrographic conditions prior to the EMT in the Ionian and Levantine basins, when the AdDW was clearly the main contributor to the EMDW (Figs. 7a and 8a). This distribution is considered a reference for the climatological state of the EM (Schlitzer et al., 1991). Indeed, the deep western and central Ionian showed a temperature of 13.30 °C and salinities lower than 38.68. On the other hand, observations conducted during the EMT phase, in 1995 and 1999 (Figs. 7b, c and 8b, c respectively), revealed the presence of Aegean waters outside the Cretan Passage, with large salinity and tempera-

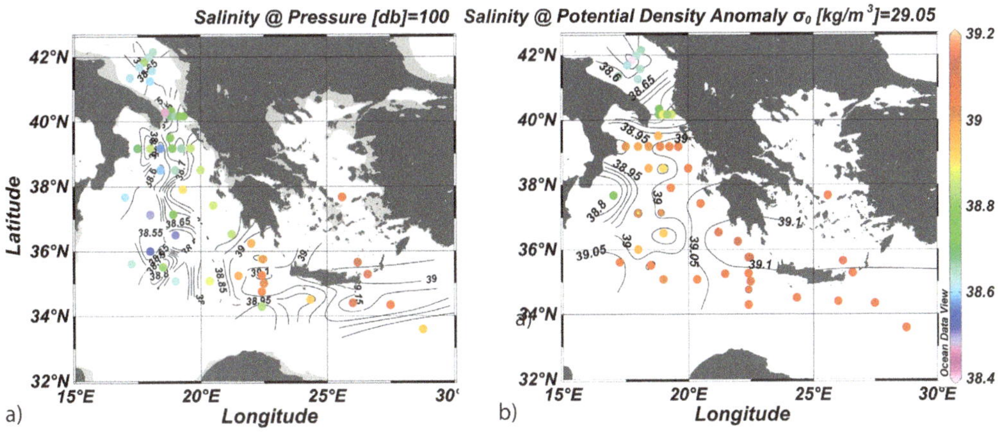

Figure 6. Horizontal distribution of salinity at 100 dbar **(a)** and on the 29.05 kg m^{-3} isopycnal **(b)**.

ture values, lying at the deepest layer of the EM. Enhanced salinity and temperature relative to 1987 (values higher than 38.84 and 13.60 °C) were already ubiquitous in early 1995 (Roether et al., 2007), with the exception of the western Ionian deep layer which was still occupied by low-salinity and low-temperature waters ($S < 38.68$, $\theta < 13.30$ °C). In 1995, waters of Aegean origin at the Cretan Passage were clearly recognizable by a dome of the ventilated bottom waters associated with $\theta > 14.0$ °C and $S > 38.78$. An analogous distribution was found in 1999 (see Theocharis et al., 2002; Manca et al., 2003) at the Cretan Arc and at the easternmost part of the Levantine basin (Figs. 7d and 8d), blocked by the bathymetry constraint of the EM Ridge. In the same years, signs of new EMDW of Adriatic origin were already evident in the western continental slope of the Ionian Sea, i.e. indicating that a major change occurred between 1999 and 2003 in the abyssal part of the Ionian basin, where AdDW returned gradually to become the principal contributor to the EMDW (Hainbucher et al., 2006; Manca et al., 2006; Rubino and Hainbucher, 2007; Bensi et al., 2013). In 2008 (Figs. 7e and 8e) only the Levantine basin was still under the influence of the warm and saline CDW that overflowed during the EMT, whilst the EMDW that resided in the Ionian abyssal layer was influenced by more recent AdDW with modified properties, being warmer and saltier ($\theta = 13.40$ °C, $S = 38.72$) than in the past. These results are consistent with the tendency toward the warming of and salinification of the AdDW pointed out by Rubino and Hainbucher (2007) and seen also in 2011 (Figs. 7f and 8f) when, in general, temperature and salinity in the abyssal layer of the Ionian Sea ($\theta \approx 13.40$ °C, $S \approx 38.75$) indicated almost everywhere a continuous relaxation of the EMT since the early 1990s.

The analysis of the upper layer (0–500 dbar) in Figs. 7 and 8 showed that the decadal variability of the Ionian and Levantine thermohaline properties is strongly related to the variability of the AW pathway, which in turn is influenced by the circulation regime in the Ionian Sea. In the last decades, different regimes alternated, leading to an increase (when AW was transported toward the Ionian basin by anticyclonic circulation) and to a decrease (when AW was transported toward the Levantine basin by cyclonic circulation) in the salinity in the Levantine basin. This variability is clearly evident in Fig. 8, where the salinity of the upper thermocline, especially in the Levantine basin, appears to oscillate on multiannual or decadal scales. A relative minimum in salinity was evident in 1987 (Fig. 8a) and was followed by an increase in 1995. At the end of 1980s the AW flow in the Ionian increased and an anticyclonic circulation up to 39° N arose (Malanotte-Rizzoli et al., 1997). This anticyclonic period lasted until 1996. The two salinity distributions in 1987 and 1995 (Fig. 8a and b, respectively) represent the limits of a period during which the salinity progressively increased.

Different mechanisms acting separately or even in combination could potentially explain part of such variability. Gačić et al. (2010, 2011) and Theocharis et al. (2014) showed that internal mechanisms seem to prevail over the atmospheric forcing in engaging the reversal of the circulation in the Ionian, which in turn modifies the position of the Atlantic–Ionian Stream responsible for the salinity distribution in the EM. In contrast, Pinardi et al. (2013) suggested that the reversal was wind driven.

The massive production and outflow of CDW was the principal manifestation of the EMT. To observe the evolution of the thermohaline properties of the water column in the CS (Fig. 9, upper panel) and in the SA (Fig. 9, lower panel), Hovmöller time diagrams of the thermohaline properties during the period 1986–2011, based on all CTD profiles sampled in the areas indicated in Fig. 1a, were obtained. Special attention was given to the density = 29.20 kg m^{-3}; this is the maximum density recorded in 1987 at the reference level of 1300 dbar in the proximity of the Kasos Strait, and it is considered the threshold density for filling the deep layers of the EM. The 900 and 800 dbar levels highlighted represent

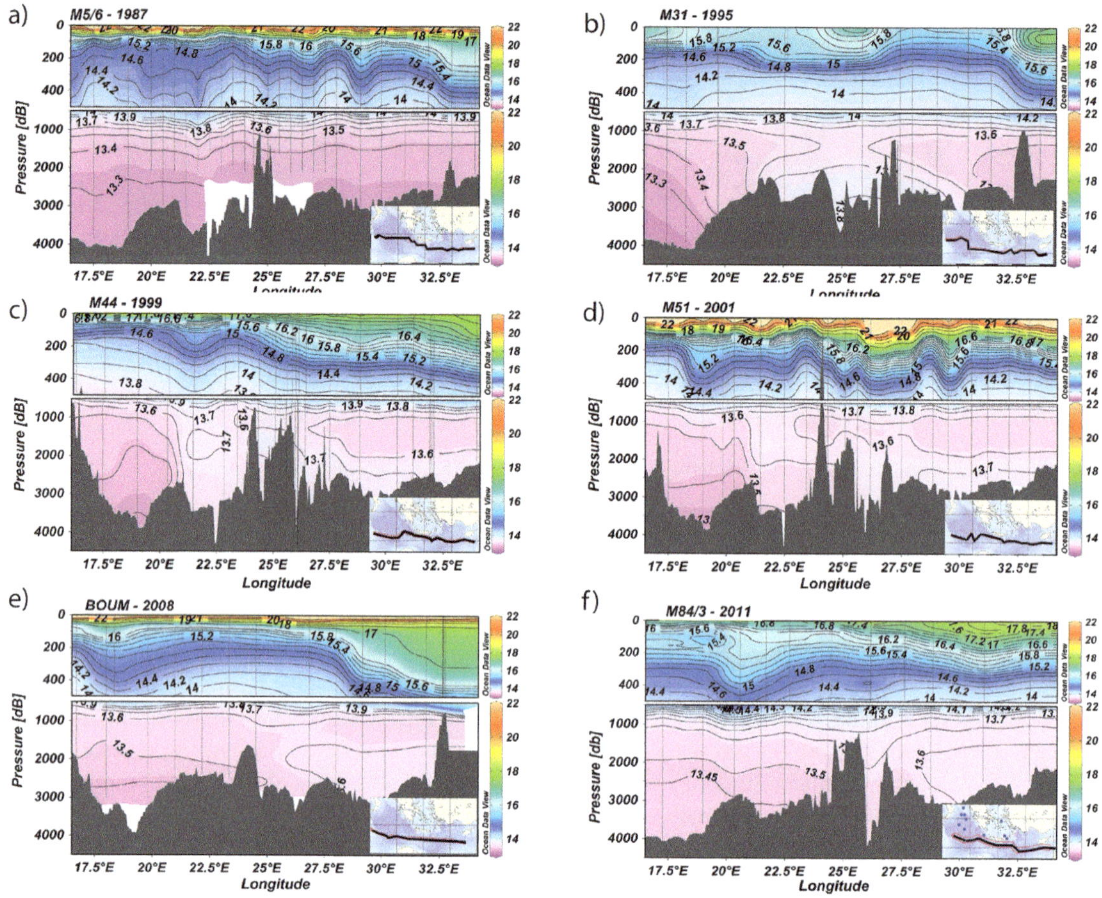

Figure 7. Sections of temperature (°C) (see inset map) in **(a)** 1987 (*Meteor* M5/6), **(b)** 1995 (*Meteor* M31/1), **(c)** 1999 (*Meteor* M44/4), **(d)** 2001 (*Meteor* M51/2), **(e)** 2008 (BOUM) and **(f)** 2011 (*Meteor* M84/3). The upper panels highlight the first 500 dbar.

the outflow levels above the Kasos and Otranto sills, respectively.

The time evolution of salinity in the CS shows, firstly, a remarkable increase in a large part of the water column (from 400 dbar to the bottom) in the period 1985–1992. As demonstrated by Theocharis et al. (1999), in the subsequent period (1992–1994) the strong decrease in temperature due to the very cold winters in 1992 and 1993 caused the accumulation of a huge amount of dense water ($\sigma_\theta > 29.20\,\mathrm{kg\,m^{-3}}$), as shown by the rapid rise of the $29.20\,\mathrm{kg\,m^{-3}}$ isopycnal (Fig. 9c, upper panel). At the same time, in the SA (Fig. 9c, lower panel) the salinity was < 38.70 from 200 dbar to the outflow depth, slightly increasing until 1990 ($S > 38.70$), and then decreasing until about 1997. The $29.20\,\mathrm{kg\,m^{-3}}$ isopycnal depth rose from 600 dbar (in 1985) to about 250 dbar (in 1992). The maximum density observed at the overflow depth was $29.25\,\mathrm{kg\,m^{-3}}$.

In the period 1992–1995, the CS density at the outflow level of the Kasos Strait remained higher than $29.25\,\mathrm{kg\,m^{-3}}$ and corresponds to the period of the most intense and continuous outflow of the new CDW into the Levantine basin (Theocharis et al., 1999). In the SA, in contrast, density rapidly decreased to values $\approx 29.20\,\mathrm{kg\,m^{-3}}$ only in the deepest portion (650–800 dbar) of the water column involved in the exchange with the Ionian Sea. In this period, the EMDW of Adriatic origin present in the Ionian Sea and in the Levantine was progressively displaced by the new denser CDW.

Our analysis confirms what was hypothesized by Roether et al. (2007) regarding the evolution of the Cretan water thermohaline properties post-EMT (from 1995) at the outflow level – i.e. that due to the slow deepening of the $29.20\,\mathrm{kg\,m^{-3}}$ isopycnal, the density levels exceeded pre-EMT values for several years. The strong CDW outflow was compensated for by an increased inflow of surface Levantine waters and by the Transitional Mediterranean Water, which in the CS was traceable by a salinity minimum (Theocharis et al., 1999) between 200 and 400 dbar in 1993–1995, progressively deepening and losing its characteristics around the second half of the 2000s. The presence of Transitional Mediterranean Water presumably inhibited further significant vertical mixing and dense water production after 1993.

As shown by Klein et al. (2000) and Cardin et al. (2011), the decrease in salinity at the intermediate level in the SA reached a minimum in 1995, after which the salinity started

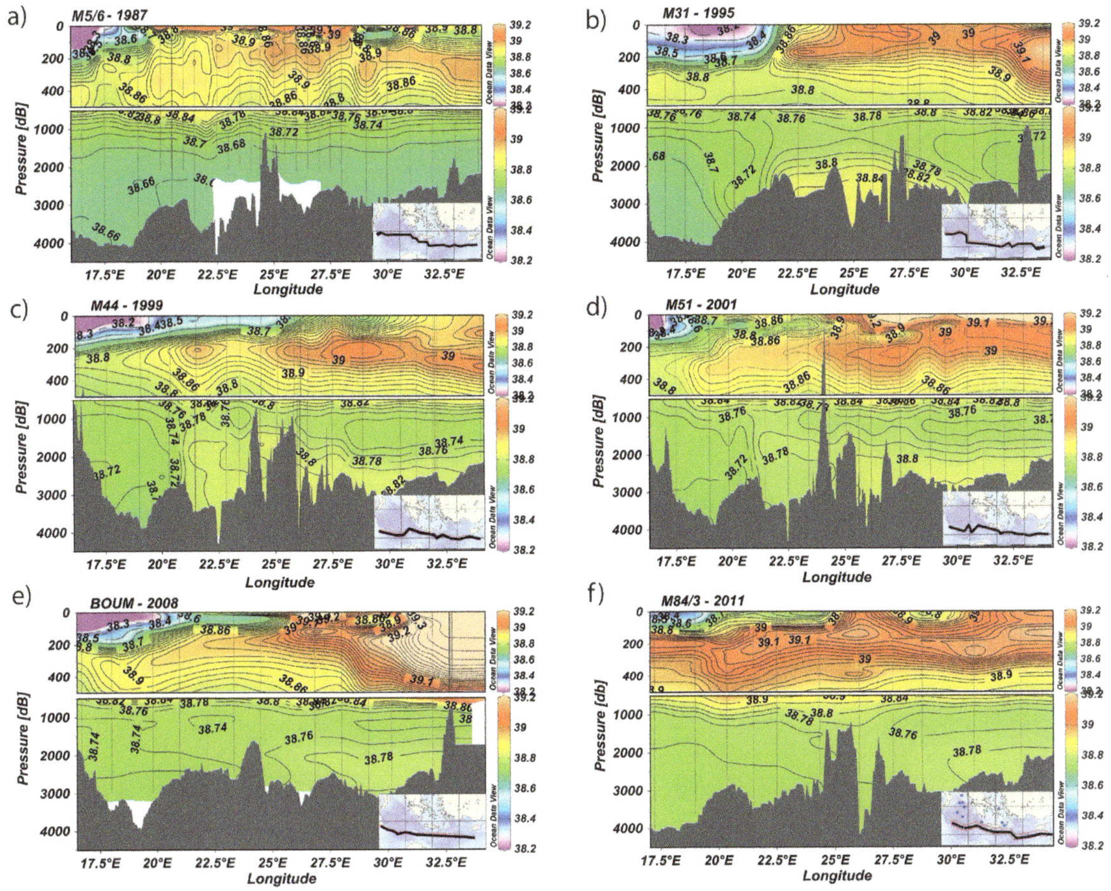

Figure 8. Sections of salinity (see inset map) in **(a)** 1987 (*Meteor* M5/6), **(b)** 1995 (*Meteor* M31/1), **(c)** 1999 (*Meteor* M44/4), **(d)** 2001 (*Meteor* M51/2), **(e)** 2008 (BOUM) and **(f)** 2011 (*Meteor* M84/3). The upper panels highlight the first 500 dbar.

to increase again until 2005. It is evident from Fig. 9 that sufficient preconditioning was established to resume the AdDW production during the first half of the 2000s. This was the result of a massive import of LIW and Cretan Intermediate Water (CIW) from the Ionian. However, it was only after 2005 that the AdDW at the overflow depth at the Otranto sill regained a density $> 29.20\,\mathrm{kg\,m^{-3}}$.

During the last part of the period studied, namely from 2005 to 2011, salinity at the intermediate level in the CS continuously increased, reaching absolute maximum values of > 39.10 in 2011. A concomitant temperature increase that prevented effective convection able to produce as much dense water to fill the deep Cretan basin was observed. In the same period in the SA the salinity of the whole water column decreased, as did the density after a relative maximum ($29.20\,\mathrm{kg\,m^{-3}}$ at 600 dbar) in 2006.

The role of the preconditioning, linked to the basin circulation, in promoting the alternation of dense water formation in the CS with that in the SA is important, but the relevance of the air–sea interaction cannot be neglected. Therefore, interannual variability in the normalized heat flux (Q_{norm}) was analysed in the two sites for the period December to March (Fig. 10). The general behaviour in the two sites was similar. Particularly noteworthy is the heat loss which occurred in the winters 1991–1992 and 1992–1993 in the CS and which triggered the active phase of the EMT. Another period of intense winter heat loss was from 2002 to 2006, for both sites. In this period, in the SA both preconditioning and heat loss were favourable to the AdDW production, whilst in the CS the presence of the above-mentioned Transitional Mediterranean Water counteracted the noticeable heat loss reported in Fig. 10. From 2007 to 2011 neither basin was subject to significant heat loss. The increased temperature in the CS and the decreased salinity in the SA made the situation in the two basins very similar in terms of the density at their own outflow depths. Also, thermohaline properties in the CS were still very different from those recorded in the pre-EMT phase.

The alternating capability of the Cretan and Adriatic seas in producing dense water by means of the internal salt redistribution between the Ionian Sea and Levantine basin has already been pointed out by Gačić et al. (2011), Theocharis et al. (2014) and Velaoras et al. (2014). Our comparative analysis also suggests that the evolution of thermohaline proper-

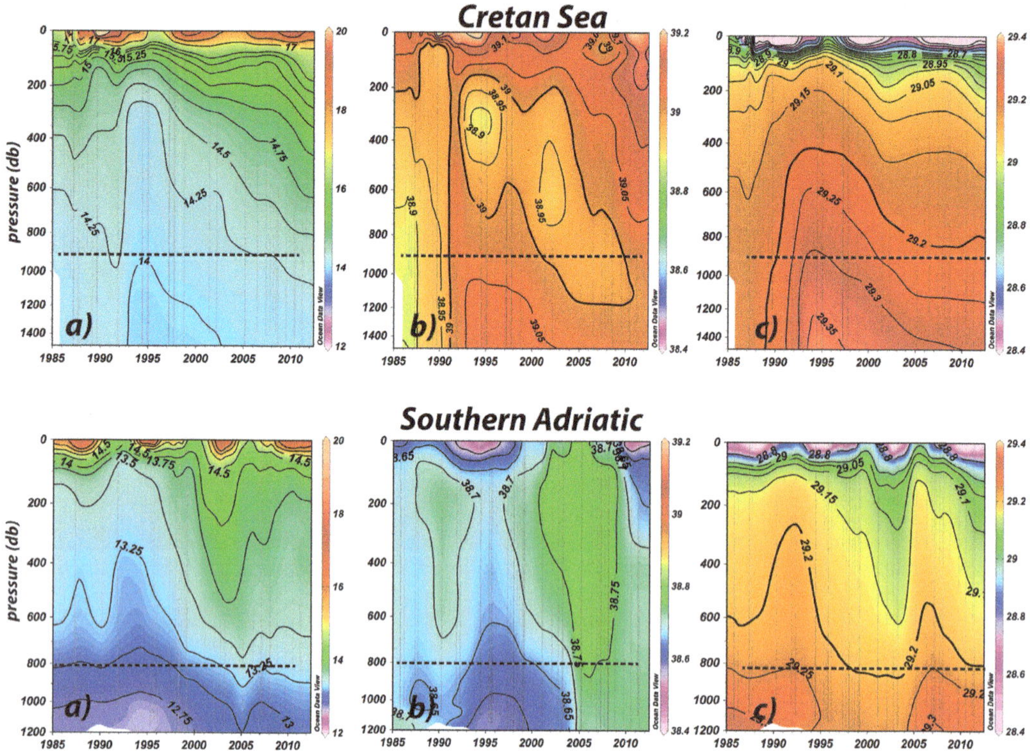

Figure 9. Hovmöller time diagrams of temperature, salinity and density for the period 1986–2011 for the Cretan Sea and Southern Adriatic. The selected areas are shown in Fig. 1a. The 900 and 800 bar levels highlighted represent the outflow level above the Kasos and Otranto sills, respectively. The y-axes are expanded at the top.

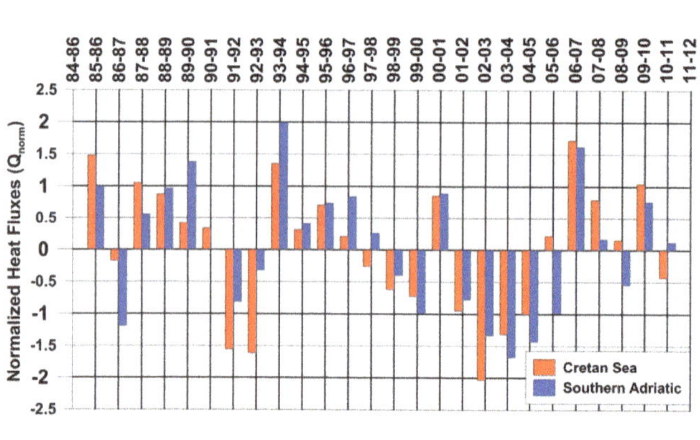

Figure 10. Normalized winter heat fluxes between 1985 and 2011 for the southern Adriatic (SA; blue) and the Cretan Sea (CS; red) with respect to their own mean. Winter period is defined as December to March. The selected areas are shown in Fig. 1a.

ties of the two potential dense water source areas is subject to the complex interactions among preconditioning, air–sea interactions, and local dynamics and circulation.

The spatial and temporal evolution of the water mass structure was also analysed through the comparison of vertical profiles of temperature and salinity within two areas: the central Ionian and the central Levantine (see Fig. 11a and b for vertical profiles and Fig. 1 for the location of the areas). This analysis includes the data from cruises already discussed in the previous paragraph. The inset graphs enhance the oceanographic characteristics for the deep layer below 1500 dbar. Salinity and temperature profiles for both areas (Fig. 11a and b, zoomed areas) are mostly indicative of changes in the water mass structure related to the EMT. In 1987, the central Levantine below 2500 dbar showed a vertical homogeneity in temperature and salinity in the ranges $\approx 13.34\,°C$ and ≈ 38.66, respectively. In 1991, a slight increase in salinity from ~ 500 dbar down to the bottom was seen. In 1995, the salinity increase was remarkable, especially below 2300 dbar where it jumped from 38.71 to ~ 38.84 because of the new dense CDW outflow (Roether et al., 1996). Its signature was tracked until 2007, when the salinity started to decrease. Meanwhile, in the central Ionian the salinity continued to increase, the abyssal layer being filled by newly formed AdDW, saltier and warmer; this was still evident in 2011. The most recent data (2008–2011) in the deep layer in the central Levantine show a salinity and temperature decrease, probably a result of the basin internal mixing (Roether et al., 2014).

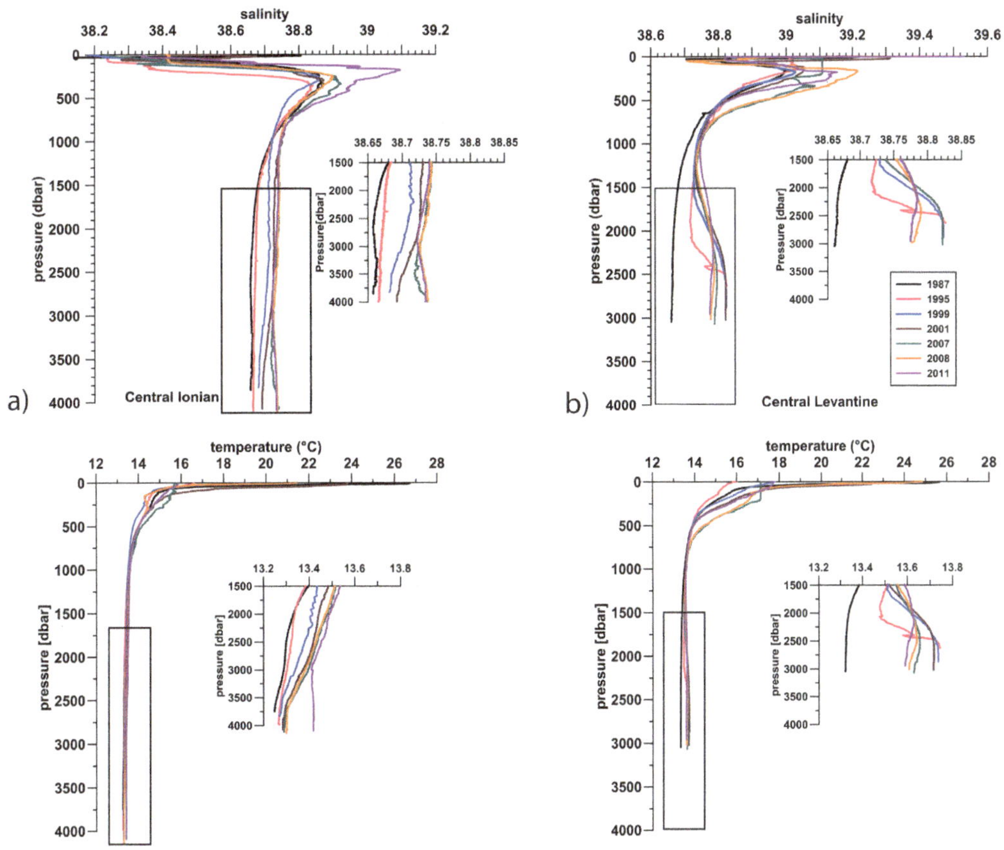

Figure 11. Vertical profiles of temperature (°C) and salinity in the central Levantine **(a)** and central Ionian **(b)**. For graphical purposes the salinity scales have the same unit length but range from 38.6 to 39.6 for the central Levantine, and from 38.2 to 39.2 for the central Ionian. The selected areas are shown in Fig. 1a.

5 Conclusions

This paper discusses data collected in 2011 during the *Meteor* M84/3 (April 2011) and the *Poseidon* P414 (June 2011) cruises in the context of the spatial and temporal evolution of the large-scale oceanographic properties of the EM during the last three decades. Additionally, we discussed the interannual variability of the preconditioning and air–sea interaction (heat fluxes) in the two main dense water source areas for the EM, i.e South Adriatic and Cretan Seas.

Our results, which are the most recent analysis for the whole EM, reveal that the thermohaline properties in the study area were still far from those observed in the pre-EMT phase (1987). The 2011 oceanographic conditions in the deep layers of the central Ionian, however, lie between the thermohaline characteristics of the EMT and those of the pre-EMT phase, indicating a possible slow return towards the latter.

The abyssal layer of the central Ionian experienced the effect of a post-EMT phase, which started in about 2003, characterized by the arrival of newly formed AdDW waters, completely different from the recent past in that they are saltier and warmer. In the following years, this new EMDW of Adriatic origin seemed to have propagated towards the abyssal Levantine basin.

This study highlights the relationship between the hydrological property distribution of the upper layer in the Levantine basin and the alternate circulation regimes in the Ionian, which modulates the salinity distribution in the EM.

The comparative analysis of the evolution of the thermohaline properties in the two potential dense-water source areas, SA and CS, demonstrates that, as far as dense-water accumulation is concerned, the alternation of the two sites is strongly influenced by the preconditioning driven by the reversal of the NIG. Nevertheless, the role of the air–sea interaction and the local dynamics should be taken into account in assessing the decadal variability of the EM thermohaline cells.

Acknowledgements. The authors would like to thank the captains and crews on the research vessels RV *Meteor* and RV *Poseidon* for their cooperation during the cruises. Special thanks to George Petihakis and the Hellenic National Oceanographic Data Centre (HNODC) for providing the CTD casts between 2005 and 2010 near the E1M3A observatory and in the Cretan Sea. Thanks go to Louis Prieur and Thierry Moutin for kindly providing the CTD of the *BOUM* cruise. The *Meteor* cruise M84/3 was supported by a grant from the Deutsch Forschungsgemeinshaft – Senatskommision für Ozeanographie (DFG), TA 311/3-1. Funding from the University of Hamburg supported the *Poseidon* P414 cruise. V. Cardin and G. Civitarese thank the CIESM for supporting their participation in the *Meteor* Cruise. The Italian Ministry of Education, University and Research (MIUR) supported this research under the RITMARE (Ricerca ITaliana per il MARE) and MedGES (Studio integrato dell'influenza dei processi fisici e dei fattori antropici sullo stato degli ecosistemi costieri e profondi del Mediterraneo Centrale) national projects. The Topic Editor is grateful to the two anonymous referees for their contributions which have helped to greatly improve the quality of this paper.

Edited by: S. Sparnocchia

References

Bensi, M., Rubino, A., Cardin, V., Hainbucher, D., and Mancero-Mosquera, I.: Structure and variability of the abyssal water masses in the Ionian Sea in the period 2003-2010, J. Geophys. Res., 118, 1–13, doi:10.1029/2012JC008178, 2013.

Bignami, F., Marullo, S., Santoleri, R., and Schiano M. E.: Long-wave radiation budget in the Mediterranean Sea, J. Geophys. Res., 100, 2501–2514, 1995.

Cardin, V. and Gačić, M.: Long-term heat flux variability and winter convection in the Adriatic Sea, J. Geophys. Res., 108, 8103, doi:10.1029/2002JC001645, 2003.

Cardin, V., Bensi, M., and Pacciaroni, M.: Variability of water mass properties in the last two decades in the Southern Adriatic Sea with emphasis on the period 2006–2009, Cont. Shelf Res., 31, 951–965, doi:10.1016/j.csr.2011.03.002, 2011.

D'Ortenzio, F. and Prieur, L.: The upper mixed layer, in: Life in the Mediterranean Sea: A look at habitat changes, edited by: Stambler, N., Nova Science Publisher, 127–156, 2012.

Gačić, M., Borzelli, G. L. E., Civitarese, G., Cardin, V., and Yari, S.: Can internal processes sustain reversals of the ocean upper circulation? The Ionian Example, Geophys. Res. Lett., 37, L09608, doi:10.1029/2010GL043216, 2010.

Gačić, M., Civitarese, G., Eusebi Borzelli, G. L., Kovačević, V., Poulain, P.-M., Theocharis, A., Menna, M., Catucci, A., and Zarokanellos, N.: On the relationship between the decadal oscillations of the Northern Ionian Sea and the salinity distributions in the Eastern Mediterranean, J. Geophys. Res., 116, C12002, doi:10.1029/2011JC007280, 2011.

Gačić, M., Civitarese, G., Kovačević, V., Ursella, L., Bensi, M., Menna, M., Cardin, V., Poulain, P.-M., Cosoli, S., Notarstefano, G., and Pizzi, C.: Extreme winter 2012 in the Adriatic: an example of climatic effect on the BiOS rhythm, Ocean Sci., 10, 513–522, doi:10.5194/os-10-513-2014, 2014.

Hainbucher, D., Rubino, A., and Klein, B.: Water mass characteristics in the deep layers of the western Ionian basin observed during May, Geophys. Res. Lett., 33, L05608, doi:10.1029/2005GL025318, 2006.

Hainbucher, D., Rubino, A., Cardin, V., Tanhua, T., Schroeder, K., and Bensi, M.: Hydrographic situation during cruise M84/3 and P414 (spring 2011) in the Mediterranean Sea, Ocean Sci., 10, 669–682, doi:10.5194/os-10-669-2014, 2014.

Hemleben, C., Roether, W., and Stoffers, P. (Eds.): Östliches Mittelmeer, Rotes Meer, Arabisches Meer, Cruise No. 31, 30 December 1994–22 March 1995, METEOR-Berichte, Universität Hamburg, 96-4, 282 pp., 1996.

Hemleben, C., Hoernle, K., Jørgensen, B.B., and Roether, W. (Eds.): Ostatlantik, Mittelmeer, Schwarzes Meer, Cruise No. 51, 12 September–28 December 2001, METEOR-Berichte, Universität Hamburg, 03-1, 225 pp., 2003.

Klein, B., Roether, W., Manca, B. B., Bregant, D., Beitzel, V., Kovačević, V., and Luchetta, A.: The large deep water transient in the Eastern Mediterranean, Deep-Sea Res. Pt. I, 46, 371–414, 1999.

Klein, B., Roether, W., Manca, B., and Theocharis, A.: The evolution of the Eastern Mediterranean Climatic transient during the last decade: the tracer viewpoint, in: The Eastern Mediterranean Transient, edited by: Briand, F., CIESM Workshop Series, vol. 10. CIESM, Monaco, 21–25, 2000.

Kondo, J.: Air-sea bulks transfer coefficients in adiabatic conditions, Bound-Lay. Meteorol., 9, 91–112, 1975.

Lascaratos, A., Roether, W., Nittis, K., and Klein, B.: Recent changes in deep water formation and spreading in the eartern Mediterranean Sea: a review, Prog. Oceanogr., 44, 5–36, 1999.

Maillard, C., Balopoulos, E., Giorgetti, A., Fichaut, M., Iona, S., Larour, M., Latrouite, A., Manca, B. B., Maudire, G., Nicolas, P., and Sanchez-Cabeza, J.-A.: An integrated system for managing multidisciplinary oceanographic data collected in the Mediterranean Sea during the basin-scale research project EU/MAST-MATER (1996–2000), J. Marine Syst., 33–34, 523–538, 2002.

Malanotte-Rizzoli, P., Manca, B., Ribera D'Alcalà, M., Theocharis, A., Bergamasco, A., Bregant, D., Budillon, G., Civitarese, G., Georgopoulos, D., Michelato, A., Sansone, E., Scarazzato, P., and Souvermezoglou, E.: A synthesis of the Ionian Sea hydrography, circulation and water mass pathways during POEM Phase I, Progr. Oceanogr., 39, 153–204, 1997.

Malanotte-Rizzoli, P., Manca, B. B., Ribera d'Alcala, M., Theocharis, A., Brenner, S., Budillon, G., and Ozsoy, E.: The Eastern Mediterranean in the 80s and in the 90s: the big transition in the intermediate and deep circulations, Dynam. Atmos. Oceans, 29, 365–395, 1999.

Molcard, A., Pinardi, N., Iskandarani, M., and Haidvogel, D. B.: Wind driven general circulation of the Mediterranean Sea simulated with a Spectral Element Ocean Model, Dynam. Atmos. Oceans, 35, 97–130, 2002.

Manca, B. B., Budillon, G., Scarazzato, P., and Ursella, L.: Evolution of dynamics in the eastern Mediterranean affecting water mass structures and properties in the Ionian and Adriatic Seas, J. Geophys. Res., 108, 8102, doi:10.1029/2002JC001664, 2003.

Manca, B. B., Burca, M., Giorgetti, A., Coatanoan, C., García, M., and Iona, A.: Physical and biogeochemical averaged vertical profiles in the Mediterranean regions: an important tool to trace the climatology of water masses and to validate incoming data from operational oceanography, J. Marine Syst., 48, 83–116, 2004.

Manca, B. B., Ibello, V., Pacciaroni, M., Scarazzato, P., and Giorgetti, A.: Ventilation of deep waters in the Adriatic and Ionian Seas following changes in thermohaline circulation of the Eastern Mediterranean, Clim. Res., 31, 239–256, 2006.

Moutin, T., Van Wambeke, F., and Prieur, L.: Introduction to the Biogeochemistry from the Oligotrophic to the Ultraoligotrophic

Mediterranean (BOUM) experiment, Biogeosciences, 9, 3817–3825, doi:10.5194/bg-9-3817-2012, 2012.

Nellen, W., Bettac, W., Roether, W., Schnack, D., Thiel, H., Weikert, H., and Zeitschel, B. (Eds.): MINDIK (Band II), Reise Nr. 5, 2 January–24 September 1987m METEOR-Berichte, Universität Hamburg, 96-2, 179 pp., 1996.

Pätzold, J., Halbach, P. E., Hempel, G., and Weikert, H. (Eds.): Östliches Mittelmeer – Nördliches Rotes Meer 1999, Cruise No. 44, 22 January–16 May 1999m METEOR-Berichte, Universität Hamburg, 00-3, 240 pp., 2000.

Pinardi, N., Zavatarelli, M., Adani, M., Coppini, G., Fratianni, C., Oddo, P., Simoncelli, S., Tonani, M., Lyubartsev, V., Dobricic, S., and Bonaduce, A.: Mediterranean Sea large-scale low-frequency ocean variability and water mass formation rates from 1987 to 2007: A retrospective analysis, Prog. Oceanogr., doi:10.1016/j.pocean.2013.11.003, in press, 2013.

Rio, M.-H., Poulain, P.-M., Pascal, A., Mauri, E., Larnicol, G., and Santoleri, R.: A mean dynamic topography of the Mediterranean Sea computed from altimetric data, in-situ measurements and a general circulation model, J. Marine Syst., 65, 484–508, 2007.

Robinson, A. R., Golnaraghi, M., Leslie, W. G., Artegiani. A., Hecht, A., Lazzoni, E., Michelato, A., Sansone. E., Theocharis, A., and Unluata, U.: The eastern Mediterranean general circulation: features, structures and variability, Dynam. Atmos. Oceans, 15, 215–240, 1991.

Roether, W. and Schlitzer, R.: Eastern Mediterranean deep water renewal on the basis of chlorofluoromethane and tritium data, Dynam. Atmos. Oceans, 15, 333–354, 1991.

Roether, W., Manca, B.B., Klein, B., Bregant, D., Georgopoulos, D., Beitzel, V., Kovačević, V., and Luchetta, A.: Recent changes in Eastern Mediterranean deep waters, Science, 271, 333–335, 1996.

Roether, W., Klein, B., Manca, B. B., Theocharis, A., and Kioroglou, S.: Transient Eastern Mediterranean deep waters in response to the massive dense-water output of the Aegean Sea in the 1990s, Prog. Oceanogr., 74, 540–571, 2007.

Roether, W., Klein, B., and Hainbucher, D.: The Eastern Mediterranean Transient: evidence for similar events previously?, in: AGU monographs: The Mediterranean Sea: Temporal Variability and Spatial Patterns, edited by: Borzelli, G. L. E., Gačić, M., Lionello, P., and Malanotte-Rizzoli, P., John Wiley & Sons, Inc., 75–83, 2014.

Rubino, A. and Hainbucher, D.: A large abrupt change in the abyssal water masses of the eastern Mediterranean, Geophys. Res. Lett., 34, L23607, doi:10.1029/2007GL031737, 2007.

Schiano, E.: Insolation over the western Mediterranean Sea: A comparison of direct measurements and Reed's formula, J. Geophys. Res., 101, 3831–3838, 1996.

Schlitzer, R., Roether, W., Hausmann, M., Junghans, H.-G., Oster, H., Johannsen, H., and Michelato, A.: Chlorofluoromethane and oxygen in the Eastern Mediterranean, Deep-Sea Res. Pt. I, 38, 1531–1551, 1991.

Tanhua, T., Hainbucher, D., Cardin, V., Álvarez, M., Civitarese, G., McNichol, A. P., and Key, R. M.: Repeat hydrography in the Mediterranean Sea, data from the *Meteor* cruise 84/3 in 2011, Earth Syst. Sci. Data, 5, 289–294, doi:10.5194/essd-5-289-2013, 2013.

Theocharis, A., Nittis, K., Kontoyannis, H., Papageorgiou, E., and Balopoulos, E.: Climatic changes in the Aegean Sea influence the eastern Mediterranean thermohaline circulation (1986–1997), Geophys. Res. Lett., 26, 1617–1620, 1999.

Theocharis, A., Klein, B., Nittis, K., and Roether, W.: Evolution and status of the Eastern Mediterranean Transient (1986–1999), Geophys. Res. Lett., 26, 1617–1620, 2002.

Theocharis, A., Krokos, G., Velaoras, D., and Korres, G.: An internal mechanism driving the alternation of the Eastern Mediterranean dense/deep water sources, in AGU monographs: The Mediterranean Sea: Temporal Variability and Spatial Patterns, edited by: Borzelli, G. L. E., Gačić, M., Lionello, P., and Malanotte-Rizzoli, P., John Wiley & Sons, Inc., 113–137, 2014.

Touratier, F. and Goyet, C.: Impact of the Eastern Mediterranean Transient on the distribution of anthropogenic CO_2 and first estimate of acidification for the Mediterranean Sea, Deep-Sea Res. Pt. I, 58, 1–15, 2011.

Velaoras, D., Krokos, G., Nittis, K., and Theocharis A.: Dense intermediate water outflow from the Cretan Sea: A salinity driven, recurrent phenomenon, connected to thermohaline circulation changes, J. Geophys. Res.-Oceans, 119, 4797–4820, doi:10.1002/2014JC009937, 2014.

4

A parameter model of gas exchange for the seasonal sea ice zone

B. Loose[1], W. R. McGillis[2], D. Perovich[3], C. J. Zappa[2], and P. Schlosser[2,4,5]

[1]Graduate School of Oceanography, University of Rhode Island, 215 South Ferry Rd., Narragansett, RI, 02882, USA
[2]Lamont-Doherty Earth Observatory of Columbia University, 61 Route 9W, Palisades, NY 10964, USA
[3]US Army Corps of Engineers Cold Regions Research and Engineering Laboratory, 72 Lyme Rd., Hanover, NH, 03755, USA
[4]Department of Earth and Environmental Sciences, Columbia University, Room 106 Geoscience Bldg., Palisades, NY 10964, USA
[5]Department of Earth and Environmental Engineering, Columbia University, 918 Seeley Mudd Bldg., Columbia University, 500 West 120th St, New York, NY 10027, USA

Correspondence to: B. Loose (brice@gso.uri.edu)

Abstract. Carbon budgets for the polar oceans require better constraint on air–sea gas exchange in the sea ice zone (SIZ). Here, we utilize advances in the theory of turbulence, mixing and air–sea flux in the ice–ocean boundary layer (IOBL) to formulate a simple model for gas exchange when the surface ocean is partially covered by sea ice. The gas transfer velocity (k) is related to shear-driven and convection-driven turbulence in the aqueous mass boundary layer, and to the mean-squared wave slope at the air–sea interface. We use the model to estimate k along the drift track of ice-tethered profilers (ITPs) in the Arctic. Individual estimates of daily-averaged k from ITP drifts ranged between 1.1 and 22 $\mathrm{m\,d^{-1}}$, and the fraction of open water (f) ranged from 0 to 0.83. Converted to area-weighted effective transfer velocities (k_{eff}), the minimum value of k_{eff} was $10^{-5}\,\mathrm{m\,d^{-1}}$ near $f = 0$ with values exceeding $k_{\mathrm{eff}} = 5\,\mathrm{m\,d^{-1}}$ at $f = 0.4$. The model indicates that effects from shear and convection in the sea ice zone contribute an additional 40 % to the magnitude of k_{eff}, beyond what would be predicted from an estimate of k_{eff} based solely upon a wind speed parameterization. Although the ultimate scaling relationship for gas exchange in the sea ice zone will require validation in laboratory and field studies, the basic parameter model described here demonstrates that it is feasible to formulate estimates of k based upon properties of the IOBL using data sources that presently exist.

1 Introduction

The regulation of air–water gas exchange occurs within a small vertical region of $\sim 100\,\mu\mathrm{m}$ at the air–water interface – a region that is too small for direct measurement of the properties governing this exchange, namely partial pressure and diffusive length scale. In addition, the air–water interface is continually being deformed by waves and surfactants and consequently the profile of turbulence and gas abundance are difficult to determine for a given surface condition (Jähne et al., 1987). This limitation at the critical scale has led to a proliferation of rate approximations based on quantities that can be more readily measured. Wind speed is the most common of these quantities, as it is an important mixing mechanism for many surface water bodies, including the ocean (Liss, 1973; Wanninkhof, 1992; Wanninkhof and McGillis, 1999; Nightingale et al., 2000; Ho et al., 2006, 2011; Sweeney et al., 2007; Wanninkhof et al., 2009). However, even for the blue water ocean, wind speed parameterizations account for 75 % or less of the observed variability in the gas exchange rate (Frew et al., 2004; Stanley et al., 2006). Wind speed parameterizations are widely used where wind is the first order mixing mechanism, and may be adequate under certain open ocean conditions.

In contrast to the open ocean, there are physical systems such as wetlands, rivers, estuaries, and the coastal surf zone where wind speed derived estimates of gas exchange are incomplete given the other mechanisms that produce aqueous turbulence. These mechanisms include tides, internal waves and boundary mixing. In these systems, it is necessary to rely on additional or alternative metrics for relating gas exchange to the underlying forcing conditions (Ho et al., 1997; Zappa et al., 2003; Frew et al., 2004).

For moderately to sparingly soluble (i.e., most) gases, the existence of an aqueous viscous sublayer at the air–sea interface is the dominant restriction to transfer across the interface. This bottleneck ranges in length scale from 20 to 200 μm (Jähne and Haubecker, 1998); turbulent eddies impinging on the free surface can reduce or collapse the viscous sublayer, leading to "bursts" of air–sea gas transfer (Rao et al., 1971). Consequently, measurement of the TKE budget close to the air-sea interface provides the best metric for determining the gas exchange rate when multiple drivers are at work.

We have argued that the sea ice covered polar oceans are another region where wind speed parameterizations are not expected to be adequate (Loose et al., 2009; Loose and Schlosser, 2011), and in this contribution we present an alternative parameter model that is based on the mechanisms that produce turbulence in the ice–ocean boundary layer (IOBL): (1) shear caused by the velocity differential between drifting ice and the water column beneath (McPhee and Martinson, 1992), (2) buoyant convection caused by heat loss and phase changes at the ocean surface (Morison et al., 1992), and (3) short period waves and wave interactions with ice floes (Wadhams et al., 1986; Kohout and Meylan, 2008) (Fig. 1). Our goal is to explore what effect these turbulent production mechanisms may have on the magnitude of k by utilizing the significant advances in theory and observation of turbulence in the IOBL, developed over the past 30 yr (McPhee, 2008). This exercise is meant to provide first-order insight into the processes by which sea ice might simultaneously produce and modulate gas exchange, and to provide a model template upon which laboratory and field studies can begin to establish firm quantitation of these processes.

In Sect. 2 we introduce the basic theory that has been developed to relate differing turbulence production mechanisms to the rate of air–sea gas exchange. Section 3 briefly reviews how to estimate turbulence production from shear in the IOBL as a function of wind speed, ice drift velocity and sea ice cover. Section 4 contains a similar review for the estimate of turbulence production by convection in the IOBL. Section 5 provides a rough estimate of k from capillary-gravity waves. Section 6 is used to present estimates of k, calculated from four ice-tethered profiler (ITP) missions. Finally we discuss the results from the ITP data and the parameter model and how they might relate to future investigations of k in the sea ice zone (SIZ).

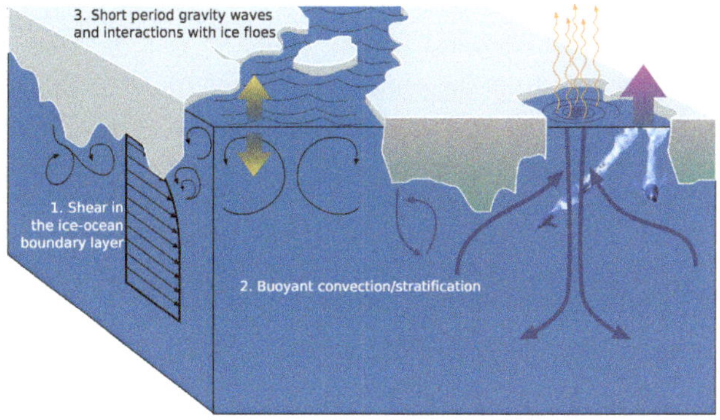

Fig. 1. A graphic illustration of three mechanisms that can lead to turbulence production and gas exchange in the ice–ocean boundary layer.

2 Methods

2.1 A note about upscaling k in the sea ice zone

The sea ice zone exists as a heterogeneous distribution of ice cover and open water, and this presents a challenge when trying to appropriately represent a gas transfer rate that takes this heterogeneity into account. Because sea ice is a porous medium (Cox and Weeks, 1983), it is possible in principle to ventilate the surface ocean of gas through the ice matrix as across the air-sea interface. To reflect this, the air–sea gas flux can be written as

$$F = k_{\text{eff}} \Delta C, \tag{1}$$

where ΔC is the molar gas differential across the air–sea interface, and k_{eff} is the "effective" gas transfer velocity,

$$k_{\text{eff}} = (1-f)k_{\text{ice}} + (f)k, \tag{2}$$

where f is the fraction of open water, $(1-f)$ is the fraction of sea ice cover, k_{ice} represents the bulk diffusive gas flux through the sea ice D/z_{ice}, and k is the open water gas transfer velocity – what we are most accustomed to working with. This representation of k_{eff} is consistent with the asymptotic homogenization of diffusive transport through a medium with rapidly varying permeability, such as the ice covered upper ocean (Dykaar and Kitandis, 1992; Jähne and Haubecker, 1998; Golden et al., 2006).

In practice, the rate of gas diffusion through cold columnar sea ice is within a factor of 10 of molecular diffusion in water (e.g., $\sim 10^{-4}\,\text{cm}^2\,\text{s}^{-1}$). Thus k_{ice} is very small compared to k – at least for cold, consolidated columnar sea ice (Gosink et al., 1976; Loose et al., 2010). In the presence of thin ice, such as grease ice, nilas or frazil ice, gas exchange may not be so slow, but solute rejection from the ice (Killawee et al., 1998; Loose et al., 2009) makes this circumstance hard to interpret. However, the distinction, between k and k_{eff} is worth clarifying for another reason. There are a variety of methods for estimating the air–sea gas transfer velocity. Based on the scale over which these methods integrate, some yield k_{eff} and others yield k or k_{ice}, and it is important to be able to put the results from these methods on a common scale. For example, tracer-based estimates of gas transfer velocity such as the radon deficit method (Smethie et al., 1985; Bender et al., 2011), and the dual-tracer method (Wanninkhof et al., 1993; Nightingale et al., 2000) will yield estimates of k_{eff}, whereas covariance flux (McGillis et al., 2001), and gradient flux (Orton et al., 2010) methods represent a much smaller spatial footprint and will yield a gas transfer velocity that will represent k_{ice} or k, but not k_{eff}. Examples of past studies that estimate k_{eff} are Fanning and Torres (1991), Loose et al. (2009), and Loose and Schlosser (2011). The study by Miller et al. (2011) would yield estimates of k_{ice} and the study by Else et al. (2011) would yield estimates of k for a polynya-type water surface.

In this study, the parameter model estimates of gas transfer are based on predictions of the aqueous turbulence and represent values of k. We will present the results as k and k_{eff}, using Eq. (2) and by assuming that k_{ice} is negligible.

2.2 TKE dissipation model

We require an estimate of k that accounts for turbulence production mechanisms that are unique to the SIZ. This can be accomplished using the relationship between TKE dissipation (ε) and gas exchange (Lamont and Scott, 1970),

$$k \propto (\nu\varepsilon)^{0.25} Sc^{-0.5}\,, \tag{3}$$

where ν is the kinematic viscosity, and Sc is the Schmidt number. For the purposes of this paper, we will assume the constant of proportionality determined by Zappa et al. (2007); 0.419. It should be noted that ε varies vertically in the water column and is likely to assume distinct values at the ice–water and air–water interfaces, therefore unique proportionality constants may exist for the SIZ. Next we assume that over the mixing timescale of the surface ocean TKE production should balance dissipation, leading to a stationary form for the kinetic energy equation (Gaspar et al., 1990),

$$u_*^2 \partial u/\partial z + \overline{b'w'} - \varepsilon = 0. \tag{4}$$

The first term represents TKE production by shear and the second represents TKE production by buoyancy flux. Monin–Obukhov similarity theory (MOST) demonstrates how to relate turbulence produced by convection and turbulence produced by shear (Lombardo and Gregg, 1989). When convection ceases and stratification sets in, MOST also provides a means for estimating the dampening effect on shear driven turbulence based on the Obukhov length scale,

$$L_0 = \frac{-u_{*0}^3}{\kappa \langle w'b' \rangle_0}, \tag{5}$$

where u_{*0} is the aqueous friction velocity at the ice–water or air–water interface, $\langle w'b' \rangle_0$ is the buoyancy flux at the ocean surface, and κ is the Von Karmann constant.

Finally, ε from Eq. (4) can be related to the turbulent friction velocity in the water (u_{*0}) by

$$\varepsilon = u_{*0}/\kappa z\,, \tag{6}$$

where z the depth scale, assuming z lies within the boundary layer. By definition, u_{*0} is a representation of the shear stress (τ) that acts on the boundary layer,

$$u_{*0} = \sqrt{\tau/\rho}, \tag{7}$$

and it is the shear stress balance that is altered by the presence of ice floes, individually and as part of the ice pack. In the next section we will review and describe the stress balance in partially ice-covered seas and then use the above equations to make estimates of k.

3 Shear in the IOBL

Sea ice floes present additional surfaces against which normal and tangential forces can operate. An individual ice floe is subject to surface drag from wind over the ice, which is transmitted to the IOBL via bottom drag, when the ice and the water beneath move relative to each other (Shirasawa and Ingram, 1991). In addition to surface stresses, an ice floe experiences form drag as it displaces water by its lateral motion (Perrie and Hu, 1997), and by wave radiation stress as gravity waves propagate from open ocean into the pack (Perrie and Hu, 1996, 1997). Steele et al. (1989) observed that as sea ice concentration decreases, the ice pack behaves more like a "flotilla of barges" where individual floes may experience form drag that exceeds surface drag, when the diameter of the floe is less than 300 times the ice draft, based on the ratio of surface and form drag coefficients.

The total shear stress exerted on the aqueous boundary layer can be estimated as a function of sea ice cover $(1-f)$, wind speed (U) and ice drift velocity (V_0). By these mechanisms the total drag on the water surface boundary layer is as follows (Steele et al., 1989):

$$\tau = (1-f)(\tau_{\text{skin-iw}} + \tau_{\text{form}}) + (f)(\tau_{\text{skin-aw}})(1-W), \tag{8}$$

where $\tau_{\text{skin-iw}}$ is the ice–water skin drag, τ_{form} is the form drag around ice floes, $\tau_{\text{skin-aw}}$ is the air–water skin friction and W is the fraction of air–water stress that is used to produce surface gravity waves (Steele et al., 1989). $W = 1.6 \times 10^{-4} \rho_w/\rho_a$ is based on the Charnock (1981) relationship.

(τ_{form}) is dependent on the profile the ice presents to the flow as it forces its way through the water. This effect was represented by Steele et al. (1989) by incorporating the freeboard (D) and length (L) of the ice floe such that form drag decreases as the characteristic dimension of the flow size increases. This behavior is meant to simulate the "wake effect" where the overall pressure gradient from the front to the back of the floe is decreased as L increases.

$$\tau_{\text{form}} = \frac{1}{2}\rho_{\text{w}}\frac{D}{L}|V_0|\,V_0, \tag{9}$$

where V_0 is the relative velocity between the water in the mixed layer and the ice floating on it (Lu et al., 2011), and ρ_{w} is the water density. Tennekes and Lumley (1972) showed that the form drag decreases approximately proportional to $L^{-0.5}$. In general, when sea ice concentration is close to 0 %, the characteristic dimension of L is small, e.g., 10 m. As ice cover increases, so does L (Toyota et al., 2006), and the relationship follows a power-law dependence of the form $N(L) \propto L^{-\alpha}$, where N is the number density of ice floes per km^2 (Rothrock and Thorndike, 1985). Toyota et al. (2006) observed that this relationship cannot be fit with a single value for α; when L is less than 40 m, the value of α is 1.15 and when L is greater than 40 m, $\alpha = 1.87$ provides the best fit. $N(L)$ can be related to the fraction of sea ice cover if we assume a characteristic shape for ice floes. The SIC was computed as the average of two "ideal" floes, one square-shaped and one circular, with sides and diameters of length L, multiplied by $N(L)$ (Fig. 2).

($\tau_{\text{skin-iw}}$), the next term in the boundary layer stress balance (Eq. 8), represents the skin friction at the ice–water interface. The ice–ocean boundary layer is often compared with the atmospheric boundary layer as both are affected by planetary rotation, both experience relatively constant stress-driven turbulence (Blackadar and Tennekes, 1968) and both tend to follow the law of the wall, which permits a complete determination of the velocity profile and turbulence closure in terms of two variables: u_{*0} and the roughness length scale z_o (McPhee, 2008),

$$u_{*0} = \frac{U(z)}{\kappa}\left(\ln\frac{z}{z_0}\right), \tag{10}$$

where $U(z)$ is the Euclidean norm velocity at depth z in the surface boundary layer. Referring to Eq. (7), the skin friction can alternately be expressed as a velocity or a stress with no loss of generality. Observations in the IOBL show that the surface boundary layer, where velocity varies linearly with the log of depth ($\sim$2 m) behaves differently from the outer layer ($\sim$70 m) (Shaw et al., 2008). Outside of the surface layer, it is necessary to account for Ekman turning on the shear stress, and this can be accomplished using the Rossby similarity theory (Blackadar and Tennekes, 1968; McPhee, 2008). To determine the surface shear stress in terms of the ice velocity it is necessary to account for the geostrophic drift at the base of IOBL because the magnitude of shear and the magnitude of rotation are interdependent (McPhee, 2008). This can be accomplished using the Rossby similarity theory (Blackadar and Tennekes, 1968; McPhee, 2008), which relates V_0 with u_{*0}, the interface friction velocity overlying a neutral geostrophic current (McPhee, 2008; Shaw et al., 2008),

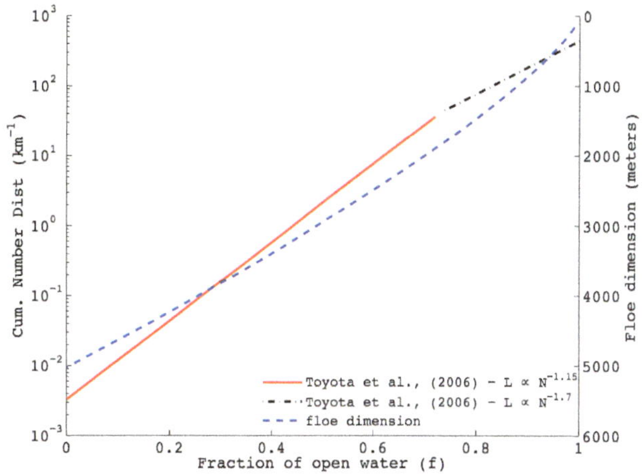

Fig. 2. The relationship between the floe number distribution (N) per km^2 (left axis) and the individual floe dimension, L (right axis), vs. the fraction (f) of open water area (or converse of the SIC) from Toyota et al. (2006). N follows a distinct power law distribution when floe size decreases below a threshold (Toyota et al., 2006).

$$\frac{V_0}{u_{*0}} = \frac{1}{\kappa}\left(\log\frac{u_*}{f z_0} - A \pm iB\right), \tag{11}$$

where V_0 and u_{*0} are the two-dimensional ice-drift and friction velocities, expressed as imaginary numbers, f is the Coriolis parameter, and A and B are matching coefficients. The term $\log\frac{u_*}{f z_0}$ is sometimes referred to as the friction Rossby number, Ro. It can be estimated with values from Eq. (10) for the nonrotating law of the wall as an initial guess of the roughness, with z approximated as the sea ice thickness.

When stratification stabilizes the surface layer, shear turbulence is damped and surface turbulence is confined to a shallower region of the IOBL. Under these conditions, the similarity terms become dependent on a stability term ($\mu_* = u_{*0}\kappa/L_0$); Ro (μ_*), $A(\mu_*)$, and $B(\mu_*)$, the solution to which is described in Sect. 4.2.3 of McPhee (2008).

The surface roughness z_0 is among the most variable and patchy IOBL parameters, and this variability is caused by the large range in ice thickness that can occur from thin ice in a lead (e.g., $\sim$cm) to deep keels underneath sea ice pressure ridges (e.g., $\sim$10 m) (Shaw et al., 2008). Estimates of z_0 span two orders of magnitude from 1 mm to 9 cm, depending on ice type (McPhee, 2008). McPhee (1992) found roughness lengths varying from 2 to 9 cm during three Arctic drift experiments, and (Shaw et al., 2008) found z_o roughness values near 10 cm using data from autonomous ocean flux buoys. Younger ice is smoother with z_0 close to 1 mm and in multiyear pack ice z_0 is about 5 cm (McPhee, 2008).

($\tau_{\text{skin-aw}}$), the final stress term in Eq. (8), refers to the shear stress exerted by the wind on the upper ocean. In the SIZ this stress occurs in the regions where water is exposed by divergent sea ice motions (i.e., leads and polynyas), or between floes in the marginal ice zone. The air–water shear stress that is transmitted into the ocean surface boundary layer can be expressed as

$$\tau_{\text{skin-aw}} = \rho_a C_d U\,|U|, \tag{12}$$

where ρ_a is the air density, C_d is the air–water drag coefficient (1.5×10^{-3}, Andreas and Murphy, 1986) and U is the 10 m wind speed.

It is instructive to visualize how the individual terms of the total stress exerted on the IOBL vary with the sea ice cover, or conversely with the fraction of open water (f). For this visualization, we use a constant $U = 10\,\text{m}\,\text{s}^{-1}$ and a constant $V_0 = 0.12\,\text{m}\,\text{s}^{-1}$, based on the average drift velocity of the autonomous ocean flux buoy #3 (http://www.oc.nps.edu/~stanton/fluxbuoy/index.html). Using these values, we have computed the individual terms in Eq. (8), $(1-f)\tau_{\text{form}}$, $(1-f)\tau_{\text{skin-iw}}$, and $(f)\tau_{\text{skin-aw}}$, for a variable ice cover ranging from $f = 0$ (100 % ice cover) to $f = 1$ (0 % ice cover), (Fig. 3). For $U = 10\,\text{m}\,\text{s}^{-1}$, the air–water stress leads to a $\tau_{\text{skin-aw}} = 0.17\,\text{N}\,\text{m}^{-2}$ and $V_0 = 0.12\,\text{m}\,\text{s}^{-1}$ results in $\tau_{\text{skin-iw}} = 0.05\,\text{N}\,\text{m}^{-2}$. Individually, these terms are proportionately modulated by the fraction of ice cover and open water, respectively, and therefore $\tau_{\text{skin-aw}}$ increases linearly as f increases and $\tau_{\text{skin-iw}}$ follows the opposite trend, achieving its maximum value when $f = 0$. In comparison, the relationship between f and τ_{form} is more complicated. We observe two tendencies, the first as τ_{form} increases quadratically from $f = 0$ to $f = 0.8$, and then decreases again from $f = 0.8$ to $f = 1$. The physical explanation for the two modes results from the competing influences of ice cover and floe dimension on the form drag. At $f = 0$, the influence of the floe dimension ($1/L \sim 10^{-4}\,\text{m}^{-1}$) strongly diminishes τ_{form}. As the ice cover decreases so does L and we observe a rapid increase in τ_{form}, as the "wake effect" of surrounding floes becomes less and each individual floe is subject to greater form drag. If we compute the gas transfer velocity from τ using Eqs. (8), (6) and then (3), we observe that shear-driven turbulence in the IOBL contributes between 1.6 and $2.0\,\text{m}\,\text{d}^{-1}$ to the overall gas transfer velocity in the sea ice zone, for a wind speed of $10\,\text{m}\,\text{s}^{-1}$ and an ice velocity of $0.12\,\text{m}\,\text{s}^{-1}$ (Fig. 3).

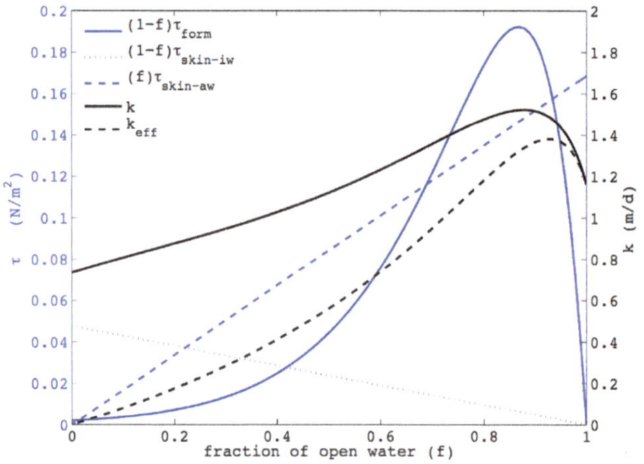

Fig. 3. A hypothetical example of the terms in the stress balance and gas transfer velocity (k) resulting from a constant RMS wind speed of $10\,\text{m}\,\text{s}^{-1}$ and an ice drift velocity of $0.12\,\text{m}\,\text{s}^{-1}$ at 75° N, 25° W. To parameterize the floe size, we used the scaling relationship of Toyota et al. (2006) to convert sea ice concentration (SIC) to a characteristic floe dimension (L).

4 Buoyant convection/stratification

As with the shear-driven turbulence, the goal is to make a determination of ε, which can then be used to estimate k. The first step is to calculate the Obukhov length scale, to determine whether convection ($L_0 < 0$) or stratification ($L_0 > 0$) characterizes the surface layer of the ocean (McPhee, 2008). In the SIZ, as contrasted with the lower latitudes, L_0 can be affected by freezing and melting of ice as well as latent and sensible and shortwave heat transfer. Here, its worth pointing out that sea ice freezing (melting) can concentrate (dilute) the gas content of the upper ocean through solute rejection (dissolution) (Killawee et al., 1998; Loose et al., 2009), and change the upper ocean air–sea gas partial pressure differential. For the purposes of this parameter model we will not consider the effect of solute rejection and dissolution.

L_0 from Eq. (5) can be rewritten as (Lombardo and Gregg, 1989)

$$L_0 = \frac{-u_{*0}^3}{\kappa J_{b0}}, \tag{13}$$

where J_{b0} is the surface buoyancy flux in $\text{W}\,\text{kg}^{-1}$ or $\text{m}^2\,\text{s}^{-3}$ equivalently. For the IOBL, the major contributions to the buoyancy flux are sensible heat transfer, sea ice freeze–melt, and the salt balance from evaporation −precipitation,

$$J_{b0} = -\frac{g\alpha}{\rho c_p}\left(J_q^{\text{SH}} + J_q^{\text{LH}}\right)(f) + J_{\text{b}}^{\text{ice}}(1-f) + g\beta(E-P)\,S_0(f), \tag{14}$$

where α is the coefficient of thermal expansion, β is the coefficient of haline contraction, ρ is the density of seawater, $E-P$ is the rate of evaporation less precipitation, S_0 is the surface seawater salinity in $\text{kg}\,\text{kg}^{-1}$, c_p is the specific heat capacity, and g is the gravitational constant. Again, the terms f and $(1-f)$ refer to the fraction of open water or ice cover, respectively. The buoyancy flux from sea ice is solved using the "three equation control interface solution" wherein the heat and salt balance are regulated by turbulent exchange across the ice–water interface and by the velocity of that interface, indicating growth or melt of sea ice (Eqs. 6.8, 6.9; McPhee, 2008), which uses T_{w}, T, and the ice–water friction velocity $\left(u_{*0} = \sqrt{\tau_{\text{skin-iw}}/\rho}\right)$ to solve for $J_{\text{b}}^{\text{ice}}$. Finally,

J_q^{SH} and J_q^{LH} are the surface sensible and latent heat flux in $\mathrm{W\,m^{-2}}$. The last term in Eq. (14) represents the surface freshening or salinification and we estimate it from the net latent heat flux, which is an NCEP Reanalysis product. J_q^{SH} and J_q^{LH} over open water can be estimated using the bulk approximation for scalar flux in the atmospheric boundary layer flux (Andreas and Murphy, 1986),

$$J_q^{\mathrm{SH}} = \rho c_p C_H (T_\mathrm{w} - T)\,|U|$$
$$J_q^{\mathrm{LH}} = \rho L_v C_H (Q_S - Q_z)\,|U|, \qquad (15)$$

where T_w is the water surface temperature, C_H – the surface drag coefficient – is approximated as 1.26×10^{-3}, and Q_s is the specific humidity at the dew point, defined by T_w. Q_z is the specific humidity at the reference height, z, which should be the same as the height of the wind speed measurement (Andreas et al., 2010).

In the absence of boundary layer flow, the TKE dissipation is directly equivalent to the heat flux: $\varepsilon = J_\mathrm{b}^0$. In a mixed condition of convection + shear, a similarity scaling for ε is chosen using L_0 to characterize the relative balance between convection and shear (Lombardo and Gregg, 1989): $L_0 \ll z$, the boundary layer thickness, indicates free convection and $L_0 \gg z$ indicates a shear-dominated regime.

5 Surface waves and interactions with sea ice

In addition to shear and convection, surface waves and their interaction with ice floes will also produce turbulence and gas exchange in the SIZ. As discussed in Sect. 2, the actions of wind-driven surface waves are the dominant mechanism by which aqueous turbulence leads to gas exchange (Jähne et al., 1987; Wanninkhof, 1992), and among these, the highest frequency gravity waves appear most effective at disturbing the viscous mass boundary layer and enhancing k (Bock et al., 1999; Frew et al., 2004, 2007). The mean squared wave slope $\langle s^2 \rangle$ of these short period gravity waves correlates well with k (Bock et al., 1999; Frew et al., 2004, 2007), however the gravity wave frequency distribution is variable and interacts strongly with sea ice (Wadhams et al., 1986; Hayes and Jenkins, 2007). To date, most studies of wave interactions with sea ice focus on much longer period waves, because they can be responsible for the breakup and melt of ice floes (Squire et al., 1995; Meylan et al., 1997; Kohout and Meylan, 2008), and relatively less attention has been paid to waves in the capillary-gravity range (40–800 $\mathrm{rad\,m^{-1}}$). Energy attenuation by wave interaction with sea ice is strongest for waves with a wave number of 1 $\mathrm{rad\,m^{-1}}$ or less (Polnikov and Lavrenov, 2007). Therefore, it might be expected that short period capillary-gravity waves are reflected with near 100 % efficiency, based on wave reflection studies (Squire and Williams, 2008).

Direct measurements of the capillary-gravity wave spectrum in the presence of sea ice are necessary to constrain the contribution of surface wave forcing of gas exchange, and we are not aware of any study that has explicitly examined the wave field in this range. In the mean time, we can apply the relationship between $\langle s^2 \rangle$ and 10 m wind speed from Frew et al. (2007) (their Fig. 1) to give an idea of how surface forcing contributes to the overall gas exchange. The choice of a parameterization based on $\langle s^2 \rangle$ instead of a relationship between wind speed and gas exchange can be justified by recognizing that Jähne et al. (1987) have shown that waves in the capillary-gravity range arise and decay within seconds, upon excitation by wind, indicating minimal fetch dependence. In contrast, it is not well known how the wind speed–gas exchange relationship varies with fetch.

In lieu of direct measurements of $\langle s^2 \rangle$ in the SIZ, we have used data from Frew et al. (2007) that relates the 10 m wind speed to $\langle s^2 \rangle_{40}^{100}$, the mean squared wave slope from 40 to 100 $\mathrm{rad\,m^{-1}}$. To capture the tendency between data $\langle s^2 \rangle_{40}^{100}$ and U_{10} we fit the Frew et al. (2007) data with a third order polynomial with fit coefficients [7.7814×10^{-6}, −0.0001624, 0.0015064, −0.0012018] and an R^2 value of 0.91 (Fig. 4). As above, U_{10} from NCEP data was interpolated onto the ITP drift track. Finally, the values of $\langle s^2 \rangle_{40}^{100}$ were used to estimate k using

$$k_\mathrm{mss} = 0.336 + 1.82 \times 10^5 \langle s^2 \rangle_{40}^{100} \qquad (16)$$

(Frew et al., 2004). The relationship between $\langle s^2 \rangle_{40}^{100}$ and gas exchange appears most robust under short fetch conditions and where bubbles are not a major contributor to the exchange rate (Frew et al., 2004, 2007). Although we do not have measurements to support this, anecdotal observations indicate that breaking waves and bubble penetration are minimal in the sea ice zone.

6 Computation of k from ITP data

We have calculated the gas transfer velocity using the parameter model for shear and convective turbulence in the IOBL, using data from ITPs together with satellite data products for the Arctic Ocean. The inputs to this parameter model (f, U, V_0, Q) and profiles of T and S, were taken from post-processed ITP missions 3, 6, 7 and 8, (http://www.whoi.edu/itp). Daily estimates of sea ice cover were derived from the Bootstrap algorithm for NIMBUS-7 and SSM/I satellite radiometers (Comiso, 2007). Daily estimates of wind speed and surface air temperature were taken from NCEP Reanalysis data provided by NOAA/OAR/ESRL PSD, Boulder, Colorado, USA, (http://www.esrl.noaa.gov/psd/). The inputs from gridded remotely sensed data were extracted by interpolating onto the recorded daily positions of the ITPs.

The calculation for k proceeded as follows.

1. To start, the equation for buoyancy flux (Eq. 14) is evaluated to determine whether convection or stratification is affecting water column turbulence. The

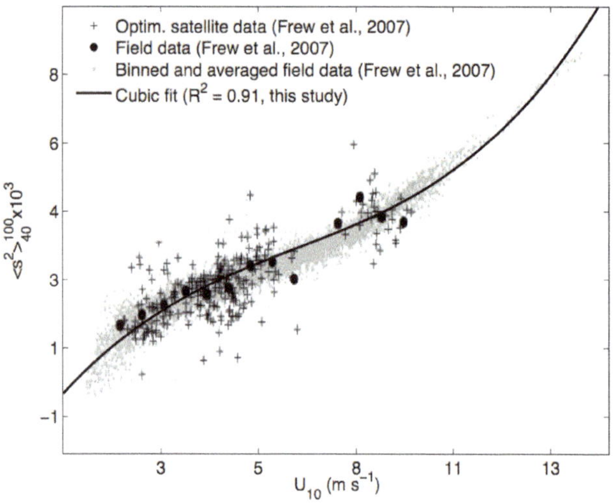

Fig. 4. Modified from Frew et al. (2007), the relationship between the 10 m wind speed and the mean-squared wave slope has been represented in this study as cubic (solid black line).

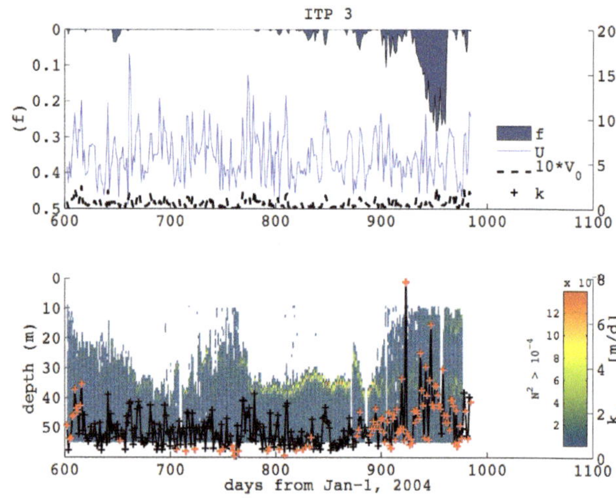

Fig. 5. (top) Wind speed (U), ice velocity (V_0), and the fraction of open water (f), calculated along the drift path of ice-tethered profiler #3 (ITP 3, http://www.whoi.edu/itp). (bottom) The gas transfer velocity (k_{shear}) that would result from shear-driven turbulence. Red "+" symbols indicate estimates of k_{shear} that have been affected by stratification. The speckled color indicates the vertical distribution of $N^2 > 10^{-4}\,s^{-2}$, giving a rough approximation of the mixed-layer depth. Some data gaps exist in the ITP profiles.

threshold for buoyant convection is determined using the Obukhov length (L_0) criteria: $H/L_0 > 1$ (Lombardo and Gregg, 1989). H is half the mixed-layer depth, and the mixed-layer depth was found by locating the maximum value of N^2 in the top 50 m of the water column – typically H was 10–15 m.

2. Shear-driven turbulence in the IOBL is determined by summing the individual terms in the stress balance (Eq. 8). To solve Eq. (9) for τ_{form}, the floe dimension, L, is estimated from the empirical relationship in Fig. 1, as follows. The floe number density, N, is first computed for the range of L from 10 to 10^4 m, using the parameterizations presented by Toyota et al. (2006). Next, the discrete values of N and L are used to compute the sea ice cover, $(1-f)$, by taking the average of square $(L \times L)$ and circular $\left(\frac{\pi}{4}L^2\right)$ floe geometry. For a discrete value of sea ice cover from the ITP drift, a value for L is computed as a table lookup. $\tau_{skin\text{-}iw}$ is computed using Eqs. (11) and (7) as described in Sect. 3: here, the similarity terms, Ro, A and B, in Eq. (11) are evaluated in terms of the buoyancy flux (J_b^0, Eq. 14). Finally, $\tau_{skin\text{-}aw}$ is computed using the Reanalysis wind speed in Eq. (12). The total shear-stress τ from Eq. (8), and resulting friction velocity (u_{*0}) is then used to compute the shear-driven TKE dissipation (ε_s, Eq. 6).

3. Finally, the aggregate total dissipation (ε) was computed as $\varepsilon = 0.84\left(0.58J_b^0 + 1.76\varepsilon_s\right)$, and k is computed from ε or ε_s using Eq. (3). In Figs. 5 and 6, k (labeled k_{shear}) was computed from ε_s, and in Figs. 7 and 8, k (labeled k_{conv}) was computed assuming $\varepsilon = J_b^0$.

We present estimates of k from shear-driven turbulence separately from convection-driven turbulence to compare their relative contributions to the overall gas transfer velocity. These separate terms are presented for two of the four ITP missions (ITP 3 and 8). The duration of ITP 3 was 786 days, beginning 24 August 2005 with f, the open water fraction, less than 0.1 until the last 100 days, when ITP 3 drifted into waters with f of nearly 0.3, near the center of the Beaufort Gyre. The maximum daily NCEP wind speed, U, was 17.2 m s^{-1} and the mean U was 5.8 m s^{-1}, whereas the maximum ice velocity, V_0 was 0.24 m s^{-1} and the average V_0 was 0.05 m s^{-1} (Fig. 5). Throughout the drift, the ice–water skin friction dominated the stress balance, because of the large fraction of ice cover. On average, the ice–water skin friction was 10 times greater than the form drag and more than 100 times greater than the air–water skin friction.

It is worth noting that between days 900 and 1000, decreased ice cover (increased f) corresponded to an increase in τ_{form} as dictated by Eq. (9). However, this effect was small compared to the coincident increase in $\tau_{skin\text{-}iw}$ that resulted from an increase in stratification. McPhee (2008) describes this effect in Sect. 4.2.3 – assuming the same boundary forcing conditions, the total shear stress in the stratified water column is the same as the neutral water column. However, stratification confines the profile of shear stress to a shallower surface layer. This effectively "concentrates" the shear stress in the surface layer, leading to an enhancement in the apparent TKE dissipation and thus gas exchange. The maximum observed values during ITP 3 all occurred during the ice retreat period and can be associated with the coincident

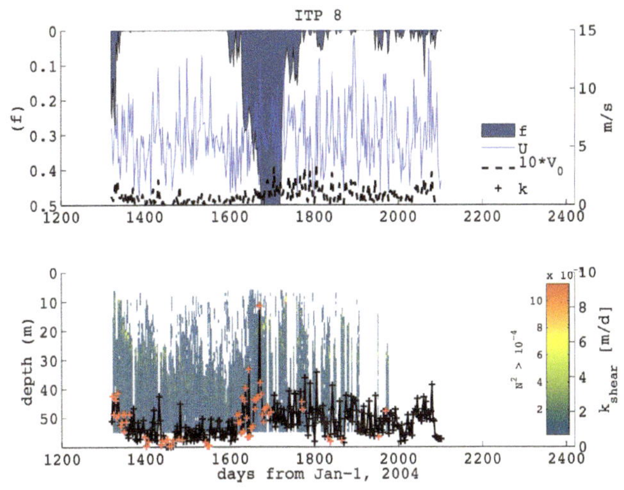

Fig. 6. (top) U, V_0, and f, calculated along the drift path of ITP 8 (http://www.whoi.edu/itp). (bottom) The k_{shear} that would result from shear-driven turbulence. Red "+" symbols indicate estimates of k_{shear} that have been altered by stratification through the stability function (Eq. 11). The speckled color indicates the vertical distribution of $N^2 > 10^{-4}\,s^{-2}$, giving a rough approximation of the mixed-layer depth. Some data gaps exist in the ITP profiles.

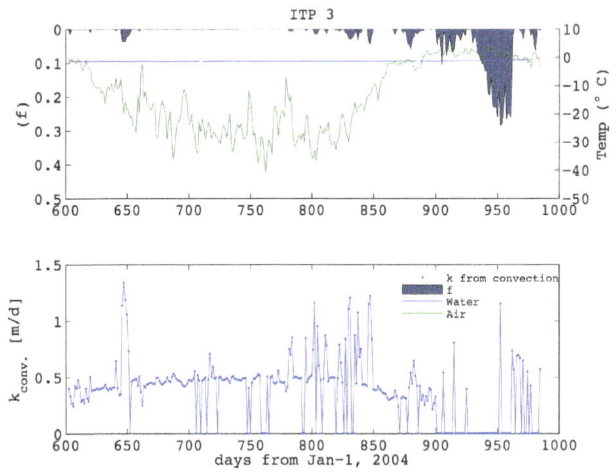

Fig. 7. (top) Surface air and water temperature and fraction of open water along the ITP 3 drift track. (bottom) Gas transfer velocity (k_{conv}) resulting from heat flux-driven convection in the IOBL.

increase in both τ_{form} and $\tau_{skin\text{-}iw}$ (Fig. 5). The mean value daily k_{shear} for ITP 3 was $1.3\,m\,d^{-1}$ and the maximum was $7.8\,m\,d^{-1}$.

A somewhat similar picture emerges for ITP 8; the ITP 8 data extended over 784 days, beginning 12 August 2007. The mean of the daily NCEP wind speed during the ITP 8 drift was also $5.7\,m\,s^{-1}$, with a maximum daily average wind speed of $14.5\,m\,s^{-1}$. However, the average ice velocity was $0.075\,m\,s^{-1}$, 25 % larger than the average drift velocity of ITP 3, and the maximum was $0.35\,m\,s^{-1}$. Overall, the variance in the drift velocity was greater for ITP 8, as compared to ITP 3. Near day 1600 (ca. 275 days after starting its drift), ITP 8 drifted into progressively opening sea ice cover, until f increased to a maximum of 0.83, around day 1700 (Fig. 6). Similar to what was observed in ITP 3, the decrease in sea ice cover produced a decrease in L, based on the parameterization in Fig. 2, resulting in an increase in τ_{form}. Likewise, stratification in the ITP profiles (apparently from ice melt) caused $\tau_{skin\text{-}iw}$ to increase coincident with the increase in f. During the increase in open water, k_{shear} achieved a local maximum of $8.0\,m\,d^{-1}$, (Fig. 6, bottom panel) on day 1705, very near the peak open water when f was 0.73 – nearly 75 % open water. The average in k_{shear} was $1.4\,m\,d^{-1}$ for ITP 8.

The parameter model estimates of k_{conv} are on average 3 times smaller as compared to the estimates of k_{shear}. These estimates depict a seasonal cycle that predictably mirrors the air–water temperature differential. This is the case for both ITP 3 and 8 (Figs. 7, 8). During the period of peak heat flux, k_{conv} reached its peak above $1.2\,m\,d^{-1}$ for ITP 3 and $1.5\,m\,d^{-1}$ for ITP 8. These peaks in k_{conv} tended to occur when $f > 0$ and when the air–water temperature differential strongly favored heat loss to the atmosphere, i.e., when fall cooling was leading to sea ice formation, but open water area yet remained (e.g., days 1700–1800, Fig. 8). Conversely, the periods of apparent decrease in sea ice cover and increase in air temperature during spring (e.g., days 1600–1700, Fig. 8), caused L_0 to increase and the convection threshold to exceed the Obukhov length criteria $H/L_0 > 1$. This causes k_{conv} to be "turned off" in the parameter model.

Having combined the TKE production from both shear and convection and using those values of ε to estimate k from $\varepsilon = 0.84\left(0.58 J_b^0 + 1.76\varepsilon_s\right)$, we find that the mean value of k from shear + convection was $1.63\,m\,d^{-1}$ and the maximum was $10.6\,m\,d^{-1}$. In general, k is a scale independent or intensive property so we would expect no strong trend between k and f. This lack of tendency is borne out in Fig. 9 (top panel), despite the areal dependence that has been introduced through Eqs. (8) and (14) for the shear and buoyancy budgets. However, there is one interesting feature between $f = 0.15$ and $f = 0.30$, with a discernable peak in k, which exceeds $4\,m\,d^{-1}$ (Fig. 9, top panel). This increase in k is the result of shear and convection enhancements during the seasonal transition periods when sea ice is in advance or retreat. As reported above, stratification during ice melt enhances k from shear by "concentrating" the net stress closer to the air–sea interface. During periods of decreasing f (e.g., fall), the effect of convection in water that is still open also leads to an enhancement in k. If k from shear and convection is converted to k_{eff} using Eq. (2), we find that the effective gas exchange across the heterogeneous ocean surface is less than $1\,m\,d^{-1}$ from $f = 0$ to $f = 0.4$, and increases almost linearly across the range from $f = 0$ to $f = 0.8$, with a value of $\sim 2\,m\,d^{-1}$ at $f = 0.8$ (Fig. 9, lower panel).

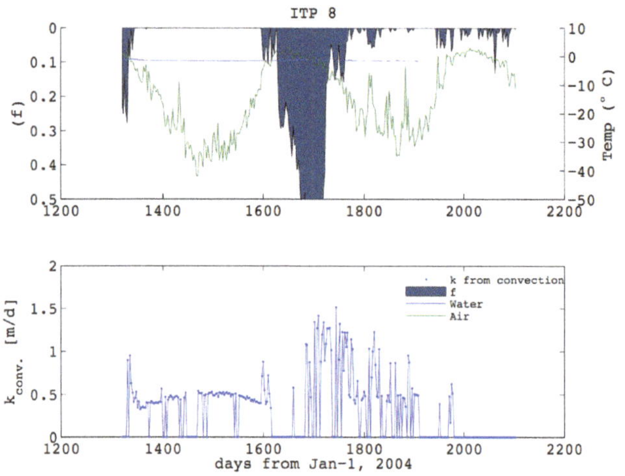

Fig. 8. (top) Surface air and water temperature and fraction of open water along the ITP 8 drift track. (bottom) Gas transfer velocity (k_{conv}) resulting from heat flux-driven convection in the IOBL.

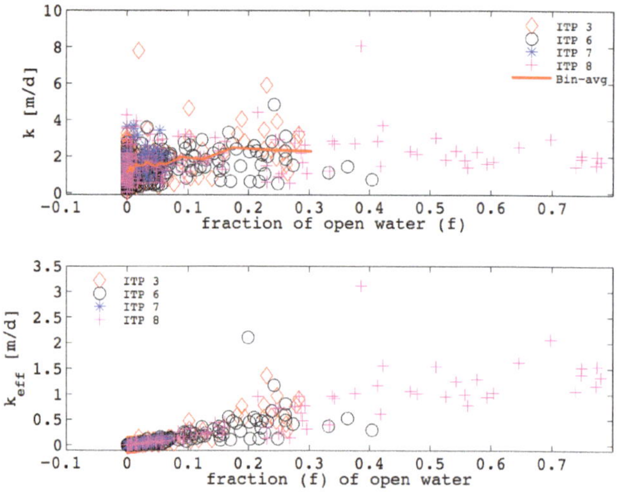

Fig. 9. (top) The gas transfer velocity, k, from shear and convection in the sea ice zone, graphed as a continuum with open water area, f. (bottom) The effective gas transfer velocity, k_{eff}, (Eq. 2) from shear and convection along the drift tracks of ITP 3, 6, 7, and 8.

To estimate the total magnitude of gas transfer in the sea ice zone, we have stacked k from the mean-squared slope relationship in Eq. (9) with the estimates of k from shear and convection (described in steps 1–6 above), displayed in Fig. 10. The $k \propto f$ line determined by the mean NCEP wind speed for the Arctic, and prior values of k_{eff} from Fanning and Torres (1991), and Loose and Schlosser (2011) have been overlaid to compare how this parameter model compares with those values. The parameter model exceeds the linear proportionality, but no values meet or exceed the median estimates of Fanning and Torres (1991). On average, the magnitude of k_{eff} from ice processes is approximately 42 % of the total magnitude of k_{eff} as can be observed in Fig. 10.

7 Discussion and conclusions

The objective of this study was to utilize the existing physical understanding of turbulence in the ice–ocean boundary layer (IOBL) to formulate a plausible parameter model for air–sea gas transfer in the seasonal sea ice zone. The model requires only six input parameters: sea ice concentration, wind speed, sea ice-drift velocity, air temperature and profiles of water temperature and salinity. These inputs are available from model output (Proshutinsky et al., 2011; Timmermans et al., 2011), from satellite data products for the polar oceans, and from Arctic Observing Network nodes, such as ITPs and autonomous ocean flux buoys. Here, we estimated the magnitude of k that results from shear and buoyancy production processes that are unique to the sea ice covered ocean. We used input data from four ITP buoy missions: ITP 3, 6, 7, and 8. The parameter model calculation is not intended to provide a definitive map of k for the SIZ as we still lack sufficient constraint to definitively produce this map. Instead it is intended to emphasize the unique mechanisms that produce ocean surface turbulence, and thus gas exchange, in the IOBL.

Several interesting observations have emerged from this simple analysis, which are worth keeping in mind as we move toward developing an accepted method to calculate k in the polar oceans. The estimates of k from ITP data predominantly occurred at low values of f; 71 % of the estimates of k were made for $f < 0.1$. This bias exists because the ITP's are purposefully situated in consolidated sea ice to ensure a long duration of the ITP mission. As ice cover becomes excessively low, the likelihood that the ITP will become dislodged and inoperable increases. This implies that marginal ice zone conditions (with $f > 0.5$) are under-represented in the data and calculations presented here, both in their likelihood of occurrence and in the gas exchange conditions that exist.

The model indicates that k becomes enhanced during the transition periods of spring and fall. In fall, convection coincides with a nonnegligible fraction of open water area resulting in enhanced gas exchange. In spring, ice melt leads to the formation a shallower mixed layer, which confines the surface stress closer to the air–sea interface, enhancing k. However, a shallower mixed layer is a smaller reservoir for gas storage; while gas transfer may be enhanced in spring, this effect on the net flux of a gas is likely more than offset by equilibration with a much smaller portion of the water column.

The characteristics of sea ice: underside roughness (z_o), floe dimension (L) and fractional ice cover $(1 - f)$ all play a direct role modulated the magnitude of k. Indeed, $(1 - f)$ modulates both the stress balance as well as the k_{eff}; consequently high resolution, high accuracy estimates of these characteristics are valuable to the parameterization of gas transfer.

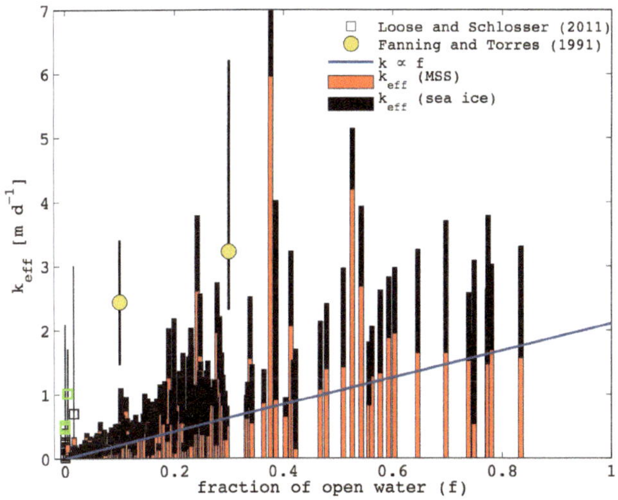

Fig. 10. The effective gas-transfer velocity (k_{eff}) plotted versus the open water fraction; adapted from Loose and Schlosser (2011), including estimates of the gas transfer velocity from the radon deficit method from Fanning and Torres (1991). The stacked bars indicate estimates of k_{eff} from sea ice zone effects (sea ice) and from surface wind waves (MSS), during the drift tracks of ITP 3, 6, 7, and 8 computed using Eq. (2) of this study.

It is known from sea ice kinematics that divergent motions cause an ephemeral fraction of the ocean surface to remain exposed, even when satellite sea ice cover is labeled as 100 % (Geiger and Drinkwater, 2005), and some estimates put this ephemeral fraction at nearly 0.1 or 10 % open water (Takahashi et al., 2009). When the fraction of ice cover exceeds 90 %, ($f < 0.1$), the fraction of open water that we need to resolve is of the same magnitude as the error in satellite ice cover estimates (Knuth and Ackley, 2006), resulting in large uncertainties in the estimate of f. In the absence of a more precise downscaled constraint on f, it would be useful to have some constraint on what the real lower limit is on open water area, if indeed it is not 0 %. The constraints on k and f in the winter when $f < 0.1$ may be particularly important as they relate to polynyas where much of the ocean's deep water is formed and properties are set.

Finally, the parameter model indicates the effect of shear and convection processes, some of which are unique to the sea ice zone, appear to contribute an additional ca. 40 % to the magnitude of air–sea gas transfer, beyond what would be predicted if a wind speed or mean-square slope parameterization is simply scaled by the fraction of open water, which has been a practice of early attempts to estimate net flux in the sea ice zone (e.g., Stephens and Keeling, 2000; Takahashi et al., 2009).

The parameter model that has been presented here can serve as a means for comparison with laboratory and field-scale investigations into the connection between ice cover, aqueous turbulence and gas exchange. At a minimum it will be necessary to investigate the effects of shear, convection, stratification and short period wind waves as individual drivers of the effective gas transfer velocity. It is unclear how turbulence and gas exchange in the SIZ transitions to turbulence and gas exchange in the open ocean. However, it is likely that a threshold exists between an upper ocean that is significantly affected by the presence of sea ice, and an ocean that is dominated by direct wind-driven turbulence. When f exceeds some critical value, wind, breaking waves and bubble entrainment are likely to exceed the shear-driven effects of ice floes that are present, but the value of this threshold in ice cover is as yet unknown.

Supplement

A copy of the Matlab routines used to calculate k_{eff} in the sea ice zone has been included as Supplement to this manuscript. A more up to date version of this code may be available at http://geotracerkitchen.org. Navigate to "Recipes".

Acknowledgements. We would like to thank Miles McPhee, Jim Ledwell, Bill Asher and one anonymous referee for their valuable comments on the manuscript. Miles McPhee provided two Matlab functions for calculating the boundary layer properties using Rossby similarity and the ice–water buoyancy flux. These functions are included with the supplemental materials. Support for this study was provided by the National Science Foundation, through award # 0944643.

Edited by: M. Hoppema

References

Andreas, E. L. and Murphy, B.: Bulk transfer coefficients for heat and momentum over leads and polynyas, J. Phys. Oceanogr., 16, 1875–1883, 1986.

Andreas, E. L., Horst, T. W., Grachev, A. A., Persson, O. G., Fairall, C. W., Gueste, P. S., and Jordan, R. E.: Parametrizing turbulent exchange over summer sea ice and the marginal ice zone, Q. J. Roy. Meteorol. Soc., 136, 927–943, 2010.

Bender, M., Kinter, S., Cassar, N., and Wanninkhof, R.: Evaluating gas transfer velocity parameterizations using upper ocean radon distributions, J. Geophys. Res., 116, C02010, doi:10.1029/2009JC005805, 2011.

Blackadar, A. K. and Tennekes, H.: Asymptotic similary in a neutral barotropic planetary boundary layer, J. Atmos. Sci., 25, 1015–1020, 1968.

Bock, E. J., Hara, T., Frew, N., and McGillis, W. R.: Relationship between air-sea gas transfer and short wind waves, J. Geophys. Res., 104, 25821–25831, 1999.

Charnock, H.: Air sea interaction. Cambridge, MA, MIT Press, 1981.

Comiso, J. C.: Bootstrap Sea Ice Concentrations from Nimbus-7 SMMR and DMSP SSM/I, in: National Snow and Ice Data Center, National Snow and Ice Data Center, Boulder, CO, 2007.

Cox, G. F. N. and Weeks, W. F.: Equations for determining the gas and brine volumes in sea-ice samples, J. Glaciol., 29, 306–316, 1983.

Dykaar, B. B. and Kitandis, P.: Determination of the effective hydraulic conductivity for a heterogeneous porous media using a numerical spectra approach 2. Results, Water Resour. Res., 28, 1167–1178, 1992.

Else, B., Papakyriakou, T., Galley, R. J., Drennan, W. M., Miller, L. A., and Thomas, H.: Eddy covariance measurements of wintertime CO_2 fluxes in an arctic polynya: Evidence for enhanced gas transfer during ice formation, J.Geophys. Res., 116, C00G03, doi:10.1029/2010JC006760, 2011.

Fanning, K. A. and Torres, L. M.: ^{222}Rn and ^{226}Ra: Indicators of sea-ice effects on air-sea gas exchange, Polar Res., 10, 51–58, 1991.

Frew, N. M., Bock, E. J., Schimpf, U., Hara, T., Haußecker, H., Edson, J. B., McGillis, W. R., Nelson, R. K., McKenna, S. P., Uz, B., and Jähne, B.: Air-sea gas transfer: Its dependence on wind stress, small-scale roughness, and surface films, J. Geophys. Res., 109, C08S17, doi:10.1029/2003JC002131, 2004.

Frew, N. M., Glover, D. M., Bock, E. J., and McCue, S. J.: A new approach to estimation of global air-sea gas transfer velocity fields using dual-frequency altimeter backscatter, J. Geophys. Res., 112, C11003, doi:10.1029/2006JC003819, 2007.

Gaspar, P., Gregoris, Y., and Lefevre, J.-M.: A simple eddy kinetic energy model for simulations of the oceanic vertical mixing: Tests at Staion Papa and Long-term upper ocean study site, J. Geophys. Res.-Oceans, 95, 16179–16193, 1990.

Geiger, C. A. and Drinkwater, M. R.: Coincident buoy- and SAR-derived surface fluxes in the western Weddell Sea during Ice Station Weddell 1992, J. Geophys. Res., 110, C04002, doi:10.1029/2003JC002112, 2005.

Golden, K. M., Heaton, A. L., Eicken, H., and Lytle, V. I.: Void bounds for fluid transport in sea ice, Mech. Mater., 38, 801–817, 2006.

Gosink, T. A., Pearson, J. G., and Kelly, J. J.: Gas movement through sea-ice, Nature, 263, 41–42, 1976.

Hayes, D. R. and Jenkins, A.: Autonomous Underwater Vehicle Measurements of Surface Wave Decay and Directional Spectra in the Marginal Sea Ice Zone, J. Phys. Oceanogr., 37, 71–83, 2007.

Ho, D. T., Bliven, L., Wanninkhof, R., and Schlosser, P.: The effect of rain on air-water gas exchange, Tellus, 49B, 149–158, 1997.

Ho, D. T., Law, C. S., Smitth, M. J., Schlosser, P., Harvey, M., and Hill, P.: Measurements of air-sea gas exchange at high wind speeds in the Southern Ocean: Implications for global parameterizations, Geophys. Res. Lett., 33, L16611, doi:10.1029/2006GL026817, 2006.

Ho, D. T., Wanninkhof, R., Schlosser, P., Ullman, D. S., Hebert, D., and Sullivan, K. F.: Towards a universal relationship between wind speed and gas exchange: Gas transfer velocities measured with $^3He/SF_6$ during the Southern Ocean Gas Exchange Experiment, J. Geophys. Res., 116, C00F04, doi:10.1029/2010JC006854, 2011.

Jähne, B. and Haubecker, H.: Air-water gas exchange, Annu. Rev. Fluid Mech., 30, 443–468, doi:10.1146/annurev.fluid.30.1.443, 1998.

Jähne, B., Münnich, K. O., Bösinger, R., Dutzi, A., Huber, W., and Libner, P.: On the parameters influencing air-water gas exchange, J. Geophys. Res., 92, 1937–1949, 1987.

Killawee, J. A., Fairchild, I. J., Tison, J.-L., Janssens, L., and Lorrain, R.: Segregation of solutes and gases in experimental freezing of dilute solutions: Implications for natural glacial systems, Geochim. Cosmochim. Ac., 62, 3637–3655, 1998.

Knuth, M. and Ackley, S. F.: Summer and early-fall sea-ice concentration in the Ross Sea: comparison of in situ ASPeCt observations and satellite passive microwave estimates, Ann. Glaciol., 44, 303–309, 2006.

Kohout, A. L. and Meylan, M. H.: An elastic plate model for wave attenuation and ice floe breaking in the marginal ice zone, J. Geophys. Res., 113, C09016, doi:10.1029/2007JC004434, 2008.

Lamont, J. C. and Scott, D. S.: An eddy cell model of mass transfer into the surface of a turbulent liquid, AIChE J., 16, 512–519, 1970.

Liss, P. S.: Processes of gas exchange across an air-water interface, Deep-Sea Res. Pt. I, 20, 221–238, 1973.

Lombardo, C. P. and Gregg, M. C.: Similiarity scaling of viscous and thermal dissipation in a convecting surface boundary layer, J. Geophys. Res., 94, 6273–6284, 1989.

Loose, B. and Schlosser, P.: Sea ice and its effect on CO_2 flux between the atmosphere and the Southern Ocean interior, J. Geophys. Res.-Oceans, 116, C11019, doi:10.1029/2010JC006509, 2011.

Loose, B., McGillis, W. R., Schlosser, P., Perovich, D., and Takahashi, T.: The effects of freezing, growth and ice cover on gas transport processes in laboratory seawater experiments, Geophys. Res. Lett., 36, L05603, doi:10.1029/2008GL036318, 2009.

Loose, B., Schlosser, P., Perovich, D., Ringelberg, D., Ho, D. T., Takahashi, T., Richter-Menge, J., Reynolds, C. M., McGillis, W. R., and Tison, J.-L.: Gas diffusion through columnar laboratory sea ice: Implications for mixed-layer ventilation of CO_2 in the seasonal ice zone, Tellus B, 63, 11, doi:10.1111/j.1600-0889.2010.00506.x, 2010.

Lu, P., Li, Z., Cheng, B., and Lepparanta, M.: A parameterization of the ice-ocean drag coefficient, J. Geophys. Res., 116, C07019, doi:10.1029/2010JC006878, 2011.

McGillis, W. R., Edson, J. B., Hare, J. E., and Fairall, C. W.: Direct covariance air-sea CO_2 fluxes, J. Geophys. Res., 106, 16729–16745, 2001.

McPhee, M.: Turbulent heat flux in the upper ocean under sea ice, J. Geophys. Res., 97, 5365–5379, 1992.

McPhee, M. G.: Air-ice-ocean interaction: Turbulent ocean boundary layer exchange, New York, Springer, 2008.

McPhee, M. and Martinson, D. G.: Turbulent mixing under drifting pack ice in the Weddell Sea, Science, 263, 218–220, 1992.

Meylan, M. H., Squire, V. A., and Fox, C.: Toward realism in modeling ocean wave behavior in marginal ice zones, J. Geophys. Res., 102, 22981–22991, 1997.

Miller, L., Papakyriakou, T., Collins, R. E., Deming, J. W., Ehn, J. K., Macdonald, R. W., Mucci, A., Owens, O., Raudsepp, M., and Sutherland, N.: Carbon dynamics in Sea Ice: A Winter Flux Time Series, J. Geophys. Res., 116, C02028, doi:10.1029/2009JC006058, 2011.

Morison, J. H., McPhee, M., Curtin, T. B., and Paulson, T. B.: The oceanography of winter leads, J. Geophys. Res., 97, 11199–11218, 1992.

Nightingale, P. D., Liss, P. S., and Schlosser, P.: Measurements of air-sea gas transfer during an open ocean algal bloom, Geophys. Res. Lett., 27, 2117–2120, 2000.

Orton, P., McGillis, W. R., and Zappa, C. J.: Sea breeze forcing of estuary turbulence and air-water CO_2 exchange, Geophys. Res. Lett., 37, L13603, doi:10.1029/2010GL043159, 2010.

Perrie, W. and Hu, Y.: Air-ice-ocean momentum exchange. Part I: Energy transfer between waves and ice floes, J. Phys. Oceanogr., 26, 1705–1720, 1996.

Perrie, W. and Hu, Y.: Air-ice-Ocean Momentum Exchange. Part II: Ice Drift, J. Phys. Oceanogr., 27, 1976–1996, 1997.

Polnikov, V. G. and Lavrenov, I. V.: Calculation of the Nonlinear Energy Transfer through the Wave Spectrum at the Sea Surface Covered with Broken Ice, Oceanology, 47, 334–343, 2007.

Proshutinsky, A., Aksenov, Y., Clement Kinney, J., Gerdes, R., Golubeva, E., Holland, E., Holloway, G., Jahn, A., Johnson, M., Popova, E., Steele, M., and Watanab, E.: Recent advances in Arctic ocean studies employing models from the Arctic Ocean Model Intercomparison Project, Oceanography, 24, 102–113, doi:10.5670/oceanog.2011.61, 2011.

Rao, K. N., Narasimha, R., and Narayanan, B.: The "bursting" phenomenon in a turbulent boundary layer, J. Fluid Mech., 48, 339–352, doi:10.1017/S0022112071001605, 1971.

Rothrock, D. A. and Thorndike, A. S.: Measuring the sea ice floe size distribution, J. Geophys. Res., 89, 6477–6486, 1985.

Shaw, W. J., Stanton, T. P., McPhee, M. G. and Kikuchi, T.: Estimates of surface roughness length in heterogeneous under-ice boundary layers, J. Geophys. Res., 113, C08030, doi:10.1029/2007JC004550, 2008.

Shirasawa, K. and Ingram, G.: Characteristics of the turbulent oceanic boundary layer under sea ice Part 1: A review of the ice-ocean boundary layer, J. Marine Syst., 2, 153–160, 1991.

Smethie, W., Takahashi, T., Chipman, D., and Ledwell, J.: Gas Exchange and CO_2 Flux in the Tropical Atlantic Ocean Determined from ^{222}Rn and pCO_2 measurements, J. Geophys. Res., 90, 7005-7022, 1985.

Squire, V. A. and Williams, T. D.: Wave propagation across sea-ice thickness changes, Ocean Model., 21, 1–11, doi:10.1016/j.ocemod.2007.10.006, 2008.

Squire, V. A., Dugan, J. P., Wadhams, P., Rottier, P. J., and Liu, A. K.: Of ocean waves and sea ice, Annu. Rev. Fluid Mech., 27, 115–168, 1995.

Stanley, R., Jenkins, W. J., and Doney, S. C.: Quantifying seasonal air-sea gas exchange processes using noble gas time-series: A design experiment, J. Marine Res., 64, 267–295, 2006.

Steele, M., Morison, J. H., and Untersteiner, N.: The Partition of Air-Ice-Ocean Momentum Exchange as a Function of Ice Concentration, Floe Size, and Draft, J. Geophys. Res., 94, 12739–12750, 1989.

Stephens, B. B. and Keeling, R. F.: The influence of Antarctic sea ice on glacial-interglacial CO_2 variations, Nature, 404, 171–174, 2000.

Sweeney, C., Gloor, E., Jacobson, A. R., Key, R. M., McKinley, G., Sarmiento, J.-L., and Wanninkhof, R.: Constraining global air-sea gas exchange for CO_2 with recent bomb 14C measurements, Global Biogeochem. Cy., 21, GB2015, doi:10.1029/2006GB002784, 2007.

Takahashi, T., Sutherland, S. C., Wanninkhof, R., Sweeney, C., Feely, R. A., Chipman, D. W., Hales, B., Friederich, G., Chavez, F., Sabine, C., Watson, A., Bakker, D. C. E., Shuster, U., Metzl, N., Yoshikawa-Inoue, H., Ishii, M., Midorikawa, T., Nojiri, Y., Körtzinger, A., Steinhoff, T., Hoppema, M., Olafsson, J., Arnarson, T. S., Tilbrook, B., Johannessen, T., Olsen, A., Bellerby, R., Wong, C. S., Delille, B., Bates, N. R., and de Baar, H. J. W.: Climatological Mean and Decadal Change in Surface Ocean pCO_2, and Net Sea-air CO_2 Flux over the Global Oceans, Deep-Sea Res. Pt. II, 56, 554–577, 2009.

Tennekes, H. and Lumley, J. L.: A first course in turbulence, Cambridge, MA, MIT Press, 1972.

Timmermans, M. L., Proshutinsky, A., Krishfield, R. A., Perovich, D. K., Richter-Menge, J. A., Stanton, T. P., and Toole, J. M.: Surface freshening in the Arctic Ocean's Eurasian Basin: An apparent consequence of recent change in the wind-driven circulation, J. Geophys. Res., 116, C00D03, doi:10.1029/2011JC006975, 2011.

Toyota, T., Takatsuji, S., and Nakayama, M.: Characteristics of sea ice floe size distribution in the seasonal ice zone, Geophys. Res. Lett., 33, L02616, doi:10.1029/2005GL024556, 2006.

Wadhams, P., Squire, V. A., Ewing, J. A., and Pascal, R. W.: The effect of the marginal ice zone on the directional wave spectrum of the ocean, J. Phys. Oceanogr., 16, 358–376, 1986.

Wanninkhof, R.: Relationship between Wind-Speed and Gas-Exchange over the Ocean, J. Geophys. Res.-Oceans, 97, 7373–7382, 1992.

Wanninkhof, R. and McGillis, W. R.: A cubic relationship between air-sea CO_2 exchangeand wind speed, Geophys. Res. Lett., 26, 1889–1892, 1999.

Wanninkhof, R., Asher, W., Weppernig, R., Chen, H., Schlosser, P., Langdon, C., and Sambrotto, R.: Gas transfer experiment on Georges Bank using two volatile deliberate tracers, J. Geophys. Res., 98, 20237–20248, 1993.

Wanninkhof, R., Asher, W. E., Ho, D. T., Sweeney, C., and McGillis, W. R.: Advances in Quantifying Air-Sea Gas Exchange and Environmental Forcing, Annu. Rev. Marine Sci., 1, 213–244, doi:10.1146/annurev.marine.010908.163742, 2009.

Zappa, C. J., Raymond, P., Terray, E., and McGillis, W. R.: Variation in surface turbulence and the gas transfer velocity over a tidal cycle in a micro-tidal estuary, Estuaries, 26, 1401–1415, 2003.

Zappa, C. J., McGillis, W. R., Raymond, P. A., Edson, J. B., Hintsa, E. J., Zemmelink, H. J., Dacey, J. W. H., and Ho, D. T.: Environmental turbulent mixing controls on air-water gas exchange in marine and aquatic systems., Geophys. Res. Lett., 34, L10601, doi:10.1029/2006GL028790, 2007.

5

Improved sea level record over the satellite altimetry era (1993–2010) from the Climate Change Initiative project

M. Ablain[1], A. Cazenave[2], G. Larnicol[1], M. Balmaseda[11], P. Cipollini[7], Y. Faugère[1], M. J. Fernandes[10,14], O. Henry[2], J. A. Johannessen[3], P. Knudsen[6], O. Andersen[6], J. Legeais[1], B. Meyssignac[2], N. Picot[12], M. Roca[8], S. Rudenko[9], M. G. Scharffenberg[4], D. Stammer[4], G. Timms[5], and J. Benveniste[13]

[1]Collecte Localisation Satellite (CLS), Ramonville Saint-Agne, France
[2]Laboratoire d'Etudes en Géophysique et Océanographie Spatiales (LEGOS), Toulouse, France
[3]Nansen Environmental and Remote Sensing Center (NERSC), Bergen, Norway
[4]University of Hamburg, Hamburg, Germany
[5]CGI, London, UK
[6]Technical University of Denmark (DTU), Lyngby, Denmark
[7]National Oceanography Centre (NOC), Southampton, UK
[8]isardSAT, Barcelona, Catalonia, Spain
[9]Helmholtz Centre Potsdam GFZ German Research Centre for Geosciences, Telegrafenberg 14473 Potsdam, Germany
[10]Faculdade de Ciências, Universidade do Porto, 4169-007 Porto, Portugal
[11]European Centre for Medium-Range Weather Forecasts (ECMWF), Reading, UK
[12]Centre National d'Etudes Spatiales (CNES), Toulouse, France
[13]European Space Agency (ESA), ESRIN, Frascati, Italy
[14]Centro Interdisciplinar de Investigação Marinha e Ambiental (CIIMAR/CIMAR), Universidade do Porto, 4050-123 Porto, Portugal

Correspondence to: M. Ablain (mablain@cls.fr)

Abstract. Sea level is one of the 50 Essential Climate Variables (ECVs) listed by the Global Climate Observing System (GCOS) in climate change monitoring. In the past two decades, sea level has been routinely measured from space using satellite altimetry techniques. In order to address a number of important scientific questions such as "Is sea level rise accelerating?", "Can we close the sea level budget?", "What are the causes of the regional and interannual variability?", "Can we already detect the anthropogenic forcing signature and separate it from the internal/natural climate variability?", and "What are the coastal impacts of sea level rise?", the accuracy of altimetry-based sea level records at global and regional scales needs to be significantly improved. For example, the global mean and regional sea level trend uncertainty should become better than 0.3 and 0.5 mm year^{-1}, respectively (currently 0.6 and 1–2 mm year^{-1}). Similarly, interannual global mean sea level variations (currently uncertain to 2–3 mm) need to be monitored with better accuracy. In this paper, we present various data improvements achieved within the European Space Agency (ESA) Climate Change Initiative (ESA CCI) project on "Sea Level" during its first phase (2010–2013), using multi-mission satellite altimetry data over the 1993–2010 time span. In a first step, using a new processing system with dedicated algorithms and adapted data processing strategies, an improved set of sea level products has been produced. The main improvements include: reduction of orbit errors and wet/dry atmospheric correction errors, reduction of instrumental drifts and bias, intercalibration biases, intercalibration between missions and combination of the different sea level data sets, and an improvement of the reference mean sea surface. We also present preliminary independent validations of the SL_cci products, based on tide gauges comparison and a sea level budget closure approach, as well as comparisons with ocean reanalyses and climate model outputs.

1 Introduction

Global warming as a result of anthropogenic greenhouse gas emissions has already shown several visible consequences, among them the increase of the earth's mean air temperature and ocean heat content, melting of glaciers, and loss of ice masses from glaciers and the Greenland and Antarctica ice sheets. Ocean warming and land ice melting in turn are causing sea level to rise, with potentially negative impacts in many low-lying regions of the world. The precise measurement of sea level changes as well as its different components, at global and regional scales, is an important issue for a number of reasons. It provides information on how the climate system and its components respond to global warming and on the relative contributions of anthropogenic forcing and natural/internal climate variability. This also allows validation of the climate models developed for projecting future changes, as the models are supposed to correctly reproduce present-day and recent-past changes. The Global Climate Observing System (GCOS) has recently defined a set of 50 climate variables (called Essential Climate Variables – ECVs) that need to be precisely monitored on the long term in order to improve our understanding of the climate system, its functioning and its response to anthropogenic forcing, as well as to provide constraints for climate modelling (GCOS, 2011). In 2010, the European Space Agency (ESA) developed a new programme, the Climate Change Initiative (CCI), dedicated to reprocessing a set of 13 ECVs currently observed from space; among them, the satellite altimetry-based sea level ECV. The objective of the CCI sea level project (called SL_cci below) was to produce a consistent and precise sea level record covering the past two decades, based on the reprocessing of all satellite altimetry data available from all missions (including the ERS-1&2 and Envisat missions, in addition to the TOPEX/Poseidon, Jason-1&2 and Geosat Follow-on (GFO) missions). During the first phase of the project, which lasted 3 years from 2011 to 2013, satellite al timetry data from seven altimeter satellites were reprocessed by the SL_cci consortium. Improved satellite orbits have been computed for all satellites except TOPEX/Poseidon and GFO using up-to-date force models and an improved reference frame realization. Updated geophysical corrections adapted to each satellite mission have been implemented after being evaluated and selected. Other improvements concern the reduction of instrumental drifts and biases (in particular for the Envisat mission), a new calculation of the mean sea surface used as reference, the method used for geographical averaging of sea surface height data, and the reduction of systematic bias between missions. The main SL_cci products computed during phase 1 consist of: (1) a global mean sea level (GMSL) time series at monthly intervals between January 1993 and December 2010, and (2) a global gridded sea level time series (resolution 0.25° × 0.25°) at the same time interval.

This paper intends to provide a global overview of the main results obtained in the frame of the SL_cci project. We firstly describe the validation protocol (Sect. 2) that has been applied to evaluate and select the algorithms and corrections used (Sect. 3) to generate the SL_cci products (described in Sect. 4). Then, Sects. 5 and 6 are focused on the assessment and the error characterization.

2 Definition of a formal validation protocol

The altimetry data processing system used to compute sea level (or the sea surface height, SSH) integrates a number of components: the altimeter range measurement (Range), the satellite orbit height (Orbit), and the instrumental and geophysical corrections. The estimation of these components needs additional information coming from different domains as orbitography (a force model) for the precise orbit determination, geodesy (geoid, mean sea surface, global isostatic adjustment (GIA), etc.), atmosphere (pressure, wind, dry and wet troposphere, etc.), and ocean (ocean tides, sea state, etc.). This information may be eventually linked together either directly or indirectly. Because of these complex interactions, sea level estimates (i.e. $\mathrm{SSH} = \mathrm{Orbit\ Range} \sum_{i=0}^{N} \mathrm{Correction}_i$) are provided with different standards. In practice, an optimized sea level calculation requires a large number of algorithms and corrections that need to be rigorously validated and regularly updated.

In the framework of the SL_cci project, we developed a new formal validation protocol which allowed us to evaluate the impact of new altimeter corrections or standards on a sea level record of climate quality, i.e. precise enough for climate studies. It consists in comparing the new altimeter corrections with corrections designed as a reference through their impact on the sea level calculation. This was done using a common set of validation diagnoses defined in such a way that they fulfil the sea level accuracy and precision requirements. The validation diagnoses are distributed into three distinct families allowing the assessment of altimetry data with complementary objectives:

1. the "global internal analyses" with the aim of checking the internal consistency of a specific mission-related altimetry system by analysing the computed sea level, its instrumental parameters (from altimeter and radiometer) and associated geophysical corrections,

2. the "global multi-mission comparisons" allowing evaluation of the coherence between two different altimetry systems through comparison of SSH data,

3. the "altimetry in situ data comparison" dedicated to the computation of the sea level differences between altimeter data and in situ sea level measurements, e.g. from

tide gauges or Argo-based steric sea level data (Valladeau et al., 2012); this 3rd approach allows for the detection of potential drifts or jumps in the long-term sea level time series.

For each family, several validation diagnoses have been defined using elementary statistical approaches (e.g. mean, standard deviation, linear regression) and data representation (e.g. global mean time series, maps, histograms, periodograms). Other tests based on altimeter correction differences, sea surface height differences at satellite track crossovers, sea level anomalies, etc. were also performed. The list of all the diagnoses and their specification is described in detail in the Product Validation Plan (PVP) report of the SL_cci project for all referenced SL_cci reports available on the SL_cci website).

The analyses of these diagnoses were performed for different spatial (global mean and regional sea level, mesoscale) and temporal scales (Fig. 1, left panel): long-term > 10 years; interannual, 2–5 years; and periodic signals (annual, semi-annual) scales. These spatio-temporal scales were chosen according to the sea level user requirements document (SL_cci User Requirements Document, 2014) presented below. This formal validation protocol allows us to determine, for each spatial and temporal scale, the level of impact (i.e. low or strong) of the new altimetry corrections on the sea level calculation (Fig. 1, right panel). For instance, if a new altimetry correction causes a GMSL trend $> 0.15\,\text{mm year}^{-1}$ (over a period > 10 years), we consider that the impact is strong, whereas if the trend effect is in the range 0.05–$0.15\,\text{mm year}^{-1}$, it is assumed low, and negligible below $0.05\,\text{mm year}^{-1}$.

Our goal is also to check whether the new altimeter corrections improved or degraded the sea level estimates for each timescale. Most of the time, it was possible to clearly detect either improvement or degradation (illustrated Fig. 1, left panel, with the symbols "+" and "−" meaning improvement and degradation). For example, increased consistency between GMSL trends derived from two different altimetry missions or from in situ measurements demonstrates that the accuracy/precision of sea level data has been improved. In only a few cases were the diagnoses inconclusive. This occurred when errors of altimetry missions were of the same order of magnitude or correlated (e.g. same error for the regional mean sea level trends). In these rare cases, thorough investigations could be conducted through a "case by case" approach. When no obvious conclusion could be reached, the sea level differences due to the new correction were then allocated to the altimetry error budget (see Sect. 6).

Thanks to this formal validation protocol, the impact of all altimeter corrections could be described through a homogeneous approach and is therefore comparable between corrections. The table presented in Fig. 1 (left panel) allows us to provide easily and quickly relevant information about the impact of each correction on the sea level products.

3 Development, validation and selection of new altimeter corrections and algorithms

In this section, we present applications of the formal validation protocol described in Sect. 2. An important output of the SL_cci project was the development of new altimetry corrections (mentioned in Sect. 2) and algorithms (e.g. for merging data from different altimetry missions). A total of 42 new corrections/algorithms were evaluated within the project using the validation protocol described above. The reference standards were those used for AVISO products (Dibarboure et al., 2011) at the beginning of the SL_cci project. In order to select the best corrections, a "selection meeting" in Toulouse in May 2012 gathered a team of international experts in satellite altimetry, not involved in the SL_cci project. The new corrections were then selected on the condition that they led to improvements in sea level calculation. In the rare cases where the new processing did not improve the results or, even worse, led to deterioration, a conservative approach was applied and finally, the former corrections were unchanged. Table 1 presents the new selected corrections for each component and altimetry missions (for detailed information, see all the "SL_cci Validation Reports"). One of the most dramatic improvements comes from the use of ERA-interim reanalyses (from the European Centre for Medium-Range Weather Forecasts – ECMWF; Dee et al., 2011) instead of operational ECMWF fields to calculate the dry tropospheric and other dynamical atmospheric corrections. Applying our validation protocol, we noted strong improvements at mesoscale and regional spatial scales, over the first altimetry decade (1993–2003) (Carrere et al., 2014; "SL_cci Validation reports, Atmospheric corrections", 2014). The GMSL error reduction (Fig. 2, top) obtained from crossover analyses is of the order of 2.5 cm on the early years of the altimetry era (1993–1995). Then, the error decreases linearly until 2004, and remains stable close to 0 during recent years. The improvement observed in the first decade (1993–2003) is stronger at high latitudes (6 cm) where the atmospheric pressure and wind fields have strong high-frequency variability. Looking at regional sea level trends (Fig. 2), significant trend differences are observed ($> 1\,\text{mm year}^{-1}$) mainly in the South Pacific Ocean below 50° S latitude.

Similarly, the model-based wet tropospheric correction was also strongly improved (until 1 cm error reduction on the GMSL) before 2002 using ERA-interim instead of ECMWF operational fields (Legeais et al., 2014). While not as good as the wet troposphere corrections derived from the on-board microwave radiometers (MWRs), the ERA-Interim wet tropospheric correction allows us to better characterize the uncertainty of wet troposphere content over the long term (Thao et al., 2014; Legeais et al., 2014). However, this was not used in the sea level calculation, where the radiometer-based corrections were preferred.

In parallel, the radiometer-based corrections have been improved using combined estimates from valid on-board MWR

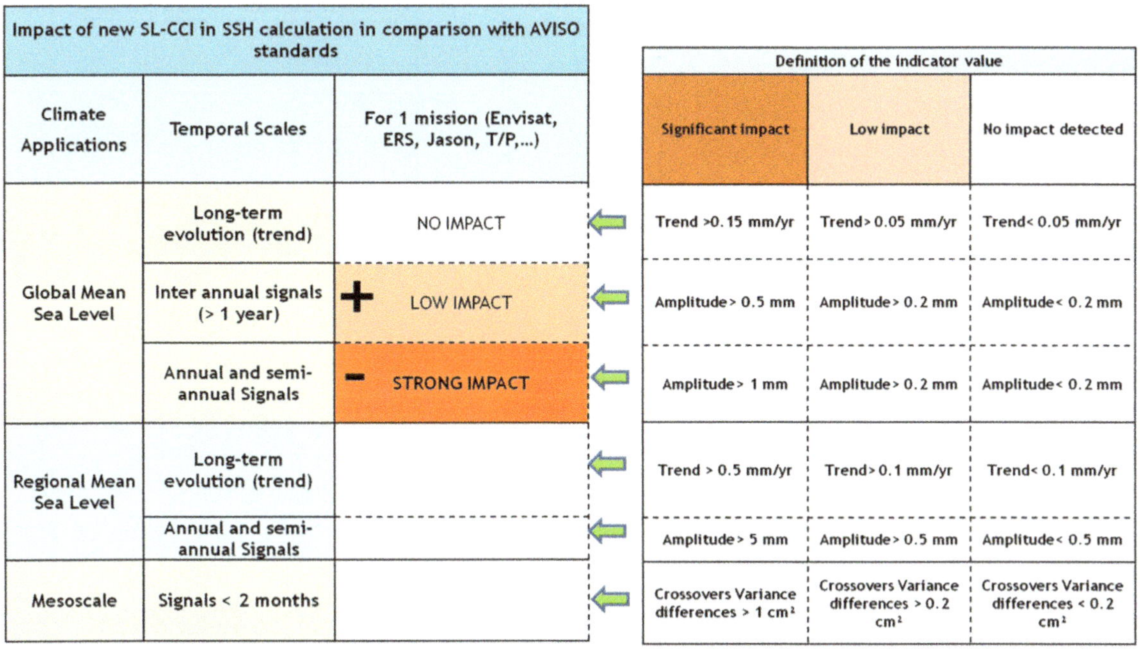

Figure 1. Definition of the temporal and spatial scales (left panel) and the indicator value table (right panel) allowing the impact characterization in sea level of new SL_cci corrections in comparison with corrections defined as reference (AVISO-2010).

Table 1. New corrections selected for the sea level calculation for the SL_cci project. The unfilled boxes indicate that the AVISO standards (release 2010) have been applied.

Corrections	ERS-1	ERS-2	Envisat	Jason-1	Jason-2	*T/P*	GFO
Orbit	Reaper combined orbit (Rudenko et al., 2012)		GDR-D CNES (Couhert et al., 2015)				
Instrumental correction			New PTR correction (Garcia and Roca, 2010)				
Sea state bias			V2.1 release		GDR-D release		
Wet troposphere	GPD corrections (Fernandes et al., 2010, 2014)						
Dry troposphere	ERA-interim based (Carrere et al., 2014)						
Dynamical atmospherical corrections	ERA-interim based (Carrere et al., 2014)						
Ocean tide	GOT 4.8 (Ray et al., 2013)						
Mean sea surface	DTU 2010 (Andersen et al., 2010)						

values, global navigation satellite systems (GNSS) measurements and ECMWF model (ERA Interim fields) in areas where the MWR measurements are degraded due to, for example, land or ice contamination or instrument malfunction (Fernandes et al., 2010, 2014). This new correction, called GNSS-derived path delay (GPD), computed for all ESA and reference missions, brings improvements mainly in coastal areas and in the polar regions. In Fig. 3, the sea level error reduction is plotted vs. the distance to the coast using the new GPD corrections instead of the reference radiometer-based corrections. For almost all missions, except Jason-2, which already benefits from an improved coastal radiometer correction (Brown et al., 2009), there is a significant SSH error reduction, close to 1 cm between 20 and 40–50 km from the coast. Improvements have also been noticed in the open ocean, especially for TOPEX data (Fernandes et al., 2014), where radiometer data gaps degrade the interpolation process. Finally, the GPD corrections have been selected for all altimeter missions because of the noted improvement in the sea level calculation at short and long timescales, mainly in coastal and polar regions.

Orbit error is the main source of the error for the long-term sea level evolution at oceanic basin scales (Couhert et al., 2015). Strong efforts have been made within the

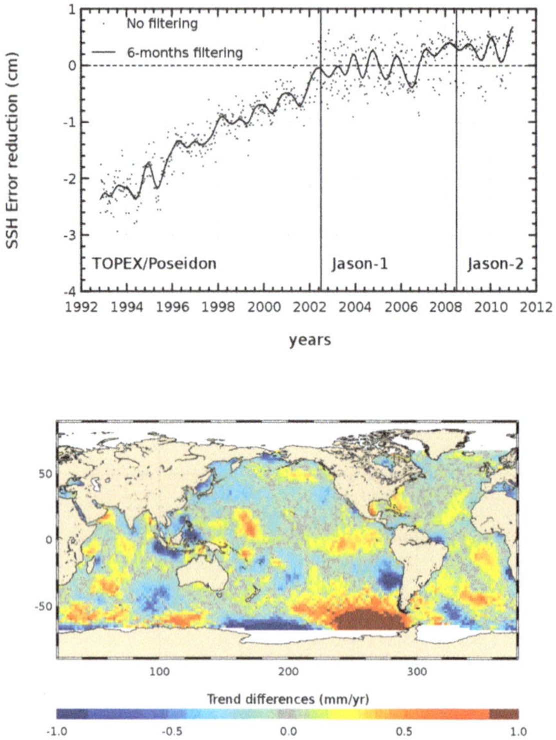

Figure 2. Evolution of the sea level error reduction applying the new dynamical atmospheric and dry troposphere corrections derived from ERA-Interim reanalyses instead of operational ECMWF fields (top) and impact on sea level regional trends (bottom).

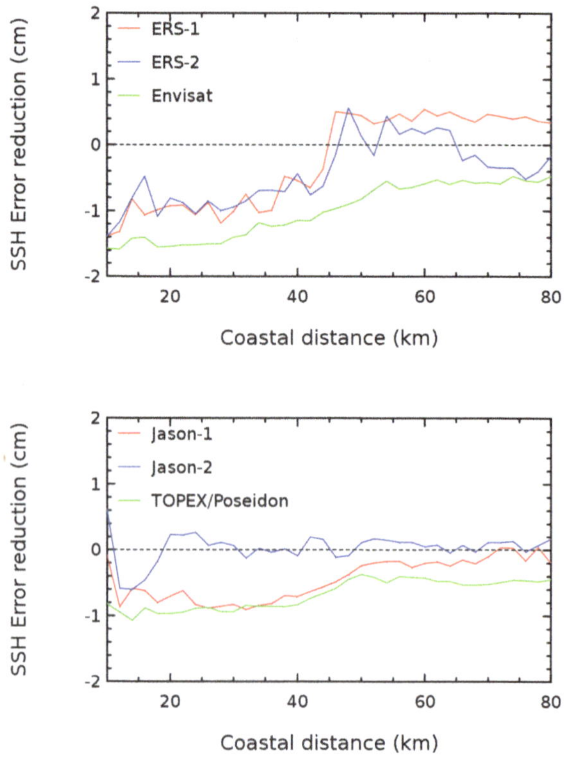

Figure 3. Evolution of the error reduction vs. the coastal distance applying the new GPD wet troposphere corrections instead of the reference radiometer-based corrections used in AVISO-2010.

SL_cci project to develop new orbit solutions (Rudenko et al., 2014) and to compare them with external solutions provided by other projects. The International Terrestrial Reference Frame (ITRF) realization (Altamimi et al., 2011) and the earth gravity field model used in the orbit computation are crucial as far as the quality of orbit solutions is concerned. After analysing all orbit solutions for all the missions, the REAPER combined orbit solutions (Rudenko et al., 2012) have been selected for ERS-1 and ERS-2, with the new CNES GDR-D orbit solutions (Couhert et al., 2015) being selected for the Jason-1, Jason-2 and Envisat missions. Strong effects were observed on the regional sea level trend, in the range of 1–2 mm year^{-1}, with large patterns at hemispheric scale when using static and time-variable earth gravity field models for orbit computation (Fig. 4). Thanks to cross-comparisons between altimetry missions (Ollivier et al., 2012) and with in situ measurements (Valladeau et al., 2012), we have demonstrated that these new orbit solutions dramatically improved the regional sea level trends. Furthermore, this inter-comparison, using different orbit solutions, provided interesting information on the orbit sensitivity to the choice of the earth gravity field model (Rudenko et al., 2014).

In addition to these major improvements, other corrections were also selected, although their impact on the sea level estimate was lower. These concern the ionospheric correction with the use of the NIC09 (New Ionosphere Climatology) model for ERS-1 (Scharroo and Smith, 2010), the GOT4.8 (Geocentric Ocean Tide) ocean tide solution (Ray et al., 2013), and the DTU10 (Danish Technical University) mean sea surface (Andersen et al., 2010) for all missions. In addition, we benefited from the reprocessing of Envisat and Jason-2 level-2 products "GDR V2.1" (Ollivier et al., 2012) and "GDR-D" (Philipps and Roinard, 2013). This allowed us to increase the data coverage (mainly for Envisat) and to improve the sea-state bias corrections along with instrumental bias and drift corrections. For the latter, the impact is strong for Envisat since a global instrumental drift of about 2 mm year^{-1} was identified and corrected in the altimeter range (Thibaut et al., 2010; Roca and Thibaut, 2009; Garcia and Roca, 2010). It is worth mentioning that the SL_cci project contributed to correcting this anomaly, while Envisat was designed not for climate studies but rather for mesoscale variability.

The last new algorithm developed and selected aims at better combining the different sea level time series from TOPEX, Jason-1, and Jason-2 at regional scale. Thanks to the verification phase between these missions, systematic geographical biases could be detected. These biases are mainly latitude-dependent, with variations close to 0.5 cm between

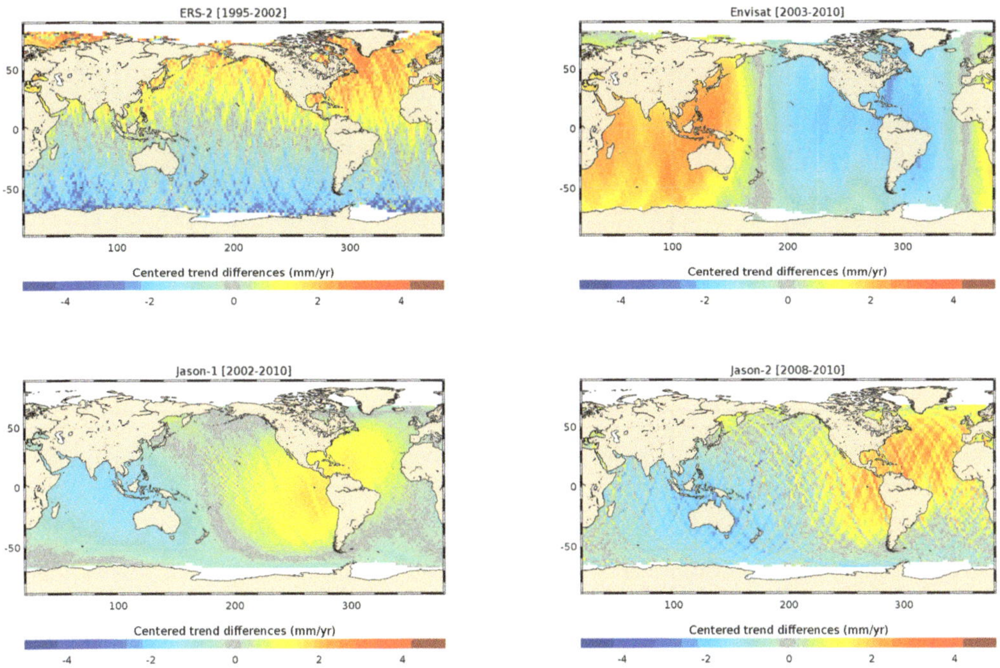

Figure 4. Impact of the new orbit solutions on the regional sea level trends for ERS-2 (Reaper combined vs. DEOS DGM-E04 orbit solutions), Envisat, Jason-1, and Jason-2 (CNES GDR-D vs. CNES GDR-C orbit solutions).

Jason-1 and Jason-2, and 1 cm between TOPEX and Jason-1. Correcting these regional and systematic sea level differences (see the SL_cci Validation Report, Regional SSH bias corrections between altimetry missions, 2014), led us to better combine these three altimetry missions and therefore better estimate the long-term sea level evolution at regional scales. The impact of these corrections on regional MSL trends plotted in Fig. 5 from 1993 to 2010 is close to ± 0.3 mm year^{-1}, with large hemispheric dependence.

4 New CCI-based sea level records

Sea level products were generated using the new altimeter corrections described in Sect. 3. The same procedure was adopted as for the SSALTO DUACS (Segment Sol Multimission Altimetrie et Orbitographie, Data Unification and Altimeter Combination System; Dibarboure et al., 2011). After calculating the along-track sea level for each of the seven missions (TOPEX/Poseidon, Jason-1, Jason-2, ERS-1, ERS-2, Envisat, and Geosat Follow-on) over the [1993, 2010] period, the main steps consisted of: combining all missions together, reducing the orbit and the long wavelength errors, computing the gridded sea level anomalies using an objective analysis approach (Ducet et al., 2000; Le Traon et al., 2003), and generating mean sea level products (e.g. GMSL time series, gridded sea level time series) dedicated for climate studies. The SL_cci products are monthly grids time series with a spatial resolution of 0.25° using a rectangular projection. The GMSL time series (also at monthly interval) is based on the geographical averaging over the oceanic domain observed by the altimetry data (82° S to 82° N) of the gridded data. Additional products (called indicators) are provided: GMSL trend, regional MSL trends, amplitudes and phases of the main periodic signals (annual, semi-annual), etc.

Access to the SL_cci products can be obtained by sending an email to: info-sealevel@esa-sealevel-cci.org. The Product User Guide (PUG, 2013) and Product Specification Document (PSD, 2013) provide further details.

Comparisons between the SL_cci product and the AVISO-2010 products (Dibarboure et al., 2011) were performed by applying the formal validation protocol described above (Sect. 2). Concerning the GMSL trend, similar values were obtained for both time series: 3.2 mm year^{-1} over the 1993–2010 time span. At the interannual timescale (highlighted by calculating the difference between the two GMSL time series (Fig. 6, top panel), small differences in the range 1-2 mm or lower are noticed, except for 1994 where a 4 mm jump is observed. This jump is due to an anomalous value of the AVISO-2010 products caused by an inadequate merging of the TOPEX data with the ERS-1 data of the non-repetitive geodetic phase (Pujol et al., 2014). The most impressive result is obtained by separating the ERS-1/ERS-2/Envisat and TOPEX/Jason-1/Jason-2 global GMSL time series using alternately the old and new altimeter corrections (Fig. 7): the trend difference between the two time series is now close to 0.6 mm year^{-1} from 1993 to 2010 instead of about 1.5 mm year^{-1} previously. This improved consistency does

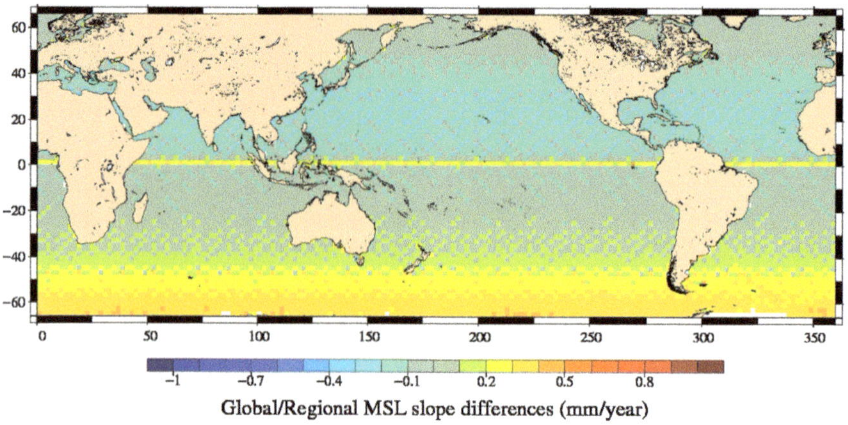

Figure 5. MSL trend differences from 1993 to 2010 between sea level maps without and with regional bias corrections for TOPEX/Jason-1 and Jason-1/Jason-2.

not have a direct impact on the GMSL trend, which depends only on the TOPEX/Jason-1/Jason-2 missions. However, this provides increased confidence in the long-term GMSL time series.

Looking at the regional sea level trend differences (Fig. 6, bottom panel), large geographically correlated structures are observed. Their amplitude is in the $\pm 2\,\mathrm{mm\,year^{-1}}$ range. They primarily result from the new orbit solutions (hemispheric effects), the new ERA-interim atmospheric fields (at high latitudes), the new wet tropospheric correction, and the geographical biases arising when linking altimetry missions together. Comparing with in situ measurements (tide gauges and Argo-based steric sea level) indicates a better consistency at the regional scale with the new SL_cci data (see SL_cci Product Validation Internal Report – PVIR, 2014). It is more difficult to detect any improvement at short spatial scales, because either the spatial or temporal sampling of in situ measurements is not good enough or because the error generated by the collocation method between the in situ and altimetry data is larger than the target signal (Couhert et al., 2015). We also examined the periodic (annual and semi-annual) sea level signals. We found differences in the order of 5 mm on average for the amplitude of the annual signal. In some regions (the tropics), the differences can reach 1 cm. While we think that the new seasonal signal is improved compared to the AVISO-2010 products, it is not possible to demonstrate this through any independent validation diagnoses. Indeed, comparisons with the in situ measurements are not accurate enough to observe such signals.

5 Validation of the temporal and spatial variations of global sea level

The SL_cci products delivered at the end of Phase 1 are currently under validation and evaluation. Two different approaches have been developed:

1. assessment of the accuracy of the SL_cci products through their use in ocean reanalyses and earth system models;

2. assessment of the global sea level budget.

In approach (1), the accuracy of the SL_cci data is evaluated by quantifying the model performances and robustness (compared to the use of a reference sea level data set, e.g. AVISO standard data) in representing a number of physical processes (e.g. the sea level drop associated with the 2011 La Niña, the Indonesian through flow, changes in the Arctic circulation, effects of monsoon on sea level, regional sea level fingerprint due to wind stress, steric sea level trend patterns).

Approach (2) consists of comparing the SL_cci GMSL and variability to (i) other GMSLs, and (ii) the sum of the climatic and non-climatic components estimated independently (changes in thermal expansion, glacier and ice sheet mass balance, and land water storage).

5.1 Assessment based on numerical ocean models

Ocean model simulations are an effective way of translating wind and heat fluxes information into sea level variations, thus providing independent verification of their contribution to sea level. Sea level from ocean-only simulations at different resolutions (1°, 0.25°) has been contrasted with along-track data and with gridded (filtered and merged) sea level maps from AVISO (Dibarboure et al., 2011) and SL_cci. The statistics of the comparison (correlation, rms error, differences in trends) were similar when using AVISO and SL_cci data. Differences between models and any observed estimations were much larger than the differences between observational products. The spatial patterns of these differences were suggestive of model error. For instance, small-scale sea level variability is much larger in observed products than in models, which is consistent with insufficient resolution in the models. In contrast, the low-frequency and large-scale vari-

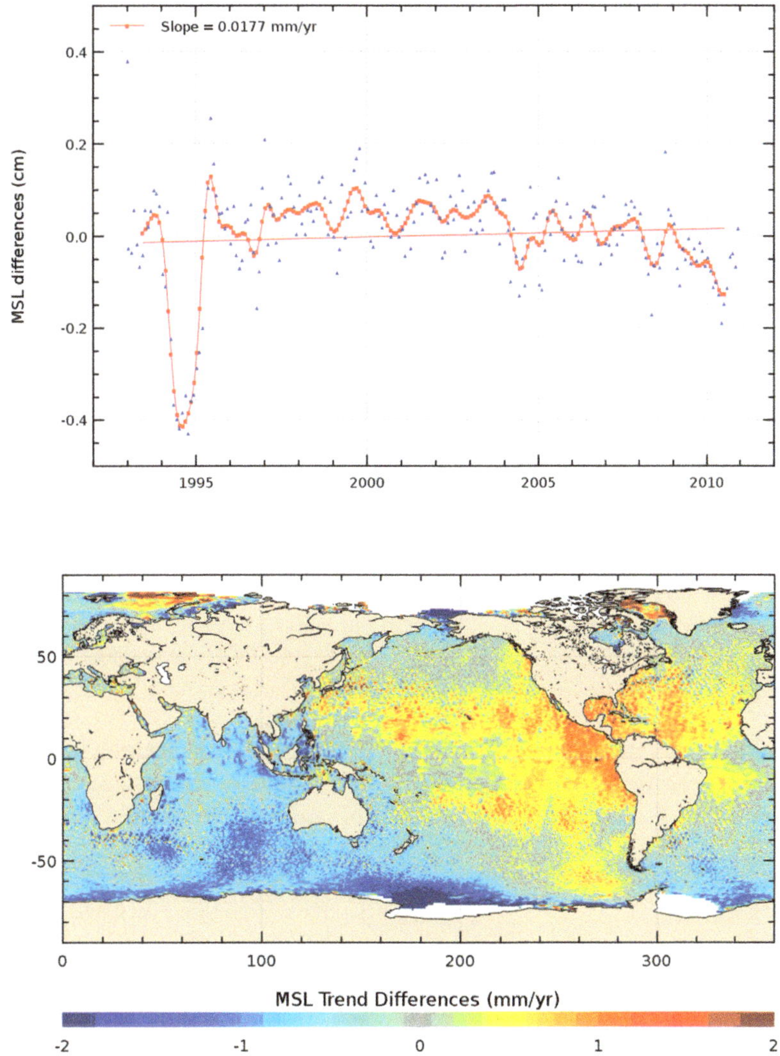

Figure 6. GMSL (top panel) and regional sea level (bottom panel) differences between the SL_cci (release 1.1) and AVISO products (release 2010).

ability is more obvious and better resolved in models. The large-scale patterns of interannual variability and trends are consistent between models and observations, but differences exist associated with the precise location of strong current systems, which models struggle to capture. This information is in itself interesting, and suggests that a large part of the sea level variability is of a dynamic nature, associated with changes in the wind-driven circulation. Both AVISO and SL_cci were useful to detect improvements in ocean model simulations due to the increased resolution.

In the Arctic Ocean the SL_cci reprocessed data reveal some distinct features of the elevated trend in sea level rise, notably in the Beaufort Sea, in the Norwegian Sea, in the Sub-Polar gyre, and in the northeast Atlantic south of the Iceland–Faroe ridge. The Beaufort Sea rise of about 6.5–7 mm year^{-1} has also been reported by Morrison et al. (2011) and Laxon et al. (2012), while the elevated feature of around 6–7 mm year^{-1}, as detected in the SL_cci field in the Lofoten Basin of the Norwegian Sea, compares rather well with the trend recovered from in situ hydrographic observations.

A first look at the three general circulation models (GCMs), NorESM (Norwegian Earth System Model), Hadley, and IPSL (Institut Pierre-Simon Laplace), reveals large individual differences in the trend of sea level change, regarding the overall trend as well as in its regional characteristic changes. The contributions to these simulated changes include the regional variability of the steric and the mass components, while there is no account of the GIA. In comparison to the SL_cci sea level change the NorESM simulations (1° resolution) yield the best agreement in the Sub-Polar gyre, in the northeast Atlantic Ocean south of the Iceland-Faroe ridge, in the Lofoten basin of the Norwegian Sea and in the Beaufort Gyre. This inter-comparison of the SL_cci trends with the trends derived from the three GCMs

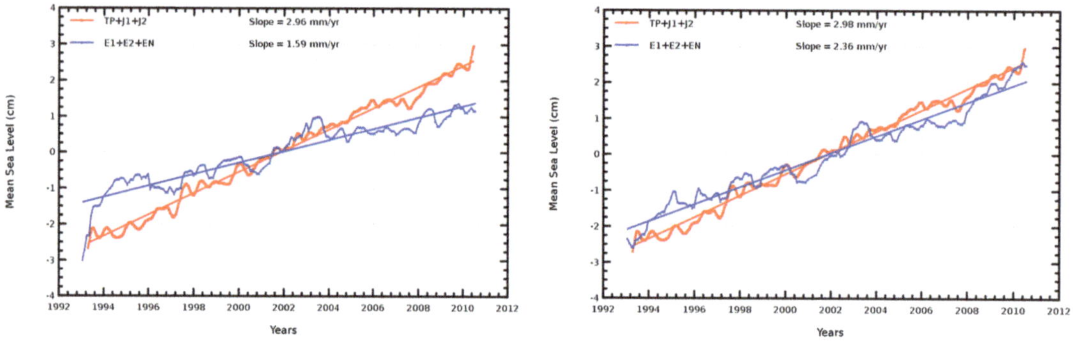

Figure 7. GMSL time series separating ERS-1/ERS-2/Envisat and TOPEX/Jason-1/Jason-2 altimeter missions using alternatively the old (AVISO-2010 standards) on left and new altimeter correction (SL_cci) on right.

can therefore provide evidence for how realistic the model simulations are with respect to the regional variability of the water masses (steric height contribution) and variability, spreading, and accumulation of freshwater discharges from melting ice sheets and glaciers (mass changes).

In summary, as was to be expected from the beginning, even ocean-only simulations are not able to identify the incremental improvement of SL_cci vs. its predecessor. Nevertheless, this validation exercise has shown that the SL_cci is a robust data set for ocean and climate models validation, and can discern verification metrics.

5.2 Assessment based on ocean data assimilation

Data assimilation methods can be very effective methods to test the quality of the input data. This approach was used here to evaluate the SL_cci products, either by direct assimilation of the product as an ocean synthesis (active mode) or by simple comparison with a reference state (passive mode), obtained by a forced ocean model combined with in situ observations, and even other sea level observations. In this way, the ocean synthesis, containing information from both the model forced with realistic atmospheric state and observations, should have less error than an ocean model simulation alone. The passive comparison can be done a posteriori (by comparing ocean reanalyses with SL_cci), or during the assimilation process, by contrasting, at the appropriate location and time, the along track altimeter data with the estimate given by an ocean model that assimilates in situ temperature and salinity.

In a first step, sea surface height fields available from the GECCO2 assimilation approach (Köhl, 2014) were compared to the AVISO products as well as to the SL_cci product. Of these two, the AVISO product was used to constrain the model, but not the SL_cci product. The comparison was performed to investigate whether the new SL_cci product is closer to the GECCO2 ocean reanalysis product, which is constrained by most of the available global data sets, than the previous AVISO data set, a test that would highlight a better consistency of the new SSH data with ocean dynamics and other ECV information. The comparisons have been performed separately for the ERS (ERS-1, ERS-2 and ENVISAT) and the TOPEX/Poseidon satellite-series (TOPEX/Poseidon, Jason-1 and Jason-2). Figure 8 shows the ratio (RMS_AVISO/RMS_SL_cci) of the rms differences between the GECCO model and the satellite time series of ERS-1, ERS-2 and ENVISAT for AVISO (RMS_AVISO) and SL_cci (RMS_SL_cci) in percent improvement at model resolution. Red indicates improvements of the SL_cci compared to the AVISO data set and blue degradation. Remarkable are the improvements in the North Atlantic, in the Indian Ocean through flow and in many parts of the ocean. The regions where SL_cci shows less skill compared to AVISO are the ones where the GECCO2 solution has adapted very well to AVISO and at the same time where the standard deviation of the data sets are very small, indicating a small signal to noise ratio in these regions. Therefore, the model might have adapted to the not as good AVISO data and thus gives less skill in comparison to the improved SL_cci data set. The improved regions (red colours) cover 62.8 % of the ocean area that had valid data for the comparison, leaving 37.2 % of the ocean area that has degraded (blue colours). Further, when averaging the ratio of RMS_AVISO/RMS_SL_cci globally, weighted by the area of each grid point, a global mean improvement of 0.91 % can be seen from the analysis on the model grid. This could demonstrate that the SL_cci has been improved in many regions.

Both AVISO and SL_cci sea levels have also been compared with the sea level from the ORAS4 ocean reanalyses (Balmaseda et al., 2013), which assimilate in situ temperature, salinity, and AVISO data along a track altimeter. Time series of standard area-averaged climate indices have been used to gain insight on the differences between the AVISO and SL_cci products. Figure 9 shows a time series of the 12 month running mean sea level anomaly differences. In the eastern Pacific (5° N–5° S, 130–90° W, left panel) both ORAS4 and SL_cci show a positive offset with respect to AVISO data after 2005 (from 2005 onwards the ocean state in ORAS4 is relatively well constrained by Argo). In addi-

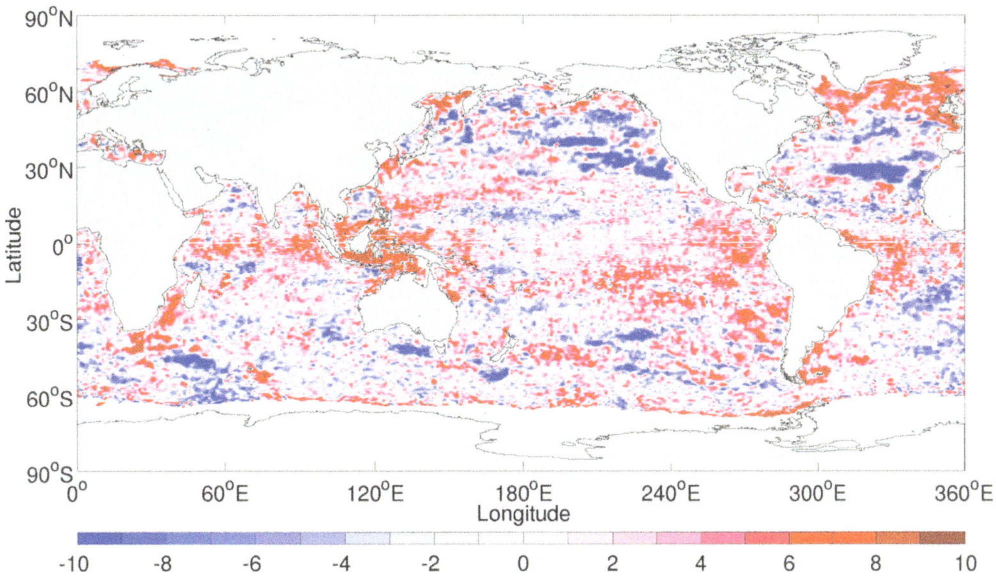

Figure 8. Ratio of the rms differences RMS_AVISO and RMS_SL_cci between the GECCO model and the satellite time series of ERS-1, ERS-2, and ENVISAT in percent improvement.

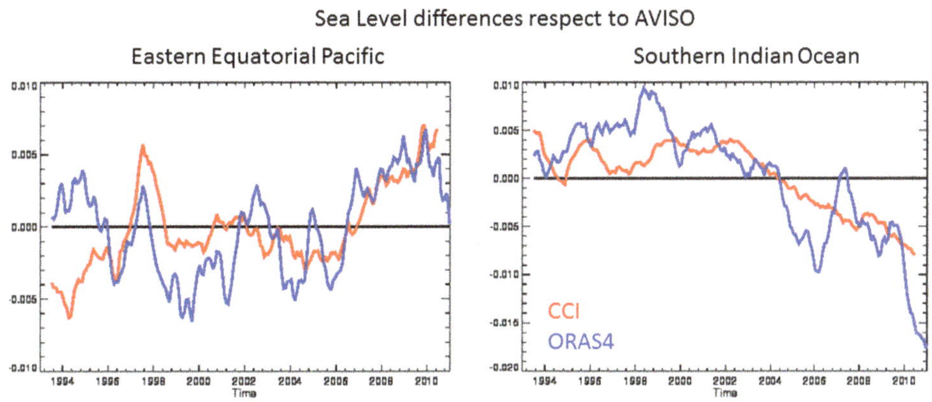

Figure 9. Differences (m) in the sea level time evolution (12 month running mean) with respect to the AVISO product of SL_cci (red) and ORAS4 (blue) . Left: eastern equatorial Pacific (5° N–5° S, 130–90° W). Right: southern Indian Ocean (30–70° S, 20–150° E). The differences in trends between SL_cci and AVISO are confirmed by ORAS4. In the eastern Pacific, both ORAS4 and SL_cci have a stronger El Niño–Southern Oscillation signature than AVISO.

tion, SL_cci and ORAS4 data consistently show stronger local maxima associated with El Niño, 1997. The precursor of this El Niño is visible in the western Pacific slightly earlier, and it is also more pronounced in SL_cci and ORAS4 than in AVISO (not shown). The right panel of Fig. 9 shows the equivalent time series for the southern Indian Ocean (30–70° S, 20–150° E), where both ORAS4 and SL_cci consistently show a negative tendency with respect to AVISO, suggesting that AVISO overestimates the sea level rise in this area. The differences in trends between SL_cci and AVISO shown in these time series are similar to those shown in Fig. 6 (bottom). The variability of the ORAS4 reanalysis agrees better with the SL_cci product than with AVISO.

5.3 Comparison of the SL_cci GMSL time series with other GMSL products

We constructed a GMSL time series by geographically averaging the SL_cci gridded data between 66° S and 66° N. A simple cosine of latitude weighting was applied to the data. As no glacial isostatic adjustment (GIA) correction was applied to the gridded data, we added the usual $+0.3\,\mathrm{mm\,year^{-1}}$ GIA trend from the SL_cci GMSL (as usually done by other processing groups). We further compared the SL_cci GMSL with altimetry-based GMSL time series computed by different processing groups (AVISO, University of Colorado (CU), NOAA (National Oceanic and Atmospheric Administration), GSFC (Goddard Space Flight Cen-

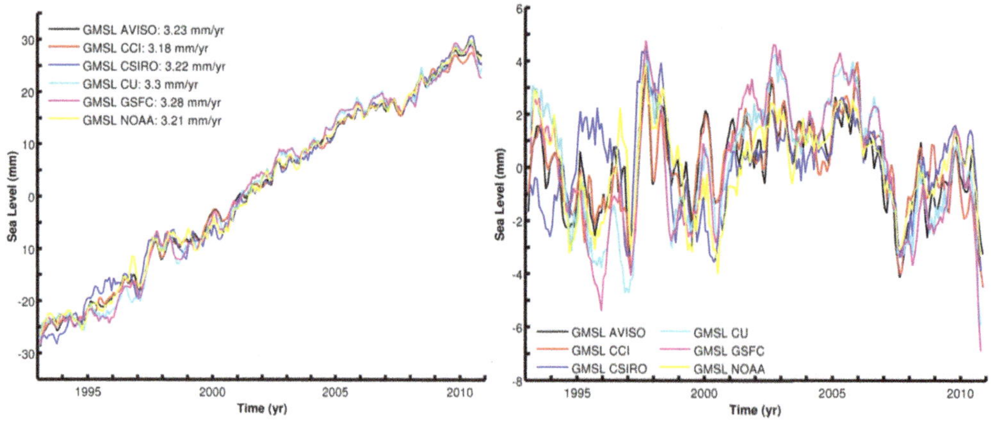

Figure 10. GMSL based on multi-mission satellite altimetry data processed by different groups (including SL_cci project). Left/right panel: with/without the global mean trend.

ter), and CSIRO (Australia's Commonwealth Scientific and Industrial Research Organisation). The results are shown in Fig. 10 (left panel). In terms of trends, all curves are very similar to each other and trend differences (< 0.2 mm year^{-1}) are fully covered by the formal error on the trend computation. However, it is interesting to note that all sea level curves differ significantly (by several mm) over an interannual timescale. This is illustrated in Fig. 10 (right panel), and is particularly noticeable during the TOPEX/Poseidon period (1993–2001), with a significant big departure of the CSIRO GMSL from other curves. The detrended SL_cci GMSL is in general close to the AVISO GMSL, although slight differences are noticed at the end of the study period.

5.4 Comparison of the SL_cci GMSL with steric and ocean mass components (sea level closure budget); interannual time scale

GMSL change is a combination of ocean mass and steric (thermal expansion) changes. We compared the GMSL computed from the SL_cci gridded product with the sum of steric and mass components over the Argo and GRACE (Gravity Recovery and Climate Experiment) operating period (since ~ 2005). Argo-based steric data used for this comparison are based on those processed by Karina von Schuckmann (von Schuckmann and Le Traon, 2011). Ocean mass has been estimated using the RL05 data from the GRACE project (average of the three products: CSR, JPL and GFZ; Chambers and Bonin, 2012). The GRACE and steric data have been averaged over the 66° S and 66° N domain. Figure 11 compares three GMSL products (AVISO, CU, and SL_cci) with the sum of steric and mass contributions over 2005–2010. Error bars of the sum "steric plus mass" time series are not shown for clarity. They are estimated to within ± 2 mm for individual monthly values. The mean trend over the study period (2005–2010) has been removed. The three GMSLs present similar variations and show reasonably good agreement with the sum of the components. Although small differences exist, the best agreement is found for the SL_cci GMSL. Correlation coefficients between the sum "steric plus mass" component and GMSL time series have also been computed. The highest correlation (of 0.65) is found with the SL_cci GMSL.

The results presented above are first attempts to validate the SL_cci products. We find some differences in terms of both global mean and regional variability with the standard products. Preliminary comparisons with the sum of the climate contributions (the sea level budget closure approach) suggest that the CCI product fits better the sum of the climatic components. However, this result is not robust considering the large uncertainties affecting the steric and mass components. Further work is needed on that matter, using different steric and ocean mass products with assessed uncertainties. For instance, the steric height from ocean reanalyses can also be used for global sea level budget closure (Balmaseda et al., 2013). This will be a topic for the CCI phase 2 activities.

6 Error budget of sea level

Although improvements were made, the SL_cci products still contain errors at different timescales. In order to inform users about these errors, we have established an error budget dedicated to the main spatio-temporal scales, i.e. global and regional, long-term (5–10 years or more), interannual (< 5 years), and seasonal (see Table 2)). For each of these, an error was determined and compared to the sea level Climate User requirements (GCOS, 2011) which have been updated in the framework of the Sea Level CCI project (Sea Level CCI User Requirement Document – URD, 2013).

Regarding the GMSL trend, an uncertainty of 0.5 mm year^{-1} was estimated over the whole altimetry era (1993–2010). This uncertainty is reduced by 0.1 mm year^{-1} compared to the previous data based on AVISO-2010

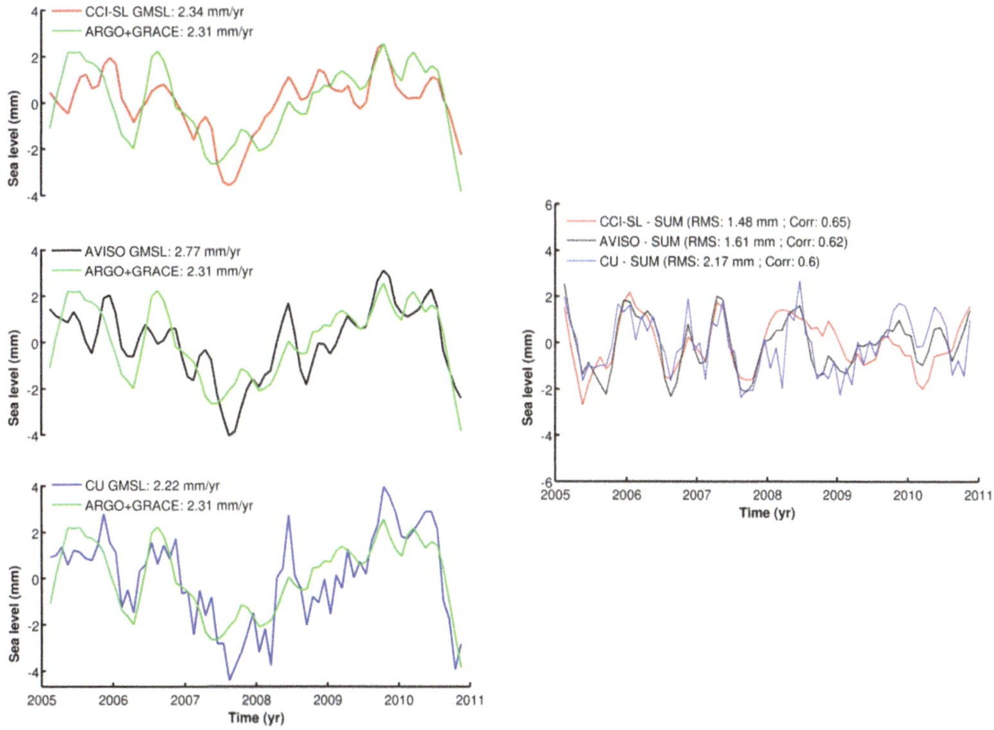

Figure 11. Sum of steric and ocean mass components based on Argo and Grace data (see Sect. 5.4) (green curve) over the January 2005–December 2010 time period and different GMSL products (left panels). Right panel: difference between the GMSL products and sum of components.

Table 2. Error budget of SL_cci products for the main climate scales.

Spatial scales	Temporal scales	Altimetry errors	User requirements
Global MSL	Long-term evolution (> 10 years)	$< 0.5\,\text{mm year}^{-1}$	$0.3\,\text{mm year}^{-1}$
	Interannual signals (< 5 years)	< 2 mm over 1 year	0.5 mm over 1 year
	Annual signals	< 1 mm	Not defined
Regional MSL	Long-term evolution (> 10 years)	$< 3\,\text{mm year}^{-1}$	$1\,\text{mm year}^{-1}$
	Annual signals	< 1 cm	Not defined

standards over 1993–2008 (Ablain et al., 2009). While small, this reduction is mainly due to a 2-year longer record as well as to the homogenization of the altimetry corrections between all the missions. The main source of the error remains the radiometer wet tropospheric correction with a drift uncertainty in the range of 0.2–$0.3\,\text{mm year}^{-1}$ (Legeais et al., 2014). To a lesser extent, the orbit error (Couhert et al., 2015) and the altimeter parameters (range, sigma-0, SWH) instabilities (Ablain et al., 2012) also add uncertainty, of the order of $0.1\,\text{mm year}^{-1}$. Notice that for these two corrections, the uncertainties are higher in the first altimetry decade (1993–2002) where TOPEX/Poseidon, ERS-1 and ERS-2 measurements display stronger errors (Ablain et al., 2013). Furthermore, imperfect links between TOPEX-A and TOPEX-B (February 1999), TOPEX-B and Jason-1 (April 2003), and Jason-1 and Jason-2 (October 2008) lead to the errors of 2 mm, 1 mm and 0.5 mm respectively (Ablain et al., 2009). They cause a GMSL trend error of about $0.15\,\text{mm year}^{-1}$ over the 1993–2010 period. Although the SL_cci project work has led to significant improvements, the remaining uncertainty of $0.5\,\text{mm year}^{-1}$ on the GMSL trend remains $0.2\,\text{mm year}^{-1}$ higher than the GCOS requirements (of $0.3\,\text{mm year}^{-1}$; see GCOS, 2011).

All sources of errors described above have also had an impact at the interannual timescale (< 5 years). Recent studies (Henry et al., 2013) have shown that the methodology applied to calculate sea level is particularly sensitive for the interannual scales (Henry et al., 2014). We estimated that the methodology uncertainty is on average ~ 2 mm over a 1-year period. Although improvements have been made, this level of error is still 1.5 mm higher than the GCOS requirement of (0.5 mm). This may have consequences on the sea level closure budget studies at the interannual timescale. For the annual signal, the amplitude error was estimated to be < 1 mm. Knowing that the annual amplitude of the GMSL is of the order of 9 mm, we can consider this error to be low. Notice that no requirement has yet been defined by GCOS for the periodic signals (at global and regional scales).

At the regional scale, the regional trend uncertainty is of the order of 2–3 mm year^{-1}. Although the orbit error has been significantly reduced for this spatial scale, it remains the main source of the error (in the range of 1–2 mm year^{-1}; Couhert et al., 2015) with large spatial patterns at hemispheric scale. The earth gravity field model errors explain an important part of these uncertainties (Rudenko et al., 2014). Furthermore, errors are higher in the first decade (1993–2002), where the earth gravity field models are less accurate due to the unavailability of the GRACE data before 2002. Additional errors are still observed; for example, for the radiometer-based wet tropospheric correction in tropical areas, other atmospheric corrections in high latitudes, and high-frequency corrections in coastal areas. The combined errors give rise to an uncertainty of 0.5–1.5 mm year^{-1}. Finally, the 2–3 mm year^{-1} uncertainty on regional sea level trends remains a significant error compared to the 1 mm year^{-1} GCOS requirement, even if this project has led to a 0.5 to 1.5 mm year^{-1} reduction (Fig. 6).

7 Conclusions and perspectives

Several groups (AVISO, University of Colorado, CSIRO, JPL (Jet Propulsion Laboratory), etc.) are currently processing satellite altimetry data to provide sea level products to user for climate applications. Within the SL_cci project, we have continued to improve the multi-mission sea level products over the altimetry era (1993–2010) through the development and computation of new corrections listed in Table 1. As far as possible, we have homogenized these corrections between all the missions in order to reduce the sources of discrepancies. Thanks to our formal validation protocol, we have been able to select the best corrections and algorithms applied in the sea level calculation. We have produced new sea level products and additional indicators over the 1993–2010 period. The SL_cci products exhibit improvements of different levels of importance for climate studies. Some of them are substantial, for instance for the estimation of the regional sea level trends, with an error reduction of 0.5–1.5 mm year^{-1} with large correlated spatial patterns. In parallel, the uncertainties of altimetry sea level have been better characterized and the sea level user requirements refined for climate applications.

The validation exercise has demonstrated that the existence of an additional good-quality sea level record has value in itself. Firstly, it clearly shows that the AVISO and SL_cci altimeter-derived sea level gridded products are robust (small uncertainty compared with the model error) and able to identify model improvements. Therefore they are a suitable data set to define metrics in the validation of ocean and climate models. SL_cci can be treated as an independent data set for verification. It has been used in the recent inter-comparison of ocean reanalyses ORAIP (Balmaseda et al., 2014; Hernandez et al., 2014). Preliminary results show that the SL_cci is closer to the ensemble mean of ocean reanalyses (a robust estimator) than its predecessor AVISO, and suggest that some ocean reanalyses that assimilate AVISO may over-fit the altimeter data. Model outputs using ocean assimilation techniques also provide independent sea level estimations that can be used to validate the SL_cci. Results obtained in the frame of the SL_cci project show that the low-frequency variability and trends of SL_cci agree better with ocean data assimilation estimators than with AVISO, especially in the Southern Ocean, the eastern Pacific, and coastal areas.

However, while a lot of improvements have been made, the user requirements are not yet reached. Remaining uncertainties are still 0.2 and 1–2 mm year^{-1} higher than the GCOS requirements for the GMSL trend and regional trends respectively. Similarly, the sea level error over a 1-year period is about 2 mm on average instead of the required 0.5 mm. Therefore it is still necessary to continue to improve the sea level time series to better understand key scientific issues, as raised in the abstract. Several ways of making improvements have already been identified and will be implemented during phase 2 of SL_cci project (January 2014 to December 2016).

For example, we plan to extend the sea level time series beyond 2010 using the same sea level corrections. By the end of year 2014, the current CCI_SL release will be extended until 2013 (included). And each subsequent year, we will extend the time series by 1 year. Additional improvements will be implemented; in particular, new orbit solutions, use of new atmospheric reanalyses based on the ERA-Clim project (Dee, 2014), new ocean tides, new radiometer-based wet troposphere corrections with improved long-term stability, etc. Furthermore, several level-2 altimetry data reprocessing activities are already planned by space agencies (CNES, NASA, ESA) for Jason-1, TOPEX/Poseidon, Envisat, and ERS missions, allowing us to benefit from homogenized data for both instrumental parameters and geophysical corrections. In addition, we intend to account for new altimeter missions already in orbit (CryoSat-2, SARAL/Altika) or to be launched in the near future (Jason-3, Sentinel-3). They are all relevant to extending the sea level time series with the same level of accuracy, and to improving coastal and high-latitude areas, which are of great interest for climate studies. Dedicated analyses will be performed in the Arctic region in order to improve sea level estimates near or under sea ice where no data are currently available. In parallel, we will continue to refine the user requirements, further developing the link with users and space agencies. This will include a quantification of the requirements for accuracy and long-term stability for climate-quality observations of sea level in the coastal zone, a key area for climate change. We also would like to refine the budget error with the new measurements and the new corrections. Lastly, to continuously answer user needs, we will produce by the end of 2016 a new, improved sea level time series covering the 1993–2015 period.

Acknowledgements. This work was performed in the framework of the ESA CCI program supported by ESA. It was also made possible thanks to the support of CNES for several years in altimetry data processing; in particular, thanks to the use of SSALTO/DUACS system developed in the framework of the SALP project (Service d'Altimétrie et de Localisation Précise). We would also like to thank all contributors to this project who have participated actively in the SL_cci project, with special recognition to S. Dinardo and B. M. Lucas in support of ESA, for their diligent reviewing of all the documents and data sets produced by the SL_cci team.

Edited by: A. Schiller

References

Ablain, M., Cazenave, A., Valladeau, G., and Guinehut, S.: A new assessment of the error budget of global mean sea level rate estimated by satellite altimetry over 1993–2008, Ocean Sci., 5, 193–201, doi:10.5194/os-5-193-2009, 2009.

Ablain, M., Philipps, S., Urvoy, M., Tran, N., and Picot, N.: Detection of Long-Term Instabilities on Altimeter Backscattering Coefficient Thanks to Wind Speed Data Comparisons from Altimeters and Models, Mar. Geodesy, 35, 42–60, doi:10.1080/01490419.2012.718675, 2012.

Ablain, M., Ollivier, A., Philipps, S., and Picot, N.: Why altimetry errors at climate scales are larger in the first decade (1993–2002)?, OSTST Boulders, October 2013, available at: http://www.aviso.altimetry.fr/fileadmin/documents/OSTST/2013/posters/Ablain_AltimetryErrorAtClimateScales.pdf (last access: 20 June 2014), 2013.

Altamimi, Z., Collilieux, X., and Métivier, L.: ITRF2008: An improved solution of the International Terrestrial Reference Frame, J. Geodesy, 85, 457–473, doi:10.1007/s00190-011-0444-4, 2011.

Andersen, O. B.: The DTU10 Gravity field and Mean sea surface (2010) Second international symposium of the gravity field of the Earth (IGFS2), 20–22 September 2010, Fairbanks, Alaska: available at: http://www.space.dtu.dk/english/~/media/Institutter/Space/English/scientific_data_and_models/global_marine_gravity_field/dtu10.ashx (last access: 20 June 2014), 2010.

Balmaseda, M. A., Mogensen, K., and Weaver, A. T.: Evaluation of the ECMWF ocean reanalysis system ORAS4, Q. J. R. Meteorol. Soc., 139, 1132–1161, doi:10.1002/qj.2063, 2013.

Balmaseda, M. A., Hernandez, F., Storto, A., Palmer, M. D., Alves, O., Shi, L., Smith, G. C., Toyoda, T., Valdivieso, M., Barnier, B., Behringer, D., Boyer, T., Chang, Y-S., Chepurin, G. A., Ferry, N., Forget, G., Fujii, Y., Good, S., Guinehut, S., Haines, K., Ishikawa Y., Keeley, S., Köhl, A., Lee, T., Martin, M., Masina, S., Masuda, S., Meyssignac, B., Mogensen, K., Parent, L., Peterson, K. A., Tang, Y. M., Yin, Y., Vernieres, G., Wang, X., Waters, J., Wedd, R., Wang, O., Xue, Y., Chevallier, M., Lemieux, J.-F., Dupont, F., Kuragano, T., Kamachi, M., Awaji, T., Caltabiano, A., Wilmer-Becker, K., and Gaillard, F.: The Ocean Reanalyses Intercomparison Project (ORA-IP), J. Operational Oceanography, submitted, 2014.

Brown, S., Desai, S., Keihm, S., and Lu, W.: Microwave radiometer calibration on decadal time scales using on-earth brightness temperature references: application to the TOPEX Microwave radiometer, J. Atmos. Oceanic Technol., 26, 2579–2591, 2009.

Carrere, L., Faugère, Y., and Ablain, M.: Improvement of pressure derived corrections for altimetry using the ERA-Interim dataset, Ocean Sci. Discuss. in preparation, 2014.

Chambers, D. P. and Bonin, J. A.: Evaluation of Release-05 GRACE time-variable gravity coefficients over the ocean, Ocean Sci., 8, 859–868, doi:10.5194/os-8-859-2012, 2012.

Couhert, A., Cerri, L., Legeais, J.-F., Ablain, M., Zelensky, N. P., Haines, B. J., Lemoine, F. G., Bertiger, W. I., Desai, S. D., and Otten, M.: Towards the 1 mm/y stability of the radial orbit error at regional scales, Adv. Space Res., 55, 2–23, doi:10.1016/j.asr.2014.06.041, 2015.

Dee, D. P.: ERA-CLIM: Final Publishable Summary Report, available at: http://www.google.fr/url?sa=t&rct=j&q=&esrc=s&source=web&cd=1&ved=0CCMQFjAA&url=httpAFFwww.era-clim.euFFR_20140228.pdf&ei=ifKsVM6bLMuAUaSfhNAJ&usg=AFQjCNHMvrEekOqszSUT_Jti_8n3oF1saw&sig2=yGNI0LhN3NBThCwNsqAC1A&bvm=bv.83134100,d.d24 (last access: 18 August 2014), 2014.

Dee, D. P., Uppala, S. M., Simmons, A. J., Berrisford, P., Poli, P., Kobayashi, S., Andrae, U., Balmaseda, M. A., Balsamo, G., Bauer, P., Bechtold, P., Beljaars, A. C. M., van de Berg, L., Bidlot, J., Bormann, N., Delsol, C., Dragani, R., Fuentes, M., Geer, A. J., Haimberger, L., Healy, S. B., Hersbach, H., Hólm, E. V., Isaksen, L., Kållberg, P., Köhler, M., Matricardi, M., McNally, A. P., Monge-Sanz, B. M., Morcrette, J.-J., Park, B.-K., Peubey, C., de Rosnay, P., Tavolato, C., Thépaut, J.-N., and Vitart, F.: The ERA-Interim reanalysis: configuration and performance of the data assimilation system, Q. J. R. Meteorol. Soc., 137, 553–597, doi:10.1002/qj.828, 2011

Dibarboure, G., Pujol, M.-I., Briol, F., Le Traon, P. Y., Larnicol, G., Picot, N., Mertz, F., and Ablain, M.: Jason-2 in DUACS: Updated System Description, First Tandem Results and Impact on Processing and Products, Mar. Geodesy, 34, 214–241, 2011.

Ducet, N., Le Traon, P. Y., and Reverdin, G.: Global high resolution mapping of ocean circulation from the combination of TOPEX/POSEIDON and ERS-1/2, J. Geophys. Res.-Ocean., 105, 19477–19498, 2000.

Fernandes, M. J., Lázaro, C., Nunes, A. L., Pires, N., Bastos, L., and Mendes, V. B.: GNSS-derived Path Delay: an approach to compute the wet tropospheric correction for coastal altimetry, IEEE Geosci. Remote Sens Lett., 7, 596–600, doi:10.1109/LGRS.2010.2042425, 2010.

Fernandes, M. J., Lázaro, C., and Nunes, A. L.: Improved Wet Path Delays for all ESA and Reference altimetric missions, Remote Sens. Environ., in preparation, 2014.

Garcia, P. and Roca, M.: ISARD_ESA_L1B_ESL_CCN_PRO_064, issue 1.b, 15 November 2010, "On-board PTR processing analysis: MSL drift differences", 2010.

GCOS: Systematic Observation Requirements for Satellite-Based Data Products for Climate (2011 Update) – Supplemental details to the satellite-based component of the "Implementation Plan for the Global Observing System for Climate in Support of the UNFCCC (2010 Update)", GCOS-154, WMO, December 2011.

Henry, O., Ablain, M., Meyssignac, B., Cazenave, A., Masters, D., Nerem, S., and Garric, G.: Effect of the processing methodology on satellite altimetry-based global mean sea level rise over the Jason-1 operating period, J. Geodesy, 88, 351–361, doi:10.1007/s00190-013-0687-3, 2014.

Hernandez, F., Ferry, N., Balmaseda, M., Chang, Y.-S., Chepurin, G., Fujii, Y., Guinehut, S., Kohl, A., Martin, M., Meyssignac, B., Parent, L., Peterson, K. A., Storto, A., Toyoda, T., Valdivieso, M., Vernieres, G., Wang, O., Wang, X., Xue, Y., and Yin, Y.: Sea level intercomparison: Inital resutls, Clivar Exchanges, 64, 18–21, 2014.

Kölh, A.: Evaluation of the GECCO2 Ocean Synthesis: Transports of Volume, Heat and Freshwater in the Atlantic, Q. J. Roy. Meteor. Soc., doi:10.1002/qj.2347, 2014.

Laxon, S. W., Giles, K. A., Ridout, A. L., Wingham, D. J., Willatt, R., Cullen, R., Kwok, R., Schweiger, A., Zhang, J., Haas, C., Hendricks, S., Krishfield, R., Kurtz, N., Farrell, S., and Davidson, M.: CryoSat-2 estimates of Arctic sea ice thickness and volume, Geophys. Res. Lett., 40, 732–737, doi:10.1002/grl.50193, 2013.

Legeais, J.-F., Ablain, M., and Thao, S.: Evaluation of wet troposphere path delays from atmospheric reanalyses and radiometers and their impact on the altimeter sea level, Ocean Sci., 10, 893–905, doi:10.5194/os-10-893-2014, 2014.

Le Traon, P. Y., Faugère, Y., Hernandez, F., Dorandeu, J., Mertz, F., and Ablain, M.: Can we merge GEOSAT Follow-On with TOPEX/POSEIDON and ERS-2 for an improved description of the ocean circulation?, J. Atmos. Ocean. Technol., 20, 889–895, 2003.

Morrison, H., Zuidema, P., Ackerman, A. S., Avramov, A., de Boer, G., Fan, J., Fridlind, A. M., Hashino, T., Harrington, J. Y., Luo, Y., Ovchinnikov, M., and Shipway, B.: Intercomparison of cloud model simulations of Arctic mixed-phase boundary layer clouds observed during SHEBA/FIRE-ACE, J. Adv. Model. Earth Syst., 3, M06003, doi:10.1029/2011MS000066, 2011.

Ollivier, A., Faugère, Y., Picot, N., Ablain, M., Femenias, P., and Benveniste, J.: Envisat Ocean Altimeter Becoming Relevant for Mean Sea Level Trend Studies, Mar. Geodesy, 35, 118–136, 2012.

Philipps, S. and Roinard, H.: Jason-2 reprocessing impact on ocean data (cycles 001 to 145), available at: http://www.aviso.altimetry.fr/fileadmin/documents/calval/validation_report/J2/Jason2ReprocessingReport-v2.1.pdf (last access: 20 June 2014), 2013.

Pujol, M. I., Faugère, Y., Ablain, M., Larnicol, G., Picot, N., and Bronner, E.: 20 years of High Resolution DUACS/Aviso Products Reprocessed, Mar. Geodesy, in preparation, 2014.

Ray, R. D.: Precise comparisons of bottom-pressure and altimetric ocean tides, J. Geophys. Res.-Ocean., 118, 4570–4584, doi:10.1002/jgrc.20336, 2013.

Roca, M. and Thibaut, P.: ISARD_CLS_SALP_PTR_TN_042, Issue 1.a, 19 November 2009, "PTR study(3) -SALP Project- ", 2009.

Rudenko, S., Otten, M., Visser, P., Scharroo, R., Schöne, T., and Esselborn, S.: New improved orbit solutions for the ERS-1 and ERS-2 satellites, Adv. Space Res., 49, 1229–1244, 2012.

Rudenko, S., Dettmering, D., Esselborn, S., Schöne, T., Förste, Ch., Lemoine, J.-M., Ablain, M., Alexandre, D., and Neumayer, K.-H.: Influence of time variable geopotential models on precise orbits of altimetry satellites, global and regional mean sea level trends, Adv. Space Res., 54, 92–118, doi:10.1016/j.asr.2014.03.010, 2014.

Scharroo, R. and Smith, W. H. F.: A global positioning system based climatology for the total electron content in the ionosphere, J. Geophys. Res., 115, A10318, doi:10.1029/2009JA014719, 2010.

SL_cci Product Validation Plan (PVP): available at: http://www.esa-sealevel-cci.org/webfm_send/101, last access: 20 June 2014.

SL_cci Product Specification Document (PSD): available at: http://www.esa-sealevel-cci.org/webfm_send/229, last access: 20 June 2014.

SL_cci Product User Guide (PUG): available at: http://www.esa-sealevel-cci.org/webfm_send/212, last access: 20 June 2014.

SL_cci User Requirement Document (URD): available at: http://www.esa-sealevel-cci.org/webfm_send/235, last access: 20 June 2014.

SL_cci Product Validation Internal Report (PVIR): available at: http://www.esa-sealevel-cci.org/webfm_send/234, last access: 20 June 2014.

SL_cci Validation Report: Executive Summary available at: http://www.esa-sealevel-cci.org/webfm_send/187, last access: 20 June 2014.

SL_cci Validation Report: Instrumental Corrections, available at: http://www.esa-sealevel-cci.org/webfm_send/175, last access: 20 June 2014.

SL_cci Validation Report: Orbit Calculation, available at: http://www.esa-sealevel-cci.org/webfm_send/176, last access: 20 June 2014.

SL_cci Validation Report: Wet Troposphere Correction, available at: http://www.esa-sealevel-cci.org/webfm_send/177, last access: 20 June 2014.

SL_cci Validation Report: Atmospherical corrections (Dynamical, Atmospherical correction, Inverse Barometer, Dry Troposphere) available at: http://www.esa-sealevel-cci.org/webfm_send/181, last access: 20 June 2014.

SL_cci Validation Report: Ionosphere correction, available at: http://www.esa-sealevel-cci.org/webfm_send/178, last access: 20 June 2014.

SL_cci Validation Report: Sea State Bias correction, available at: http://www.esa-sealevel-cci.org/webfm_send/179, last access: 20 June 2014.

SL_cci Validation Report: Ocean tides correction, available at: http://www.esa-sealevel-cci.org/webfm_send/180, last access: 20 June 2014.

SL_cci Validation Report: Regional SSH bias corrections between altimetry missions, available at: http://www.esa-sealevel-cci.org/webfm_send/182, last access: 20 June 2014.

SL_cci Validation Report: Impact of altimeter satellite constellation, available at: http://www.esa-sealevel-cci.org/webfm_send/183, last access: 20 June 2014.

SL_cci Validation Report: Mapping Methods, available at: http://www.esa-sealevel-cci.org/webfm_send/184, last access: 20 June 2014.

SL_cci Validation Report: High latitudes areas, available at: http://www.esa-sealevel-cci.org/webfm_send/185, last access: 20 June 2014.

SL_cci Validation Report: Coastal areas, available at: http://www.esa-sealevel-cci.org/webfm_send/186, last access: 20 June 2014.

Thao, S., Eymard, L., Obligis, E., and Picard, B.: Trend and Variability of the Atmospheric Water Vapor: a Mean Sea Level issue, J. Atmos. Ocean. Technol., 31 1881–1901, 2014.

Thibaut, P.: PTR position and drift analysis – Theoretical Approach", RA-2 Quality Working Group presentation, Reading, UK, May 2010.

Valladeau, G., Legeais, J. F., Ablain, M., Guinehut, S., and Picot, N.: Comparing Altimetry with Tide Gauges and Argo Profiling Floats for Data Quality Assessment and Mean Sea Level Studies, Mar. Geodesy, 35, 42–60, 2012.

von Schuckmann, K. and Le Traon, P.-Y.: How well can we derive Global Ocean Indicators from Argo data?, Ocean Sci., 7, 783–791, doi:10.5194/os-7-783-2011, 2011.

An alternative method for correcting fluorescence quenching

L. Biermann[1], C. Guinet[2], M. Bester[3], A. Brierley[4], and L. Boehme[1]

[1]Sea Mammal Research Unit, Scottish Oceans Institute, St. Andrews, UK
[2]Centre National de la Recherche Scientifique, Centre d'Etudes Biologiques de Chizé, Villiers en Bois, France
[3]Mammal Research Institute, Department of Zoology and Entomology, University of Pretoria, Pretoria, South Africa
[4]Pelagic Ecology Research Group, Scottish Oceans Institute, St. Andrews, UK

Correspondence to: L. Biermann (lb66@st-andrews.ac.uk)

Abstract. Under high light intensity, phytoplankton protect their photosystems from bleaching through non-photochemical quenching processes. The consequence of this is suppression of fluorescence emission, which must be corrected when measuring in situ yield with fluorometers. We present data from the Southern Ocean, collected over five austral summers by 19 southern elephant seals tagged with fluorometers. Conventionally, fluorescence data collected during the day (quenched) were corrected using the limit of the mixed layer, assuming that phytoplankton are uniformly mixed from the surface to this depth. However, distinct deep fluorescence maxima were measured in approximately 30 % of the night (unquenched) data. To account for the evidence that chlorophyll is not uniformly mixed in the upper layer, we propose correcting from the limit of the euphotic zone, defined as the depth at which photosynthetically available radiation is ~ 1 % of the surface value. Mixed layer depth exceeded euphotic depth over 80 % of the time. Under these conditions, quenching was corrected from the depth of the remotely derived euphotic zone Z_{eu}, and compared with fluorescence corrected from the depth of the density-derived mixed layer. Deep fluorescence maxima were evident in only 10 % of the day data when correcting from mixed layer depth. This was doubled to 21 % when correcting from Z_{eu}, more closely matching the unquenched (night) data. Furthermore, correcting from Z_{eu} served to conserve non-uniform chlorophyll features found between the 1 % light level and mixed layer depth.

1 Introduction

Monitoring distribution and abundance of primary producers in the marine environment is useful for understanding larger biological and physical processes. Phytoplankton are predominantly single-celled, green microscopic organisms, present at variable concentrations in every ocean (Falkowski and Kolber, 1995; Behrenfeld et al., 2009). They are the first stepping stone in transferring energy into the marine ecosystem. When conditions allow for growth, their collective impact is such that they are able to change the spectral properties of the water (McClain, 2009). This can be exploited to measure abundance and distribution using their photosynthetic pigment, chlorophyll *a* (Chl *a*) as a marker (Morel and Prieur, 1977; Gordon and Morel, 1983; Falkowski et al., 1998; Henson et al., 2010).

Satellite-derived ocean colour is the most comprehensive data set available for monitoring surface Chl *a* concentration. However, the limitations of remote sensing are significant (Dierssen and Smith, 2000; McClain, 2009). Perhaps most significantly, satellite sensors can only see ocean colour from the surface to one optical depth, providing little information on the vertical structure of the water column (Morel and Berthon, 1989). Despite innovative algorithms to improve data collected by satellites, limitations are not likely to be resolved with the next generation of ocean colour sensors. Continued collection of in situ data on phytoplankton distribution and abundance is thus essential (Johnson et al., 2009).

Fluorescence has been widely used as a relatively inexpensive, non-invasive method for quantifying Chl *a* since the 1960s (Lorenzen, 1966; Lorenzen and Jeffrey, 1980; Cullen and Eppley, 1981; Falkowski and Kolber, 1995; Xing et al., 2011; Lavigne et al., 2012). Chlorophyll pigments packaged inside phytoplankton cells re-radiate ~ 2 % of light energy as fluorescence. Active fluorescence can thus be measured by a fluorometer delivering voltage output equivalent to 460 nm (excitation in the blue) and measuring resultant fluorescence in the 620–715 nm range (detection in the red). Assuming that the measured yield is proportional to the abundance of photosynthesising phytoplankton, relative values of fluorescence can be used to quantify primary biomass. However, yield (and thus proportionality) is affected by several factors, including the intensity of sunlight each cell is exposed to (Behrenfeld and Boss, 2006).

To regulate high sunlight intensity, phytoplankton employ the mechanism of non-photochemical quenching. This process helps cells protect themselves in environments where light energy absorption exceeds the capacity for light utilisation (Müller et al., 2001; Behrenfeld et al., 2009). During periods of high light stress, shallow-mixed phytoplankton in the upper euphotic zone protect their photosystems from bleaching by emitting excess energy as heat (Milligan et al., 2012). In this state of photo-protection, photosynthesis is inhibited and fluorescence yield drops (Müller et al., 2001). However, deep-mixed phytoplankton are shaded by surface biomass and protected by light-attenuating properties of the water itself. Thus, during high sunlight intensity, deeply mixed phytoplankton will fluoresce while shallow-mixed biomass will be invisible to a fluorometer. If uncorrected, quenched fluorescence yields will generate values under-representative of phytoplankton abundance at the surface. Avoiding such under-representation is particularly important for studies comparing satellite-derived surface Chl *a* with in situ fluorescence-derived Chl *a*.

On the vertical scale, quenched fluorescence profiles are also problematic in that they mimic the shape of true deep fluorescence maxima (DFM) (Mignot et al., 2011). Sub-surface fluorescence maxima can be indicative of deep chlorophyll maxima (DCM). These features either form due to differences in Chl *a* packaging, or when the bulk of phytoplankton biomass settles to depths where both nutrients and light are sufficient (Cullen and Eppley, 1981; Holm-Hansen and Hewes, 2004; Mellard et al., 2011). Because DCM are found below one optical depth, abundance and distribution of such features cannot be measured by satellite (McClain, 2009; Charrassin et al., 2010). However, vertical distributions of phytoplankton play important roles in the organisation of pelagic trophic food webs (Mellard et al., 2011). DCM in particular play potentially under-reported roles in net primary production and carbon fixation (Fairbanks et al., 1982; Estrada et al., 1993; Claustre et al., 2008). It is therefore useful to try to accurately identify and map these features.

In previous work on glider data, correcting for quenching involved using backscatter (Sackmann et al., 2008) or surface light intensity (Todd et al., 2009), each measured simultaneously to fluorescence. For autonomous platforms collecting only fluorescence, salinity and temperature data in remote locations, this is not possible. Using the depth of the density-derived mixed layer, Xing et al. (2011) corrected quenched fluorescence data collected by Argo floats in Pacific, Atlantic and Mediterranean offshore zones. This mixed layer depth (MLD) method was then extended to fluorescence data collected by tagged southern elephant seals (*Mirounga leonina*) off Kerguelen Island in the Southern Ocean (Xing et al., 2012; Guinet et al., 2013). Following the method of Xing et al. (2012), the maximum fluorescence yield within the density-derived mixed layer would be representative of the whole, assuming homogeneity. How robust is this assumption for the waters of the Southern Ocean?

The upper layer of the Southern Ocean tends to be deeply mixed, even over austral summers, and DCM are thought to be rare (discussed in Holm-Hansen and Hewes, 2004). However, ship-based studies undertaken in different regions of the Southern Ocean have shown that mixing and settling patterns within and between functional types are dynamic (Holm-Hansen and Hewes, 2004; Mengesha et al., 1998; Quéguiner, 2013; Sangrà et al., 2014). This is especially true for motile (flagellated) and seasonally successive (heavily silicified) species (Mengesha et al., 1998; Quéguiner, 2013). Furthermore, even within homogenously mixed layers of biomass, Chl *a* packaging has been shown to vary with depth, generating DCM (and DFM) independent of biomass (Behrenfeld and Boss, 2006). Reliance on the assumption of homogeneity is, therefore, possibly problematic.

In this paper, we present an alternative method to account for quenching without relying on homogeneity in the density-derived upper layer. Non-photochemical quenching is corrected using the limit of the euphotic zone (Z_{eu}), defined as the depth at which downward photosynthetic available radiation (PAR) is at ~ 1 % of the surface value (Shang et al., 2011; Soppa et al., 2013). At this level, light should be sufficient for photosynthesis (Ryther and Menzel, 1959; Morel and Berthon, 1989; Saulquin et al., 2013; Palmer et al., 2013) but too weak to cause quenching (Alderkamp et al., 2011; Ross et al., 2011). This method is applied to fluorescence data collected by 19 animal-borne fluorometer, conductivity, temperature and depth satellite-relayed data loggers (FCTD-SRDLs) deployed in the Southern Ocean over several austral summers.

2 Methods

2.1 In situ data

Between 2009 and 2013 18 adult female southern elephant seals from Kerguelen and one from Marion Island were equipped with FCTD-SRDLs (Sea Mammal Research Unit, University of St Andrews, UK, Scotland) before they undertook their post-breeding foraging migration over the austral summer (Fig. 1). Following established tagging protocols

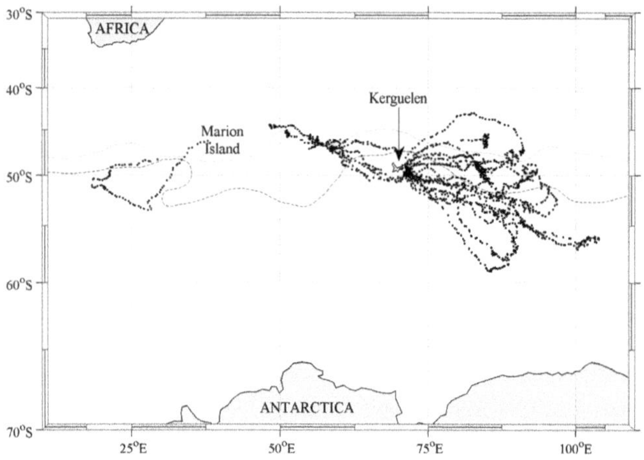

Figure 1. Tracks of 19 southern elephant seals tagged with FCTD-SRDLs over austral summers from 2009 to 2013 in black. The general positions of the subantarctic front and the polar front are shown in green and blue, respectively (Orsi and Harris, 2014). The incomplete track seen off Marion Island is due to the FCTD-SRDL battery failing after 55 days.

(Bester, 1988; McIntyre et al., 2010), the seal on Marion Island was immobilised with ketamine using a remote injection method and the tag was glued to the fur on the head with quick-setting epoxy resin. Seals on Kerguelen were anesthetised with an intravenous injection of tiletamine and zolazepam 1 : 1 and tags were glued to the fur on the head using a two component industrial epoxy (Jaud et al., 2012).

The FCTD-SRDL instrument records behavioural data as well as in situ pressure, temperature, salinity and fluorescence (Charrassin et al., 2010; Xing et al., 2012). At-sea data were relayed via the Argos satellite system (http://www.argos-system.org). Locations were provided by Service Argos based on Doppler shift measurements and data were downloaded from the Sea Mammal Research Unit Instrumentation Group's website (http://www.smru.st-andrews.ac.uk/). Detailed information on the hardware and software of the CTD-SRDL is described by Boehme et al. (2009), and on-board data processing is described comprehensively by Fedak et al. (2002).

Fluorescence is recorded by a Cyclops 7 fluorometer (Turner Designs, CA, USA), which delivers a voltage output proportional to the fluorescence detected in a wavelength between 620 and 715 nm. The Cyclops instrument is programmed to measure fluorescence every 2 s during the ascent (upcast), from 180 m to the surface. Due to the tight limits on data transfer through the Argos satellite system, data have to be compressed (Boehme et al., 2009). Fluorescence yields were therefore binned into 10 m vertical intervals. The reading for 175 m is thus a weighted mean of all fluorescence readings taken between 180 and 170 m. This is the deepest bin available – deep and dark enough to anticipate an absence of live, photosynthesising phytoplankton (Guinet et al., 2013). However, instruments do not return readings of zero at these depths. This dark count is an offset value added by the manufacturers, and is useful because at very low signal, readings are indistinguishable from noise. The offset value increases the signal to noise ratio, which must be removed during data processing (Lavigne et al., 2012). We therefore calculated the median of readings at 175 m and subtracted this from all measurements collected by the same tag (Xing et al., 2011). This is not only useful for making the fluorescence values more representative, it also serves to reduce variability between tags (Xing et al., 2012). The resulting values are considered proportional to fluorescence, and are termed relative fluorescence units (RFU).

2.2 Satellite data

The depth of Z_{eu} reflects the limit where PAR is 1 % of its surface value. In this study, we use an estimate of the euphotic zone Z_{eu} based on an algorithm by Lee et al. (2007). These estimates are derived from measurements of the water-leaving radiance by the Moderate Imaging Spectrometer (MODIS) as 4 km×4 km monthly composites (http://oceancolor.gsfc.nasa.gov).

To optimise temporal and spatial coverage in a region as cloudy as the Southern Ocean, values of L3 8 day Z_{eu} were used whenever available, and values from monthly composites were then added to fill in the blanks. However, despite the broad coverage in time, gaps were still present in space. For each fluorometer profile the closest Z_{eu} value in space was extracted from the gridded data set. If no value was available, Z_{eu} was interpolated linearly from the nearest seal position with an associated Z_{eu} value.

2.3 Correcting for quenching

Quenching occurs in surface waters during periods of high light stress. Despite daily variability attributed to cloud (changing light intensity), and differences in phytoplankton concentrations between years and regions, suppression of fluorescence is a ubiquitous feature. Fitting a non-linear model (sin function) to surface fluorescence as a function of time of day showed that quenching around midday was significant ($n = 1267$, $R^2 = 0.26$, $P < 0.001$) (Supplement Fig. S1). Data were thus categorised as day: possibly quenched (from sunrise to sunset, with the sun 8 degrees above the horizon) or night: unquenched. To illustrate the quenching effect without interannual and regional variability, day yields were divided by the preceding night yields to generate ratios. Data were then binned by hour and plotted using box plots (Fig. 2).

Before processing, a small proportion of profiles with maximum fluorescence yields below 0.15 RFU were removed from the data set. These low signals tend to fall within the noise, and the resulting vertical profiles cannot be meaningfully interpreted.

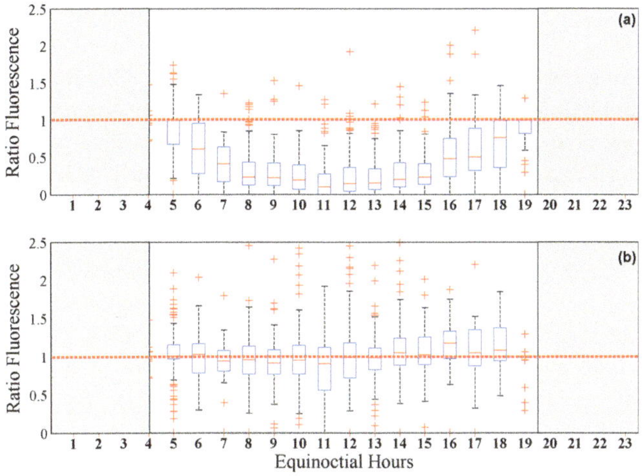

Figure 2. Ratio of day to night surface fluorescence yield (first 15 m) per local hour from data collected by 19 animal-borne fluorometers over five austral summers. Box plots show the distribution of day to night ratios in **(a)**, representing quenched to unquenched yield, respectively. The horizontal red line at 1 indicates the 1 : 1 ratio. Ratios of fluorescence yield decrease from around 06:00 until around 18:00 LT, then recover to the 1 : 1 ratio line. Corrected ratios of fluorescence in **(b)** do not follow the same trend of deviation from the 1 : 1 ratio line at 1. Flanking grey sections represent times of the day where the sun is less than 6 degrees below the horizon.

The night: unquenched data were then normalised and each profile was interrogated for maximum values below the near-surface yield. To ensure a deep maximum value is not within the measurement error for each tag, a 95 % confidence interval for the error was calculated for each instrument using bootstrapping (1000 bootstrap samples). The upper confidence limit was set as a threshold value, subtracted from the deep maximum and added to the near-surface value. Only differences outside of this threshold were flagged as potential DFM, and these profiles were then visually inspected. Vertical profiles conforming to the shape of DFM in this data set were considered true DFM, independent of quenching (Fig. S2).

For the day: quenched data, we compare the results of fluorescence quenching correction from two depths. The first is a density-derived MLD calculated from the concurrent CTD measurements of the FCTD-SRDL. Here, MLD is defined as the depth where the vertical density gradient equals or exceeds a threshold value of $\Delta\rho = 0.03\,\mathrm{kg\,m^{-3}}$; as employed by Xing et al. (2011, 2012) (de Boyer Montégut et al., 2004; Kara et al., 2000). The MLD in the region of interest tends to remain deeper than 50 m, even over the austral summer (Dong et al., 2008; Fig. 3). The second depth is the remotely derived Z_{eu} using Lee's algorithm (2007). Separately, maximum fluorescence yield between MLD and the surface, or Z_{eu} and the surface, were then extended to fill in suppressed yield (Fig. 3).

For over 80 % of the day data set, mixed layer depth was deeper than Z_{eu}. However, the small proportion (~ 20 %) of waters with MLD shallower than Z_{eu} proved problematic for correcting quenching. In these instances, correcting from MLD generates still-quenched values. Conversely, correcting from Z_{eu} would mask shallow but potentially true subsurface features. Where MLD shoaled above Z_{eu}, fluorescence data were thus flagged and removed from the data set.

3 Results

The at-sea phases of the 19 tagged southern elephant seals (Fig. 1) tended to start in November and end in late January of the following year, meaning data were collected over the height of the austral summer. In the top 10–15 m of the upper mixed layer across all years of sampling, suppression of daytime fluorescence yield was significant (Fig. 2a).

To correct for quenching in deeply mixed waters, surface values collected between dawn and dusk were filled in with maximum values from the depth of the mixed layer, or the depth of Z_{eu}. The difference between these methods is illustrated in Fig. 3, where individual normalised fluorescence profiles are treated with the two correction schemes. Corrected surface fluorescence yields are overestimated using the depth of the mixed layer, but perhaps more importantly, vertical complexity is lost. Correcting from Z_{eu} serves to conserve unusual (not homogenous) dynamics on the vertical scale. This is illustrated in Fig. 3b, where phyotoplankton may either be settling into different layers (floristic shifts) or where Chl *a* packaging has become measurably different between deep-mixed and shallow-mixed layers (photoacclimation).

The difference between the two methods is also evident looking at fluorescence sections in Fig. 4. Uncorrected vertical profiles in the top row (Fig. 4a, d and g) illustrate how evident suppression of daytime fluorescence yield is in the surface layers (greyscale bar with daylight hours in white and black for night). Quenching is corrected using MLD in the second row (Fig. 4b, e and h) and Z_{eu} in the third (Fig. 4c, f and i). The distinct subsurface fluorescence feature in (a) is within the mixed layer, and is thus masked when corrected from this depth (b). However, correcting from Z_{eu} conserves vertical complexity while still correcting quenching at the surface (c). The deep fluorescence feature in Fig. 4d also has a maximal yield below euphotic depth but within the mixed layer. This deep signal is masked when correcting from MLD (e) but remains distinct when using Z_{eu} (f). Finally, the dynamics in Fig. 4g are lost when correcting from the depth of the mixed layer (h) but conserved when correcting from Z_{eu} (i). The night data from this example in particular corroborates complex Chl *a* dynamics with depth.

Unquenched fluorescence profiles collected at night support the evidence that DCM should be accounted for in this oceanic regime. Of the 19 seals tagged with FCTD-SRDLs

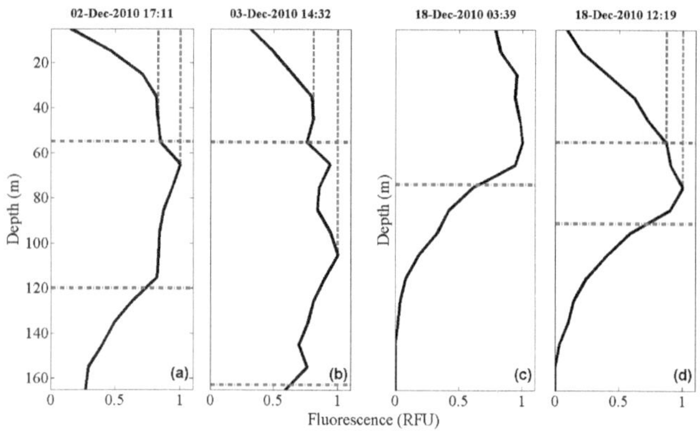

Figure 3. Normalised vertical profiles of fluorescence (solid black line) corrected using either the depth of the mixed layer (dashed red line) or the 1 % light level, Z_{eu} (dashed blue line). The fluorescence in profile **(c)** was measured at night and is unquenched (mixed layer is shown by the dashed red line but Z_{eu} is absent at night). Profiles **(a)**, **(b)** and **(d)** were collected at varying times during the day. For each profile, quenched fluorescence is corrected using either Z_{eu} (depth and corresponding fluorescence yield shown with dashed blue line) or MLD (depth and corresponding fluorescence yield shown with dashed red line).

between 2009 and 2013, 18 instruments measured DFM in the night (unquenched) data. For the combined night data set, approximately 30 % of the 797 profiles showed DFM (mean = 13; standard deviation = 9) (Fig. S3). Correcting with Z_{eu} doubled the number of DFM conserved in the day data (21 %) compared with MLD (10 %), but was still unable to fully match the percentage found in the night data.

4 Discussion

The phenomenon of non-photochemical quenching is well described and appears to be ubiquitous across oceans and seasons (Sackmann et al., 2008; Milligan et al., 2012). The depth where light levels are sufficient for photosynthesis but too weak to generate quenching is key for correcting in situ fluorescence data; ensuring fluorescence yield is representative of vertical Chl *a* concentrations. However, when fluorescence data are collected autonomously, this depth cannot always be measured in space and time.

For this study, removing quenched fluorescence would mean discarding 1352 of the 2149 profiles collected by 19 animal-borne tags over several austral summers in the Southern Ocean. Losing over 60 % of the data collected from such an undersampled region is simply not viable. Thus, two proxy depths are compared for quenching correction. The first is a density-derived MLD, calculated from CTD data measured concurrently by the tag (Xing et al., 2012; Guinet et al., 2013). The second is the depth of the remotely derived euphotic zone, Z_{eu} (Lee et al., 2007).

During daylight hours, a proportion of the euphotic zone (defined as the surface to the 1 % light level) will be light-saturated relative to photosynthetic efficiency (Behrenfeld and Falkowski, 1997). Shallow-mixed phytoplankton are exposed to a range of super-saturating light intensities, while deep-mixed phytoplankton will be exposed to low-levels of light intensity that are suboptimal for photosynthesis in the absence of photoadaptation.

Cells throughout any given euphotic zone adapt to variations in the light field through a range of physiological strategies (Arrigo et al., 2010; Kropuenske et al., 2009, 2010). These include non-photochemical quenching for high light stress, and a range of dark adaptation strategies below a saturating light threshold. Photosynthesis typically saturates at light intensities of ~200 mmol photons m^{-2} s^{-1} (van der Poll et al., 2009) and below ~100 mmol photons m^{-2} s^{-1} cells must initiate dark adaptation strategies (Garcia and Purdie, 1992). By definition, such dark-adapted phytoplankton at the base of the euphotic zone are not experiencing high light stress. However, light is still sufficient for photosynthesis. Indeed, Hameedi (1978), Platt et al. (1982) and Lee et al. (2010) have shown that shade-acclimated phytoplankton from the 1 % light level are more productive than phytoplankton from the 50 % light level (Alderkamp et al., 2011; Palmer et al., 2013).

At noon on a cloudless summer's day, the Southern Ocean can receive surface irradiance of up to 2000 mmol photons m^{-2} s^{-1} (Ross et al., 2008). Light saturation would be expected at 10 % surface light levels, while dark-adapted phytoplankton would be found at ~5 % of the maximal expected surface irradiance. Therefore, at the 1 % light level, phytoplankton are likely dark adapted (i.e. not exhibiting non-photochemical quenching) and actively photosynthesizing.

Composites of MODIS Z_{eu} are provided as evaluative products and these were used to determine euphotic depth, where light intensity would be too low to cause light stress and suppression of fluorescence yield. Lee's algorithm for Z_{eu} provided accurate and reliable estimates of Z_{eu} in open waters of the China Sea (Morel, 1988; Morel et al., 2007; Lee et al., 2007; Shang et al., 2011). We assume that this method can also reliably estimate the limit of the euphotic zone in the open waters of the Southern Ocean. However, validation with in situ light data at the high latitudes would be beneficial for measuring the accuracy of this product.

Combinations of 8 day and monthly Z_{eu} products were used to account for the gaps in coverage. Satellites cannot see through clouds, and the high latitudes are persistently cloudy. Despite the large temporal difference, 8 day and monthly products showed excellent agreement ($R^2 = 0.82$, $p < 0.0001$). This is less likely to be true for coastal or near-coastal regimes, where dynamics change on finer scales. For this reason, we would recommend using daily data combined with 8 day composites in these areas, or any region where cloud cover is less of a persistent problem. As a compro-

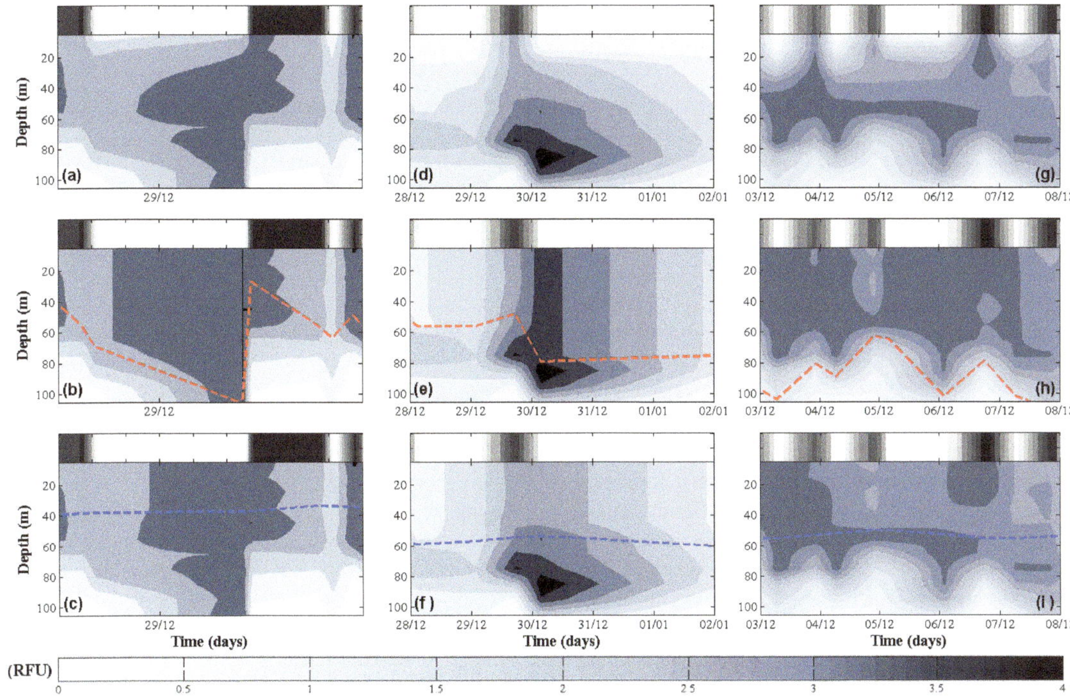

Figure 4. Sections of fluorescence corrected using either the depth of the mixed layer (dashed red line) or the depth of the 1 % light level, Z_{eu} (dashed blue line). The greyscale bar attached to the top of each plot illustrates when fluorescence data were collected relative to the time of day, with black for night and white for day. The top row of vertical profiles **(a)**, **(d)** and **(g)** show sections of uncorrected fluorescence data collected by different instrumented seals over austral summers. The fluorescence data in the second row **(b)**, **(e)** and **(h)** have been corrected using mixed layer depth. The last row of fluorescence data **(c)**, **(f)** and **(i)** have been corrected using euphotic depth. Conditions where MLD shoals above Z_{eu} cannot be corrected and have been filtered from the day data set, but not from the night (unquenched).

mise, generating 16-day composites using SeaDAS 7.0 (Fu et al., 1998) would address the temporal compromise as well as fill in some of the gaps. However, this step requires custom processing, rather than the use of freely available and pre-existing products. This may limit wider application of the method. Our aim is to create a method that is easily applied and universally useful for autonomously collected fluorescence data in regions where waters tend to be deeply mixed.

Surface waters of the Antarctic are deeply mixed and deep chlorophyll maxima were previously thought to be rare (Holm-Hansen and Hewes, 2004). However, while the vertical distribution of phytoplankton is controlled by turbulence and density, it is also driven by nutrient and light gradients (Navarro et al., 2006). Cell motility has been shown to prevail over dispersion by turbulence (Durham et al., 2013). When ideal levels of light and nutrients change with depth over time, so too will the vertical distribution of biomass in the water column (evolutionary stable strategy depth) (Mellard et al., 2011). Indeed, 19 of the 19 seals tagged with FCTD-SRDLs between 2009 and 2013 returned DFM in the night data. These in situ fluorescence data sets add to growing evidence that DCM may be regionally and seasonally persistent in the Southern Ocean (Holm-Hansen et al., 2005; Charrassin et al., 2010; Quéguiner, 2013).

Distinct DCM are only a part of the story, however. Heterogeneous phytoplankton patterns within layers of homogenous density are a common feature of the North Atlantic, resulting from slow mixing rates (relative to growth rates) during periods without deep convective mixing (Taylor and Ferrari, 2011). Although we have no information on mixing or turbulence, finding deep maxima in 30 % of the night data points to moderate or even weak turbulence in the upper layer (Ross et al., 2011; Franks, 2014). Correcting from Z_{eu} serves to conserve unusual (not homogenous) dynamics between the 1 % light level and MLD. From a biological perspective, this vertical information may provide useful insights into mixing and settling patterns of different phytoplankton species (floristic shifts), or differences in chlorophyll packaging in the same species (photoacclimation). However, the fact that DCM and non-uniform phytoplankton patterns persist over the five summers recorded may also be interesting from a purely physical perspective.

The problem of phytoplankton not being uniformly distributed in the mixed layer was commented on by Xing et al. (2011), but not addressed until now. The limitations of our own correction scheme would be improved with testing of the accuracy of the remotely sensed Z_{eu} product in Southern Ocean waters with in situ light data. Furthermore, until we are able to apply our quenching method to fluorescence data outside of the high latitudes, we are only able to suggest that using Z_{eu} improves the current method in the deeply mixed waters of the Southern Ocean.

Acknowledgements. This work received funding from the MASTS pooling initiative (The Marine Alliance for Science and Technology for Scotland) and their support is gratefully acknowledged. MASTS is funded by the Scottish Funding Council (grant reference HR09011) and contributing institutions. A number of field assistants were involved in the deployment of tags on Marion and we are particularly grateful to Mia Wege. FCTD-SRDL tags and fluorescence data were generously provided by CEBC-CNRS as part of the research program supported by CNES-TOSCA, ANR-VMC, Total Foundation and IPEV. The Mammal Research Institute, University of Pretoria, covered Marion Island Argos transmission costs and we thank Trevor McIntyre for facilitating this. We are grateful to Sandy Thomalla and other reviewers for significantly improving this manuscript. Special thanks to Drew Lucas, Mick Wu and Hayley Evers-King for their assistance with the editing, statistics and bio-optics, respectively.

Edited by: M. Hoppema

References

Alderkamp, A. C., Garcon, V., de Baar, H. J., and Arrigo, K. R.: Short-term photoacclimation effects on photoinhibition of phytoplankton in the Drake Passage (Southern Ocean), Deep Sea Res. Pt. I, 58, 943–955, doi:10.1016/j.dsr.2011.07.001, 2011.

Arrigo, K. R., Mills, M. M., Kropuenske, L. R., van Dijken, G. L., Alderkamp, A. C., and Robinson, D. H.: Photophysiology in two major Southern Ocean phytoplankton taxa: photosynthesis and growth of Phaeocystis antarctica and Fragilariopsis cylindrus under different irradiance levels, Integr. Comp. Biol., 50, 950–966, doi:10.1093/icb/icq021, 2010.

Behrenfeld, M. J. and Boss, E.: Beam attenuation and chlorophyll concentration as alternative optical indices of phytoplankton biomass, J. Mar. Res., 64, 431–451, doi:10.1357/002224006778189563, 2006.

Behrenfeld, M. J. and Falkowski, P. G.: Photosynthetic rates derived from satellite-based chlorophyll concentration, Limnol. Oceanogr., 42, 1–20, doi:10.4319/lo.1997.42.1.0001, 1997.

Behrenfeld, M. J., Westberry, T. K., Boss, E. S., O'Malley, R. T., Siegel, D. A., Wiggert, J. D., Franz, B. A., McClain, C. R., Feldman, G. C., Doney, S. C., Moore, J. K., Dall'Olmo, G., Milligan, A. J., Lima, I., and Mahowald, N.: Satellite-detected fluorescence reveals global physiology of ocean phytoplankton, Biogeosciences, 6, 779–794, doi:10.5194/bg-6-779-2009, 2009.

Bester, M. N.: Chemical restraint of Antarctic fur seals and southern elephant seals, South Afr. J. Wildl. Res., 18, 57–60, 1988.

Boehme, L., Lovell, P., Biuw, M., Roquet, F., Nicholson, J., Thorpe, S. E., Meredith, M. P., and Fedak, M.: Technical Note: Animal-borne CTD-Satellite Relay Data Loggers for real-time oceanographic data collection, Ocean Sci., 5, 685–695, doi:10.5194/os-5-685-2009, 2009.

Charrassin, J. B., Roquet, F., Park, Y. H., Bailleul, F., Guinet, C., Meredith, M., , Nicholls, K., Thorpe, S., Tremblay, Y., Costa, D., Göbel, M., Muelbert, M., Bester, M. N., Plötz, J., Bornemann, H., Timmermann, R., Hindell, M., Meijers, A., Coleman, R. C., Field, I. C., McMahon, C., Rintoul, S., Sokolov, S., Fedak, M., Lovell, P., Biuw, M., Kovacs, K., and Lydersen, C.: New insights into Southern Ocean physical and biological processes revealed by instrumented elephant seals, Proceedings of OceanObs 09: Sustained Ocean Observations and Information for Society (Vol. 2), edited by: Hall, J., Harrison, D. E., and Stammer, D., Venice, Italy, 21–25 September 2009, doi:10.5270/OceanObs09.cwp.15, ESA Publication WPP-306, 2010.

Claustre, H., Sciandra, A., and Vaulot, D.: Introduction to the special section bio-optical and biogeochemical conditions in the South East Pacific in late 2004: the BIOSOPE program, Biogeosciences, 5, 679–691, doi:10.5194/bg-5-679-2008, 2008.

Cullen, J. J. and Eppley, R. W.: Chlorophyll maximum layers of the Southern-California Bight and possible mechanisms of their formation and maintenance, Oceanologica Acta, 4, 23–32, 1981.

de Boyer Montégut, C., Madec, G., Fischer, A. S., Lazar, A., and Iudicone, D.: Mixed layer depth over the global ocean: An examination of profile data and a profile-based climatology, J. Geophys. Res.-Ocean., (1978–2012), 109, C12003, doi:10.1029/2004JC002378, 2004.

Dierssen, H. M. and Smith, R. C.: Bio-optical properties and remote sensing ocean color algorithms for Antarctic Peninsula waters, J. Geophys. Res.-Ocean., (1978–2012), 105, 26301–26312, doi:10.1029/1999JC000296, 2000.

Dong, S., Sprintall, J., Gille, S. T., and Talley, L.: Southern Ocean mixed?layer depth from Argo float profiles, J. Geophys. Res.-Oceans, 113, C06013, doi:10.1029/2006JC004051, 2008.

Durham, W. M., Climent, E., Barry, M., De Lillo, F., Boffetta, G., Cencini, M., and Stocker, R.: Turbulence drives microscale patches of motile phytoplankton, Nature communications, 4, 2148, doi:10.1038/ncomms3148, 2013.

Estrada, M., Marrase, C., Latasa, M., Berdalet, E., Delgado, M., and Riera, T.: Variability of deep chlorophyll maximum characteristics in the Northwestern Mediterranean, Marine Ecol. Progr. Ser., 92, 289–289, 1993.

Fairbanks, R. G., Sverdlove, M., Free, R., Wiebe, P. H., and Bé, A. W.: Vertical distribution and isotopic fractionation of living planktonic foraminifera from the Panama Basin, Nature, 298, 841–844, 1982.

Falkowski, P. G. and Kolber, Z.: Variations in Chlorophyll Fluorescence Yields in Phytoplankton in the World Oceans, Austr. J. Plant Physiol., 22, 341–355, 1995.

Falkowski, P. G., Barber, R. T., and Smetack, V.: Biogeochemical controls and feedbacks on ocean primary productivity, Science, 281, 200–206, 1998.

Fedak, M., Lovell, P., McConnell, B., and Hunter, C.: Overcoming the constraints of long range radio telemetry from animals: getting more useful data from smaller packages, Integr. Comparat. Biol., 42, 3–10, doi:10.1093/icb/42.1.3, 2002.

Franks, P. J.: Has Sverdrup's critical depth hypothesis been tested? Mixed layers vs. turbulent layers, ICES J. Mar. Sci., doi:10.1093/icesjms/fsu175, online firstm, 2014.

Fu, G., Baith, K. S., and McClain, C. R.: SeaDAS: The SeaWiFS Data Analysis System, Proceedings of The 4th Pacific Ocean Remote Sensing Conference, Qingdao, China, 28–31 July, 73–79, 1998.

Garcia, V. M. T. and Purdie, D. A.: The influence of irradiance on growth, photosynthesis and respiration of Gyrodinium cf. aureolum, J. Plankton Res., 14, 1251–1265, 1992.

Gordon, H. R. and Morel, A.: Remote Sensing of Ocean Color for Interpretation of Satellite Visible Imagery: A Review, (Springer-Verlag, New York), 1983.

Guinet, C., Xing, X., Walker, E., Monestiez, P., Marchand, S., Picard, B., Jaud, T., Authier, M., Cotté, C., Dragon, A. C., Diamond, E., Antoine, D., Lovell, P., Blain, S., D'Ortenzio, F., and Claustre, H.: Calibration procedures and first dataset of Southern Ocean chlorophyll a profiles collected by elephant seals equipped with a newly developed CTD-fluorescence tags, Earth Syst. Sci. Data, 5, 15–29, doi:10.5194/essd-5-15-2013, 2013.

Hameedi, M. J.: Aspects of water column primary productivity in the Chukchi Sea during summer, Marine Biol., 48, 37–46, 1978.

Henson, S. A., Sarmiento, J. L., Dunne, J. P., Bopp, L., Lima, I., Doney, S. C., John, J., and Beaulieu, C.: Detection of anthropogenic climate change in satellite records of ocean chlorophyll and productivity, Biogeosciences, 7, 621–640, doi:10.5194/bg-7-621-2010, 2010.

Holm-Hansen, O. and Hewes, C. D.: Deep chlorophyll *a* maxima (DCMs) in Antarctic waters, Polar Biol., 27, 699–710, doi:10.1007/s00300-004-0641-1, 2004.

Holm-Hansen, O., Kahru, M., and Hewes, C. D.: Deep chlorophyll a maxima (DCMs) in pelagic Antarctic waters. II. Relation to bathymetric features and dissolved iron concentrations, Marine Ecol. Progr. Ser., 297, 71–81, doi:10.3354/meps297071, 2005.

Jaud, T., Dragon, A. C., Garcia, J. V., and Guinet, C.: Relationship between Chlorophyll a Concentration, Light Attenuation and Diving Depth of the Southern Elephant Seal Mirounga leonina, PLoS one, 7, e47444, doi:10.1371/journal.pone.0047444, 2012.

Johnson, K. S., Berelson, W. M., Boss, E. S., Chase, Z., Claustre, H., Emerson, S. R., Gruber, N., Kortzinger, A., Perry, M. J., and Riser, S. C.: Observing Biogeochemical Cycles at Global Scales with Profiling Floats and Gliders: Prospects for a Global Array, Oceanography, 22, 216–225, doi:10.5670/oceanog.2009.81, 2009.

Kropuenske, L. R., Mills, M. M., van Dijken, G. L., Bailey, S., Robinson, D. H., Welschmeyer, N. A., and Arrigo, K. R.: Photophysiology in two major Southern Ocean phytoplankton taxa: photoprotection in Phaeocystis antarctica and Fragilariopsis cylindrus, Limnol. Oceanogr., 54, 1176, doi:10.1111/j.1529-8817.2010.00922.x, 2009.

Kropuenske, L. R., Mills, M. M., Van Dijken, G. L., Alderkamp, A. C., Mine Berg, G., Robinson, D. H., Welschmeyer, N. A., and Arrigo, K. R.: Strategies And Rates Of Photoacclimation In Two Major Southern Ocean Phytoplankton Taxa: Phaeocystis Antarctica (Haptophyta) And Fragilariopsis Cylindrus (Bacillariophyceae) 1, J. Phycol., 46, 1138–1151, 2010.

Lavigne, H., D'Ortenzio, F., Claustre, H., and Poteau, A.: Towards a merged satellite and in situ fluorescence ocean chlorophyll product, Biogeosciences, 9, 2111–2125, doi:10.5194/bg-9-2111-2012, 2012.

Lee, S. H., Stockwell, D., and Whitledge, T. E.: Uptake rates of dissolved inorganic carbon and nitrogen by under-ice phytoplankton in the Canada Basin in summer 2005, Polar Biol., 33, 1027–1036, 2010.

Lee, Z., Weidemann, A., Kindle, J., Arnone, R., Carder, K. L., and Davis, C.: Euphotic zone depth: Its derivation and implication to ocean-color remote sensing, J. Geophys. Res.-Oceans (1978–2012), 112, 223–227, doi:10.1016/0011-7471(66)91102-8, 2007.

Lorenzen, C. J.: A method for the continuous measurement of in vivo chlorophyll concentration, Deep Sea Res. Oceanogr. Abstr., 13, 223–227, 1966.

Lorenzen, C. J. and Jeffrey, S. W.: Determination of chlorophyll in seawater, Unesco tech. pap. Mar. Sci., 35, 1–21, 1980.

McClain, C. R.: A Decade of Satellite Ocean Colour Observations, Ann. Rev. Marine Sci., 1, 19–22, 2009.

McIntyre, T., de Bruyn, P. J. N., Ansorge, I. J., Bester, M. N., Bornemann, H., Plötz, J., and Tosh, C. A.: A lifetime at depth: vertical distribution of southern elephant seals in the water column, Polar Biol., 33, 1037–1048, 2010.

Mellard, J. P., Yoshiyama, K., Litchman, E., and Klausmeier, C. A.: The vertical distribution of phytoplankton in stratified water columns, J. Theor. Biol., 269, 16–30, doi:10.1016/j.jtbi.2010.09.041, 2011.

Mengesha, S., Dehairs, F., Fiala, M., Elskens, M., and Goeyens, L.: Seasonal variation of phytoplankton community structure and nitrogen uptake regime in the Indian Sector of the Southern Ocean, Polar Biol., 20, 259–272, 1998.

Mignot, A., Claustre, H., D'Ortenzio, F., Xing, X., Poteau, A., and Ras, J.: From the shape of the vertical profile of in vivo fluorescence to Chlorophyll-a concentration, Biogeosciences, 8, 2391–2406, doi:10.5194/bg-8-2391-2011, 2011.

Milligan, A. J., Aparicio, U. A., and Behrenfeld, M. J.: Fluorescence and nonphotochemical quenching responses to simulated vertical mixing in the marine diatom Thalassiosira weissflogii, Mar. Ecol. Progr. Ser., 448, 67–78, doi:10.3354/meps09544, 2012.

Morel, A.: Optical modeling of the upper ocean in relation to its biogenous matter content (case I waters), J. Geophys. Res.-Ocean., 93, 10749–10768, 1988.

Morel, A. and Berthon, J. F.: Surface pigments, algal biomass profiles, and potential production of the euphotic layer: Relationships reinvestigated in view of remote-sensing applications, Limnol. Oceanogr., 34, 1545–1562, 1989.

Morel, A. and Prieur, L.: Analysis of variations in ocean color, Limnol. Oceanogr., 22, 709–722, 1977.

Morel, A., Huot, Y., Gentili, B., Werdell, P. J., Hooker, S. B., and Franz, B. A.: Examining the consistency of products derived from various ocean color sensors in open ocean (Case 1) waters in the perspective of a multi-sensor approach, Remote Sens. Environ., 111, 69–88, doi:10.1016/j.rse.2007.03.012, 2007.

Müller, P., Li, X. P., and Niyogi, K. K.: Non-photochemical quenching. A response to excess light energy, Plant Physiol., 125, 1558–1566, 2001.

Navarro, G., Ruiz, J., Huertas, I. E., García, C. M., Criado-Aldeanueva, F., and Echevarría, F.: Basin-scale structures governing the position of the deep fluorescence maximum in the Gulf of Cádiz, Deep Sea Res. Pt II, 53, 1261–1281, 2006.

Orsi, A. H. and Harris, U.: Locations of the various fronts in the Southern Ocean Australian Antarctic Data Centre-CAASM Metadata, 2001, available at: https://data.aad.gov.au/aadc/metadata/metadata_redirect.cfm?md=/AMD/AU/southern_ocean_fronts, 15, 1835–1852, updated 2014.

Palmer, M. A., van Dijken, G. L., Mitchell, B. G., Seegers, B. J., Lowry, K. E., Mills, M. M., and Arrigo, K. R.: Light and nutrient control of photosynthesis in natural phytoplankton populations from the Chukchi and Beaufort seas, Arctic Ocean, Limnol. Oceanogr., 58, 2185–2205, 2013.

Platt, T., Harrison, W. G., Irwin, B., Horne, E. P., and Gallegos, C. L.: Photosynthesis and photoadaptation of marine phytoplankton in the Arctic, Deep Sea Res. Pt A, 29, 1159–1170, 1982.

Quéguiner, B.: Iron fertilization and the structure of planktonic communities in high nutrient regions of the Southern Ocean. Deep Sea Res. Pt II, 90, 43–54, doi:10.1016/j.dsr2.2012.07.024, 2013.

Ross, O. N., Moore, C. M., Suggett, D. J., MacIntyre, H. L., and Geider, R. J.: A model of photosynthesis and photo-protection based on reaction center damage and repair, Limnol. Oceanogr., 53, 1835, doi:10.1029/2008JC005086, 2008.

Ross, O. N., Geider, R. J., Berdalet, E., Artigas, M. L., and Piera, J.: Modelling the effect of vertical mixing on bottle incubations for determining in situ phytoplankton dynamics. I. Growth rates, Marine Ecol. Progr. Ser., 435, 33–45, 2011.

Ryther, J. H. and Menzel, D. W.: Light adaptation by marine phytoplankton, Limnol. Oceanogr., 4, 492–497, 1959.

Sackmann, B. S., Perry, M. J., and Eriksen, C. C.: Seaglider observations of variability in daytime fluorescence quenching of chlorophyll-a in Northeastern Pacific coastal waters, Biogeosciences Discuss., 5, 2839–2865, doi:10.5194/bgd-5-2839-2008, 2008.

Sangrà, P., García-Muñoz, C., García, C. M., Marrero-Díaz, A., Sobrino, C., Mouriño-Carballido, B., Aguiar-González, B., Henríquez-Pastene, C., Rodríguez, S. A., Lubian, L., Hernández, M., Hernández-León, S., Vázquez, E., and Estrada-Allis, S. N.: Coupling between upper ocean layer variability and size-fractionated phytoplankton in a non-nutrient-limited environment, Marine Ecol. Progr. Ser., 499, 35–46, 2014.

Saulquin, B., Hamdi, A., Gohin, F., Populus, J., Mangin, A., and d'Andon, O. F.: Estimation of the diffuse attenuation coefficient K(dPAR) using MERIS and application to seabed habitat mapping, Remote Sens. Environ., 128, 224–233, 2013.

Shang, S., Lee, Z., and Wei, G.: Characterization of MODIS-derived euphotic zone depth: Results for the China Sea, Remote Sens. Environ., 115, 180–186, 2011.

Soppa, M. A., Dinter, T., Taylor, B. B., and Bracher, A.: Satellite derived euphotic depth in the Southern Ocean: Implications for primary production modelling, Remote Sens. Environ., 137, 198–211, doi:10.1016/j.rse.2013.06.017, 2013.

Taylor, J. R. and Ferrari, R.: Shutdown of turbulent convection as a new criterion for the onset of spring phytoplankton blooms, Limnol. Oceanogr., 56, 2293–2307, doi:10.4319/lo.2011.56.6.2293, 2011.

Todd, R. E., Rudnick, D. L., and Davis, R. E.: Monitoring the greater San Pedro Bay region using autonomous underwater gliders during fall of 2006, J. Geophys. Res.-Ocean., 114, 483–495, doi:10.1029/2008JC005086, 2009. 2009.

van de Poll, W. H. and Buma, A. G. J.: Does ultraviolet radiation affect the xanthophyll cycle in marine phytoplankton?, Photochem. Photobiol. Sci., 8, 1295–1301, doi:10.1039/b904501e, 2009.

Xing, X., Morel, A., Claustre, H., Antoine, D., D'Ortenzio, F., Poteau, A., and Mignot, A.: Combined processing and mutual interpretation of radiometry and fluorimetry from autonomous profiling Bio-Argo floats: Chlorophyll a retrieval, J. Geophys. Res.-Oceans, (1978–2012), 116, doi:10.1029/2010JC006899, 2011.

Xing, X., Claustre, H., Blain, S., D'Ortenzio, F., Antoine, D., Ras, J., and Guinet, C.: Quenching correction for in vivo chlorophyll fluorescence acquired by autonomous platforms: A case study with instrumented elephant seals in the Kerguelen region (Southern Ocean), Limnol. Oceanogr. Methods, 10, 483–495, doi:10.4319/lom.2012.10.483, 2012.

7

The CO_2 system in the Mediterranean Sea: a basin wide perspective

M. Álvarez[1], H. Sanleón-Bartolomé[1], T. Tanhua[2], L. Mintrop[3], A. Luchetta[4], C. Cantoni[4], K. Schroeder[5], and G. Civitarese[6]

[1]IEO – Instituto Español de Oceanografía, Apd. 130, A Coruña, 15001, Spain
[2]GEOMAR Helmholtz Centre for Ocean Research Kiel, Düsternbrooker Weg 20, 24105 Kiel, Germany
[3]MARIANDA, Tulpenweg 28, 24145 Kiel, Germany
[4]CNR ISMAR – Consiglio Nazionale delle Ricerche, Istituto di Scienze Marine, viale R. Gessi 2, 34123, Trieste, Italy
[5]CNR ISMAR – Arsenale, Tesa 104, Castello 2737/F, 30122 Venice, Italy
[6]OGS – Istituto Nazionale di Oceanografia e di Geofisica Sperimentale, Borgo Grotta Gigante 42/c, 34018 Sgonico, Trieste, Italy

Correspondence to: M. Álvarez (marta.alvarez@co.ieo.es)

Abstract. The Mediterranean Sea (MedSea) is considered a "laboratory basin" being an ocean in miniature, suffering dramatic changes in its oceanographic and biogeochemical conditions derived from natural and anthropogenic forces. Moreover, the MedSea is prone to absorb and store anthropogenic carbon due to the particular CO_2 chemistry and the active overturning circulation. Despite this, water column CO_2 measurements covering the whole basin are scarce. This work aims to be a base-line for future studies about the CO_2 system space-time variability in the MedSea combining historic and modern CO_2 cruises in the whole area. Here we provide an extensive vertical and longitudinal description of the CO_2 system variables (total alkalinity – TA, dissolved inorganic carbon – DIC and pH) along an East-West transect and across the Sardinia-Sicily passage in the MedSea from two oceanographic cruises conducted in 2011 measuring CO_2 variables in a coordinated fashion, the RV Meteor M84/3 and the RV Urania EuroFleets 11, respectively. In this sense, we provide full-depth and length CO_2 distributions across the MedSea, and property-property plots showing in each sub-basin post-Eastern Mediterranean Transient (EMT) situation with regard to TA, DIC and pH.

The over-determined CO_2 system in 2011 allowed performing the first internal consistency analysis for the particularly warm, high salinity and alkalinity MedSea waters. The CO_2 constants by Mehrbach et al. (1973) refitted by Dickson and Millero (1987) are recommended. The sensitivity of the CO_2 system to the atmospheric CO_2 increase, DIC and/or TA changes is evaluated by means of the Revelle and buffer factors.

1 Introduction

The Mediterranean Sea (MedSea hereafter) is a particular system with respect to its physical and biogeochemical oceanography (e.g. Tanhua et al., 2013a). The MedSea is a land-locked sea exporting warm and salty intermediate water into the Atlantic Ocean through the shallow and narrow Strait of Gibraltar affecting the global thermohaline circulation (e.g. McCartney and Mauritzen, 2001). Apart from the shallow open thermohaline cell encompassing the whole Mediterranean basin, two closed deep overturning cells are active in the western and eastern basins, respectively. Additionally, a complex upper layer circulation is present including significant permanent and quasi-permanent eddies.

One of the most relevant biogeochemical characteristics of the MedSea, especially the eastern basin, is the anomalous high Nitrogen to Phosphorous ratio (e.g. Ribera d'Alcalà et al., 2003; Krom et al., 2005; Pujo-Pay et al., 2011). This fact along with the oligotrophic gradient from west to east could be attributed to the difference in the allochtonous nutrient sources in terms of quantity and quality (Ribera d'Alcalà et al., 2003; Ludwig et al., 2010; Schroeder et al., 2010; Krom et al., 2010) and also to the specific hydrodynamic features of the MedSea (Robinson et al., 2001; Millot and Taupier-Letage, 2005; Huertas et al., 2012).

With a limited exchange with the ocean and its specific circulation, the MedSea is considered a "laboratory basin" (e.g. Robinson and Golnaraghi, 1994; Bergamasco and Malanotte-Rizoli, 2010) where processes occurring on a global scale can be approximated to smaller time and space

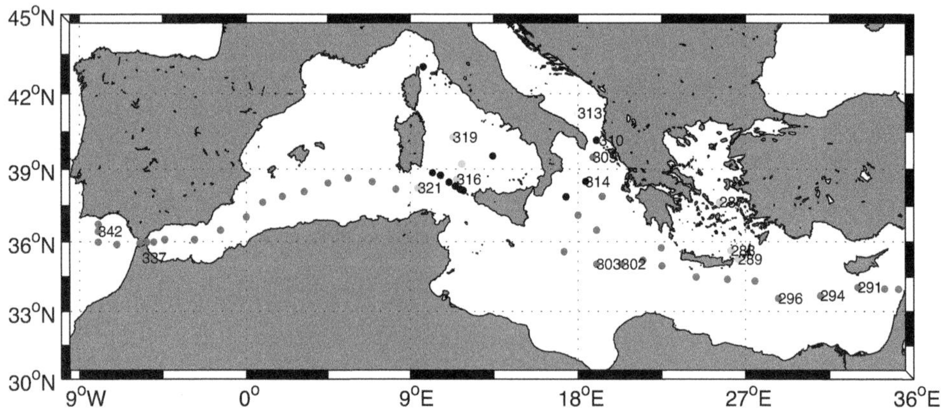

Fig. 1. Map of the Mediterranean Sea with the stations from the Meteor M84/3 (colour dots) and Urania EF11 (black dots) cruises where CO_2 measurements were taken. The colour code for the M84/3 stations corresponds to the data showed in Figs. 12–14.

scales. This fact mostly mentioned for general circulation studies can also be applied for biogeochemical studies, eutrophication, hypoxia, acidification, etc.

The MedSea represents only 0.8 % of the global oceanic surface and despite the general sparseness of water column CO_2 measurements (Álvarez, 2012) it has been identified as an important anthropogenic carbon storage where the column inventory is much higher than in the Atlantic or Pacific oceans (Schneider et al., 2010; Lee et al., 2011). The reasons for this are the intrinsic physico-chemical characteristics of the MedSea where warm and high alkalinity waters, thus low Revelle factor, are prone to absorb CO_2 from the atmosphere and be transported to the interior by the active overturning circulation. With regard to interior CO_2 data, the published literature discussed specific cruises, in the eastern MedSea sub-basins (e.g. Schneider et al., 2010; Luchetta et al., 2010; Krasakopoulu et al., 2011) and the western MedSea (e.g. Brunet et al., 1984; Claustre et al., 2001; Copin-Montegut, 1993; Millero et al., 1979). Cruises covering the eastern and western MedSea are few: the TRANSMED cruise in 2007 (Rivaro et al., 2010), BOUM in 2008 (Touratier and Goyet, 2012), ours in 2011 and the not yet published MedSeA 2013 cruise (medseaoceancruise.wordpress.com).

This paper is a contribution to the Special Issue "Physical, chemical and biological oceanography of the Mediterranean Sea". In particular, it focuses on CO_2 system analysis performed on board RV Meteor (M84/3 cruise) and RV Urania (EuroFleets 11 cruise), both sampled the MedSea in April 2011 (Tanhua et al., 2013a) (Fig. 1). Besides characterising the particular chemistry of the CO_2 in the MedSea, we will present and discuss vertical distributions and property-property plots in order to present a quasy-synoptic picture of the post-EMT oceanographic situation with regard to the CO_2 system. For those not familiar, briefly, the EMT was a dramatic event changing the hydrography and the large-scale circulation of deep waters in the eastern MedSea, which changed the formation area from the Adriatic to the Aegean Sea in the 1990s (e.g. Roether et al., 1996, 2007).

2 Data and methods

2.1 Meteor M84/3 cruise

During the Meteor M84/3 cruise in April 2011 (Fig. 1) (Tanhua et al., 2013a, b) Dissolved Inorganic Carbon (DIC), pH and Total Alkalinity (TA) were measured in all stations and depths in order to have an over-determined CO_2 system. Details about the sampling, analysis and quality control are given in Tanhua et al. (2013b). The bottle data is available at the Carbon Dioxide Information Analysis Center (CDIAC) (Tanhua et al., 2012).

The DIC content was determined using a SOMMA instrument. Samples were collected in borosilicate bottles according to standard operation protocol. The precision of the analysis was determined to $\pm 0.6\,\mu\text{mol kg}^{-1}$ by titration of several bottles filled from the same Niskin bottle. The accuracy was determined to be $2.5\,\mu\text{mol kg}^{-1}$ by analysing a total of 42 bottles of certified reference material (CRM, Andrew Dickson, Scripps, CA, USA, batch 108); the DIC of this batch is certified at $2022.70 \pm 0.7\,\mu\text{mol kg}^{-1}$. Measurements of the CRMs were also used to daily correct the temporal drift in the coulometer cell; this correction was never larger than $3\,\mu\text{mol kg}^{-1}$.

pH was measured in cylindrical optical glass 10-cm path-length cells with a spectrophotometric procedure (Clayton and Byrne, 1993). pH is reported in the total scale at 25 °C (pH25T). The precision was estimated to ±0.0012 by measuring replicates from the same Niskin bottle. Regarding the accuracy, the theoretical pH25T value for the 108 CRM batch using the dissociation constants from Mehrbach et al. (1973) refitted by Dickson and Millero (1987) is 7.8782. Replicate pH measurements along the cruise on CRMs are lower than the theoretical value by 0.0962 ± 0.0012 pH units. We will discuss this point in Sect. 3.2.

Samples for TA determination were collected in 600 mL borosilicate bottles and were analysed within one day after sampling. The TA samples were analysed following the double end point potentiometric technique by Pérez and Fraga (1987) and Pérez et al. (2000). Measurements of CRM were performed in order to control the accuracy of the TA measurements; the pH as measured by the electrode was corrected to obtain the closest mean TA to the CRM; the pH correction was never larger than 0.05. The precision of the TA measurements was better than $0.6\,\mu$mol kg^{-1} determined from duplicate Niskin bottles, and the accuracy of the TA measurements (i.e. including sampling errors) is determined to be $1\,\mu$mol kg^{-1}. Additionally, the daily drift of the system was determined by regular measurements of a substandard (surface seawater); this drift was always very low.

Other data used in this study from the M84/3 cruise are the hydrographic variables from CTD (Conductivity Temperature Depth) casts, and the bottle data for dissolved oxygen measured by a Winkler potentiometric technique and inorganic nutrients measured with a Quattro auto-analyser. See Tanhua et al. (2013b) for more details. The Apparent Oxygen Utilisation (AOU) is calculated using the saturation equation from Benson and Krause (1984).

2.2 Urania EF11 cruise

During the EF11 survey, samples for DIC, pH and TA were collected and measured on land. Unfortunately, fewer stations (15 for TA and pH, and 9 for DIC) than originally planned (Tanhua et al., 2013b) were sampled due to the geopolitical situation in the area. Two inter-calibration stations with the M84/3 cruise in the Ionian Sea were occupied (Fig. 1). All samples were carefully sampled in borosilicate glass bottles, poisoned with mercuric chloride, sealed with grease (silicone) and stored in the dark at a constant temperature of about 12 °C, until the analyses in the laboratories on shore (ISMAR for TA and pH and GEOMAR for DIC).

In ISMAR pH was measured as described in the previous section, with the spectrophotometric procedure described by Clayton and Byrne (1993), and reported at 25 °C and on the Total scale. The precision was estimated to be ± 0.003 measuring replicates from the same Niskin bottle. Regarding the accuracy, the theoretical pH25T value for the CRM batch 100, using the dissociation constant from Mehrbach et al. (1973) refitted by Dickson and Millero (1987), is 7.907. Our pH replicates on CRMs batch 100 are lower than the theoretical value by 0.019 ± 0.003 pH units. We further checked the accuracy against Tris Buffer Solution Synthetic Seawater batch 13 (distributed by Dr. Dickson), with certified pH corresponding to 8.093 pH25T. Our replicates on Tris are lower than the theoretical value by 0.018 ± 0.004. As the two different accuracies match very well, this correction was applied to the EF11 pH values.

TA was determined by potentiometric titration in an open cell with a difference derivative readout (Hernandez-Ayon et al., 1999). The precision of the TA measurements was better than $2.3\,\mu$mol kg^{-1} determined from duplicate Niskin bottles. The accuracy of the TA measurements on CRM (batch 117) is determined to be $0.8\,\mu$mol kg^{-1}, as our results yielded a mean value of $2240.0\,\mu$mol kg^{-1}, comparable to the certified value $2339.2\,\mu$mol kg^{-1}. The daily drift of the system was determined by regular measurements of substandard seawater; this drift was always very low.

The DIC was analysed at GEOMAR with the same methodology and equipment as in M84/3. Precision and accuracy of measurements are the same as those reported for the M84/3 cruise, see above.

3 Consistency of CO_2 analysis

3.1 Inter-cruise consistency

A crossover analysis was performed between the M84/3 and EF11 CO_2 data using the methodology applied to the CARINA database (Tanhua et al., 2010). We studied two areas: the Tyrrhenian Sea (4 stations each cruise) and the Ionian Sea (3 stations from M84/3 and 2 from EF11) (Fig. 1). We compared the DIC, pH and TA data below 1000 dbars: DIC difference is less than $2\,\mu$mol kg^{-1} in both sites; EF11 TA is about $11\,\mu$mol kg^{-1} higher and was therefore reduced by this amount, and pH25T differences are quite noisy and less than 0.01. Therefore, EF11 CO_2 data were not included in the following analysis of the internal consistency of the CO_2 system, Sect. 3.2.

Using the preferred set of constants reported on in Sect. 3.2, we compared measured DIC with that calculated from pH and TA using option 4 and the CO2SYS programme for ®MATLAB (see next section). After rejecting some outliers, the mean difference between measured and calculated DIC is $-0.4 \pm 10\,\mu$mol kg ^{1}. In summary, no adjustments were done to the EF11 DIC and pH measurements, whereas the TA was reduced by $11\,\mu$mol kg^{-1} as a consequence of the comparison between M84/3 and EF11 data.

The EF11 CO_2 data will be used to discuss the CO_2 system properties in water masses of the passage between Sardinia and Sicily, Sect. Sardinia Sicily passage.

3.2 Internal consistency analysis for the CO_2 system in the MedSea

There is a wealth of literature regarding the internal consistency of CO_2 analysis in seawater and its dissociation constants, see the review by Millero (2007), but to the best of our knowledge the last study about this issue was published in 2002 compiling up to 6000 measurements from NSF/JGOFS, NOAA/OACES and DOE/WOCE cruises in the Atlantic, Pacific, Southern and Indian oceans (Millero et al., 2002), but without any CO_2 data from the MedSea.

Table 1. Mean values and standard deviation of the residuals (measured minus calculated) for alkalinity (TA), Dissolved Inorganic Carbon (DIC) and pH at the total scale and referred to 25 °C (pH25T) calculated with different thermodynamic constants for CO_2 according to the CO2SYS programme options. The sulphate constant from Dickson (1990) and total borate concentration from Khoo et al. (1977) were used in all cases.

Constants	TA	DIC	pH25T
1	-16.1 ± 4.0	14.7 ± 3.6	0.024 ± 0.005
2	-11.1 ± 3.7	10.2 ± 3.4	0.016 ± 0.005
3	-5.0 ± 3.5	4.6 ± 3.2	0.007 ± 0.005
4	0.1 ± 3.3	-0.1 ± 3.0	-0.0003 ± 0.005
5	-3.7 ± 3.5	3.4 ± 3.2	0.005 ± 0.005
6	6.3 ± 3.2	-5.9 ± 3.0	-0.009 ± 0.005
7	-3.6 ± 3.5	3.3 ± 3.2	0.005 ± 0.005
9	17.3 ± 3.2	-16.1 ± 3.0	-0.027 ± 0.005
10	0.6 ± 3.3	-0.6 ± 3.0	-0.001 ± 0.005
11	-4.3 ± 3.6	3.9 ± 3.3	0.006 ± 0.005
12	-10.2 ± 4.4	9.3 ± 4.0	0.015 ± 0.006
13	2.8 ± 3.3	-2.6 ± 3.1	-0.004 ± 0.005
14	1.5 ± 3.4	-1.4 ± 3.1	-0.002 ± 0.005

(1) Roy (1993); (2) Goyet and Poisson (1989); (3) Hansson (1973a, b) refit by Dickson and Millero (1987); (4) Mehrbach et al. (1973) refit by Dickson and Millero (1987); (5) Hansson (1973a, b) and Mehrbach et al. (1973) refit by Dickson and Millero (1987); (6) GEOSECS choice (Takahashi et al., 1982); (7) Peng et al. (1987); (9) Cai and Wang (1998); (10) Lueker et al. (2000); (11) Mojica-Prieto and Millero (2002); (12) Millero et al. (2002); (13) Millero et al. (2006); (14) Millero (2010).

Despite the lack of surface or column partial pressure CO_2 (pCO_2) measurements, this study presents high-quality DIC, TA and pH data covering the whole MedSea that can be used to perform the first internal consistency analysis in this area for CO_2 measurements.

By using the consensus programme for seawater CO_2 calculations CO2SYS, configured for ®MATLAB by van Heuven et al. (2011), we tested the internal consistency of our measurements for the various CO_2 constants using all options, except that for pure water only (option 8). We also changed the sulfate constants (Dickson, 1990; Khoo, 1977) and the parameterization of borate as a function of salinity (Uppström, 1974; Lee et al., 2010). The inorganic nutrients concentrations were also taken into account.

The option yielding minimum differences between the calculated and measured variable for the three combinations (TA-pH, TA-DIC and DIC-pH) (Table 1 and Fig. 2) is obtained by using option 4 [CO_2 constants from Mehrbach et al., 1973; refitted by Dickson and Millero, 1987], the sulphate constant from Dickson (1990) and the parameterization of borate from Uppström (1974). Using this borate function and the sulphuric constant from Khoo (1977) yields practically the same results (not shown). Using option 13 (Millero et al., 2006), the more recent borate function by Lee et al. (2010) and either sulphate constant (Khoo, 1977 or Dickson, 1990) low residuals are also obtained (not shown).

By using our preferred set of constants (see above) our pH, TA and DIC measurements along the M84/3 cruise are internally consistent to -0.0003 ± 0.005 for pH, $0.1 \pm 3.3\,\mu\text{mol kg}^{-1}$ for TA and $-0.1 \pm 3\,\mu\text{mol kg}^{-1}$ for DIC. This combination of constants is also the preferred option in global ocean CO_2 data synthesis projects such as GLODAP (Key et al., 2004) and CARINA (Key et al., 2010). At the light of these results, the bias of 0.0962 ± 0.0012 in the pH25T measurements obtained after CRM measurements for pH during M84/3 was disregarded.

The effect of uncertainties in pK_1 and pK_2 using the pH – TA combination is identical to those observed using pH – DIC. As summarized in Lee et al. (1997) when pH and TA (pH and DIC) are used as input parameters, calculated DIC (TA) depends on pK_2, which is also dependent upon the DIC/TA ratio; the lower the ratio the higher the sensitivity to pK_2. However, only a reliable pK_1 is needed to calculate pCO_2 from the pH–TA (pH–DIC) combination; in this case the lower the DIC/TA ratio the lower the sensitivity to pK_1. Calculated pH from the TA–DIC input combination relies on a good estimation of $pK_1 - pK_2$ (the K_1/K_2 ratio). The DIC/TA ratio is an indicative of the concentration of carbonates; the lower the ratio the higher the $[CO_3]^{2-}$ concentration, which is also sensitive to pK_2. Lacking measured pCO_2 data we are not able to check the dependence on K_1 in the calculation of pCO_2 (pH-DIC or pH-TA inputs) or pH (pCO_2-DIC or pCO_2-TA inputs).

Having the former information in mind, a closer look at the pH, TA and DIC residuals (measured minus calculated values) reveals some insights and peculiarities about the internal consistency of CO_2 measurements in the high TA (low DIC/TA) and salinity MedSea waters. Figure 3 shows the longitudinal distribution of the residuals in the MedSea pointing to a decreasing trend eastwards for TA (Fig. 3a) and on the contrary an increasing trend eastwards for DIC and pH (Fig. 3b and c) which seem to be related with the eastwards salinity increase (Fig. 3 in Tanhua et al., 2013a), DIC/TA ratio decrease (Fig. 10h) or $[CO_3]^{2-}$ increase. The pH residuals are a function of $pK_1 - pK_2$, but mostly dependent on K_2, while the TA or DIC residuals are a function of pK_2 (Lee et al., 1997; Millero, 2007). The relation between pH and TA residuals is linear (Fig. 4), but two regression lines are detected depending on salinity. Samples with salinity higher than 36.1 have a higher slope ($-0.0015 \pm 4.5 \times 10^{-6}$, $r^2 = 0.9937$, $n = 725$) than fresher ones ($-0.0020 \pm 2.9 \times 10^{-5}$, $r^2 = 0.9945$, $n = 28$). Including measured pH, TA and DIC data from five CARINA cruises (Key et al., 2010) between the Strait of Gibraltar and 24° W and 40° N to 24° N the same tendency is detected (data with salinity < 36.1, slope $= -0.0022 \pm 8.7 \times 10^{-6}$, $r^2 = 0.9919$, $n = 671$; data with salinity ≥ 36.1, slope $= -0.0016 \pm 6.7 \times 10^{-6}$, $r^2 = 0.9977$, $n = 182$). This result points to the need for a revision of the K_2 parameterization for waters with high salinity and low DIC/TA ratio, as in the MedSea, as the CO_2 chemistry is more sensitive to K_2.

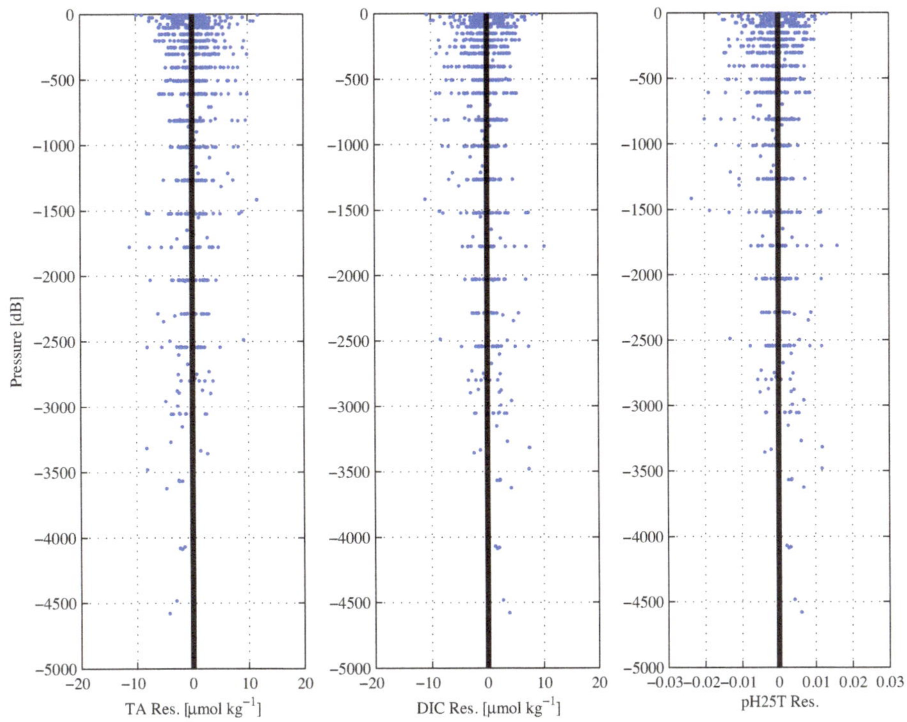

Fig. 2. Vertical distribution of the residual (measured minus calculated) values for TA ($\mu mol\,kg^{-1}$), DIC ($\mu mol\,kg^{-1}$) and pH25T using option 4 in CO2SYS, see Table 1.

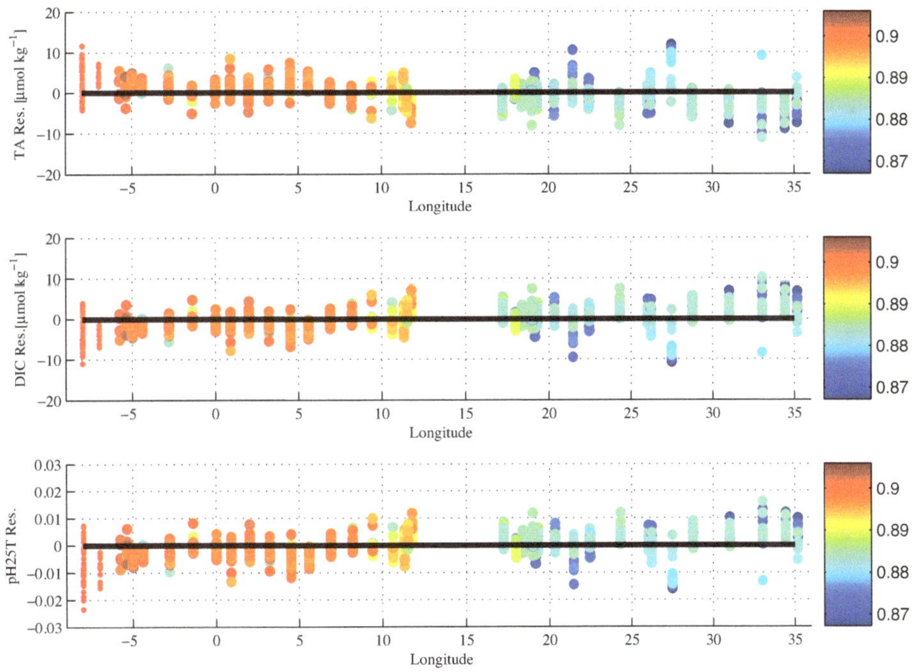

Fig. 3. Same as in Fig. 2 but plotted vs. longitude. The colour code corresponds to the DIC/TA ratio. Two stations to the west of the Strait of Gibraltar were coded as red because DIC/TA ratio > 0.94.

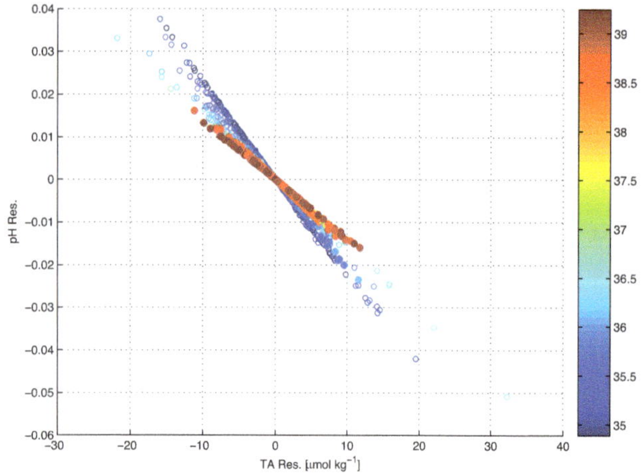

Fig. 4. Relationship between the TA and pH25T residuals for the M84/3 data (filled symbols) and some CARINA cruises (open symbols). The colour code stands for salinity.

In summary, the preferred option for CO_2 constants for global studies on internal consistency (CO_2 constants from Mehrbach et al., 1973; refitted by Dickson and Millero, 1987; sulphate constant from Dickson, 1990; total borate from Uppström, 1974) also applies for MedSea waters. However, further studies are needed, with an additional fourth measurement, either pCO_2 or $[CO_3]^{2-}$ (Byrne and Yao, 2008) to confirm the internal consistency of K_1 and K_2 with the measurements in the peculiar MedSea waters.

4 Distribution of CO_2 species in the Mediterranean Sea

The general aim of this section is to provide a large-scale distribution of measured and derived CO_2 properties along the full length of the Mediterranean Sea and corresponding sub-basins, being complementary to the physical (temperature and salinity) and chemical (dissolved oxygen and inorganic nutrients) variables presented by Tanhua et al. (2013a). Water mass characterisation will be commented using vertical distributions and property-property plots for the different sub-basins.

4.1 Vertical distributions

4.1.1 Measured variables

Meteor M84/3 cruise data

Vertical distributions of TA (Fig. 5), DIC (Fig. 6) and pH25T (Fig. 7) clearly distinguish several water masses in the MedSea, from surface to bottom:

i. Levantine Surface Water (LSW): this water formed by intensive heating and evaporation, has the largest salinity and temperature of the entire MedSea (> 17 °C, > 38.9), clearly seen in the surface layer of the Levantine Basin (Tanhua et al., 2013a). LSW has TA values around 2610 μmol kg^{-1}, DIC around 2270 μmol kg^{-1} and pH25T around 8.03 (Figs. 5–7).

ii. Atlantic Water (AW): enters through the Strait of Gibraltar and flows eastward, it is mainly detected by a low salinity signal in the upper 50–100 dbars (Fig. 3 in Tanhua et al., 2013a) from the Strait of Gibraltar to the Levantine Basin (36.2–38.9). AW is also characterised by a low TA signal evolving from values < 2560 μmol kg^{-1} in the western basin to < 2600 μmol kg^{-1} in the Levantine Basin and Ionian Sea (Fig. 5). The same trend was seen in DIC evolving from 2250 μmol kg^{-1} in the western to 2270 μmol kg^{-1} in the Levantine Basin (Fig. 6), pH25T increased from 8 to 8.014 eastwards (Fig. 7).

The MedSea is considered one of the most oligotrophic areas in the world (Azov, 1991) with low phytoplankton biomass and primary production decreasing eastward, confirmed by in situ measurements (Ignatiades et al., 2009; Moutin and Raimbalut, 2002; Siokou-Frangou et al., 2010), satellite data (Bosc et al., 2004; Volpe et al., 2007; D'Ortenzio and Ribera d'Alcalà, 2009) and modelling studies (Crise et al., 1999; Lazzari et al., 2012). The period of the M84/3 cruise, April, coincides with the end of the winter-spring bloom (D'Ortenzio and Ribera d'Alcalà, 2009; Lazzari et al., 2012). Surface Chlorophyll *a* (Chl *a*) and primary production data during the M84/3 cruise were presented in Rahav et al. (2013) with a clear westward increasing gradient, and values ranging from 0 to 0.4 μg L^{-1} for Chl *a*, and 0.2 to 15 μgC L^{-1} d^{-1} for primary production.

The gathered biogeochemical data during the M84/3 cruise showed that the upper 50 dbars are supersaturated in oxygen (AOU < 0) with more negative values in the western basin where higher primary production was reported (Rahav et al., 2013). While in the western basin inorganic nutrients were still available, in the eastern basin they were depleted, especially phosphate (see corresponding figures in Tanhua et al., 2013a). For waters with pressure < 60 dbars, there was no clear relationship between AOU and either DIC, TA or pH25T, except in the western basin where pH25T related with AOU (pH25T $= -0.0015 \pm 0.00008 \cdot$ AOU $+ 7.97 \pm 0.001$, $r^2 = 0.83$ $p < 0.001$), pointing to a pH modulation by primary production.

iii. Levantine Intermediate Water (LIW): this warm and high salinity water is formed in the Levantine Basin, mainly in the Rhodes Gyre area. During the M84/3 cruise LIW was clearly detected by the salinity maximum around 200 dbars in the eastern basin (Tanhua et al., 2013a), with TA > 2620 μmol kg^{-1},

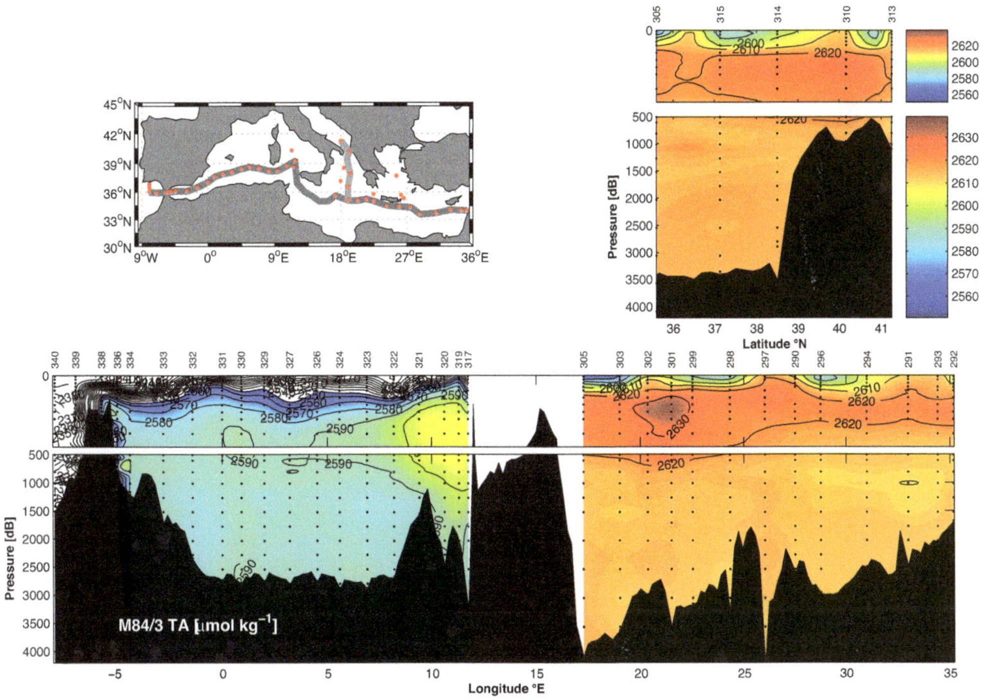

Fig. 5. Sections of TA (μmol kg^{-1}) in the Mediterranean Sea from the Meteor cruise M84/3 in April of 2011. The top right panel is a meriodional section from the Adriatic Sea to the Ionian Sea (light gray line on the map) and the lower panel is the zonal section from the coast of Lebanon in the Eastern Mediterranean Sea to through the Strait of Gibraltar (dark gray line on the map). The depth scale and the colour scale are identical all both panels. The top 500 dbars in each section are slightly expanded. No stations are shown in the Atlantic due to the very different TA encountered there as compared to the Mediterranean Sea.

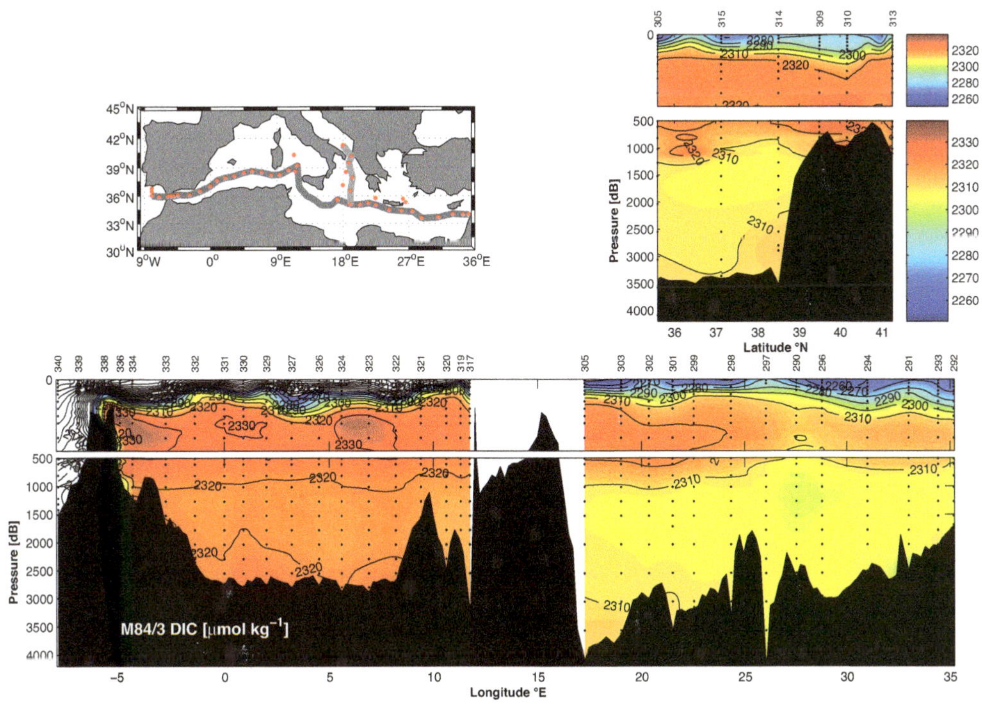

Fig. 6. As in Fig. 5 but for DIC (μmol kg^{-1}).

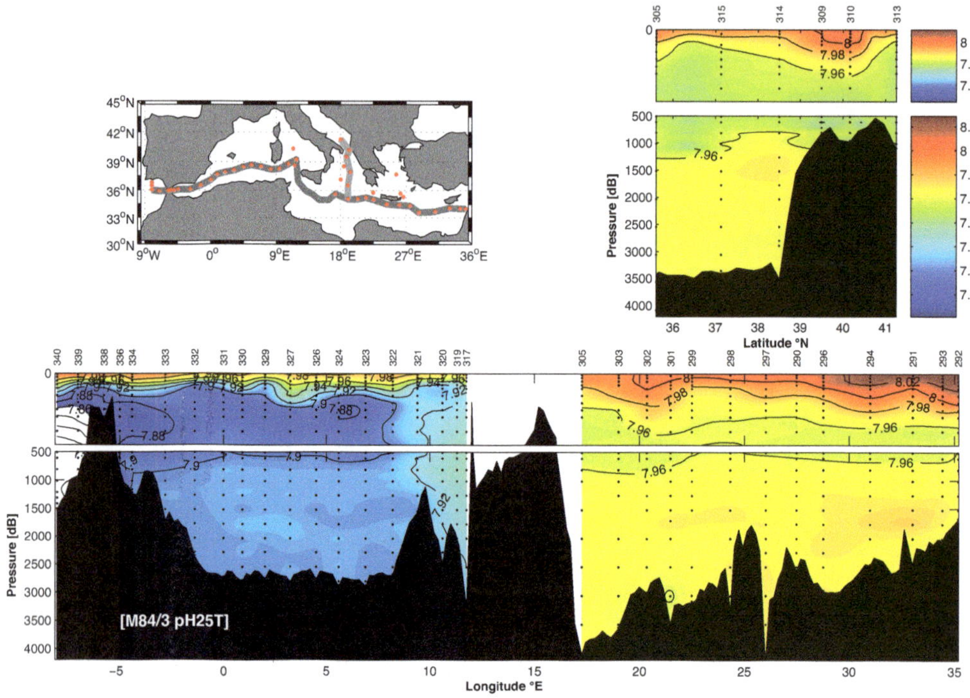

Fig. 7. As in Fig. 5 but for pH25T.

DIC > 2310 μmol kg^{-1}, but no clear signal in pH which varies between 7.96 and 7.98 (Figs. 5, 6 and 7). In the western basin LIW was found around 500 dbars as a maximum in salinity, corresponding to TA ($\approx$ 2590 μmol kg^{-1}) (Fig. 5), maximum in DIC ($\approx$ 2330 μmol kg^{-1}) (Fig. 6) and minimum pH25T ($\approx$ 7.9) (Fig. 7). In the eastern basin, the maximum layer of remineralization, indicated by the oxygen minimum (see Fig. 6 in Tanhua et al., 2013a), was located below LIW while in the western basin, both coincided.

iv. Eastern Mediterranean Deep Water (EMDW): along the M84/3 cruise track in the eastern MedSea three types of EMDW could be distinguished mainly using salinity and AOU (Tanhua et al., 2013a) and tracers such as CFC-12 (Stöven and Tanhua, 2013). Around 1000 dbar in the Levantine Basin a layer with higher salinity and AOU maxima related with pre-EMT-EMDW with an Adriatic Sea origin. In this basin, below 1500 dbars a more saline water and higher in oxygen pointed to EMDW with an Aegean Sea origin produced during the EMT, although more mixed compared to that found in the Ionian Sea. Here salinity and AOU were lower. The third EMDW is that found in the near-bottom layer of the Ionian Sea, it was produced after the EMT (> 2001) and has an Adriatic Sea origin (Hainbucher et al., 2013).

The upper panels in Figs. 5, 6 and 7 show the distribution of CO_2 species from the centre of the Ionian Sea to the Adriatic Pit. Here, below 3000 dbars, salinity increased in about 0.01 units (not shown), and TA as well (not resolved by the contour or the colour scale in Fig. 5) and DIC (Fig. 6) increased towards the bottom by about 4 μmol kg^{-1}. Correspondingly, pH25T decreased by about 0.002 (note the greenish colour in the bottom of the upper panel in Fig. 7). The tracers distributions, CFC-12 and SF_6 (Stöven, 2011), also increased towards the bottom. These results point to a recent arrival of EMDW formed in the Adriatic Sea as commented previously.

v. Western Mediterranean Deep Water (WMDW): being fresher and colder than EMDW, the deep layers below 1500 dbars in the western MedSea during M84/3 were quite homogeneous in the physical (potential temperature and salinity) properties. A closer look showed two distinct WMDW varieties: a younger one, close to the bottom, as noted by the CFC-12 distribution (Stöven, 2011) and saltier than the older WMDW, immediately above 2000 dbars (Hainbucher et al., 2013). Bottom (younger) WMDW also had higher DIC values by about 7 μmol kg^{-1}, lower pH25T values by about 0.004, and higher TA values by about 4 μmol kg^{-1}, i.e. within the accuracy limits of the latter two variables (Figs. 5–7).

An outstanding feature in the vertical distribution of CO_2 variables was the pH25T minimum (Fig. 7) around 500 dbars in the Levantine and Ionian basins. In the Levantine basin this minimum was coincident with the oxygen minimum (Fig. 6 in Tanhua et al., 2013a) and a DIC maximum (Fig. 6). Whereas in the Ionian Sea the pH25T minimum/DIC maximum (500 dbars) was shallower than the oxygen minimum that was situated around 1000 dbars. All these extreme layers differ from the LIW layer, the salinity maximum was shallower in both basins, Levantine and Ionian. However, in the western basin, the LIW signal was coincident with the layer of highest remineralization as indicated by the DIC maxima and the oxygen/pH25T minimum.

During the M84/3 cruise we sampled the interior of the Tyrrhenian Sea: in the upper layer low salinity (Tanhua et al., 2013a) and TA values (Fig. 5) pointed to AW. Around 400 dbars the maximum in salinity, TA, DIC and pH25T (Figs. 5, 6 and 7) were associated with LIW entering trough the Sicily channel; Tyrrhenian Deep Water (TDW) formed from the mixing of LIW and WMDW was detected below the TA maximum associated with LIW and bottom water with lower TA and pH25T and higher DIC values (Figs. 5, 6 and 7). This bottom water corresponds to WMDW that has started to cascade into the Tyrrhenian Sea from the western basin in 2011, also higher in tracer concentrations (Stöven and Tanhua, 2013).

As a reference for future studies and calculations in Sect. 4.1.2, Table 2 shows the mean values and standard deviation for salinity and different CO_2 species in the MedSea sub-basins and Atlantic waters west of the Strait of Gibraltar as sampled in 2011 during the Meteor M84/3 cruise by different depth intervals.

The direct comparison of our CO_2 data with previously published CO_2 data, either DIC, TA or pH, in order to identify and quantify trends is out of the scope of our manuscript. The available data for CO_2 variables in the MedSea is quite heterogeneous (see review in Álvarez, 2012, and below). A detail study about CO_2 temporal changes in the interior MedSea will be the aim of a forthcoming work. We do however make a rough comparison in order to detect possible biases in the historic dataset as compared to ours and vice versa. Qualitatively, our TA and DIC data in the western MedSea seems to be 5–10 and 40–70 $\mu mol\,kg^{-1}$ lower, respectively, than those shown in Millero et al. (1979) from a cruise in 1976. This result is counterintuitive due to the expected increase in DIC due to input of anthropogenic carbon (C_{ANT}). However, the data from Millero et al. (1979) originates before the use of certified reference materials (CRMs), so that the accuracy of these early data can be questioned. Our TA and DIC are however more comparable to those from the Almofront cruise in 1991 (Copin-Montegut, 1993). The data shown by Touratier and Goyet (2009) from the DYFAMED site in the Ligurian Sea below 1000 dbar, corresponding to WMDW, are lower than our WMDW values by about 5 $\mu mol\,kg^{-1}$ for both DIC and TA.

The reference cruise for the CO_2 system in the eastern and part of the western MedSea is the M51/2 cruise (R/V Meteor 2001) available at CDIAC; the TA data was presented in Schneider et al. (2007) and the DIC in Schneider et al. (2010). The TA data are comparable, but our DIC are roughly 10 $\mu mol\,kg^{-1}$ higher, which could be a real signal considering the 10 yr between the measurements and the uptake of anthropogenic carbon.

The VECTOR cruise in 2007 sampled 8 stations in the eastern and western MedSea for pH and TA, whereas DIC was calculated (Rivaro et al., 2010). Examining data from the western basin below 700 dbars, the TA data are comparable to each other whereas the DIC and pH values reported by Rivaro et al. (2010) are higher than ours by 10 $\mu mol\,kg^{-1}$ and by 0.01, respectively. In the eastern basin, our data below 700 dbars for pH and TA are comparable but our DIC seems 20 $\mu mol\,kg^{-1}$ higher. More recent CO_2 data are reported from the BOUM cruise in 2008 (Pujo-Pay et al., 2011; Touratier and Goyet, 2012); their DIC data seem higher than ours by about 10 $\mu mol\,kg^{-1}$ in both the eastern and western basins.

Sardinia Sicily passage

The vertical distributions of physical properties in the Sardinia Sicily passage, already analysed in the last decade of last century (Astraldi et al., 2002), indicate that the passage is a crucial chokepoint where almost all MedSea waters can be intercepted (see the review by Astraldi et al., 2002). The Sardinia Sicily transect, for its local character, was not included in the overall MedSea description done by Tanhua et al. (2013a). The main water masses crossing the passage have been identified according to their physical characteristics using potential temperature and salinity (Fig. 8a and b), AOU, (Fig. 8c) and CO_2 variables (Fig. 8d–f). On the eastern part (close to Sicily) only inflowing waters can be observed, from surface to bottom: (1) a fresh vein of AW (0–200 dbars), coming directly from Gibraltar through the Algerian sub-basin, was evident from the low salinity ($\ll 38.00$) (Fig. 8b) and supported by the lowest TA values (2525–2575 $\mu mol\,kg^{-1}$) (Fig. 8d); this inflowing AW was not in equilibrium with atmospheric oxygen (AOU between 20–40 $\mu mol\,kg^{-1}$, Fig. 8c), pH25T (≈ 7.950, Fig. 8f) and DIC values (2225–2285 $\mu mol\,kg^{-1}$, Fig. 8e) were the lowest observed along the Sect. 2) Below AW, the inflow of the saltiest water (> 38.75) in the transect, LIW, coming from the eastern MedSea, after crossing the Sicily Channel, the high AOU (50–60 $\mu mol\,kg^{-1}$) and DIC ($\approx 2300\,\mu mol\,kg^{-1}$) values aside of the low pH25T (< 7.940) highlight the organic matter remineralization processes occurring within the LIW; TA values as high as 2605–2610 $\mu mol\,kg^{-1}$were observed. (3) Below LIW, there was a thin layer of dense transitional Eastern Mediterranean Deep Water (tEMDW, i.e. the upper layer of the deep water filling the Eastern Mediterranean, which is able to cross the channel); tEMDW is cooler and

Table 2. Mean values and standard deviation for salinity and CO_2 species in the different MedSea sub-basins and west of the Strait of Gibraltar as sampled during the M84/3 cruise. Several vertical intervals are considered: surface (0–150 dbars in the East and 0–200 dbars in the West MedSea), intermediate (East: 150–500 and West: 200–500 dbars), deep (East: 500–2500 and West: 600–2000 dbars) and bottom (East: > 2500 and West: > 2000 dbars). IS stands for in situ values. Ar stands for aragonite and Ca for calcite.

SURFACE	Salinity	TA μmol kg^{-1}	DIC μmol kg^{-1}	pH25T	CO_2 IS μmol kg^{-1}	Ω−Ar IS	Ω−Ca IS
Aegean	39.20 ± 0.01	2628 ± 2	2291 ± 4	8.009 ± 0.006	13.1 ± 0.2	3.7 ± 0.1	5.6 ± 0.1
Levantine	39.05 ± 0.13	2612 ± 10	2277 ± 15	8.010 ± 0.017	13.1 ± 0.6	3.6 ± 0.1	5.5 ± 0.2
Ionian	38.73 ± 0.21	2607 ± 16	2289 ± 17	7.985 ± 0.015	13.5 ± 0.6	3.4 ± 0.1	5.1 ± 0.2
Adriatic	38.70 ± 0.12	2616 ± 1	2298 ± 16	7.980 ± 0.023	14.0 ± 0.9	3.5 ± 0.2	5.3 ± 0.2
Tyrrhenian	38.32 ± 0.29	2570 ± 26	2294 ± 32	7.929 ± 0.021	16.2 ± 1.0	3.1 ± 0.1	4.7 ± 0.2
Western	37.79 ± 0.51	2522 ± 46	2250 ± 54	7.930 ± 0.033	15.9 ± 1.6	3.0 ± 0.2	4.6 ± 0.2
Atlantic	36.75 ± 0.89	2423 ± 83	2166 ± 83	7.927 ± 0.038	14.6 ± 1.7	2.7 ± 0.2	4.1 ± 0.3
INTERM.	Salinity	TA μmol kg^{-1}	DIC μmol kg^{-1}	pH25T	CO_2 IS μmol kg^{-1}	Ω−Ar IS	Ω−Ca IS
Aegean	39.14 ± 0.02	2636 ± 2	2318 ± 8	7.975 ± 0.01	14.3 ± 0.4	3.4 ± 0.1	5.1 ± 0.2
Levantine	39.04 ± 0.11	2621 ± 4	2310 ± 10	7.970 ± 0.016	14.6 ± 0.6	3.3 ± 0.1	5.0 ± 0.2
Ionian	39.00 ± 0.08	2627 ± 6	2321 ± 5	7.962 ± 0.012	14.9 ± 0.5	3.3 ± 0.1	5.0 ± 0.2
Adriatic	38.79 ± 0.13	2620 ± 6	2320 ± 7	7.954 ± 0.015	15.2 ± 0.6	3.2 ± 0.1	4.9 ± 0.2
Tyrrhenian	38.70 ± 0.03	2603 ± 3	2324 ± 2	7.926 ± 0.005	16.4 ± 0.3	3.0 ± 0.1	4.5 ± 0.1
Western	38.50 ± 0.08	2586 ± 8	2326 ± 7	7.892 ± 0.010	17.7 ± 0.5	2.8 ± 0.1	4.2 ± 0.1
Atlantic	36.03 ± 0.66	2372 ± 55	2172 ± 42	7.828 ± 0.038	19.7 ± 1.7	2.2 ± 0.2	3.4 ± 0.3
DEEP	Salinity	TA μmol kg^{-1}	DIC μmol kg^{-1}	pH25T	CO_2 IS μmol kg^{-1}	Ω−Ar IS	Ω−Ca IS
Aegean	39.05 ± 0.02	2633 ± 2	2323 ± 2	7.964 ± 0.003	14.5 ± 0.2	2.7 ± 0.2	4.1 ± 0.4
Levantine	38.78 ± 0.02	2613 ± 2	2306 ± 3	7.967 ± 0.004	14.5 ± 0.3	2.7 ± 0.2	4.1 ± 0.4
Ionian	38.77 ± 0.06	2615 ± 3	2309 ± 5	7.965 ± 0.004	14.6 ± 0.3	2.8 ± 0.2	4.2 ± 0.3
Adriatic	38.75 ± 0.04	2618 ± 3	2326 ± 5	7.940 ± 0.008	15.6 ± 0.3	2.9 ± 0.1	4.4 ± 0.2
Tyrrhenian	38.57 ± 0.05	2595 ± 5	2318 ± 2	7.923 ± 0.008	16.3 ± 0.3	2.6 ± 0.2	3.8 ± 0.2
Western	38.48 ± 0.01	2587 ± 2	2319 ± 2	7.906 ± 0.003	16.6 ± 0.2	2.5 ± 0.1	3.7 ± 0.2
Atlantic	36.10 ± 0.25	2390 ± 16	2201 ± 6	7.795 ± 0.026	21.0 ± 1.0	1.9 ± 0.1	2.9 ± 0.2
BOTTOM	Salinity	TA μmol kg^{-1}	DIC μmol kg^{-1}	pH25T	CO_2 IS μmol kg^{-1}	Ω−Ar IS	Ω−Ca IS
Aegean	–	–	–	–	–	–	–
Levantine	38.75 ± 0.01	2613 ± 2	2308 ± 1	7.969 ± 0.001	12.9 ± 0.2	2.0 ± 0.2	2.9 ± 0.3
Ionian	38.73 ± 0.001	2615 ± 2	2310 ± 2	7.965 ± 0.003	14.3 ± 0.1	2.1 ± 0.2	3.1 ± 0.3
Adriatic	–	–	–	–	–	–	–
Tyrrhenian	38.51 ± 0.01	2589 ± 3	2317 ± 1	7.916 ± 0.005	16.3 ± 0.2	2.1 ± 0.1	3.1 ± 0.2
Western	38.48 ± 0.001	2587 ± 2	2321 ± 2	7.906 ± 0.002	16.7 ± 0.1	2.1 ± 0.1	3.1 ± 0.1
Atlantic	–	–	–	–	–	–	–

fresher than LIW (< 13.8 °C and < 38.65), as well as for the higher DIC, TA (2318 and 2612 μmol kg^{-1}, respectively, Fig. 8d and e) and inorganic nutrient values (not shown) and the lower pH25T (< 7.930, Fig. 8f); given its high density tEMDW cascades at greater depths to the interior of the Tyrrhenian Sea, contributing to the formation of the Tyrrhenian Deep Water, TDW, and provoking an intense mixing that dilutes its original physical properties (tEMDW is significantly warmer and saltier than TDW), in contrast the chemical properties of these two deep water masses remain very similar (Fig. 8).

On the western part (close to Sardinia) there are both inflowing and outflowing water masses, from surface to bottom: (1) close to Sardinia the outflowing AW has circulated through the Tyrrhenian Sea, becoming saltier (38.00 < salinity < 38.20, note the depth decrease towards Sardinia of the 38 isoline, Fig. 8b) and even less saturated in oxygen (AOU 35–40 μmol kg^{-1}, note the depth decrease towards Sardinia of the 20 isoline, Fig. 8c) than the inflowing AW on the Sicilian side; CO_2 variables are also slightly different (2550 < TA < 2575 μmol kg^{-1}; 2270 < DIC < 2285 μmol kg^{-1}; 7.940 < pH25T < 7.950, note the sloping isolines in Fig. 8d–f). (2) Below the outflowing AW, an outflowing LIW vein can be noted, which in turn, after its path through the whole Tyrrhenian Sea has become colder (≈ 14.0 °C) and less salty (< 38.68) than the inflowing LIW on the eastern edge of the section, due to the mixing; this fact also lowers TA (≈ 2600 μmol kg^{-1},

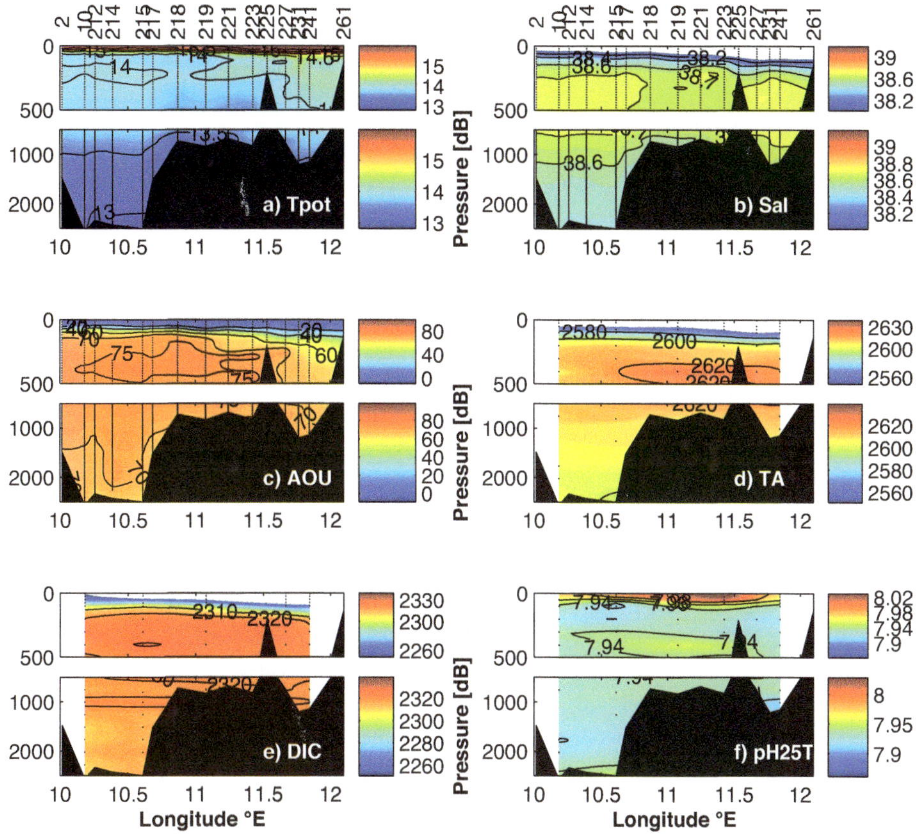

Fig. 8. Vertical distributions across the Sardinia Sicily passage using EF11 data of **(a)** potential temperature (°C), **(b)** salinity, **(c)** Apparent Oxygen Utilisation (AOU, μmol kg^{-1}), **(d)** TA (μmol kg^{-1}) **(e)** DIC (μmol kg^{-1}) and **(f)** pH25T. Potential temperature, salinity and AOU are derived from CTD data, CTD-oxygen was calibrated with bottle data.

Fig. 8d) and increases DIC ($\approx 2310\,\mu$mol kg^{-1}, Fig. 8e) while the pH25T (Fig. 8f) remains similar; outflowing LIW is also characterised by an absolute AOU maximum ($65 < \text{AOU} < 73\,\mu$mol kg^{-1}, Fig. 8c) due to ageing. (3) In the central part of passage, over the plateau, there is an inflow of "old LIW" (Astraldi et al., 2002) (identified by its salinity maximum, but less salty than the outflowing LIW), also identified by higher AOU (55–70 μmol kg^{-1}, Fig. 8c) than the inflowing LIW (50–60 μmol kg^{-1}), while differences in CO_2 variables cannot be distinguished. (4) In deep layers (600–1900 dbars) the properties of LIW progressively become those of TDW, exiting the Tyrrhenian Sea, pH25T show the lowest values (< 7.930, Fig. 8f), AOU and DIC the highest ones ($72 < \text{AOU} < 74\,\mu$mol kg^{-1}; $2315 < \text{DIC} < 2320\,\mu$mol kg^{-1}, Fig. 8c and e) and TA values ($> 2605\,\mu$mol kg^{-1}, Fig. 8d) are not particularly characterising. (5) At the bottom and squeezed on the eastern wall in the deep canyon, there is an inflowing vein of Western Mediterranean Deep Water, which can be distinguished from the TDW because is fresher (< 38.52) and colder (< 13.4 °C), has slightly lower AOU ($< 70\,\mu$mol kg^{-1}), higher pH25T values (> 7.940) but similar DIC ($> 2318\,\mu$mol kg^{-1}) and TA (2605–2610 μmol kg^{-1}) values.

4.1.2 Derived variables

We used measured TA, DIC and inorganic nutrients data to calculate the degree of saturation (Ω) for calcite and aragonite using the preferred set of constants (see Sect. 3.2). The whole MedSea is supersaturated ($\Omega \gg 1$) with respect to both calcite and aragonite throughout the whole water column. Figure 9 shows the vertical distribution of Ω, a clear longitudinal gradient is evident for surface, intermediate and deep waters with increasing values eastwards and a clear separation between the western, eastern MedSea basins and Atlantic waters to the west of the Strait of Gibraltar. The Ω values reported here compare well with those found by Millero et al. (1979) and Schneider et al. (2007). According to the classification by Kleypas et al. (1999) only the eastern basin in the upper 200 dbars presents adequate to optimal conditions for the development of coral reefs with $\Omega\text{-Ar} > 3.5$.

Several projects, finished (e.g. EPOCA) or on going (e.g. BIOACID, MedSeA), try to decipher the combined effect of acidification and warming on the highly adapted MedSea

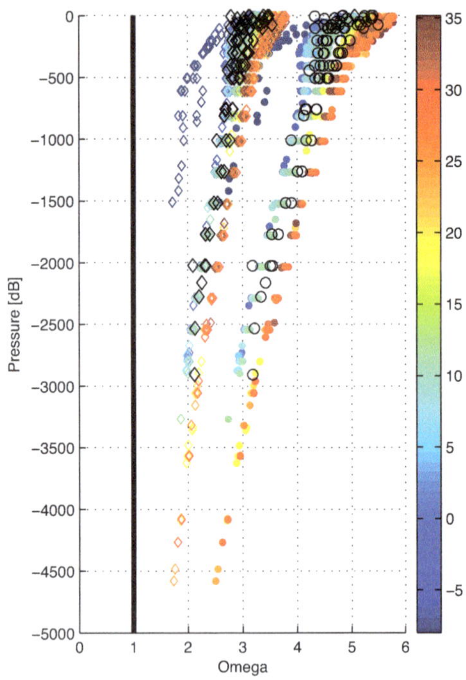

Fig. 9. Vertical distribution of the carbonate saturation state for calcite (circles) and aragonite (diamonds). Data from the EF11 cruise are shown as black circles (saturation of calcite) and diamonds (aragonite). The colour code stands for longitude.

calcareous and non-calcareous organisms. As a result of being too small and too complex in its circulation and biogeochemistry, the MedSea is usually not included in global projections models (e.g. Ricke et al., 2013) and a specific ocean-biogeochemistry model needs to be used. Here we provide a basin wide perspective for the CO_2 chemistry in the MedSea as a basis for future studies and projections.

In order to quantify the ability of MedSea waters to resist changes in the CO_2 system we calculated the buffer factors formulated by Egleston et al. (2010). The six factors proposed in their Table 1 quantify the sensitivity of aqueous CO_2 $[CO_2]$ (γi), protons [H+] (βi) and the carbonate saturation state Ω (ωi) to changes in DIC and TA (subindex i) when the other parameter is kept constant. The calculations were performed with the output from CO2SYS at in situ conditions with the preferred set of constants and measured DIC and TA as inputs. The buffer factors have dimensions of $\text{mol kg}^{-1} \times 10^{-3}$ of seawater (mmol kg^{-1} hereinafter). See the formulation in Table 3 as several editing errors were detected in the original equations from Egleston et al. (2010).

The buffer factors γi express the fractional change in CO_2 when DIC changes at constant TA (air-sea CO_2 exchange) or when TA changes at constant DIC (additions of strong acid or base). The traditional Revelle factor, R, is directly calculated as $R = \text{DIC}/\gamma_{\text{DIC}}$. The buffer factor β_{TA} expresses the fractional change in hydrogen ion concentration or activity when TA changes at constant DIC (addition of a strong base or acid). The factor β_{DIC} is identical to $-\gamma_{\text{TA}}$. Since $[Ca^{2+}]$ and the solubility products of carbonate are invariant to DIC and TA changes, the buffer factors ωi account for the fractional change in $[CO_3^{2-}]$ when DIC or TA changes when correspondingly TA or DIC is constant. A natural process just inducing strong acid/base additions is difficult to imagine, but a cargo disaster could be possible, unfortunately. A change in natural processes (primary production/remineralization of organic matter, formation/dissolution of carbonates, freshwater balance) usually affect both TA and DIC and therefore, the sensitivity of $[CO_2]$, pH and $[CO_3^{2-}]$ to a change in those processes can be calculated using a combination of buffer factors (Egleston et al., 2010).

The values of γ_{DIC} and β_{DIC} are positive while ω_{DIC} is negative because adding CO_2 to seawater increases $[CO_2]$ and $[H^+]$ but decreases $[CO_3^{2-}]$. Likewise, γ_{TA} and β_{TA} are negative while ω_{TA} is positive because adding a strong base to seawater decreases $[CO_2]$ and $[H^+]$ but increases $[CO_3^{2-}]$. Low buffer values imply low buffering capacity for a given change in DIC or TA. Minimum absolute values for the six buffer factors are found in waters with similar DIC and TA (DIC/TA ratio ≈ 1) and pH of about 7.5.

The six buffer factors, R and the DIC/TA ratio are shown in Fig. 10 and their mean values by sub-basin in Table 3. As the DIC/TA ratio approaches one the buffer capacities are reduced since $[CO_2]$ is more sensitive to temperature changes at this DIC/TA ratio. From the vertical distributions in Fig. 10 we can conclude that:

- The DIC/TA ratio decreased eastwards in the MedSea; being low compared to Atlantic waters. The highest ratios, i.e. higher $[CO_2]$ sensitivity to temperature, were associated with the Mediterranean Overflow, DIC/TA ratio > 0.9 (Fig. 10h).

- The absolute values of the buffer factors increased eastwards in the MedSea, therefore, the CO_2 system in the eastern basin is less sensitive to changes in DIC and/or TA than the western basin. Atlantic waters, to the west of the Strait of Gibraltar, would be more sensitive than waters in the MedSea.

- The absolute buffering capacities distribute as $|\omega_{\text{DIC}}| > \omega_{\text{TA}} > |\beta_{\text{TA}}| > (\beta_{\text{DIC}} = -\gamma_{\text{TA}}) > \gamma_{\text{DIC}}$. That is, the sensitivity of CO_2 concentration to changes in DIC at constant TA (air-sea CO_2 exchange) is the highest, followed by the change in pH due to the air-sea CO_2 exchange (equivalent to the change in CO_2 due to a strong acid /base addition), then changes in pH due to strong acid/base additions. Being $|\omega_{\text{DIC}}|$ and ω_{TA} the highest buffer values means that MedSea waters are able to buffer changes in the saturation state due to additions of a strong acid or base and the air-sea CO_2 exchange more easily than Atlantic waters.

Table 3. Mean values and standard deviation of several buffer factors in the different MedSea sub-basins and west of the Strait of Gibraltar as sampled during the M84/3 cruise. Same vertical intervals as in Table 2. The buffer factors have units of mmol kg^{-1} and Revelle is dimensionless.

SURFACE	γ_{DIC}	β_{DIC}	ω_{DIC}	γ_{TA}	β_{TA}	ω_{TA}	Revelle
Aegean	0.23 ± 0.01	0.28 ± 0.01	−0.37 ± 0.01	−0.28 ± 0.01	−0.31 ± 0.01	0.35 ± 0.01	9.9 ± 0.1
Levantine	0.23 ± 0.01	0.28 ± 0.01	−0.36 ± 0.01	−0.28 ± 0.01	−0.31 ± 0.01	0.35 ± 0.01	9.9 ± 0.2
Ionian	0.22 ± 0.01	0.26 ± 0.01	−0.34 ± 0.01	−0.26 ± 0.01	−0.29 ± 0.01	0.32 ± 0.01	9.8 ± 0.2
Adriatic	0.23 ± 0.01	0.27 ± 0.01	−0.35 ± 0.01	−0.27 ± 0.01	−0.30 ± 0.01	0.33 ± 0.01	10.2 ± 0.3
Tyrrhenian	0.21 ± 0.01	0.25 ± 0.01	−0.31 ± 0.01	−0.25 ± 0.01	−0.27 ± 0.01	0.30 ± 0.01	10.8 ± 0.2
Western	0.21 ± 0.01	0.25 ± 0.01	−0.31 ± 0.01	−0.25 ± 0.01	−0.27 ± 0.01	0.29 ± 0.01	10.8 ± 0.4
Atlantic	0.19 ± 0.01	0.22 ± 0.01	−0.27 ± 0.02	−0.22 ± 0.01	−0.24 ± 0.01	0.26 ± 0.02	10.3 ± 0.4
INTERM.	γ_{DIC}	β_{DIC}	ω_{DIC}	γ_{TA}	β_{TA}	ω_{TA}	Revelle
Aegean	0.23 ± 0.01	0.28 ± 0.01	−0.35 ± 0.01	−0.28 ± 0.01	−0.30 ± 0.01	0.33 ± 0.01	10.2 ± 0.1
Levantine	0.22 ± 0.01	0.27 ± 0.01	−0.34 ± 0.01	−0.27 ± 0.01	−0.30 ± 0.01	0.33 ± 0.01	10.3 ± 0.2
Ionian	0.22 ± 0.01	0.27 ± 0.01	−0.34 ± 0.01	−0.27 ± 0.01	−0.29 ± 0.01	0.32 ± 0.01	10.4 ± 0.1
Adriatic	0.22 ± 0.01	0.27 ± 0.01	−0.33 ± 0.01	−0.27 ± 0.01	−0.29 ± 0.01	0.32 ± 0.01	10.5 ± 0.2
Tyrrhenian	0.21 ± 0.01	0.26 ± 0.01	−0.32 ± 0.01	−0.26 ± 0.01	−0.28 ± 0.01	0.30 ± 0.01	10.8 ± 0.1
Western	0.21 ± 0.01	0.24 ± 0.01	−0.30 ± 0.01	−0.24 ± 0.01	−0.26 ± 0.01	0.28 ± 0.01	11.2 ± 0.1
Atlantic	0.18 ± 0.01	0.21 ± 0.01	−0.24 ± 0.02	−0.21 ± 0.01	−0.22 ± 0.01	0.23 ± 0.02	12.2 ± 0.5
DEEP	γ_{DIC}	β_{DIC}	ω_{DIC}	γ_{TA}	β_{TA}	ω_{TA}	Revelle
Aegean	0.22 ± 0.01	0.27 ± 0.01	−0.34 ± 0.01	−0.27 ± 0.01	−0.30 ± 0.01	0.32 ± 0.01	10.3 ± 0.1
Levantine	0.22 ± 0.01	0.27 ± 0.01	−0.34 ± 0.01	−0.27 ± 0.01	−0.29 ± 0.01	0.32 ± 0.01	10.4 ± 0.1
Ionian	0.22 ± 0.01	0.27 ± 0.01	−0.34 ± 0.01	−0.27 ± 0.01	−0.29 ± 0.01	0.32 ± 0.01	10.4 ± 0.1
Adriatic	0.22 ± 0.01	0.26 ± 0.01	−0.33 ± 0.01	−0.26 ± 0.01	−0.28 ± 0.01	0.31 ± 0.01	10.6 ± 0.1
Tyrrhenian	0.21 ± 0.01	0.25 ± 0.01	−0.31 ± 0.01	−0.25 ± 0.01	−0.27 ± 0.01	0.30 ± 0.01	10.9 ± 0.1
Western	0.21 ± 0.01	0.24 ± 0.01	−0.30 ± 0.01	−0.24 ± 0.01	−0.26 ± 0.01	0.29 ± 0.01	10.9 ± 0.1
Atlantic	0.17 ± 0.01	0.20 ± 0.01	−0.23 ± 0.01	−0.20 ± 0.01	−0.21 ± 0.01	0.22 ± 0.01	12.6 ± 0.3
BOTTOM	γ_{DIC}	β_{DIC}	ω_{DIC}	γ_{TA}	β_{TA}	ω_{TA}	Revelle
Aegean	–	–	–	–	–	–	–
Levantine	0.20 ± 0.01	0.24 ± 0.01	−0.30 ± 0.01	−0.24 ± 0.01	−0.26 ± 0.01	0.29 ± 0.01	9.4 ± 0.1
Ionian	0.22 ± 0.01	0.27 ± 0.01	−0.33 ± 0.01	−0.27 ± 0.01	−0.29 ± 0.01	0.32 ± 0.01	10.4 ± 0.1
Adriatic	–	–	–	–	–	–	–
Tyrrhenian	0.21 ± 0.01	0.25 ± 0.01	−0.31 ± 0.01	−0.25 ± 0.01	−0.27 ± 0.01	0.29 ± 0.01	11.0 ± 0.1
Western	0.21 ± 0.01	0.25 ± 0.01	−0.30 ± 0.01	−0.25 ± 0.01	−0.27 ± 0.01	0.29 ± 0.01	11.1 ± 0.1
Atlantic	–	–	–	–	–	–	–

$$\gamma_{DIC} = DIC - \frac{TA_C^2}{S}$$

$$\gamma_{TA} = \frac{TA_C^2 - DIC \times S}{TA_C}$$

$$\beta_{DIC} = \frac{DIC \times S - TA_C^2}{TA_C}$$

$$\beta_{TA} = \frac{TA_C^2}{DIC} - S$$

$$\omega_{DIC} = DIC - \frac{TA_C \times (2 \times [CO_2] + [HCO_3^-])}{P}$$

$$\omega_{TA} = TA_C - \frac{DIC \times P}{2 \times [CO_2] + [HCO_3^-]}$$

$$S = [HCO_3^-] + 4 \times [CO_3^{2-}] + \frac{[H^+] \times [B(OH)_4^-]}{K_{hb} + [H^+]} + [H^+] + [OH^-]$$

$$P = [HCO_3^-] - \frac{[H^+] \times [B(OH)_4^-]}{K_{hb} + [H^+]} - [H^+] - [OH^-]$$

$$TA_C = [HCO_3^-] + 2 \times [CO_3^{2-}]$$

$$DIC = [HCO_3^-] + [CO_3^{2-}] + [CO_2]$$

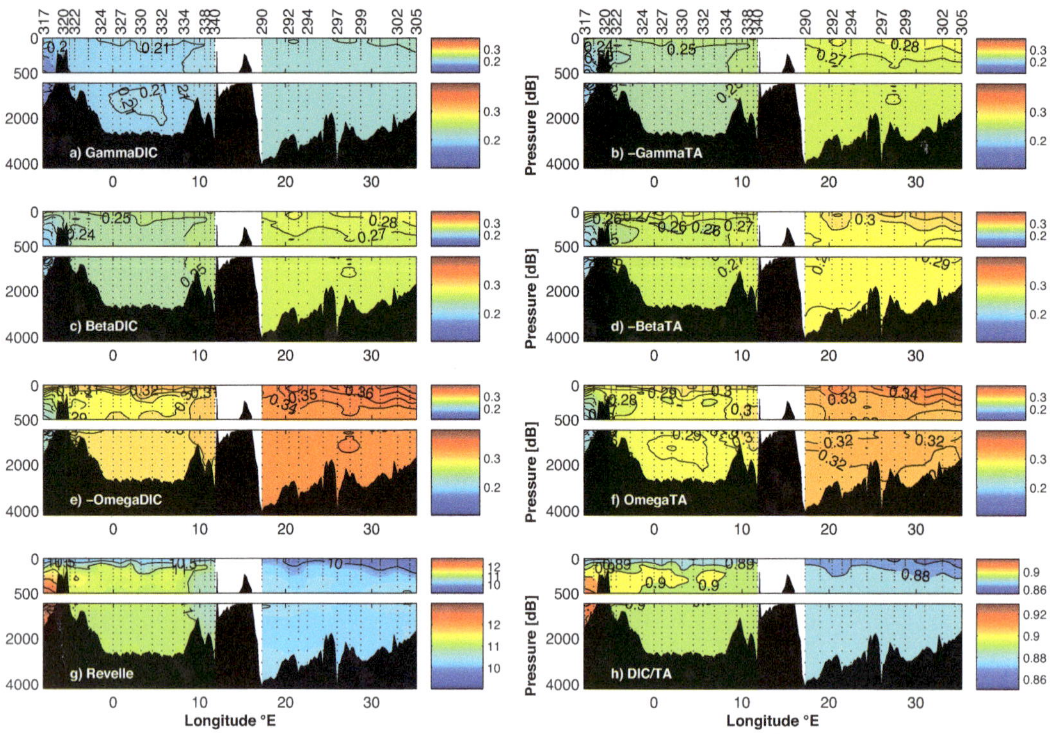

Fig. 10. Vertical distribution in the Mediterranean Sea from the Meteor cruise M84/3 of the buffer factors **(a)** γ_{DIC}, **(b)** $-\gamma_{\mathrm{TA}}$, **(c)** β_{DIC}, **(d)** $-\beta_{\mathrm{TA}}$, **(e)** $-\omega_{\mathrm{DIC}}$, **(f)** ω_{TA}, all in $\mathrm{mmol\,kg^{-1}}$, **(g)** Revelle factor and **(h)** DIC/TA ratio, both dimensionless.

- By examining the Revelle factor it is obvious that for a given change in CO_2 the relative change in DIC for MedSea waters is higher than in the adjacent Atlantic waters. That means that MedSea waters have the ability to store more anthropogenic carbon (C_{ANT}) than Atlantic waters. Within the MedSea the eastern basin is more prone to absorb more C_{ANT} for a given CO_2 increase than the western basin.

The impact of the three main anthropogenic processes altering the surface CO_2 chemistry in MedSea waters is given in Table 4, which shows absolute changes in aqueous CO_2 ($[CO_2]$), pH and the aragonite saturation (Ω-Ar) from the initial surface values in Table 2; the different sub-basins are treated separately. A C_{ANT} input through air-sea exchange (affecting DIC at constant TA) increasing DIC by $10\,\mu\mathrm{mol\,kg^{-1}}$ will lead to a increase of $[CO_2]$ from 0.56 to $0.78\,\mu\mathrm{mol\,kg^{-1}}$, a pH decrease of around 0.016 units in MedSea waters, less than the pH decrease in Atlantic waters (0.020) for the same perturbation; the saturation state is reduced by about 0.1 units, but a reduction of at least 0.9 is required for deep waters to become under-saturated in aragonite. A hypothetical decrease in calcification, resulting in a DIC increase of $10\,\mu\mathrm{mol\,kg^{-1}}$ and double in TA, will compensate the pH and Ω decrease due to the input of C_{ANT} and roughly half the corresponding increase in $[CO_2]$. Within the MedSea changes in $[CO_2]$ and pH increase westwards being smaller than in Atlantic waters. An increase in salinity of 0.03 will cause minor changes in the CO_2 chemistry.

The buffer factors here provided are useful to predict changes in the CO_2 chemistry as a function of DIC and/or TA changes due to different processes, natural and/or anthropogenic. However, it is relevant to relate the increase of C_{ANT} in the ocean to the increasing partial pressure of CO_2 (pCO_2) in the atmosphere instead of using a DIC increase as reference. Table 5 shows the absolute changes in DIC, pH and Ω-Ar from initial states in Table 2 as a consequence of prescribed pCO_2 increases. Note that a pCO_2 increase of $25\,\mu\mathrm{atm}$ causes changes in DIC by about $10\,\mu\mathrm{mol\,kg^{-1}}$, as already presented in Table 4. The main result in Table 5 is that as the Revelle factor in warm, salty and high TA waters is lower and the other buffer factors distribute as $|\omega| > |\beta| > \gamma_{\mathrm{DIC}}$ (Table 3), the change in DIC and Ω-Ar for a given pCO_2 increase is higher in the MedSea, but that for pH is lower than in the Atlantic. Therefore, MedSea waters are more (less) sensitive to DIC and Ω-Ar (pH) changes induced from the pCO_2 increase in the atmosphere (C_{ANT} increase).

4.2 Water mass characterisation

Here we present property-property plots for general physical and chemical variables, typical potential temperature-salinity, AOU-salinity and silicate concentration (SiO_2)-salinity (Figs. 11 and 13) and the CO_2 variables, TA, DIC and

Table 4. Absolute associated changes in aqueous [CO_2] (μmol kg^{-1}), pH (no units) and aragonite saturation (no units) state due to several processes related to global change: a DIC increase of 10 μmol kg^{-1} at constant TA (CO_2 air-sea exchange), a decrease in calcification of 10 μmol kg^{-1} in DIC and double in TA, a salinity increase of 0.03. Surface values from Table 2 and 3 are taken to perform the calculations.

	Air-sea CO_2 gas exchange ΔDIC = + 10 μmol kg^{-1}		
	Δ[CO_2]	ΔpH	ΔOmega
Aegean	0.56	−0.016	−0.10
Levantine	0.57	−0.016	−0.10
Ionian	0.62	−0.017	−0.07
Adriatic	0.62	−0.016	−0.10
Tyrrhenian	0.77	−0.018	−0.10
Western	0.76	−0.018	−0.10
Atlantic	0.78	−0.020	−0.10
	Calcification decrease ΔDIC = +10 μmol kg^{-1} ΔTA = +20 μmol kg^{-1}		
	Δ[CO_2]	ΔpH	ΔOmega
Aegean	−0.36	0.011	0.11
Levantine	−0.36	0.011	0.11
Ionian	−0.40	0.012	0.11
Adriatic	−0.40	0.012	0.11
Tyrrhenian	−0.52	0.013	0.11
Western	−0.52	0.013	0.11
Atlantic	−0.53	0.015	0.11
	Salinity increase		
	Δ[CO_2]	ΔpH	ΔOmega
Aegean	0.01	0.0001	0.004
Levantine	0.01	0.0001	0.004
Ionian	0.01	0.0001	0.002
Adriatic	0.01	0.0001	0.004
Tyrrhenian	0.01	0.0001	0.003
Western	0.01	0.0001	0.003
Atlantic	0.01	0.0001	0.003

pH25T vs. salinity, SiO_2 and AOU (Figs. 12 and 14). We separated the western and eastern basins showing with colours the different sub-basins comprised. Some special characteristics, mainly in deep waters, are presented in more detail with figures in the Supplement. The companion paper by Hainbucher et al. (2013) gives more details about the physical and hydrodynamical variability.

4.2.1 Aegean Sea (stations 287, 288 and 289)

Three stations during the M84/3 cruise were located in the Aegean Sea: 287 in the Chios Basin (central Aegean Sea), 289 in the eastern Cretan basin and 289 in the Kasos Strait

Table 5. Absolute associated changes in DIC concentration, pH and aragonite saturation state due several partial pressure of CO_2 in the atmosphere increase. The initial conditions are the mean 2011 values (Table 2).

	ΔDIC in μmol kg^{-1}			
	25 μatm	100 μatm	200 μatm	300 μatm
Aegean	13.1	52.4	104.8	157.2
Levantine	13.0	52.1	104.2	156.3
Ionian	12.9	51.6	103.1	154.7
Adriatic	11.9	47.8	95.5	143.3
Tyrrhenian	9.7	38.8	77.7	116.5
Western	9.8	39.1	78.2	117.2
Atlantic	10.8	43.3	86.6	129.9
	$\Delta\Omega$− Ar			
	25 μatm	100 μatm	200 μatm	300 μatm
Aegean	−0.13	−0.52	−1.05	−1.57
Levantine	−0.13	−0.52	−1.04	−1.56
Ionian	−0.13	−0.52	−1.03	−1.55
Adriatic	−0.12	−0.48	−0.96	−1.43
Tyrrhenian	−0.10	−0.38	−0.76	−1.14
Western	−0.10	−0.38	−0.77	−1.15
Atlantic	−0.11	−0.43	−0.85	−1.28
	ΔpH			
	25 μatm	100 μatm	200 μatm	300 μatm
Aegean	−0.020	−0.088	−0.199	−0.349
Levantine	−0.021	−0.089	−0.200	−0.350
Ionian	−0.022	−0.094	−0.215	−0.383
Adriatic	−0.019	−0.083	−0.186	−0.321
Tyrrhenian	−0.017	−0.073	−0.160	−0.269
Western	−0.017	−0.074	−0.164	−0.278
Atlantic	−0.022	−0.095	−0.216	−0.384

(both in the southern Aegean Sea). In the upper 100 dbars high salinity identifies LSW (Levantine Surface Water) entering the Aegean Sea through the eastern Cretan arc straits (Fig. 11a). This LSW was also identified by the highest pH25T values measured in the M84/3 cruise (> 8.01, Fig. 12h) and also high TA and DIC values (≈ 2620 and ≈ 2280 μmol kg^{-1}, respectively, Fig. 12a and d). At intermediate levels, 100 to 250 dbars, CIW (Cretan Intermediate Water) presented the highest TA values for the M84/3 cruise (≈ 2645 μmol kg^{-1}, Fig. 12a and d). Deep layers below 2500 dbars were occupied by CDW (Cretan Deep Water), warm, salty and very dense (Fig. 11a), also identified by the distinct SiO_2-salinity relation (Fig. 11c), practically constant DIC (≈ 2320 μmol kg^{-1}) and TA (≈ 2630 μmol kg^{-1}) values vs. SiO_2 (Fig. 12b and e).

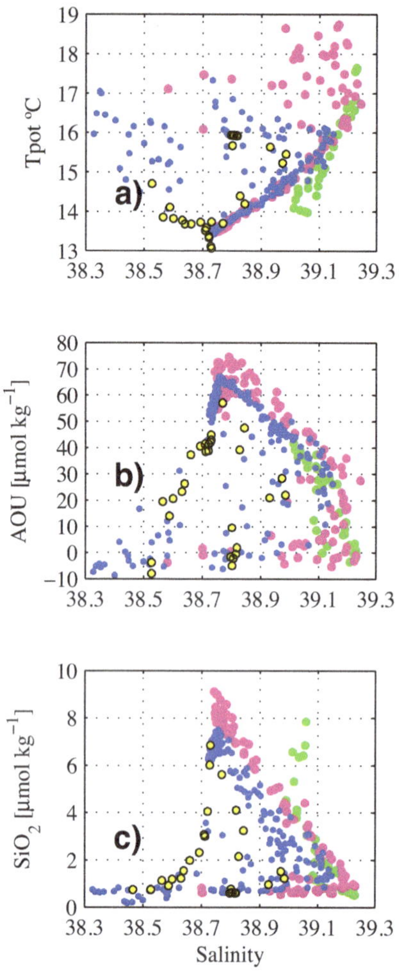

Fig. 11. Property-property plots for the M84/3 data in the eastern basin: Aegean Sea (green dots), Levantine basin (magenta), Ionian Sea (blue) and Adriatic Sea (yellow). **(a)** Potential temperature (°C) vs. salinity, **(b)** Apparent Oxygen Utilisation (AOU in μmol kg^{-1}) vs. salinity and (c) silicate concentration (SiO_2 in μmol kg^{-1}) vs. salinity. The station positions are shown in Fig. 1.

4.2.2 Levantine Basin (from station 290 to 298)

In the upper water column, < 100 dbars, AW and LSW could be distinguished using salinity (Fig. 11a), LSW presented salinity > 38.9 and AW from 38.58 to 38.9 (Fig. 11a). TA, DIC and pH25T values much higher in LSW could also differentiate them (Fig. 12a, d and g). Interestingly, LSW presented pH25T values higher, and AW lower than 8.01 for waters with AOU < 0 (Fig. 12i).

At intermediate depths, 100–400 dbars, the main water mass was LIW (Levantine Intermediate Water), In addition to the typical scorpion tail relation between salinity and potential temperature, LIW was recognized by SiO_2 minima associated with a salinity maximum (Fig. 11c) and high TA ($\approx 2630\,\mu$mol kg^{-1}, Fig. 12a) values.

In the deep layers below 1000 dbars within the Levantine basin, the salinity minimum and then a pronounced temperature-salinity inversion (temperature and salinity increase) typically detected during the EMT (Roether et al., 2007) was still detected in this basin (Hainbucher et al., 2013) (red lines in Fig. S1). Water in the inversion is EMDW with an Aegean Sea origin (Aeg-EMDW) produced during the EMT.

Adriatic EMDW (Ad-EMDW) to the west of the East Mediterranean Ridge was differentiated from Aeg-EMDW mainly with AOU and SiO_2 (Fig. 11b and c), both being lower in the Ad-EMDW, which also contained higher transient tracer values (Stöven and Tanhua, 2013), indicating a more recent origin. It is difficult to separate them using TA, DIC and pH25T, both EMDW presented similar values, Ad-EMDW was slightly higher in DIC (around $5\,\mu$mol kg^{-1}) and TA (around $4\,\mu$mol kg^{-1}) and lower in pH25T (0.005 pH units) than Aeg-EMDW (Fig. 12c, f and i) (blue and black lines in Fig. S1).

The salinity minimum around 1000 dbars and below the temperature-salinity inversion previously commented provoke a sort of hook in the relationship between DIC – TA – pH25T vs. AOU or SiO_2 (red lines in Fig. S1) for deep waters in this basin. DIC and TA decrease with decreasing SiO_2 or AOU until their corresponding maxima (salinity minimum) and then towards the bottom DIC and TA increase with decreasing AOU (increasing salinity). pH is the opposite, it decreases with AOU increasing until the AOU maximum (salinity minimum) and then pH increases and AOU decreases (Fig. S1).

4.2.3 Ionian Sea (from station 299 to 309, 314 and 315)

Two upper water masses could be differentiated in the Ionian Sea, AW (< 17 °C, < 38.9) and LSW (> 16 °C, > 38.8) (Fig. 11a). During M84/3 LSW in the Ionian Sea was found in the stations to the western part of the Hellenic Trench (Sts. 299, 300, 301) (Fig. 1). Below, two intermediate water masses, LIW and CIW could be distinguished. LIW with temperature and salinity maxima around 200 dbars was also associated with a TA and DIC maximum (Fig. 12a and d). The mixing between LIW and AW produced an exponential relationship between TA and DIC vs. salinity (Fig. 12a and d). Deviations from this relationship towards lower TA and DIC values corresponded to LSW in this basin. The other intermediate water detected, CIW, mixed with LIW south of Crete, in the eastern basin was found in stations 299, 300 and 301 to the west of the Antikithera strait. This water mass with also very high salinity, presented an even higher TA than LIW (Fig. 12a, b and c). For example in the case of the TA vs. SiO_2 plot (Fig. 12b) CIW was detected as a deviation towards higher values from the typical TA – SiO_2 relationship. CIW was also seen as a hump between 0 to $30\,\mu$mol kg^{-1} AOU over the LIW data (Fig. 12c).

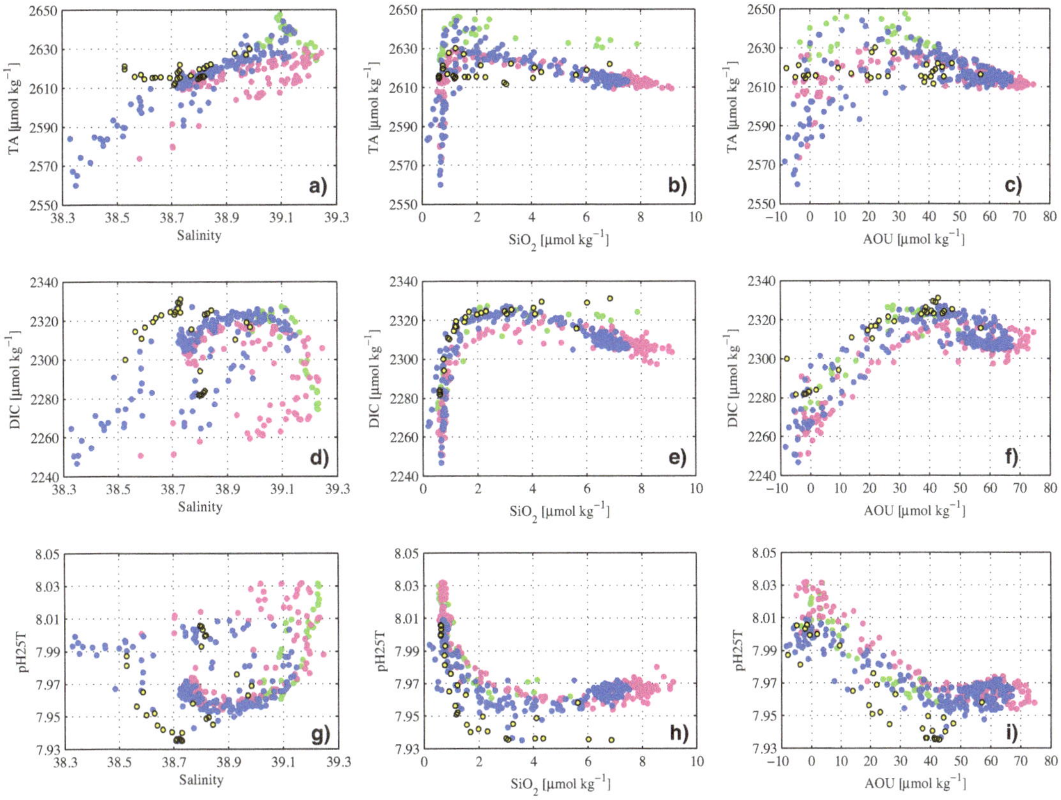

Fig. 12. Property-property plots for the M84/3 data in the eastern basin: Aegean Sea (green dots), Levantine basin (magenta), Ionian Sea (blue) and Adriatic Sea (yellow). First row: alkalinity (TA in μmol kg^{-1}) vs. **(a)** salinity, **(b)** silicate concentration (SiO_2 in μmol kg^{-1}) and **(c)** AOU (μmol kg^{-1}). Second row: Dissolved Inorganic Carbon (DIC in μmol kg^{-1}) vs. **(d)** salinity, **(e)** SiO_2 and **(f)** AOU. Third row: pH25T vs. **(d)** salinity, **(e)** SiO_2 and **(f)** AOU. The station positions are shown in Fig. 1.

In the Ionian Sea, Post-EMT-EMDW occupied the layer below 1500 dbars, it presented lower AOU and SiO_2 values than EMT Aeg-EMDW (Fig. 11b and c) in the Levantine basin. The recent replenishment from the Adriatic Sea, 2001 onwards, was detected as a temperature-salinity increase in bottom waters of the central Ionian Sea (Rubino and Hainbucher, 2007) (green and cyan lines in Fig. S2). These younger bottom waters in the Ionian Sea explain the DIC increase (Fig. 6), AOU and pH25T (Fig. 7) decrease towards the bottom of the Ionian Sea (compare red line with the blue and cyan lines in Fig. S2).

The hook structure in the relationships DIC – TA – pH vs. AOU for deep waters in the Levantine basin (Fig. S1) was here eroded as is the temperature-salinity inversion associated with EMT Aeg-EMDW. The AOU maximum in the Ionian Sea was no longer as clearly associated with the salinity minimum as in the Levantine basin. Below 500 dbars DIC and TA decrease with increasing AOU, from the AOU maximum TA practically stays constant and DIC increases. The very recent input of bottom Ad-EMDW is noticed in the pH vs. AOU relationship, from 500 dbars pH decreases with AOU increasing, from the AOU maximum pH increases (AOU decreases) but close to the bottom both pH and AOU decrease (Fig. S2).

4.2.4 Adriatic Sea (stations 310 and 313)

Only two stations were sampled in the Adriatic Sea for CO_2 variables, one in the Adriatic Pit (313) and one to the east of the Otranto Strait (310). One of the densest deep waters during the M84/3 cruise was sampled in the Adriatic Pit at 1100 dbars (Fig. 11a), this water remains confined as the Otranto Strait is only 800 m deep. Waters in the Adriatic presented high TA values despite a moderate salinity (Fig. 12a), they were out of the general linear relationship between TA and salinity due to the high TA of the rivers discharging in the area (Luchetta et al., 2010; Cantoni et al., 2012; Copin-Montegut, 1993). The relationship TA vs. SiO_2 was also different for Adriatic waters (Fig. 12b), being lower than the general trend in the eastern MedSea. Additionally DIC and pH in the Adriatic had a distinct signal, high values for DIC and low values for pH in waters with salinity lower than 38.7 (Fig. 12d and g).

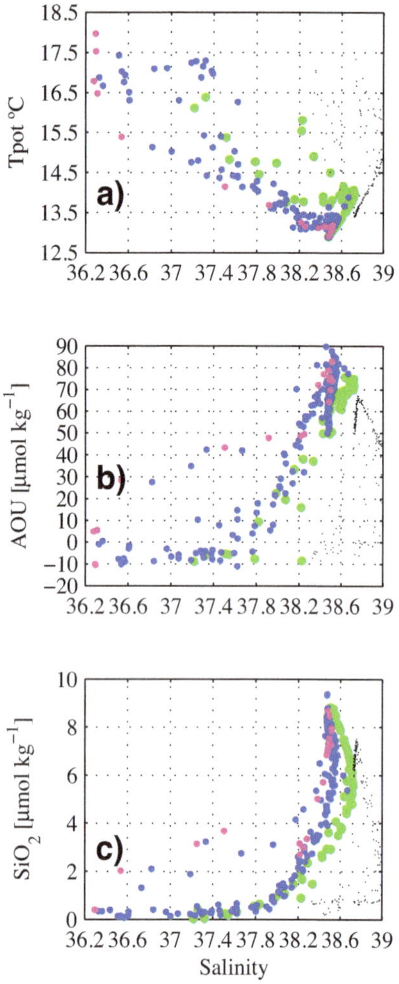

Fig. 13. Property-property plots for the M84/3 data in the western basin: Tyrrhenian Sea (green dots), Western basin (blue) and Strait of Gibraltar (magenta). Light black dots correspond to Ionian Sea data. **(a)** Potential temperature (°C) vs. salinity, **(b)** Apparent Oxygen Utilisation (AOU in μmol kg^{-1}) vs. salinity and **(c)** silicate concentration (SiO_2 in μmol kg^{-1}) vs. salinity. The station positions are shown in Fig. 1.

The current overflow from the Adriatic stood out in the pH – AOU plot (Figs. 14i and 7, upper panel): pH data for AOU values between 40 and 45 μmol kg^{-1} in stations 309 and 310 located in the Otranto Strait, deviated towards 7.94 pH25T values, corresponding to waters around 500 dbars in the Adriatic Pit.

4.2.5 Tyrrhenian Sea (from station 316 to 321)

This basin acts as a connector between the eastern and the western MedSea, here intermediate waters entering from the east mix with deep waters entering from the west (e.g. Hopkins, 1988; Astraldi et al., 1996). This fact was noticed in the property-property plots shown in Fig. 13 were the points corresponding to the Tyrrhenian Sea clearly connected the western and eastern MedSea waters.

In the surface layer, AW with salinity lower than 38 was detected in all the Tyrrhenian stations except 319 in the centre of the basin (Fig. 13a). The freshest AW was seen in station 316, the nearest to Sicily, here AW comes directly from Gibraltar through the Algerian sub-basin. AW was also distinguished by low AOU and SiO_2 values close to zero (Fig. 13b and c). The minimum TA and DIC associated with AW are 2466 and 2178 μmol kg^{-1}, respectively (Fig. 14a and d). In the centre of the Tyrrhenian Sea the upper 52 dbars were occupied by probably modified AW with 38.5 > salinity > 38 affected by evaporation or mixed with the salty intermediate water entering through the Strait of Sicily. This water also presented a different relationship between salinity and pH25T compared to unmixed AW (Fig. 14g).

Maximum salinity values in the intermediate layer between 250 and 500 dbars pointed to LIW entering from the eastern basin (Fig. 13a). As LIW flows into the Tyrrhenian Sea remineralization processes increase its AOU, DIC and lower pH25T, with extreme values found in St. 319 (cyan line in Fig. S3) in the interior of this basin and at St. 321 (magenta line in Fig. S3) in the Sardinia Channel where aged LIW flows out of the Tyrrhenian Sea into the western basin.

At St. 316 (red line in Fig. S3), the nearest one to Sicily in the Sardinia Sicily passage, the intermediate layer was occupied by a fresher LIW (38.66–38.68) with higher AOU and DIC (lower pH25T and TA) values. It might be LIW mixed with AW returning to the western basin although this is an area where LIW is supposed to enter the Tyrrhenian Sea.

Deep layers in the Tyrrhenian are occupied by TDW with a typical linear relationship in the temperature-salinity plot (Fig. 13a). No hook in the temperature-salinity diagram to lower salinities was detected in bottom waters of the Tyrrhenian Sea, this hook would show the input of recently formed WMDW as detected in the bottom of the western basin. In the Tyrrhenian Sea bottom WMDW stands out from TDW by the sharp decrease in AOU and pH25T (Fig. 7) and increase in DIC (Fig. 6) for example with salinity (red dots in Fig. S4).

4.2.6 Western basin (from station 322 to 336)

The upper layer in this basin was occupied by AW very fresh ($\ll$ 38) and practically 100 % saturated in oxygen (AOU $\approx$ 0 μmol kg^{-1}) and with no SiO_2 (Fig. 13b and c). In the intermediate layer the salinity maximum associated with LIW evolves from 38.67 at St. 322 to 38.51 at St. 336. There was a marked reduction in salinity (and TA) when LIW entered the western basin passing from 38.71 at St. 321 in the Sardinia Channel, 38.67 at St. 322 (right west of the channel) and then 38.51 at St. 323 in the western basin. The change in the other chemical variables, SiO_2, AOU, DIC or pH25T was less acute.

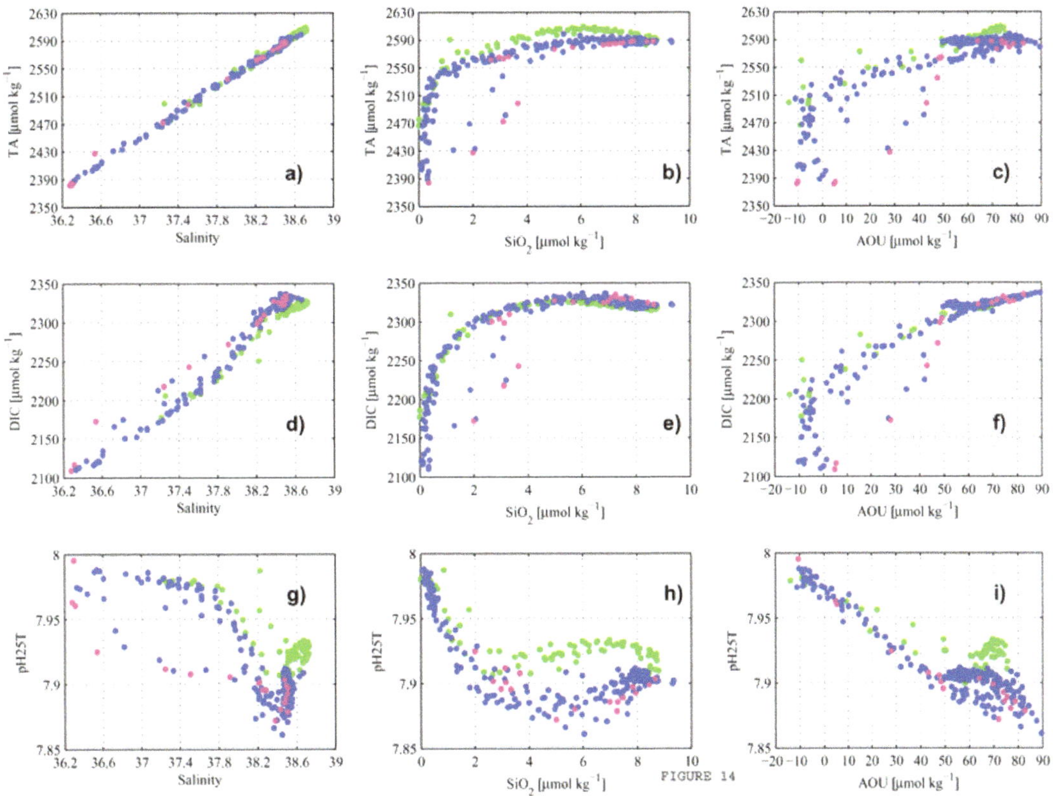

Fig. 14. Property-property plots for the M84/3 data in the western basin: Tyrrhenian Sea (green dots), Western basin (blue) and Strait of Gibraltar (magenta). First row: alkalinity (TA in μmol kg^{-1}) vs. **(a)** salinity, **(b)** silicate concentration (SiO_2 in μmol kg^{-1}) and **(c)** AOU (μmol kg^{-1}). Second row: Dissolved Inorganic Carbon (DIC in μmol kg^{-1}) vs. **(d)** salinity, **(e)** SiO_2 and **(f)** AOU. Third row: pH25T vs. **(d)** salinity, **(e)** SiO_2 and **(f)** AOU. The station positions are shown in Fig. 1.

When in the western basin, LIW was associated with a maximum in AOU (Fig. 13b), DIC (Fig. 14d) and minimum in pH25T (Fig. 14g), so it was the layer of maximum remineralization of organic matter, contrary to the eastern basin where this layer was right below LIW.

The near bottom layer in the western basin was occupied by a recently formed WMDW with higher salinity, so the temperature-salinity plots present a sort of hook structure (Hainbucher et al., 2013). This new WMDW presented higher DIC and lower AOU than the older WMDW (Figs. 14f and S5).

4.2.7 Atlantic waters (from station 337 to 344)

Due to the dramatic scale change, Figs. 13 and 14 show only the data up to St. 338 (corresponding to the Strait of Gibraltar). All the samples collected to the west of this Strait are shown in Fig. S6.

At the Strait of Gibraltar (St. 337 and 338) the intermediate LIW outflowing presented a salinity of 38.48 (TA $\approx$ 2584 μmol kg^{-1}, DIC $\approx$ 2331 μmol kg^{-1}, pH25T $\approx$ 7.89, AOU $\approx$ 75.3 μmol kg^{-1}) (Figs. 14 and S6). The outflow was only detected at stations 342 and 344 (pink lines in Fig. S6) in the Portuguese continental slope, with a salinity of 36.50 (TA $\approx$ 2409 μmol kg^{-1}, DIC $\approx$ 2193 μmol kg^{-1}, pH25T $\approx$ 7.86, AOU $\approx$ 58.2 μmol kg^{-1}). In the centre of the Gulf of Cádiz (339 and 340, black and green lines respectively in Fig. S6), the outflow was mixed with the upper low salinity central waters and presented a salinity of 36.20 (TA $\approx$ 2392 μmol kg^{-1}, DIC $\approx$ 2202 μmol kg^{-1}, pH25T $\approx$ 7.79, AOU $\approx$ 93 μmol kg^{-1}).

5 Summary

Two coordinated cruises sampled the Mediterranean Sea for CO_2 variables (TA, DIC and pH) in 2011: the RV Urania EF11 (Sardinia Sicily passage) and the RV Meteor M84/3 (East-West full-length transect).

The over-determined and high-quality CO_2 data collected in the M84/3 cruise allowed performing the first internal consistency analysis for CO_2 data in the peculiar warm, salty and low DIC/TA ratio MedSea waters. Using the ®MATLAB version for the consensus CO2SYS programme the preferred set of constants is option 4 [CO_2 constants from Mehrbach et al. (1973) refitted by Dickson and Millero (1987)] along with the sulphate constant from Dickson (1990) and the parameterization of borate from Uppström (1974). Using this combination our pH, TA and DIC measurements from the M84/3 cruise are internally consistent to -0.0003 ± 0.005 for pH, $0.1 \pm 3.3\,\mu\text{mol kg}^{-1}$ for TA and $-0.1 \pm 3\,\mu\text{mol kg}^{-1}$ for DIC. Despite this encouraging result further studies are needed including a fourth variable: $p\text{CO}_2$ in order to verify the internal consistency of K_1 (only a reliable K_1 is needed to calculate $p\text{CO}_2$ from the input of pH – TA or pH – DIC) and $[\text{CO}_3^{2-}]$ in order to revise the K_2 parameterization in low DIC/TA ratio waters where the CO_2 system is more sensitive to K_2. Using the combination of pH – TA – DIC only a reliable K_2 is needed in order to calculate bicarbonate, carbonate and aqueous CO_2 (Park, 1969). Particularly comparing measured (Byrne and Yao, 2008) and calculated $[\text{CO}_3^{2-}]$ could be helpful to revise K_2.

The Meteor M84/3 data allowed a post-EMT, full length and depth description of CO_2 variables in the MedSea. This study aims to be base-line for future data and model studies dealing with the effects of natural (climate) and human (warming, acidification, etc...) driven changes in the CO_2 species and ancillary variables associated.

As a general view, the eastern MedSea presents higher TA (2560–2644 μmol kg^{-1}) and pH25T (7.935–8.032) and less variable DIC (2247–2331 μmol kg^{-1}) than the western basin with TA (2388–2608 μmol kg^{-1}), pH25T (7.861–7.988) and DIC (2110–2336 μmol kg^{-1}). The saturation for calcite and aragonite in both basins is well above 1, for the eastern basin Ω-Ca vary 2.5–5.78 and Ω-Ar 1.73–3.80 and for the western basin Ω-Ca 2.73–5.25 and Ω-Ar 1.86–3.44.

The CO_2 chemistry in the warm, salty and high TA waters in the MedSea compared to the Atlantic is in general less sensitive to temperature changes because of the low DIC/TA ratio. MedSea waters are more resistant to direct changes in DIC and/or TA due to natural and/or anthropogenic processes, since the calculated buffer factors are higher in MedSea waters, showing an increasing westwards gradient. The CO_2 chemistry in the western basin would be more sensitive than in the eastern basin.

However, from the point of view of an increasing $p\text{CO}_2$ in the atmosphere, that increases indirectly DIC and keeps TA constant in the ocean, waters in the MedSea have a lower Revelle factor than in the Atlantic, and therefore are able to store more DIC for a given $p\text{CO}_2$ increase than in the Atlantic but the changes in pH (Ω) would be lower (higher) than in the Atlantic.

As a summary the water masses encountered according to depth are:

i. Levantine Surface Water (LSW): warm and salty (> 38.9) surface water (upper 100 dbars) formed by intensive heating and evaporation in the Levantine basin, found during the M84/3 cruise in this basin and the Aegean and Ionian Seas. LSW was characterised by high TA ($\approx 2610\,\mu$mol kg^{-1}) and DIC ($\approx 2270\,\mu$mol kg^{-1}). In April 2011 LSW presented pH25T > 8.01 associated with AOU < 0 pointing to active photosynthetic activity.

ii. Atlantic Water (AW): it is the upper part (< 125 dbars) of the open estuarine cell in the MedSea, enters through the Strait of Gibraltar and flows to the Levantine basin while suffering evaporation and heating. The main CO_2 variable showing the presence of AW is TA, which evolved from very low values in the Strait of Gibraltar ($\approx 2390\,\mu$mol kg^{-1}) to high values ($\approx 2610\,\mu$mol kg^{-1}) in the Levantine basin. Evaporation processes and mixing with the salty Levantine Intermediate Water (LIW) induced a practically linear relation between TA and salinity. Another linear relationship was found between pH25T and AOU for the western MedSea in April 2011 pointing to a biological modulation of pH.

iii. Levantine Intermediate Water (LIW): main intermediate water mass in the MedSea (125–500 dbars), is the lower part of the open estuarine cell in the MedSea. Typically LIW is identified by a scorpion tail in the temperature-salinity diagram associated with a salinity maximum. The main CO_2 variable identifying LIW in any MedSea sub-basin was an intermediate TA maximum evolving from $\approx 2620\,\mu$mol kg^{-1} in the Levantine and Ionian basins to $\approx 2600\,\mu$mol kg^{-1} in the Tyrrhenian Sea, $\approx 2590\,\mu$mol kg^{-1} in the western basin, and $\approx 2584\,\mu$mol kg^{-1} overflowing in the Strait of Gibraltar, and finally $\approx 2409\,\mu$mol kg^{-1} in the Mediterranean Water located in the Gulf of Cadiz, that then circulates into the North Atlantic.

In the Levantine and Ionian basins LIW was found shallower (≈ 250 dbars) than in the western basin (≈ 450 dbars). In the eastern MedSea LIW was situated above the layer of maximum organic matter mineralization while in the western MedSea they coincided and therefore was associated with a maximum in DIC and AOU and minimum in pH.

iv. Cretan Intermediate Water (CIW): formed in the Cretan Sea, this intermediate (250 dbars) water was characterised by even higher TA (2630–2645 μmol kg^{-1}) than LIW and detected in the Aegean Sea and the Ionian Sea west of the Antikithera strait.

v. Cretan Deep Water (CDW): produced by convection in the Cretan Sea, was clearly identified apart of the high density, by a distinct relation between SiO_2 – salinity, high constant TA (2630 μmol kg^{-1}) and DIC (2320 μmol kg^{-1}) values with varying SiO_2 (2–4 μmol kg^{-1}). This water remained confined in the Aegean Sea.

vi. Pre-EMT Eastern Mediterranean Deep Water (EMDW): this water mass, of an Adriatic Sea origin, was still detected in the Levantine basin by a salinity minimum around 1000 dbars, it presented a minimum in tracer values as CFC-12 or SF_6, but also a minimum in pH because was the layer of organic matter remineralization as also indicated by the maximum in AOU and inorganic nutrients.

vii. EMT-EMDW: below the salinity minimum in the Levantine basin temperature and salinity increase producing a hook in the temperature-salinity diagram which is associated with the input of EMDW produced in the Aegean Sea during the EMT. Due to a more recent origin it presents lower AOU, slightly higher DIC and pH values than pre-EMT EMDW.

viii. Post-EMT-EMDW: bottom layers in the Hellenic Trench (Levantine basin) and deep layers (> 1000 dbars) in the Ionian Sea are occupied by EMDW with an Adriatic origin produced after the EMT. The hook in the temperature-salinity diagram was no longer detected and a less salty post-EMT-EMDW occupied the deep layer in the Ionian Sea. This water was also identified by lower AOU and pH values compared to the saltier EMT-EMDW.

Bottom layers in the Ionian Sea, below about 3000 dbars, were occupied by recent overflow waters produced in the Adriatic Sea after 2001. It presented a clear signature in DIC ($> 2310\,\mu\text{mol kg}^{-1}$), AOU ($< 55\,\mu\text{mol kg}^{-1}$) and pH ($< 7.967$).

ix. Waters in the Adriatic Sea: surface waters in this basin presented a distinct TA vs. salinity relationship with values around $2610\,\mu\text{mol kg}^{-1}$ for salinity < 38.7, higher than the general TA–salinity relation mainly due to the high TA input through Adriatic rivers. The overflow from the Adriatic Sea was mainly detected in the pH vs. AOU relationship, for AOU values between 40–45 $\mu\text{mol kg}^{-1}$ pH deviated to 7.94, those are waters from around 500 dbars in the Adriatic Pit spilling into the Ionian Sea.

x. Waters in the Tyrrhenian Sea: the property-property plots showed that this basin is a connector between the eastern and western basins, where LIW mixes with deep waters from the west producing Tyrrhenian Deep Water (TDW). During 2011 no recently formed Western Mediterranean Deep Water (WMDW) was detected in the bottom of the Tyrrhenian Sea.

The water mass exchange between the eastern and western MedSea is highlighted using data from the Urania EF11 cruise in the Sardinia Sicily passage: AW mainly enters the Tyrrhenian on the eastern part of the passage, close to Sicily, and exits this basin on the Sardinian side, becoming saltier and also higher in TA; the same circulation scheme is done by LIW, which becomes colder and saltier after circulating in the Tyrrhenian, TA is decreased and DIC increased.

xi. Western Mediterranean Deep Water (WMDW): this water occupied layers below 1000 dbars in the western basin of the MedSea, with respect to CO_2 variables it presented DIC ($2318–2320\,\mu\text{mol kg}^{-1}$) higher and TA ($2584–2588\,\mu\text{mol kg}^{-1}$) and pH25T (7.90–7.91) lower than EMDW. In the bottom layer a recent input, after 2005, of WMDW was detected with higher salinity and temperature, while DIC (AOU) clearly increased (decreased) towards the bottom.

xii. In the Strait of Gibraltar LIW presented in 2011 a salinity of 38.48 and TA $\approx 2584\,\mu\text{mol kg}^{-1}$, DIC $\approx 2331\,\mu\text{mol kg}^{-1}$, pH25T ≈ 7.89, AOU $\approx 75.3\,\mu\text{mol kg}^{-1}$; once mixed with central waters in the Gulf of Cadiz, Mediterranean Water sampled in the Portuguese slope was detected with a salinity of 36.50, TA $\approx 2409\,\mu\text{mol kg}^{-1}$, DIC $\approx 2193\,\mu\text{mol kg}^{-1}$, pH25T ≈ 7.86, AOU $\approx 58.2\,\mu\text{mol kg}^{-1}$.

The CO_2 and ancillary data collected in 2011 highlight the need for a sustained programme monitoring the temporal evolution of water masses in the MedSea. This small but relevant marginal sea is a perfect laboratory basin to detect the effect of natural and anthropogenic driven changes in the physics and CO_2 chemistry. Special emphasis should be given to CO_2 variables as MedSea waters are especially sensitive to pCO_2 increases in the atmosphere: they are prone to absorb more DIC for given pCO_2 increase, this DIC increase in the surface is rapidly transported to depth with the overturning circulation, and therefore pH and carbonate saturation conditions are altered, with consequences still unknown for the well adapted organisms.

Acknowledgements. The authors want to thank the captains and crews on the research vessels R/V Meteor and R/V Urania for the excellent cooperation during the campaigns. The Meteor cruise M84/3 was supported by a grant from the Deutsche Forschungsgemenschaft- Senatskommision für Ozenographie (DFG), and from a grant from the DFG; TA 317/3-1. The participation of G. C. was partly supported by the Mediterranean Science Commission (CIESM). The authors are grateful to C. Colmenero-Modrón, I. Hueso-Zabaleta, F. Rozada and I. Buns for the technical assistance during CO_2, oxygen and inorganic nutrients sampling and analysis. Our special thanks to X. A. Álvarez-Salgado for providing some equipment. Funding for CO_2 measurements was provided by MICINN (MedSea Repeat, CTM2010-12244-E). This is a HOTMIX (CTM2011-30010-C02) contribution.

Edited by: M. Hoppema

References

Álvarez, M.: The CO_2 system observations in the Mediterranean Sea: past, present and future, in: CIESM, Designing Med-SHIP: a Program for repeated oceanographic surveys, N° 43 in CIESM Workshop monographs, edited by: Briand, F., 164 pp., Monaco, 2012.

Astraldi, M., Gasparini, G. P., Sparnocchia, S., Moretti, S., and Sansone, E.: The characteristics of the Mediterranean water masses and the water transport in the Sicily Channel at longtime scales, in: Dynamics of Straits and Channels, edited by: Briand, F., vol. 2, CIESM Science Series, Monaco, 95–118, 1996.

Astraldi, M., Gasparini, G. P., Vetrano, A., and Vignudelli, S.: Hydrographic characteristics and interannual variability of water masses in the central Mediterranean: a sensitivity test for long-term changes in the Mediterranean Sea, Deep Sea Res. I, 49, 661–680, 2002.

Azov, I.: The Mediterranean Sea, a marine desert?, Mar. Pollut. Bull., 23, 225–232, 1991.

Benson, B. B. and Krause Jr, D.: The concentration and isotopic fractionation of oxygen dissolved in freshwater and seawater in equilibrium with the atmosphere, Limnol. Oceanogr., 29, 620–632, 1984.

Bergamasco, A. and Malanotte-Rizzoli, P.: The circulation of the Mediterranean Sea: a historical review of experimental investigations, Adv. Oceanogr. Limnol., 1, 11–28, doi:10.1080/19475721.2010.491656, 2010.

Bosc, E., Bricaud, A., and Antoine, D.: Seasonal and interannual variability in algal biomass and primary production in the Mediterranean Sea, as derived from 4 years of SeaWiFS observations, Global Biogeochem. Cy., 18, GB1005, doi:10.1029/2003GB002034, 2004.

Brunet, C., Poisson, A., Lebel, J., and Porot, V.: Alcalinité totale, carbone inorganique, calcium, densité, in: Propriétés Hydrologiques et Chimiques des Eaux du Bassin Occidental de la Méditerranée. Campagne Mediprod IV-15 Octobre–17 Novembre 1981, Résultats des Campagnes à la Mer, 26, 89–93, Centre National pour l'Exploitation des Océans, Brest, France, 1984.

Byrne, R. H. and Yao, W. S.: Procedures for measurement of carbonate ion concentrations in seawater by direct spectrophotometric observations of Pb(II) complexation, Mar. Chem., 112, 128–135, 2008.

Cai, W.-J. and Wang, Y.: The chemistry, fluxes and sources of carbon dioxide in the estuarine waters of the Satilla and Altamaha Rivers, Georgia, Limnol. Oceanogr., 43, 657–668, 1998.

Cantoni, C., Luchetta, A., Celio, M., Cozzi, S., Raicich, F., and Catalano, G.: Carbonate system variability in the Gulf of Trieste (North Adriatic Sea), Estuarine, Coast. Shelf Sci., 115, 51–62, doi:10.1016/j.eccs.2012.07.006, 2012.

Claustre, H., Morel, A., Hooker, S. B., Babin, M., Antoine, D., Oubelkheir, K., Bricaud, A., Leblanc, K., Quéguiner, B., and Maritorena, S.: Is desert dust making oligotrophic waters greener?, Geophys. Res. Lett., 29, doi:10.1029/2001GL014056, 2001.

Clayton, T. D. and Byrne, R. H.: Spectrophotometric seawater pH measurements: total hydrogen ion concentration scale concentration scale calibration of m-cresol purple and at-sea results, Deep-Sea Res. Pt. I, 40, 2115–2129, 1993.

Copin-Montegut, C.: Alkalinity and carbon budgets in the Mediterranean Sea, Global Biogeochem. Cy., 7, 915–925, 1993.

Crise, A., Allen, J. I., Baretta, J., Crispi, G., Mosetti, R., and Solidoro, C.: The Mediterranean pelagic ecosystem response to physical forcing, Prog. Oceanogr., 44, 219–243, 1999.

Dickson, A. G.: Thermodynamics of the dissociation of boric acid in synthetic seawater from 273.15 to 318.15 K, Deep-Sea Res., 37, 755–766, 1990.

Dickson, A. G. and Millero, F. J.: A comparison of the equilibrium constants for the dissociation of carbonic acid in seawater media, Deep-Sea Res., 34A, 1733–1743, 1987.

D'Ortenzio, F. and Ribera d'Alcalà, M.: On the trophic regimes of the Mediterranean Sea: a satellite analysis, Biogeosciences, 6, 139–148, doi:10.5194/bg-6-139-2009, 2009.

Egleston, E. S., Sabine, C. L., and Morel, F. M. M.: Revelle revisited: Buffer factors that quantify the response of ocean chemistry to changes in DIC and alkalinity, Global Biogeochem. Cy., 24, GB1002, doi:10.1029/2008GB003407, 2010.

Goyet, C. and Poisson, A.: New determination of carbonic acid dissociation constants in seawater as a function of temperature and salinity, Deep-Sea Res., 36, 1635–1654, 1989.

Hainbucher, D., Rubino, A., Cardin, V., Tanhua, T., Schroeder, K., and Bensi, M.: Hydrographic situation during cruise M84/3 and P414 (spring 2011) in the Mediterranean Sea, Ocean Sci. Discuss., 10, 2399–2432, doi:10.5194/osd-10-2399-2013, 2013.

Hansson, I.: A new set of acidity constants for carbonic acid and boric acid in sea water, Deep-Sea Res., 20, 4611–478, 1973a.

Hansson, I.: The determination of dissociation constants of carbonic acid in synthetic sea water in the salinity range of 20–40°/‰ and temperature range of 5–30 °C, Acta Chemica Scandanavia, 27, 931–944, 1973b.

Hernández-Ayón, J. M., Belli, S. L., and Zirino A.: pH, alkalinity, and total CO_2 in coastal seawater by potentio-metric titration with a difference derivative readout, Anal. Chim. Acta, 394, 101–108, 1999.

Hopkins, T. S.: Recent observations on the intermediate and deep water circulation in the Southern Tyrrhenian Sea, Oceanol. Acta, 9, 41–50, 1988.

Huertas, I. E., Ríos, A. F., García-Lafuente, J., Navarro, G., Makaoui, A., Sánchez-Román, A., Rodriguez-Galvez, S., Orbi, A., Ruíz, J., and Pérez, F.F.: Atlantic forcing of the Mediterranean oligotrophy, Global Biogeochem. Cy., 26, GB2022, doi:10.1029/2011GB004167, 2012.

Ignatiades, L., Gotsis-Skretas, O., Pagou, K., and Krasakopoulou, E.: Diversification of phytoplankton community structure and related parameters along a large-scale longitudinal east-west transect of the Mediterranean Sea, J. Plankton. Res., 31, 411–428, 2009.

Key, R. M., Kozyr, A., Sabine, C. L., Lee, K., Wanninkhof, R., Bullister, J. L., Feely, R. A., Millero, F. J., Mordy, C., and Peng, T.-H.: A global ocean carbon climatology: Results from Global Data Analysis Project (GLODAP), Global Biogeochem. Cy., 18, GB4031, doi:10.1029/2004GB002247, 2004.

Key, R. M., Tanhua, T., Olsen, A., Hoppema, M., Jutterström, S., Schirnick, C., van Heuven, S., Kozyr, A., Lin, X., Velo, A., Wallace, D. W. R., and Mintrop, L.: The CARINA data synthesis project: introduction and overview, Earth SySt. Sci. Data, 2, 105–121, doi:10.5194/essd-2-105-2010, 2010.

Khoo, K. H., Ramette, R. W., Culberson, C. H., and Bates, R. G.: Determination of hydrogen ion concentrations in seawater from

5 to 40 °C: standard potentials at salinities from 20 to 45°/‰, Anal. Chemist., 49, 29–34, 1977.

Kleypas, J. A., McManus, J. W., and Meñez, L. A. B.: Environmental limits to coral reef development: where do we draw the line?, Am. Zool., 39, 146–159, 1999.

Krasakopoulou, E., Souvermezoglou, E., and Goyet, C.: Anthropogenic CO_2 fluxes in the Otranto strait (E. Mediterranean) in February 1995, Deep-Sea Res. I, 58, 1103–1114, 2011.

Krom, M. D., Woodward, E. M. S., Herut, B., Kress, N., Carbo, P., Mantoura, R. F. C., Spyres, G., Thingstad, T. F., Wassmann, P., Wexels-Riser, C., Kitidis, V., Law, C. S., and Zodiatis, G.: Nutrient cycling in the south east Levantine basin of the eastern Mediterranean: results from a phosphorus starved system, Deep Sea Res. II, 52, 2879–2896, 2005.

Lazzari, P., Solidoro, C., Ibello, V., Salon, S., Teruzzi, A., Béranger, K., Colella, S., and Crise, A.: Seasonal and inter-annual variability of plankton chlorophyll and primary production in the Mediterranean Sea: a modelling approach, Biogeosciences, 9, 217–233, doi:10.5194/bg-9-217-2012, 2012.

Lee, K., Millero, F. J., and Wanninkhof, R.: The carbon dioxide system in the Atlantic Ocean, J. Geophys. Res., 102, 15693–15707, doi:10.1029/97JC00067, 1997.

Lee, K., Kim, T.-W., Byrne, R. H., Millero, F. J., Feely, R. A., and Liu, M.-M.: The universal ratio of boron to chlorinity for the North Pacific and North Atlantic Oceans, Geochim. Cosmochim. Acta, 74, 1801–1811, 2010.

Lee, K., Sabine, C. L., Tanhua, T., Kim, T. W., Feely, R. A., and Kim, H. C.: Roles of marginal seas in absorbing and storing fossil fuel CO_2, Energy Environ. Sci., 4, 1133–1146, doi:10.1039/C0EE00663G, 2011.

Luchetta, A., Cantoni, C., and Catalano, G.: New observations of CO_2 induced acidification in the Norterhn Adriatic Sea over the last quarter century, Chem. Ecol., 26, 1–17, doi:10.1080/02757541003627688, 2010.

Ludwig, W., Bouwman, A. F., Dumont, E., and Lespina, F.: Water and nutrient fluxes from major Mediterranean and Black Sea rivers: Past and future trends and their implications for the basin-scale budgets, Global Biogeochem. Cy., 24, GB0A13, doi:10.1029/2009GB003594, 2010.

Lueker, T. J., Dickson, A. G., and Keeling, C. D.: Ocean pCO_2 calculated from dissolved inorganic carbon, alkalinity, and equations for K_1 and K_2: validation based on laboratory measurements of CO_2 in gas and seawater at equilibrium, Mar. Chem., 70, 105–119, 2000.

McCartney, M. S. and Mauritzen, C.: On the origin of the warm inflow to the Nordic Seas, Prog. Oceanogr., 51, 125–214, 2001.

Mehrbach, C., Culberson, C. H., Hawley, J. E., and Pytkowitcz, R. M.: Measurements of the apparent dissociation constants of carbonic acid in seawater at atmospheric pressure, Limnol. Oceanogr., 18, 897–907, 1973.

Millero, F. J.: The marine inorganic carbon cycle, Chem. Rev., 107, 308–341, doi:10.1021/cr0503557, 2007.

Millero, F. J.: Carbonate constants for estuarine waters, Mar. Freshw. Res., 61, 139–142, 2010.

Millero, F., Morse, J., and Chen, C. T.: The carbonate system in the western Mediterranean Sea, Deep-sea Res., 26, 1395–1404, 1979.

Millero, F. J., Pierrot, D., Lee, K., Wanninkhof, R., Feely, R. Sabine, C. L., Key, R. M., and Takahashi, T.: Dissociation constants for carbonic acid determined from field measurements, Deep-Sea Res. I, 49, 1705–1723, 2002.

Millero, F. J., Graham, T. B., Huang, F., Bustos-Serrano, H., and Pierrot, D.: Dissociation constants of carbonic acid in seawater as a function of salinity and temperature, Mar. Chem., 100, 80–94, 2006.

Millot, C. and Taupier-Letage, I.: Circulation in the Mediterranean Sea, in: The Mediterranean Sea, the Handbook of Environmental Chemistry, Springer, Berlin/Heidelberg, 29–66, 2005.

Mojica-Prieto, F. J. and Millero, F. J.: The values of $pK1 + pK2$ for the dissociation of carbonic acid in seawater, Geochim. Cosmochim. Acta, 66, 2529–2540, 2002.

Moutin, T. and Raimbault, P.: Primary production, carbon export and nutrients availability in western and eastern Mediterranean Sea in early summer 1996 (MINOS cruise), J. Marine. Syst., 33, 273–288, 2002.

Park, P. K.: Oceanic CO_2 system: an evaluation of ten methods of investigation, Limnol. Oceanogr., 2, 179–186, 1969.

Peng, T. H., Takahashi, T., Broecker, W. S., and Olafsson, J.: Seasonal variability of carbon dioxide, nutrients and oxygen in the northern North Atlantic surface water: Observations and model, Tellus, 39B, 439–458, 1987.

Pérez, F. F. and Fraga, F.: A precise and rapid analytical procedure for alkalinity determination, Mar. Chem., 21, 169–182, 1987.

Pérez, F. F., Ríos, A. F., Rellán, T., and Álvarez, M.: Improvements in a fast potentiometric seawater alkalinity determination, Cienc. Mar., 26, 463–478, 2000.

Pujo-Pay, M., Conan, P., Oriol, L., Cornet-Barthaux, V., Falco, C., Ghiglione, J.-F., Goyet, C., Moutin, T., and Prieur, L.: Integrated survey of elemental stoichiometry (C, N, P) from the western to eastern Mediterranean Sea, Biogeosciences, 8, 883–899, doi:10.5194/bg-8-883-2011, 2011.

Rahav, E., Herut, B., Levi, A., Mulholland, M. R., and Berman-Frank, I.: Springtime contribution of dinitrogen fixation to primary production across the Mediterranean Sea, Ocean Sci., 9, 489–498, doi:10.5194/os-9-489-2013, 2013.

Ribera d'Alcalà, M., Civitarese, G., Conversano, F., and Lavezza, R.: Nutrient ratios and fluxes hint at overlooked processes in the Mediterranean Sea, J. Geophys. Res., 108, 8106, doi:10.1029/2002JC001650, 2003.

Ricke, K. L., Orr, J. C., Schneider, K., and Caldeira, K.: Risks to coral reefs from ocean carbonate chemistry changes in recent earth system model projections, Environ. Res. Lett., 8, 034003, doi:10.1088/1748-9326/8/3/034003, 2013.

Rivaro, P., Messa, R., Massolo, S., and Frache, R.: Distributions of carbonate properties along the water column in the Mediterranean Sea: Spatial and temporal variations, Mar. Chem., 121, 236–245, 2010.

Robinson, A. R. and Golnaraghi, M.: The Physical and Dynamical Oceanography of the Mediterranean Sea, in: Ocean Processes in Climate Dynamics: Global and Mediterranean Examples, edited by: Malanotte-Rizzoli, R. and Robison, A. R., NATO ASI Series Volume 419, 255–306, 1994.

Robinson, A. R., Leslie, W. G., Theocharis, A., and Lascaratos, A.: Mediterranean Sea Circulation, Encyclopedia of Ocean Sciences, Academic Press, 1689–1706, 2001.

Roether, W., Klein, B., Manca, B. B., Theocharis, A., and Kioroglou, S.: Transient Eastern Mediterranean deep waters in response to the massive dense-water output of the Aegean Sea in the 1990s, Progr. Ocean., 74, 540–571, 2007.

Roether, W., Manca, B. B., Klein, B., Bregant, D., Georgopoulos, D., Beitzel, V., Kovacevich, V., and Lucchetta, A.: Recent changes in Eastern Mediterranean deep waters, Science, 271, 333–335, 1996.

Roy, R. N., Roy, L. N., Vogel, K. M., Porter-Moore, C., Pearson, T., Good, C. E., Millero, F. J., and Campbell, D. M.: The dissociation constants of carbonic acid in seawater at salinities 5 to 45 and temperatures 0 to 45 °C, Mar. Chem., 44, 249–267, 1993.

Rubino, A. and Hainbucher, D.: A large abrupt change in the abyssal water masses of the eastern Mediterranean, Geophys. Res. Lett., 34, L23607, doi:10.1029/2007GL031737, 2007.

Schneider, A., Wallace, D. W. R., and Kortzinger, A.: Alkalinity of the Mediterranean Sea, Geophys. Res. Lett., 34, L15608, doi:10.1029/2006GL028842, 2007.

Schneider, A., Tanhua, T., Körtzinger, A., and Wallace, D. W. R.: High anthropogenic carbon content in the eastern Mediterranean, J. Geophys. Res., 115, C12050, doi:10.1029/2010JC006171, 2010.

Schroeder, K., Gasparini, G. P., Borghini, M., and Delfanti, R.: Biogeochemical tracers and fluxes in the Western Mediterranean Sea, spring 2005, J. Mar. Systems, 80, 8–24, 2010.

Siokou-Frangou, I., Christaki, U., Mazzocchi, M. G., Montresor, M., Ribera d'Alcalá, M., Vaqué, D., and Zingone, A.: Plankton in the open Mediterranean Sea: a review, Biogeosciences, 7, 1543–1586, doi:10.5194/bg-7-1543-2010, 2010.

Stöven, T.: Ventilation processes of the Mediterranean Sea based on CFC-12 ad SF6 measurements, Diploma, Leibniz-Institut für Meereswissenschaften der Mathematischnatur- wissenschaftlichen Fakultät, Christian-Albrecht-Universit at zu Kiel, Kiel, 2011.

Stöven, T. and Tanhua, T.: Ventilation of the Mediterranean Sea constrained by multiple transient tracer measurements, Ocean Sci. Discuss., 10, 1647–1705, doi:10.5194/osd-10-1647-2013, 2013.

Takahashi, T., Williams, R. T., and Bos, D. L.: Carbonate chemistry, in: GEOSECS Pacific Expedition, edited by: Broecker, W. S., Spencer, D. W., and Craig, H., Volume 3, Hydrographic Data 1973–1974, National Science Foundation, Washington, D.C., 77–83, 1982.

Tanhua, T., van Heuven, S., Key, R. M., Velo, A., Olsen, A., and Schirnick, C.: Quality control procedures and methods of the CARINA database, Earth SySt. Sci. Data, 2, 35–49, doi:10.5194/essd-2-35-2010, 2010.

Tanhua, T., Álvarez, M., and Mintrop, L.: Carbon Dioxide, Hydrographic, and Chemical Data Obtained During the R/V Meteor MT84_3 Mediterranean Sea Cruise (5 April–28 April 2011), available at: http://cdiac.ornl.gov/ftp/oceans/CLIVAR/Met_84_3_Med_Sea/, Carbon Dioxide Information Analysis Center, Oak Ridge National Laboratory, US Department of Energy, Oak Ridge, Tennessee, doi:10.3334/CDIAC/OTG.CLIVAR_06MT20110405, 2012.

Tanhua, T., Hainbucher, D., Schroeder, K., Cardin, V., Álvarez, M., and Civitarese, G.: The Mediterranean Sea system: a review and an introduction to the special issue, Ocean Sci., 9, 789–803, doi:10.5194/os-9-789-2013, 2013a.

Tanhua, T., Hainbucher, D., Cardin, V., Álvarez, M., Civitarese, G., McNichol, A. P., and Key, R. M.: Repeat hydrography in the Mediterranean Sea, data from the Meteor cruise 84/3 in 2011, Earth SySt. Sci. Data, 5, 289–294, doi:10.5194/essd-5-289-2013, 2013b.

Touratier, F. and Goyet, C.: Decadal evolution of anthropogenic CO_2 in the north western Mediterranean Sea (at the Dyfamed site) from the mid-1990's to the mid-2000's, Deep-Sea Res. I, 56, 1708–1716, 2009.

Touratier, F., Guglielmi, V., Goyet, C., Prieur, L., Pujo-Pay, M., Conan, P., and Falco, C.: Distributions of the carbonate system properties, anthropogenic CO_2, and acidification during the 2008 BOUM cruise (Mediterranean Sea), Biogeosciences Discuss., 9, 2709–2753, doi:10.5194/bgd-9-2709-2012, 2012.

Uppström, L. R.: The boron/chloronity ratio of deep-sea water from the Pacific Ocean, Deep-Sea Res., 21, 161–162, 1974.

van Heuven, S., Pierrot, D., Rae, J. W. B., Lewis, E., and Wallace, D. W. R.: MATLAB Program Developed for CO2 System Calculations. ORNL/CDIAC-105b. Carbon Dioxide Information Analysis Center, Oak Ridge National Laboratory, U.S. Department of Energy, Oak Ridge, Tennessee, doi:10.3334/CDIAC/otg.CO2SYS_MATLAB_v1.1, 2011.

Volpe, G., Santoleri, R., Vellucci, V., Ribera d'Alcalà, M., Marullo, S., and D'Ortenzio, F.: The colour of the Mediterranean Sea: Global versus regional bio-optical algorithms evaluation and implication for satellite chlorophyll estimates, Remote Sens. Environ., 107, 625–638, 2007.

8

Temporal variations of zooplankton biomass in the Ligurian Sea inferred from long time series of ADCP data

R. Bozzano[1], E. Fanelli[2], S. Pensieri[1], P. Picco[2], and M. E. Schiano[3]

[1]National Research Council of Italy (CNR-ISSIA) – Via de Marini 6, 16149, Genova, Italy
[2]Marine Environment Research Centre (ENEA) – Santa Teresa, Pozzuolo di Lerici 19032, Italy
[3]National Research Council of Italy (CNR-ISMAR) – Via de Marini 6, 16149, Genova, Italy

Correspondence to: R. Bozzano (roberto.bozzano@cnr.it)

Abstract. Three years of 300 kHz acoustic doppler current profiler data collected in the central Ligurian Sea are analysed to investigate the variability of the zooplankton biomass and the diel vertical migration in the upper thermocline. After a pre-processing phase aimed at avoiding the slant range attenuation, hourly volume backscattering strength time series are obtained. Despite the lack of concurrent net samples collection, different migration patterns are identified and their temporal variability examined by means of time–frequency analysis. The effect of changes in the environmental condition is also investigated. The highest zooplankton biomasses are observed in April–May just after the peak of surface primary production in March–April. The main migration pattern found here points to a "nocturnal" migration, with zooplankton organisms occurring deeper in the water column during the day and shallower at night. Also, twilight migration is highlighted during this study. The largest migrations are recorded in November–December, corresponding to lowest backscattering strength values and they are likely attributable to larger and more active organisms (i.e. euphausiids and mesopelagic fish). The results suggest further applications of the available historical acoustic doppler current profiler time series.

1 Introduction

The Acoustic Doppler Current Profiler (ADCP) is a widely used instrument to monitor the marine currents. Time series of these measurements span from a few days up to several years and are available for many coastal and open ocean sites. However, ADCP data may also be usefully employed to measure biological variables as pointed out by Flagg and Smith (1989) and Plueddemann and Pinkel (1989) at the end of the 1980s, who showed that the acoustic backscatter signal was correlated with the zooplankton biomass. Since then, several biological investigations have been carried out using ADCP observations (Rippeth and Simpson, 1998; Pinot and Jansá, 2001; Jiang et al., 2007; Ashjian et al., 2002, 1994; van Haren, 2007). Unfortunately, ADCP data are more qualitative than quantitative because the conversion from backscatter intensity to equivalent zooplankton biomass, usually obtained by direct comparison against coincident data from net samples, is somewhat problematic, and the resulting relationship provides only rough estimates (Pinot and Jansá, 2001; Ashjian et al., 2002; Fielding et al., 2004). Nevertheless, the instrument, ensuring long-term autonomous monitoring, offers the opportunity to study the zooplankton distribution and its variability on many different temporal and spatial scales. Thus, its data supply important information required in marine ecology research that cannot be satisfactorily obtained using the classical observational methodology based on discrete net sampling.

Furthermore, ADCP data may help in reconstructing the diel vertical migration (DVM) of the zooplankton which is probably the biggest animal migration, in terms of biomass, on the planet (Hays, 2003). Since zooplankton represents the trophic link between primary producers (i.e. phytoplankton in the photic zone) and the higher trophic levels up to top predators, the comprehension of their migratory patterns and biomass distribution is of crucial importance in understanding the pelagic ecosystem functioning.

In this context, the paper analyses the echo intensity and the vertical velocity data obtained from an ADCP moored in the open Ligurian Sea from September 2003 to February 2006 in terms of variations in the zooplankton biomass and DVM. These patterns are discussed, taking into account the results achieved by several different previous studies carried out in the Ligurian Sea, one of the most dynamically active regions in the Mediterranean.

Although the experiment did not include specific biological measurements, the aim of this work is to highlight the usefulness of long-term ADCP data series to enhance current knowledge on zooplankton, especially their migratory patterns.

The paper is organized into the following parts: Sect. 2 describes the investigated area, the data and the methodologies used in the study. The analysis of the zooplankton behaviour and its variability, with particular attention to the characteristic patterns of the DVM and the influence of some environmental variables, is given in Sect. 3. The results are discussed in Sect. 4.

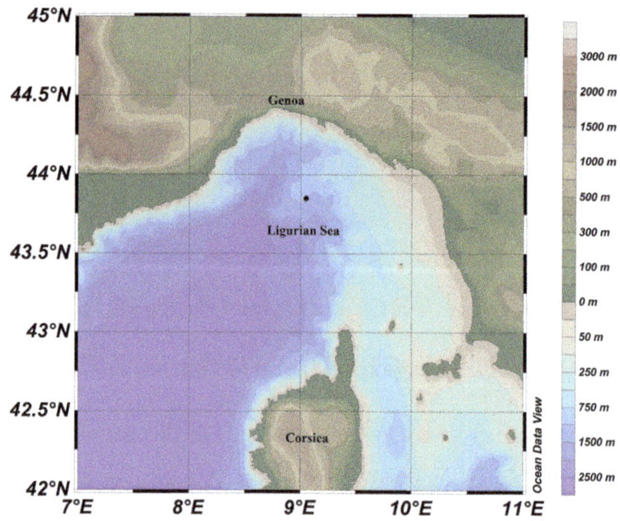

Fig. 1. Bathymetry and horography of the Ligurian region. Black dot corresponds to the mooring position.

2 Materials and methods

2.1 Main features of the investigated area

The Ligurian Sea is a 3000 m-deep basin surrounded in the north by the Alps and limited by Corsica to the south (Fig. 1). These topographic constraints as well as the thermal contrast between land and sea give rise to specific local effects that influence the general circulation of both atmosphere and ocean, causing a strong variability in the upper ocean. The general circulation of the Ligurian Basin is characterized by a permanent basin-wide cyclonic circulation involving both the surface Modified Atlantic Water (MAW) and the lower Levantine Intermediate Water (LIW) (Crepon et al., 1982; Millot, 1999); it also shows important seasonal variability (Astraldi and Gasparini, 1992). The currents are generally weaker in the summer than during the winter and the contribution from the Tyrrhenian Sea is strongly reduced in summer. Significant inter-annual variability is observed (Vignudelli et al., 2000; Birol et al., 2010), as well as an intense mesoscale activity with marked seasonal variations (Taupier-Letage and Millot, 1986; Sammari et al., 1995).

Furthermore, due to the interplay of these particular climatic, oceanographic and physiographic factors, the area is highly productive and hosts a rich and complex ecosystem. This is also sustained by vertical mixing and coastal upwelling, generated by the prevailing northwesterly wind, which pumps deep nutrients and other organic substances contributed by rivers into the euphotic zone where they fertilize growing phytoplankton populations. Hence, the area attracts several cetacean species and is part of the "Cetacean Sanctuary" protected area, established to preserve the richness and variety of cetaceans living here with more than eight species, including the fin whale *Balaenoptera physalus*.

All these issues make the Ligurian Basin a meaningful research site for both physicists and biologists.

Several previous studies were conducted in the Ligurian Sea to determine the composition and biomass distribution of zooplankton communities (Licandro and Ibanez, 2000; Licandro and Icardi, 2009; Tarling et al., 2001; Andersen and Sardou, 1992; Sardou and Andersen, 1996). Although a comprehensive study which analyses the zooplankton composition at a seasonal scale in the whole Ligurian Basin is lacking, a rough reconstruction of species composition and dominance throughout the year can be done.

The mesozooplankton is mostly dominated by copepods (64 % of the total number of taxa); as a general pattern, in the entire Mediterranean Sea the bulk of the copepod population is concentrated in the epipelagic layer above 100 m depth with abundances decreasing sharply thereafter (Mazzocchi et al., 2007; di Carlo et al., 1984; Weikert and Trinkaus, 1990; Brugnano et al., 2011). Seasonal variations are observed in the composition of the zooplankton population with the dominance of different species throughout the year, such as *Clausocalanus* spp. and *Fritillaria* spp. in winter, and *Oithona helgolandica* and *Temora stylifera* in autumn (Licandro and Ibanez, 2000). Among the macrozooplankton/micronekton, Tarling et al. (2001) report a dominance of euphausiids (mainly *Meganyctiphanes norvegica* and *Nematoscelis megalops*) and pteropods (almost exclusively *Cavolinia inflexa*) in September, while mostly *Nematoscelis megalops* is found in the Ligurian central zone in May (Andersen and Sardou, 1992). A further study by Sardou and Andersen (1996) shows that *Meganyctiphanes norvegica* on the coastal side of the Ligurian Front has peaks of abundance in February and again in August, as also confirmed by McGehee et al. (2004).

In addition, due to the particular hydrographic conditions of the area, three main assemblages may be defined: one linked to the divergence zone of the basin, one associated to the periphery of the divergence and the latter with the eastern continental shelf (Licandro and Icardi, 2009). Different mesozooplanktonic taxa describe each assemblage, being copepods of Scolecithricidae family and appendicularians of genus *Fritillaria* dominant in the divergence, while *Clausocalanus furcatus* and *Paracalanus* spp. mainly inhabit the eastern continental shelf. The third assemblage is mostly characterized by ostracods and the copepods *Neocalanus gracilis* and *Clausocalanus paululus*. Each assemblage is identified by different biomass values, being least in the divergence zone (0.8–1.4 mg m^{-3}), greater at the periphery of the divergence, and at its maximum in shallower waters on the eastern continental shelf (1.7–4.2 mg m^{-3}) close to the study area (Licandro and Icardi, 2009).

Overall, in the Mediterranean Sea, zooplankton dynamic is characterized by late winter and summer blooms which are ubiquitous in the different basins and are preceded by phytoplanktonic blooms related to environmental factors including stratification (Furnestin, 1960; Margalef, 1985). In the northwestern Mediterranean Sea the late winter zooplankton bloom is generally delayed to spring (April–May) since the peak of surface primary production is generally recorded in February–March (Andersen, 2001a; Fanelli et al., 2009).

2.2 Environmental conditions during the period 2003–2006

The study was carried out in three periods related to the recovery and the re-positioning at sea of the mooring (Table 1). Some unusual marine conditions occurred during each of the three periods. Particularly, the first one took place at the end of summer 2003, which was among the hottest of the last century. The anomalous warming affected the whole of Europe and, above all, the central Mediterranean and was characterized by persistent calm weather conditions from May to the end of September 2003 (Sparnocchia et al., 2006). Several mass-mortality episodes were observed along Mediterranean coasts (Garrabou et al., 2009), as the strong vertical stratification prevented vertical mixing, thus reducing the oxygen contribution from the atmosphere to the deeper layers of the ocean, and allowing intense warming of the surface waters. Only at the end of October 2003, after several strong wind events due to the passage of lows, were the upper layers completely mixed and normal conditions were again established. The second experiment was carried out during the severe winter of 2004–2005 when very intense events of deep water formation took place in the northwestern Mediterranean Sea, despite significant warming and salinification of the entire water column which occurred in this winter (Font et al., 2007). The third experiment was characterized by an anomalous long period (from April 2005 to February 2006) of very weak currents. Indeed, the recorded values did not exceed 30 cm s^{-1} until November when they increased during the passage of the only low system of the overall period and reached the maximum in the first days of January 2006. Moreover, winter 2005–2006 showed an abrupt change of the water mass physical characteristics (temperature and salinity) linked to deep water convection which occurred in 2006 (Smith et al., 2008). Even current direction showed a very uncommon pattern, with several episodes of eastward currents.

This significant inter-annual variability in the circulation was also evident in the horizontal currents measured by the used ADCP data (Fig. 2). Analysis for inter-annual variability was performed on the data collected from October to February since this period was available for all three deployments.

During the period 2003–2004 currents were mainly directed towards north-northwest and very few events of weak southward currents were recorded. On average, the mean hourly velocities were greater than 15 cm s^{-1}, sometimes with peaks in the surface layers exceeding 60 cm s^{-1}. On the contrary, during the period 2005–2006, the average velocity was less than 10 cm s^{-1} at all depths and the maximum values never rose above 45 cm s^{-1}. The prevailing current was to the west and occurrences of southward currents increased.

The anomalous behaviour of the currents in the last period, both in velocity and direction, may be associated with the persistence of a high pressure system over the Ligurian Sea, inducing an extraordinary long period of calm wind condition. The anomalous conditions were also confirmed by the presence of saltier, warmer and shallower-than-normal intermediate water in the Ligurian Basin (compared with previous years) that was responsible in the first days of February 2006 for the unusual mixed layer depth of 2000 m, in a basin where it is commonly a few hundred metres deep, and for dense water formation from February to April 2006 (Martín et al., 2010).

2.3 Experimental setup

An upward-looking RDI Sentinel 300 kHz ADCP was operating in the central Ligurian Sea (43°47.77′ N, 9°02.85′ E) from 13 September 2003 until 22 February 2006 to investigate the upper layer dynamics in the area (Picco et al., 2010).

During this period, the mooring was recovered for maintenance and redeployed twice. Each time, the water depth and hence instruments depths were inevitably slightly different, also the ADCP temporal resolution was refined for the final deployment. The whole observation period was divided into three different phases (Table 1).

The vertical displacement of the mooring was checked using pressure measurements from a CTD device installed close to the ADCP and using the distance of the ADCP to water surface computed following Visbeck and Fischer (1995).

Table 1. Experimental configuration of the ADCP during the three deployments, from September 2003 to February 2006.

Period	No. of samples	Sample frequency	Depth	Bin size
13 Sep 2003–24 May 2004	6120	1 h	58 m	8 m
23 Sep 2004–15 Apr 2005	4943	1 h	100 m	8 m
19 Apr 2005–22 Feb 2006	14934	30 min	80 m	8 m

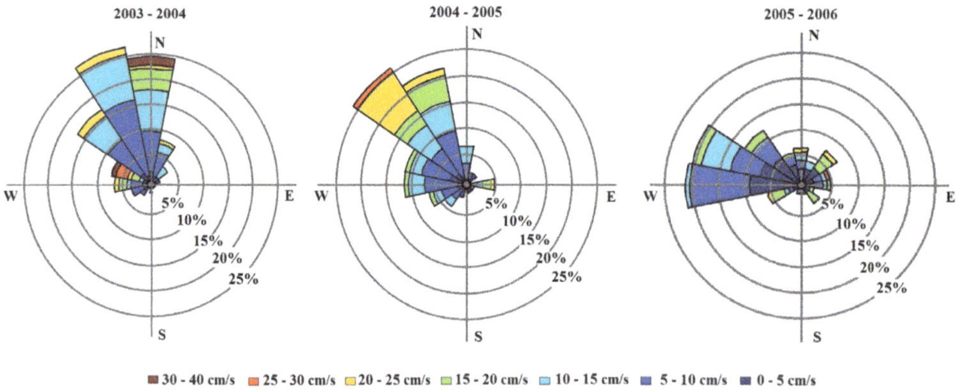

Fig. 2. Distribution of the horizontal currents recorded at 40 m depth from October to February during the three deployments.

The mooring in all three periods showed good stability, indeed pitch and roll data never exceeded 2.5°, well below the 9° limit defined by the manufacturer for tilt compensation. Pitch and roll data showed a standard deviation less than than 1° for all three periods, satisfying the requirements for good velocities data (RD Instruments, 2007).

All observations were quality checked; data were considered valid only (a) if characterized by at least three beam solutions, (b) if satisfying the constraints of maximum range established by Gordon (1996) and (c) if the threshold of error velocity was not exceeded.

In all three periods the percentage of data rejected was less than 0.1 %. The final overall ADCP data set consists of 11 063 hourly data and 14 934 samples with a sampling time of 30 min (Table 1).

Several supplementary data were used. Particularly, the sunrise/sunset times at the mooring location were computed by using the air–sea toolbox developed at Woods Hole Science Center (http://woodshole.er.usgs.gov/operations/sea-mat/air_sea-html/index.html), while surface wind data on the mooring site were obtained from the ERA-Interim (Dee et al., 2011) database products from the European Centre for Medium-Range Weather Forecasts (ECMWF).

In addition, values of net primary production (NPP) obtained from the vertically generalized production model based on MODIS and SeaWIFS measurements (http://www.science.oregonstate.edu/ocean.productivity) were used as proxies of surface primary production in the area (Behrenfeld and Falkowski, 1997). Considering the normal delay of about 1 month between peaks of surface production and the increase in zooplankton biomass (Truscott and Brindley, 1994) values of NPP may help to interpret the ADCP observed profiles.

2.4 Volume backscattering strength computation

The measure of the echo intensity scattered by the ocean is usually given in terms of volume backscattering strength (Sv hereafter) in dB re $(4\pi\ \mathrm{m})^{-1}$. Following Deines (1999), Sv is computed for each depth cell along each of the four beams through Eq. (1):

$$\mathrm{Sv} = C + 10\log_{10}((T_x + 273.16)R^2) - L_{\mathrm{DBM}} - P_{\mathrm{DBM}} + 2\alpha R + K_{\mathrm{c}}(E - E_{\mathrm{r}}). \quad (1)$$

C (−143.5 dB) is the instrumental constant of RDI profilers for Workhorse Sentinel ADCP 300 kHz. T_x is the ADCP internal temperature in °C. R is the slant range defined in Eqs. (3) and (6). L_{DBM} is the $10\log_{10}$ of the transmit pulse length in m (8.21 m for all three deployments). P_{DBM} is the $10\log_{10}$ of the transmit power in Watt, defined by RDI for the ADCP model here used as 14 dBW. E is the echo intensity provided by the ADCP. E_{r} (40 counts) is the echo reference value when there is no signal and it is specifically determined for each instrument. K_{c} is the factor for converting to dB unit the raw echo data provided by the ADCP and it is defined through Eq. (2) (RD Instruments, 2007):

$$K_{\mathrm{c}} = \frac{127.3}{T_x + 273}. \quad (2)$$

α is the sound absorption coefficient for the seawater that is variable with depth, computed according to the Ainslie

(Ainslie, 1998) formula, a simple expression which takes into account the contribution of boric acid, magnesium sulfate and pure water, and depends on temperature and salinity. These last two parameters are obtained from the climatological MED6 database, a gridded monthly mean of in situ measurements of temperature and salinity for the entire Mediterranean and the near-North Atlantic area at 0.25° resolution (Brankart and Pinardi, 2001). Data from the MED6 database are selected in a 0.25°, square centred on 9.375° E, 43.75° N, the grid point closest to the mooring.

R is the slant range to the scattering layers being measured along the beam in metres expressed by Eq. (3):

$$R = \frac{B + (\frac{L+D}{2}) + [(n-1)D] + \frac{D}{4}}{\cos\theta}\frac{C'}{C_1}, \quad (3)$$

where B is the blank in metres (1.76 m for all deployments), L is the transmit pulse length in metres, D is the depth cell length in metres (8 m for all deployments), n is the depth cell number of the scattering layer being measured, θ is the beam angle in degrees (20°), C_1 is the speed of sound in the water used by the ADCP (set to 1475.1 m s^{-1}), while C' is the average sound speed from the transducer to the range cell that depends on the depth (computed by means of Mackenzie formula; (MacKenzie, 1981)) using the same MED6 data previously extracted and interpolated at a vertical resolution of 1 m from 0 up to 100 m. To be used in Eq. (1), the slant range should not be less than $\pi R_0/4$, where R_0 is the Rayleigh distance defined by RDI for the Workhorse Sentinel 300 kHz ADCP as 0.98 m. According to Gordon (1996), the maximum range of acceptable data $R_{\max}$ should satisfy Eq. (4):

$$R_{\max} = H\cos\theta, \quad (4)$$

where H is the distance of the instrument to the surface. In all three deployments both requirements were satisfied. The backscatter strength value Sv was obtained by averaging the results of Eq. (1) applied to the four available beams. The averaged Sv profiles for the three deployments at different depths are shown in Fig. 3. The results show that these values can be influenced by a low signal to noise ratio, as pointed out by Gostiaux and van Haren (2010), who suggest modifying Eq. (1) as

$$\mathrm{Sv} = 20\log_{10}(R) + 2\alpha R - A + 10\log_{10}(10^{K_c E/10} - 10^{K_c E_r/10}) \quad (5)$$

and introducing the ADCP transmit lag L_a in Eq. (3), thus obtaining Eq. (6):

$$R = \frac{B + (\frac{L+D+L_a}{2}) + [(n-1)D] + \frac{D}{4}}{\cos\theta}. \quad (6)$$

In order to determine the constant A, a best linear fit between the Sv values obtained by the Gostiaux and van Haren

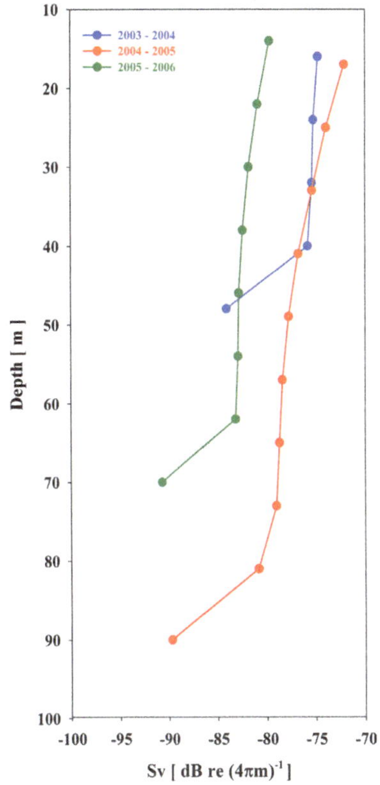

Fig. 3. Average backscatter strength profiles for the three deployments.

equation and the Sv values calculated by the Deines formula is performed using only the bins that satisfy the test $K_c(E - E_r) > 10$. In the 2003–2004 data set, the signal to noise ratio is very high for almost all acquired data, so that the improvement of the Gastiaux and van Haren equation is limited to a quite constant little shift for all layers. On the contrary, in the 2004–2005 data set, the differences between the backscattering strength values computed using the two methods increase with the increasing of the slant range. These differences become more marked for the data acquired in the third deployment, while disagreement is also found for the first two bins.

Furthermore, the closest bin to the transducer is characterized by a very weak signal and consequently the ADCP profile shows a high gradient on the second bin: this behaviour is due to the transient time needed by the instrument to transmit and receive the signal, thus during this gap of time the acquired data can be erroneous (Lane et al., 1999).

Taking into account the previous issues, the used Sv values have been computed using Eq. (5), discarding the bin closest to the transducer, as well as the one closest to the sea surface since it may be strongly affected by the atmosphere (Schott, 1989).

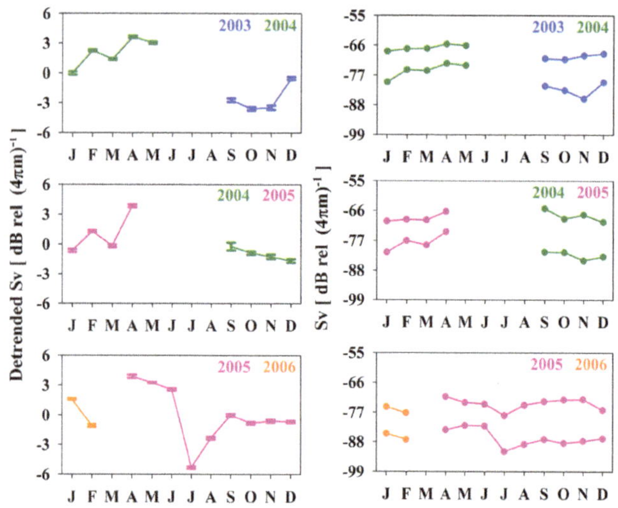

Fig. 4. Backscatter strength at 40 m depth during the three deployments: (left) detrended monthly averaged time series with error bars corresponding to uncertainties in the average estimation; (right) monthly maximum- and minimum values. Data from year 2003 in blue, 2004 in green, 2005 in pink, and 2006 in orange. Time axis aligned from January to December.

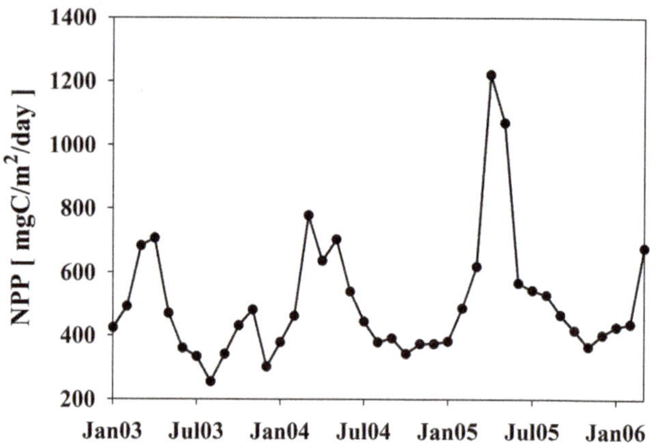

Fig. 5. Monthly averaged net primary production values recorded in the study area during the sampling period.

3 Results

3.1 Seasonal variability

The monthly mean backscattering strength values computed for each time series at the common depth of about 40 m are shown in Fig. 4. Although none of the time series covers an entire year and not all months are complete, the three distributions show similar seasonal behaviour with an increase in spring and lower values during the fall months. A relative maximum is also detected in January–February, while an abrupt drop, with very low values, is recorded in July 2005.

The seasonal NPP values show a pronounced interannual variability (Fig. 5). In 2003 the mean annual NPP cycle is characterized by two peaks: the major one in April (707 mg C/m^2 day^{-1}) and the other in November (481 mg C/m^2 day^{-1}). In 2004 there is a unique peak of NPP from March to May (mean NPP = 706 $\pm$ 70 mg C/m^2 day^{-1}), while in 2005 an extraordinarily high NPP peak is recorded from April (1221 mg C/m^2 day^{-1}) to May (1070 mg C/m^2 day^{-1}), but high NPP values persist until August (on average 547 $\pm$ 42 mg C/m^2 day^{-1}).

The pattern of backscattering strength values is consistent with NPP trends with an expected delay of about 1 month between the peak of surface primary production and the response by the zooplankton community. Indeed, the peak of zooplankton biomass is recorded in April–May (Fig. 4) after the peak of NPP in March–April (Fig. 5).

In Fig. 4, apart from the generally high values observed from April to June (common to the three deployments), which are consistent with the main peak of NPP in temperate areas, the second greatest value for the third deployment occurred in January 2006. This fully agrees with the increase in NPP values recorded in December 2005 and the recognized delay between phytoplankton production and the successive zooplankton biomass increase, as previously explained. On the other hand the low Sv value found in July 2005, the only year in which this period was examined, was in agreement with the almost complete disappearance of *Meganyctiphanes norvegica* from the upper water layers, as observed by Andersen (2001b) in June in the western Ligurian Sea. This species goes deeper at the end of its reproductive season (end of spring) and lives at 500 m and even 1000 m, being the deepest sampled species (Franqueville, 1971; Sardou and Andersen, 1996). In addition, Huntley and Brooks (1982) reported that when food is scarce in the surface waters, the usual DVM performers cease vertical migration until food concentrations suffice to support it and this likely occurs in July, far from the peak of NPP in surface, when food availability is low.

The observed ADCP time series is also in agreement, in terms of seasonal dynamic, with the distribution of monthly mean Continuous Plankton Recorder (CPR) data collected in the northwestern Mediterranean Sea in the period 1977–1999 (Licandro and Icardi, 2009). This ADCP time series is substantially uniform down the water column, at least to 30 m depth, while a greater variability is noted below this depth and between the ADCP time series of the three deployments.

3.2 Daily mean cycle

The first experiment starts at the end of the anomalous warm summer 2003 and the measures are limited to the upper 50 m of the sea. From the beginning of the observational period in September 2003 to the first half of October, the mean daily backscatter strength values are high, particularly in the layers above 30 m depth (Fig. 6). Starting from mid-October until the first days of December 2003, with the exception

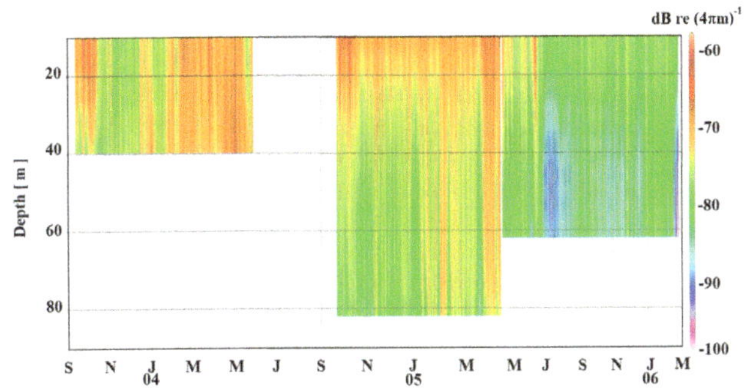

Fig. 6. Time series of daily mean backscatter strength data recorded during the three deployments. Year is indicated below January label.

of early November, they undergo a significant decrease and slightly higher values are recorded only sporadically below 30 m. From mid-December 2003, the signal begins to grow and the whole observed water column becomes substantially homogeneous. From mid-April the measured values are again high, even if only on a few occasions, reaching those recorded in October. A weak reduction appears in the surface layer shortly before the recovery of the mooring at the end of May 2004.

The mooring configuration adopted in the second experiment allows investigation down to 80 m depth. The mean daily backscatter strength data collected above 35 m depth show a trend similar to the one of the previous period 2003–2004: starting from mid-October 2004 the values decrease until December, when they start to grow slowly again to reach (after a short period of attenuation in March 2005) the highest values in April. However, in the upper 30 m layers the Sv values remain high the whole time and the homogenization of the examined water column occurs only sporadically below 25 m depth until January 2005, when strong signals begin to be registered even in the deepest layers. At the end of March, after a short period of decrease, which affects the water column up to almost the surface, the Sv values are large at all depths.

Summer data are available only for the third experiment, in a long period of exceptionally calm conditions of both marine and atmospheric dynamics. From mid-June to the end of August the recorded Sv values are very small, especially below 30 m depth. A significant reduction of total zooplankton biomass in the summer months is also shown in the CPR data (Licandro and Icardi, 2009). However, except for a few events of short duration, the mean daily backscatter remains weak until mid-February, when a significant further drop is observed just before the end of the measurements.

Despite the significant inter-annual differences, a fairly constant characteristic in the time series of the monthly mean hourly Sv data is the presence of a marked daily cycle, with low values during the daylight and high ones in the night (Fig. 7). In correspondence with the sudden changes in the Sv signal, negative peaks in the early morning and positive

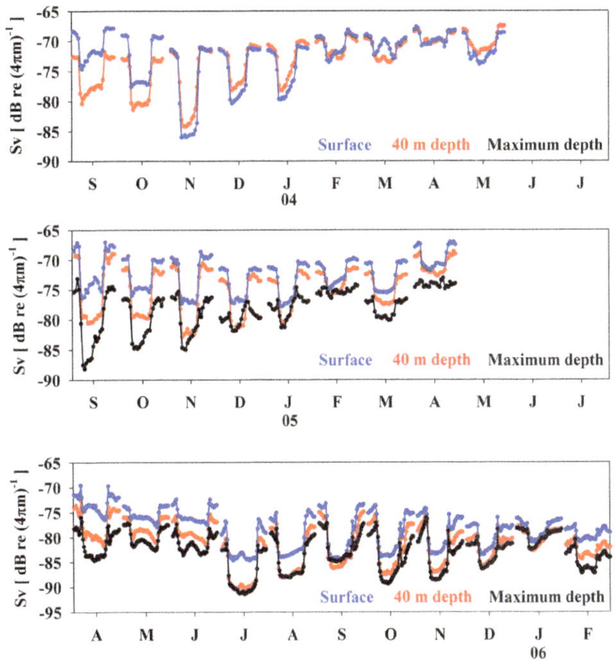

Fig. 7. Monthly mean daily cycle of backscatter strength values at different depths. Year is indicated below January label.

ones in the afternoon are found in the monthly time series of the vertical velocity mean hourly values (Fig. 8).

A noteworthy agreement results between the time of sunrise and sunset at the mooring position, and the time at which the lowest and highest daily values of both vertical velocity and Sv hourly changes occur, especially for the third period when the used sampling time is set at 30 min. It should be noted that significant values of vertical velocity are recorded almost exclusively for a short time around dusk and dawn. This characteristic suggests that these velocities are related mainly to displacements of the scatter elements rather than to vertical motions of the water masses.

3.3 Diel vertical migration

Results mentioned above can be ascribed to the diel vertical migration performed by several species of the zooplankton population. This is a vertical movement, generally involving a 24 h cycle, the causes of which are not yet fully understood (Ringelberg, 2010). Three main patterns have been identified: "normal" or "nocturnal" DVM involves animals moving deeper in the water column during the day and shallower at night. A less common behaviour is exhibited by a slow descent following arrival at the surface at dusk, and a subsequent second ascent to the surface towards the end of the night, prior to the dawn descent ("twilight migration"). Other species or live stages undergo "reverse migration", where the zooplankton ascend at dawn and descend at dusk (Heywood, 1996; Jiang et al., 2007; Cisewski et al., 2010).

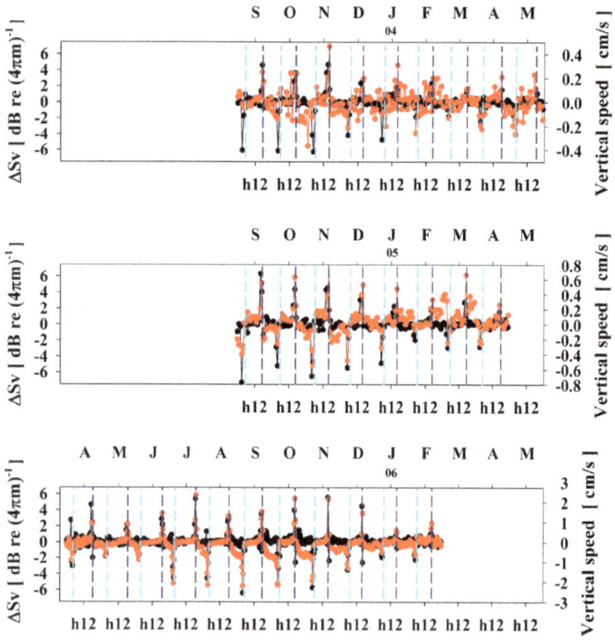

Fig. 8. Monthly mean daily cycle of temporal derivative of backscatter strength and vertical speed at 40 m depth for the three deployments: (top) 2003–2004, (middle) 2004–2005, and (bottom) 2005–2006. Vertical dashed lines correspond to hour of sunrise (cyan) and sunset (blue) of each month. Time axis aligned from April of one year to May of the following year. Year is indicated below January label.

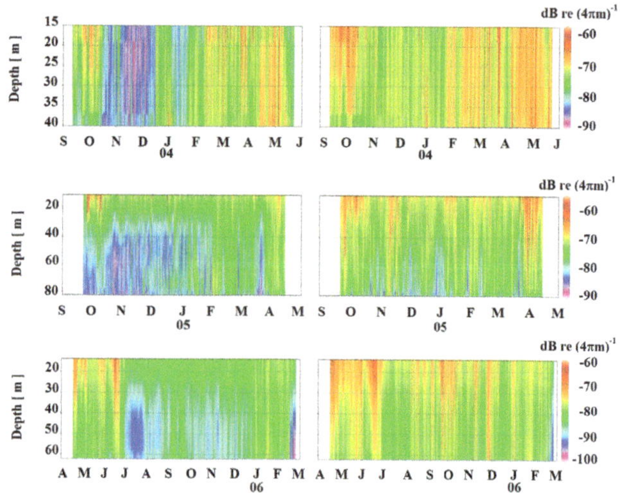

Fig. 9. Time series of backscatter strength profile for data collected (left) between sunrise and sunset and (right) between sunset and sunrise in the three deployments. Year is indicated below January label.

The backscatter strength shows significant differences both in time and with depth. To better investigate these patterns, two different data sets for each experiment are obtained, separating the observations taken between sunrise and sunset (Fig. 9, on the left) from those collected between sunset and sunrise (Fig. 9, on the right).

During the first experiment, the measured data show a quite uniform vertical distribution of the backscattering strength. These values are for most of the examined period rather large during both day and night. Only in two periods, between October and December 2003 and for a few days in January 2004, the diurnal values undergo a significant reduction, denoting fewer scatter organisms during daylight. In the second and third experiments, such small values are never observed above 30 m depth. In this upper layer the backscatter strength signal remains large with small differences between day and night. Below 40 m depth a strong reduction is often observed during daylight, while in the night the signal increases and has a tendency to homogenize the whole observed water column.

3.4 Spectral analysis

The spectral analysis performed on the three time series at the shared depth of 40 m does not differ significantly among samples. It is characterized by a dominant 24 h peak and a series of minor peaks at the higher harmonics. It confirms the predominance of a signal with a 24 h cycle corresponding to the so-called "normal" or "nocturnal" pattern (Fig. 10). The secondary 12 h peak may be attributable to a different DVM pattern, the so-called twilight migration (Cushing, 1951).This is sometimes distinguished in the hourly Sv values, especially during the third experiment that has 30 min temporal resolution (Fig. 11). Twilight migrations were found in more than 80 % of records of northwestern Atlantic zooplankton DVM (Ashjian et al., 1994). The observed behavioural patterns have different interpretations, with hunger-satiation and escapes from predators (i.e. krill) as the most plausible causes (Tarling et al., 1999).

Time–frequency analysis provides evidence for the temporal evolution of the amplitude of each signal and allows for the identification of the periods when a signal characterizing a specific DVM pattern is prevailing. The spectrograms are obtained using a 240 h sliding window with 216 h overlap between each sample. For all three periods, the time evolution of the amplitude of the 12 h and 24 h harmonics at different depths is given in Fig. 12.

The three distributions show that both 12 h and 24 h cycles are particularly intense between November and December, when the backscattering strength values are least. Furthermore, during the first experiment the maximum of the spectrum for the 24 h harmonic is obtained for the surface layer, and it is found at 40 m depth for the 12 h harmonic. On the contrary, during the other two experiments, the spectrum for both harmonics is weak at the surface. The 24 h harmonic time series shows the maximum amplitude at 40 m depth, while the 12 h harmonic shows it at the deepest layers.

These results may suggest the presence of different types of zooplankton organisms, some of which migrate according

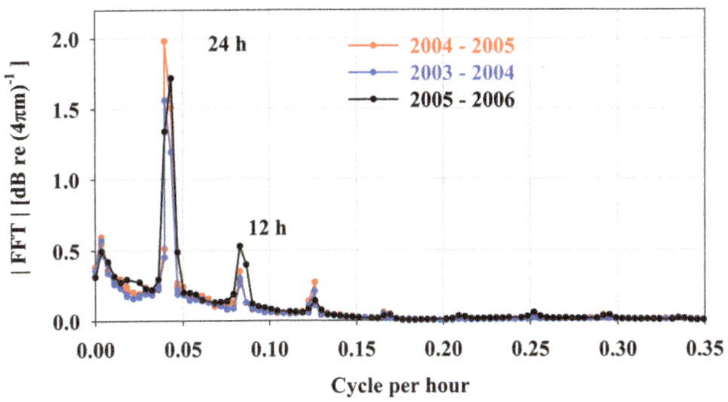

Fig. 10. Power spectrum for the three times series of the backscatter strength at 40 m depth.

to their own specific DVM pattern, while others are stationary, as observed in other areas like in the Irish Sea where generally more than 60 % of the zooplankton community does not perform significant DVM (Irigoien et al., 2004). In the Mediterranean Sea few mesozooplanktonic species show considerable DVM. Thus, taking into account that the widest migrations appear when the backscatter is weak, the more marked DVM signals may be ascribed to some large and more active organisms, likely macrozooplankton and micronekton (i.e. euphausiids, small mesopelagic fish) which are quite abundant in late autumn in other areas of the northwestern Mediterranean (McGehee et al., 2004; Olivar et al., 2012). Among them, the euphausiid *Meganyctiphanes norvegica* is one of the most widespread in the Ligurian Sea (Tarling et al., 2001; Andersen, 2001a). Its vertical migration is usually wide and has a 24 h cycle, however, depending on multiple factors (i.e. reproduction, moon phase, etc.), it can also take place on a 12 h cycle.

3.5 Vertical velocities analysis

Data obtained by ADCP are considered to be an average of a backscattering field consisting of both migrating and stationary organisms (Plueddemann and Pinkel, 1989). Depending on the prevalence of one or another organism, the DVM signal may be more or less detectable by ADCP measurements. Each migrating species has its own behaviour that may change during the different stages of life and due to environmental conditions. Different studies provided accurate estimates of vertical velocity of zooplankton population, which generally varied (depending on the frequency used for the estimates) between 1 and 8 cm s^{-1}. Vertical velocity is greater during ascent (5–8 cm s^{-1}) than during descent (3–4 cm s^{-1}) and with increasing depth (Heywood, 1996; Smith et al., 2008).

Although the ADCP measurements do not give the true value of the zooplankton vertical speed but, generally, a lesser one (Plueddemann and Pinkel, 1989; Tarling et al., 2001), the analysis of the recorded mean hourly vertical velocity may help to investigate if part of the observed

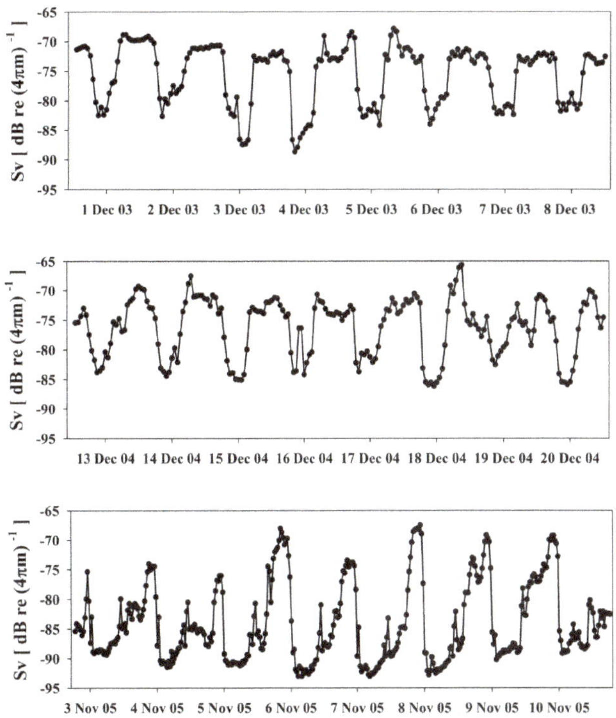

Fig. 11. Subset of backscatter strength hourly data at 40 m depth in presence of twilight diel vertical migration pattern for the three deployments. Label at 12:00 UTC.

variability could be ascribed to a change of the proportion between the different genotypes making up the observed zooplankton population.

Almost all vertical velocity data have values close to zero. Values greater than ±0.5 cm s^{-1} are recorded mainly in conjunction with changes of Sv or during strong wind events.

As a result, the analysis is carried out on the time series of the maximum and minimum daily values (Fig. 13). The time series of maximum and minimum mean hourly vertical velocity show that values tend to increase from the first to the third experiment and going from the surface to the deepest layers. In fact, during the first deployment more than 94 % of the values are between ±1.5 cm s^{-1}, of which 40 % fall in the range ±0.5 cm s^{-1}. During the second deployment, the percentage in the range ±1.5 cm s^{-1} is still greater than 90 %, but far fewer values are between ±0.5 cm s^{-1} and the decrease is even more evident in the deepest layer, where they are less than 10 %. During the third deployment the recorded values are larger at all depths and fewer than 10 % are in the interval ±0.5 cm s^{-1}.

Although during the first experiment the extreme values of the vertical velocity are small, the analysis of the daily data shows that, under strong wind conditions lasting a few days, peaks of up to 5 cm s^{-1} and more are recorded, mainly in the surface layer whilst they decrease with depth. These large values are never reached in the other two periods, despite the occurrence of several episodes with even stronger

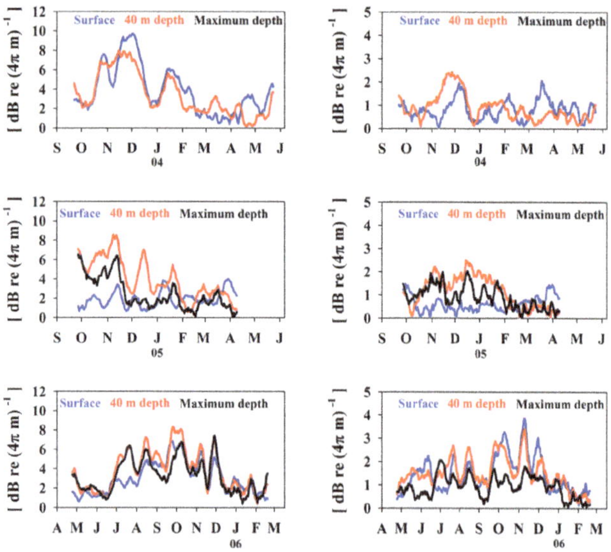

Fig. 12. (left) Time evolution of the amplitude of the 24 h and (right) 12 h harmonics at different depths. Year is indicated below January label.

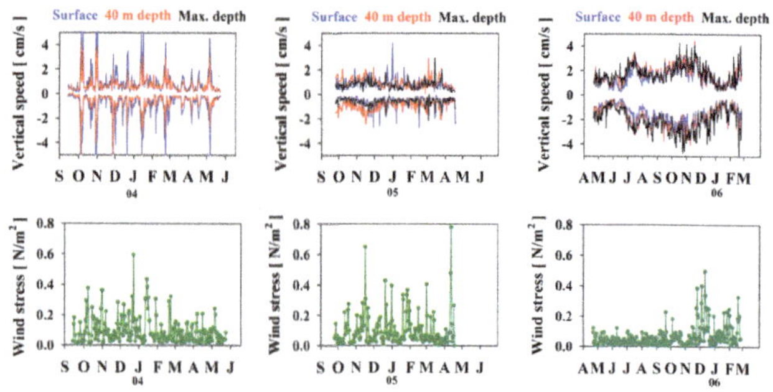

Fig. 13. (upper panel) Maximum and minimum vertical velocity and (lower panel) wind stress at different depths. Year is indicated below January label.

wind. Furthermore, the first period is also characterized by a different distribution of the vertical velocities. This does not show any particular seasonal pattern, but rather single episodes whose amplitude is less in spring. On the contrary, a weak seasonal trend may be detected for the data collected during the second and third periods. Particularly, the extreme values of the vertical velocity increase from late summer to November, when they start again to decrease, reaching minimum values in January. During the period 2005–2006, an increase and subsequent decrease is also observed between the middle of June and the end of August.

4 Discussion and conclusions

Three years of acoustic backscatter and vertical velocities data collected by a 300 kHz ADCP in the central Ligurian Sea are analysed to investigate the zooplankton dynamics. Even based on only one frequency and without net samples, the analysis of backscatter variability at different timescales allows for the identification of different zooplankton migration patterns and, from these, to infer about its presence and composition in the area.

At seasonal scale, the biomass follows the NPP signal with a delay of about 1 month, having higher values in April and May and a secondary maximum in January or February; lower values are generally observed in autumn.

The prevailing vertical migration pattern is the 24 h cycle performed by the zooplankton swimming upward at sunset and downward at sunrise. A second pattern having a 12 h cycle is also identified, the twilight migration, in particular in the measurements with 30 min temporal resolution.

Furthermore, the analysis of both Sv and vertical velocity data suggests that changes in composition of zooplankton population may occur during the three years of continuous monitoring.

Although no biological sampling was performed during the experiments, the results of several studies made in adjacent areas (Licandro and Ibanez, 2000) and the zooplankton composition there reported can help in interpreting the findings of this ADCP data analysis. Among the mesozooplanktonic species abundantly found in those studies, such as *Clausocalanus* spp., *Paracalanus* spp. and *Oithona* spp. (McGehee et al., 2004; Licandro and Ibanez, 2000), no one species shows a strong diel vertical migration (Andersen, 2001b). Indeed, according to Brugnano et al. (2011), only species of the Scolecithricidae family show significant DVM in the area, confirming previous information about the presence of a few strong migrants in the Mediterranean Sea. Thus, it may be supposed that the species mainly responsible for the strong Sv signal found in some periods of this study, at times of small biomass, are ascribable to the macroplanktonic/micronektonic component. Particularly, the area is dominated by the euphausiid *Meganyctiphanes norvegica* (Tarling et al., 2001; Andersen, 2001a) which attains its maximum abundance values in August–September. This species is known to perform wide vertical migration (Kaartvedt, 2010) and it could be responsible for the maximum amplitude found in ADCP data recorded in September 2005.

Furthermore, other previous investigations (Boucher et al., 1987; Licandro and Ibanez, 2000; McGehee et al., 2004; de Puelles and Molinero, 2008; Raybaud et al., 2008; Licandro and Icardi, 2009) point out that the Ligurian Sea is characterized by different zooplankton populations whose distribution is related to the main hydrological features of this basin. The importance of Mediterranean circulation dynamics in the determination of different zooplankton associations was also found in the Gulf of Trieste (Cataletto, 1995) and in the Gulf of Naples (Carrada et al., 1980).

Water masses and marine circulation of the whole Mediterranean Sea, and particularly of the Ligurian Basin, underwent major changes over the three years of the study. The contemporaneous measurements of sea currents show a significant modification in the study area, with an anticlockwise rotation from north to west and a decrease of intensity (Fig. 2). This may lead to the dominance of different zooplankton associations, related to changes in current intensity and direction, as observed by Licandro and Ibanez (2000) in a long-term study in the Gulf of Tigullio, an area adjacent to the ADCP mooring position.

Although more qualitative than quantitative, the results of this study clearly show the skill of ADCP to highlight some characteristics of the zooplankton population that the usual biological observations hardly fail to grasp. Particularly, they show the important role of the time at which the discrete biological sampling is carried out. Thus, in the future, the joint use of long-term continuous monitoring by ADCP and periodic net samplings may be a good observational strategy for deepening the zooplankton knowledge. Nevertheless, at present, the re-examination of backscatter signals of many long time series of ADCP data that have been collected to estimate horizontal and vertical oceanic currents can also contribute to the biological monitoring of the oceans, even in the absence of corresponding in situ direct observations.

Acknowledgements. The work has been partially supported by the Flagship Project RITMARE – The Italian Research for the Sea – coordinated by the National Research Council of Italy and funded by the Italian Ministry of Education, University and Research within the National Research Program 2011–2013 and by the Italian National Program of Antarctic Research (PNRA) under grant 2010/A4.01. The authors acknowledge the ECMWF organization for providing the ERA-Interim data used in this study and Stefano Salon from OGS (Trieste, Italy) for providing the climatological sound speed profiles.

Edited by: J. M. Huthnance

References

Ainslie, M. A.: A simplified formula for viscous and chemical absorption in sea water, J. Acoust. Soc. Am., 103, 1671–1672, 1998.

Andersen, V.: Zooplankton Community During the Transition from Spring Bloom to Oligotrophy in the Open NW Mediterranean and Effects of Wind Events. 1. Abundance and Specific Composition, J. Plankton Res., 23, 227–242, 2001a.

Andersen, V.: Zooplankton Community During the Transition from Spring Bloom to Oligotrophy in the Open NW Mediterranean and Effects of Wind Events. 2. Vertical Distributions and Migrations, J. Plankton Res., 23, 243–261, 2001b.

Andersen, V. and Sardou, J.: The diel migrations and vertical distributions of zooplankton and micronekton in the Northwestern Mediterranean Sea. 1. Euphausiids, mysids, decapods and fishes, J. Plankton Res., 14, 1129–1154, 1992.

Ashjian, C. J., Smith, S. L., Flagg, C. N., Mariano, A. J., and Beherens, W. J.: The influence of a Gulf Stream meander on the distribution of zooplankton biomass in the Slope Water, the Gulf Stream, and the Sargasso Sea, described using a shipboard acoustic Doppler current profiler, Deep Sea Res.-Pt. I, 41, 23–50, 1994.

Ashjian, C. J., Smith, S. L., Flagg, C. N., and Idrisi, N.: Distribution, annual cycle, and vertical migration of acoustically derived biomass in the Arabian Sea during 1994–1995, Deep Sea Res.-Pt. II, 49, 2377–2402, 2002.

Astraldi, M. and Gasparini, G. P.: The seasonal characteristics of the circulation in the north Mediterranean basin and their relationship with the atmospheric-climatic conditions, J. Geophys. Res., 97, 9531–9540, 1992.

Behrenfeld, M. J. and Falkowski, P. G.: Photosynthetic rates derived from satellite-based chlorophyll concentration, Limnol. Oceanogr., 42, 1–20, 1997.

Birol, F., Cancet, M., and Estournel, C.: Aspects of the seasonal variability of the Northern Current (NW Mediterranean Sea) observed by altimetry, J. Marine Syst., 81, 297–311, 2010.

Boucher, J., Ibanez, F., and Prieur, L.: Daily and seasonal variations in the spatial distribution of zooplankton populations in relation to the physical structure in the Ligurian Sea Front, J. Marine Res., 45, 133–173, 1987.

Brankart, J.-M. and Pinardi, N.: Abrupt Cooling of the Mediterranean Levantine Intermediate Water at the Beginning of the 1980s: Observational Evidence and Model Simulation, J. Phys. Oceanogr., 31, 2307–2320, 2001.

Brugnano, C., D'Adamo, R., Fabbrocini, A., Granata, A., and Zagami, G.: Zooplankton responses to hydrological and trophic variability in a Mediterranean coastal ecosystem (Lesina Lagoon, South Adriatic Sea), Chem. Ecol., 27, 461–480, 2011.

Carrada, G. C., Hopkins, T. S., Bonaduce, G., Ianora, A., Marino, D., Modigh, M., Ribera, D. M., and Scotto, C. B.: Variability in the Hydrographic and Biological Features of the Gulf of Naples, Mar. Ecol., 1, 105–120, 1980.

Cataletto, B.: Eleven years of time-series analysis on the net-zooplankton community in the Gulf of Trieste, Ices J. Mar. Sci., 52, 669–678, 1995.

Cisewski, B., Strass, V. H., Rhein, M., and Krägefsky, S.: Seasonal variation of diel vertical migration of zooplankton from ADCP backscatter time series data in the Lazarev Sea, Antarctica, Deep Sea Res.-Pt. I, 57, 78–94, 2010.

Crepon, M., Wald, L., and Monget, J. M.: Low-frequency waves in the Ligurian Sea during December 1977, J. Geophys. Res., 87, 595–601, 1982.

Cushing, D. H.: The vertical migration of planktonic crustacea, Biol. Rev., 26, 158–192, 1951.

de Puelles, M. L. F. and Molinero, J. C.: Decadal changes in hydrographic and ecological time-series in the Balearic Sea (western Mediterranean), identifying links between climate and zooplankton, Ices J. Mar. Sci., 65, 311–317, 2008.

Dee, D. P., Uppala, S. M., Simmons, A. J., Berrisford, P., Poli, P., Kobayashi, S., Andrae, U., Balmaseda, M. A., Balsamo, G., Bauer, P., Bechtold, P., Beljaars, A. C. M., van de Berg, L., Bidlot, J., Bormann, N., Delsol, C., Dragani, R., Fuentes, M., Geer, A. J., Haimberger, L., Healy, S. B., Hersbach, H., Hólm, E. V., Isaksen, L., Kållberg, P., Köhler, M., Matricardi, M., McNally, A. P., Monge-Sanz, B. M., Morcrette, J.-J., Park, B.-K., Peubey, C., de Rosnay, P., Tavolato, C., Thépaut, J.-N., and Vitart, F.: The

ERA-Interim reanalysis: configuration and performance of the data assimilation system, Q. J. Roy. Meteorol. Soc., 137, 553–597, 2011.

Deines, K. L.: Backscatter estimation using Broadband acoustic Doppler current profilers, in: Working Conference on Current Measurement, 1999.

di Carlo, B., Ianora, A., Fresi, E., and Hure, J.: Vertical zonation patterns for Mediterranean copepods from the surface to 3000 m at a fixed station in the Tyrrhenian Sea, J. Plankton Res., 6, 1031–1056, 1984.

Fanelli, E., Cartes, J. E., Rumolo, P., and Sprovieri, M.: Food-web structure and trophodynamics of mesopelagic–suprabenthic bathyal macrofauna of the Algerian Basin based on stable isotopes of carbon and nitrogen, Deep Sea Res.-Pt. I, 56, 1504–1520, 2009.

Fielding, S., Griffiths, G., and Roe, H. S. J.: The biological validation of ADCP acoustic backscatter through direct comparison with net samples and model predictions based on acoustic-scattering models, Ices J. Mar. Sci., 61, 184–200, 2004.

Flagg, C. and Smith, S.: On the use of the acoustic Doppler current profiler to measure zooplankton abundance, Deep Sea Res.-Pt. A, 36, 455–474, 1989.

Font, J., Puig, P., Salat, J., Palanques, A., and Emelianov, M.: Sequence of hydrographic changes in NW Mediterranean deep water due to the exceptional winter of 2005, Sci. Mar., 71, 339–346, 2007.

Franqueville, C.: Macroplancton profond (Invertébrés) de la Méditerranée nord-occidentale, Tethys, 3, 11–56, 1971.

Furnestin, M. L.: Zooplancton du Golfe du Lion et de la côte orientale de Corse, Rev. Trav. Inst. Pêches Marit., 24, 153–252, 1960.

Garrabou, J., Coma, R., Bensoussan, N., Bally, M., Chevaldonné, P., Cigliano, M., Diaz, D., Harmelin, J., Gambi, M., Kersting, D., Ledoux, J., Lejeusne, C., Linares, C., Marschal, C., Pérez, T., Ribes, M., Romano, J., Serrano, E., Teixido, N., Torrents, O., Zabala, M., Zuberer, F., and Cerrano, C.: Mass mortality in Northwestern Mediterranean rocky benthic communities: Effects of the 2003 heat wave, Glob. Change Biol., 15, 1090–1103, 2009.

Gordon, R.: Acoustic Doppler Current Profilers Principles of Operation: A Practical Primer, RD Instruments, 1996.

Gostiaux, L. and van Haren, H.: Extracting Meaningful Information from Uncalibrated Backscattered Echo Intensity Data, J. Atmos. Ocean. Tech., 27, 943–949, 2010.

Hays, G. C.: A review of the adaptive significance and ecosystem consequences of zooplankton diel vertical migrations, Hydrobiologia, 503, 163–170, 2003.

Heywood, K. J.: Diel vertical migration of zooplankton in the Northeast Atlantic, J. Plankton Res., 18, 163–184, 1996.

Huntley, M. and Brooks, E.: Effects of age and food availability on diel vertical migration of Calanus pacificus, Mar. Biol., 71, 23–31, 1982.

Irigoien, X., Conway, D., and Harris, R.: Flexible diel vertical migration behaviour of zooplankton in the Irish Sea, Marine Ecology-progress Series, 267, 85–97, 2004.

Jiang, S., Dickey, T. D., Steinberg, D. K., and Madin, L. P.: Temporal variability of zooplankton biomass from ADCP backscatter time series data at the Bermuda Testbed Mooring site, Deep Sea Res.-Pt. I, 54, 608–636, 2007.

Kaartvedt, S.: Diel Vertical Migration Behaviour of the Northern Krill (Meganyctiphanes norvegica Sars), Vol. 57, 255–275, doi:10.1016/B978-0-12-381308-4.00009-1, 2010.

Lane, A., Knight, P. J., and Player, R. J.: Current measurement technology for near-shore waters, Coast. Eng., 37, 343–368, 1999.

Licandro, P. and Ibanez, F.: Changes of zooplankton communities in the Gulf of Tigullio (Ligurian Sea, Western Mediterranean) from 1985 to 1995. Influence of hydroclimatic factors, J. Plankton Res., 22, 2225–2253, 2000.

Licandro, P. and Icardi, P.: Basin scale distribution of zooplankton in the Ligurian Sea (north-western Mediterranean) in late autumn, Hydrobiologia, 617, 17–40, 2009.

MacKenzie, K. V.: Nine-term equation for sound speed in the oceans, J. Acoust. Soc. Am., 70, 807–812, 1981.

Margalef, R.: Environmental Control of the Mesoscale Distribution of Primary Producers and Its Bearing to Primary Production in the Western Mediterranean, Plenum Press, 1985.

Martín, J., Miquel, J.-C., and Khripounoff, A.: Impact of open sea deep convection on sediment remobilization in the western Mediterranean, Geophys. Res. Lett., 37, L13604, doi:10.1029/2010GL043704, 2010.

Mazzocchi, M. G., Christou, E. D., Capua, I. D., de Puelles, M. F., Fonda-Umani, S., Molinero, J. C., Nival, P., and Siokou-Frangou, I.: Temporal variability of Centropages typicus in the Mediterranean Sea over seasonal-to-decadal scales, Prog. Oceanogr., 72, 214–232, 2007.

McGehee, D., Demer, D., and Warren, J.: Zooplankton in the Ligurian Sea: Part I. Characterization of their dispersion, relative abundance and environment during summer 1999, J. Plankton Res., 26, 1409–1418, 2004.

Millot, C.: Circulation in the Western Mediterranean Sea, J. Mar. Syst., 20, 423–442, 1999.

Olivar, M., Bernal, A., Molí, B., na, M. P., Balbín, R., Castellón, A., Miquel, J., and Massutí, E.: Vertical distribution, diversity and assemblages of mesopelagic fishes in the western Mediterranean, Deep Sea Res.-Pt. I, 62, 53–69, 2012.

Picco, P., Cappelletti, A., Sparnocchia, S., Schiano, M. E., Pensieri, S., and Bozzano, R.: Upper layer current variability in the Central Ligurian Sea, Ocean Sci., 6, 825–836, doi:10.5194/os-6-825-2010, 2010.

Pinot, J. and Jansá, J.: Time variability of acoustic backscatter from zooplankton in the Ibiza Channel (western Mediterranean), Deep Sea Res.-Pt. I, 48, 1651–1670, 2001.

Plueddemann, A. and Pinkel, R.: Characterization of the patterns of diel migration using a Doppler sonar, Deep Sea Res.-Pt. I, 36, 509–530, 1989.

Raybaud, V., Nival, P., Mousseau, L., Gubanova, A., Altukhov, D., Khvorov, S., Ibañez, F., and Andersen, V.: Short term changes in zooplankton community during the summer-autumn transition in the open NW Mediterranean Sea: species composition, abundance and diversity, Biogeosciences, 5, 1765–1782, doi:10.5194/bg-5-1765-2008, 2008.

RD Instruments: WinRiver User's Guide, available at: http://www.rdinstruments.com/smartlink/wr/support_docs/WINRIV2UG.PDF (last access: 19 February 2014), 2007.

Ringelberg, J.: Diel Vertical Migration of Zooplankton in Lakes and Oceans: Causal Explanations and Adaptive Significances, Springer Netherlands, 2010.

Rippeth, T. and Simpson, J.: Diurnal signals in vertical motions on the Hebridean Shelf, 1998.

Sammari, C., Millot, C., and Prieur, L.: Aspects of the seasonal and mesoscale variabilities of the Northern Current in the western Mediterranean Sea inferred from the PROLIG-2 and PROS-6 experiments, Deep Sea Res.-Pt. I, 42, 893–917, 1995.

Sardou, J. E. M. and Andersen, V.: Seasonal abundance and vertical distributions of macroplankton and micronekton in the Northwestern Mediterranean Sea, Oceanol. Ac., 19, 645–656, 1996.

Schott, F.: Measuring winds from underneath the ocean surface by upward looking acoustic Doppler current profilers, J. Geophys. Res.-Oceans, 94, 8313–8321, 1989.

Smith, R. O., Bryden, H. L., and Stansfield, K.: Observations of new western Mediterranean deep water formation using Argo floats 2004–2006, Ocean Sci., 4, 133–149, doi:10.5194/os-4-133-2008, 2008.

Sparnocchia, S., Schiano, M. E., Picco, P., Bozzano, R., and Cappelletti, A.: The anomalous warming of summer 2003 in the surface layer of the Central Ligurian Sea (Western Mediterranean), Ann. Geophys., 24, 443–452, doi:10.5194/angeo-24-443-2006, 2006.

Tarling, G., Matthews, J., David, P., Guerin, O., and Buchholz, F.: The swarm dynamics of northern krill (Meganyctiphanes norvegica) and pteropods (Cavolinia inflexa) during vertical migration in the Ligurian Sea observed by an acoustic Doppler current profiler, Deep Sea Res.-Pt. I, 48, 1671–1686, 2001.

Tarling, G. A., Buchholz, F., and Matthews, J.: The effect of lunar eclipse on the vertical migration behaviour of Meganyctiphanes norvegica (Crustacea: Euphausiacea) in the Ligurian Sea, J. Plankton Res., 21, 1475–1488, 1999.

Taupier-Letage, I. and Millot, C.: General hydrodynamical features in the Ligurian Sea inferred from the DYOME experiment, Oceanol. Ac., 9, 119–131, 1986.

Truscott, J. E. and Brindley, J.: Equilibria, Stability and Excitability in a General Class of Plankton Population Models, Philos. T. R. Soc. A, 347, 703–718, 1994.

van Haren, H.: Echo intensity data as a directional antenna for observing processes above sloping ocean bottoms, Ocean Dynam., 57, 135–149, 2007.

Vignudelli, S., Cipollini, P., Astraldi, M., Gasparini, G. P., and Manzella, G.: Integrated use of altimeter and in situ data for understanding the water exchanges between the Tyrrhenian and Ligurian Seas, J. Geophys. Res.-Oceans, 105, 19649–19663, 2000.

Visbeck, M. and Fischer, J.: Sea Surface Conditions Remotely Sensed by Upward-Looking ADCPs, J. Atmos. Ocean.-Tech., 12, 141–149, 1995.

Weikert, H. and Trinkaus, S.: Vertical mesozooplankton abundance and distribution in the deep Eastern Mediterranean Sea SE of Crete, J. Plankton Res., 12, 601–628, 1990.

Is coccolithophore distribution in the Mediterranean Sea related to seawater carbonate chemistry?

A. Oviedo[1], P. Ziveri[1,2,5], M. Álvarez[3], and T. Tanhua[4]

[1]Institute of Environmental Science and Technology (ICTA), Universitat Autònoma de Barcelona (UAB), 08193 Bellaterra, Spain
[2]Earth & Climate Cluster, Department of Earth Sciences, FALW, Vrije Universiteit Amsterdam, FALW, HV1081 Amsterdam, the Netherlands
[3]IEO – Instituto Español de Oceanografia, Apd. 130, A Coruña 15001, Spain
[4]GEOMAR Helmholtz-Zentrum für Ozeanforschung Kiel, Marine Biogeochemistry, Düsternbrooker Weg 20, 24105 Kiel, Germany
[5]ICREA, Institució Catalana de Recerca i Estudis Avançats, Barcelona, Spain

Correspondence to: A. Oviedo (angelamaria.oviedo@uab.cat)

Abstract. The Mediterranean Sea is considered a "hot spot" for climate change, being characterized by oligotrophic to ultra-oligotrophic waters and rapidly increasing seasurface temperature and changing carbonate chemistry. Coccolithophores are considered a dominant phytoplankton group in these waters. As marine calcifying organisms they are expected to respond to the ongoing changes in seawater carbonate chemistry. We provide here a description of the springtime coccolithophore distribution in the Mediterranean Sea and relate this to a broad set of in situ-measured environmental variables. Samples were taken during the R/V *Meteor* (M84/3) oceanographic cruise in April 2011, between 0 and 100 m water depth from 28 stations. Total diatom and silicoflagellate cell concentrations are also presented. Our results highlight the importance of seawater carbonate chemistry, especially [CO_3^{2-}] but also [PO_4^{3-}] in unraveling the distribution of heterococcolithophores, the most abundant coccolithophore life phase. Holo- and heterococcolithophores respond differently to environmental factors. For instance, changes in heterococcolithophore assemblages were best linked to the combination of [CO_3^{2-}], pH, and salinity ($\rho = 0.57$), although salinity might be not functionally related to coccolithophore assemblage distribution. Holococcolithophores, on the other hand, showed higher abundances and species diversity in oligotrophic areas (best fit, $\rho = 0.32$ for nutrients), thriving in nutrient-depleted waters. Clustering of heterococcolithophores revealed three groups of species sharing more than 65 % similarities. These clusters could be assigned to the eastern and western basins and deeper layers (below 50 m), respectively. In addition, the species *Gephyrocapsa oceanica*, *G. muellerae*, and *Emiliania huxleyi* morphotype B/C are spatially distributed together and trace the influx of Atlantic waters into the Mediterranean Sea. The results of the present work emphasize the importance of considering holo- and heterococcolithophores separately when analyzing changes in species assemblages and diversity. Our findings suggest that coccolithophores are a main phytoplankton group in the entire Mediterranean Sea and can dominate over siliceous phytoplankton. They have life stages that are expected to respond differently to the variability in seawater carbonate chemistry and nutrient concentrations.

1 Introduction

Marine phytoplankton constitutes about 1–2 % of the global biomass among primary producers (Falkowski, 1994); however, it contributes to ~46 % of the primary production in a global scale (Field et al., 1998). Coccolithophores represent ~10 % of global phytoplankton biomass (Tyrrell and Young, 2009). They play an important role in biogeochemical cycles, contributing to both the organic and inorganic carbon pumps through photosynthesis and calcification. The latter is the main process controlling PIC : POC (rain ratio). For instance, in the eastern Mediterranean Sea, they are the main contributor to the inorganic carbon pump ($CaCO_3$ production and flux) throughout the year (Knappertsbusch, 1993; Ziveri et al., 2000; Malinverno et al., 2003).

Most studies looking at coccolithophore assemblages and distribution take into account parameters such as nutrients, photosynthetically active radiation (PAR), temperature, salinity, and oxygen (e.g., Young, 1994; Ziveri et al., 1995; Hagino et al., 2000; Takahashi and Okada, 2000; Haidar and Thierstein, 2001; Cortés et al., 2001; Ignatiades et al., 2009). Parameters related to the seawater carbonate system have only recently been considered due to their importance for calcification and the ongoing and projected changes directly related to the rapidly increasing atmospheric pCO_2. Carbonate chemistry parameters have been suggested as drivers of coccosphere morphology modification in field samples (Beaufort et al., 2008, 2011; Meier et al., 2014; Triantaphyllou et al., 2010), *Emiliania huxleyi* blooms (Merico et al., 2006; Tyrrell et al., 2008), and changes in coccolithophore assemblage composition (Charalampopoulou et al., 2011). Although it is not clear why coccolithophores calcify, calcification is an energy-consuming process for coccolithophores (Brand, 1994; Balch, 2004), maintained by natural selection over millions of years, that changes the carbonate chemistry of their surrounding media. It is therefore plausible that the availability of the necessary resources for carrying out calcification (i.e., HCO_3^- and CO_3^{2-}) should facilitate coccolithophore's growth in the ocean. In this context, it is important to understand how marine calcifying organisms could respond to the rapid accumulation of atmospheric CO_2 and their interaction with the ocean's carbonate chemistry (Kroeker et al., 2013).

The Mediterranean Sea provides an ideal ground to explore the factors controlling coccolithophore distribution because of the well-known large gradient in physicochemical parameters. It has a negative fresh-water balance, with evaporation exceeding precipitation. Surface water temperature, salinity, total alkalinity (TA), and $[CO_3^{2-}]$ increase towards the eastern basin. The Mediterranean Sea is one of the most nutrient-poor regions of the global ocean (Dugdale and Wilkerson, 1988), with a trophic status ranging from mesotrophic in the northwest to extremely oligotrophic in the east (Krom et al., 1991; Berman et al., 1984; Berland et al., 1988; Yacobi et al., 1995; Psarra et al., 2000). The spatial distribution of the phytoplankton community along an east–west transect shows that coccolithophores can dominate along the Mediterranean Sea, in the Levantine, Ionian, and Tyrrhenian basins (Ignatiades et al., 2009). Ocean acidification, warming, and changes in nutrient availability are expected to significantly alter primary production rates, as well as the overall plankton community structure. Studies on coccolithophores distribution in the Mediterranean Sea are mostly regional (Dimiza et al., 2008; Malinverno et al., 2003), losing part of the abovementioned gradients. Additionally, carbonate chemistry parameters were not available in older studies focussed on the distribution of coccolithophores in the Mediterranean Sea. Thus comparisons between the different basins are scarce and the influence of carbonate chemistry parameters on actual coccolithophore assemblages remains therefore uncertain.

The present work investigates the regional and vertical distribution of living coccolithophores in the Mediterranean Sea with respect to in situ-measured environmental parameters and with focus on those of the carbonate chemistry. It provides a description of the late springtime coccolithophore assemblage's composition and distribution in the Mediterranean Sea, with a basin resolution that has not been assessed before, and along physical and chemical gradients.

2 Material and methods

2.1 Sampling

A detailed water sampling was conducted during the M84/3 cruise from 6 to 28 April 2011 onboard the R/V *Meteor* (Tanhua et al., 2013a). Here we investigate a subset of 81 samples from 28 stations collected between 0 and 100 m water depth. Samples were taken using a carousel of 24 Niskin bottles interfaced with the SBE911 CTD system (SeaBird) that provided the hydrographical data for in situ temperature, salinity, and dissolved oxygen of the seawater samples. Samples collected at < 1 m water depth were obtained by filling a 5 L plastic container with surface water. Between 1.5 and 4.5 L of water was gently filtered onto acetate cellulose membranes (Millipore, 0.45 μm pore size, 47 mm diameter). Membrane filters were oven-dried at 40 °C for ~ 12 h and stored in sealed Petri dishes. Figure 1 shows the location of all sampled stations during the cruise trajectory. Table 1 shows the stations coordinates, dates, and sampled depths. The bottle data can be found at http://cdiac.ornl.gov/ftp/oceans/CLIVAR/Met_84_3_Med_Sea/.

2.2 Phytoplankton analyses

A portion of each filter was placed on aluminum stubs and gold-coated using an EMITECH K550X sputter coater. The quantification and identification of the main phytoplankton groups and coccolithophore species were performed by JEOL-JSM 6300 and ZEISS-EVO MA10 scanning electron microscopes (SEMs). Transects of 5–15 mm on the filter, corresponding to an average of 2.3 mL of seawater, were observed at 3000x and phytoplankton groups quantified as coccolithophores, diatoms, and silicoflagellates (*Dictyocha* spp.). A subset of five random samples was analyzed under a light microscope for estimation of coccolithophore and diatom cell abundance at 1000x. This was done in order to check whether counts at higher magnification were biased towards smaller cells. This comparison showed that

Table 1. Location, date, and water depth from which the samples were collected.

Station	Month/day/year	Longitude (° E)	Latitude (° N)	Bot. depth (m)	Sampled depth (m) 100	50	25	5	0
287	4/6/2011	25.60	37.67	829	X*		X	X	
288	4/7/2011	26.22	35.65	2293			X	X	
291	4/8/2011	33.00	34.07	2474	X	X	X	X	
292	4/9/2011	35.17	33.99	1681				X	
293	4/9/2011	34.42	34.00	2034				X	
294	4/10/2011	31.00	33.70	2437	X	X	X	X	X
296	4/11/2011	28.77	33.58	2934			X		
297	4/11/2011	26.02	34.40	4210	X		X	X	
298	4/12/2011	24.33	34.50	3287			X	X	
299	4/12/2011	22.50	35.00	3117				X	
302	4/13/2011	20.35	35.07	2968	X	X	X	X	X
305	4/14/2011	17.25	35.60	4440			X	X	
306	4/15/2011	19.00	36.50	3445		X	X	X	
307	4/15/2011	19.30	37.90	3305				X	X
308	4/15/2011	19.00	38.50	3462	X	X		X	X
309	4/16/2011	18.80	39.50	805	X	X	X		
313	4/16/2011	18.00	41.25	1105			X	X	
316	4/19/2011	11.50	38.60	1665			X	X	
319	4/20/2011	11.30	40.30	2880				X	X
320	4/20/2011	10.61	38.75	2490	X			X	
321	4/20/2011	9.40	38.25	1565	X	X	X	X	
324	4/21/2011	5.60	38.65	2845	X	X	X	X	X
329	4/23/2011	2.00	37.90	2730	X	X	X	X	X
331	4/23/2011	0.00	37.05	2704	X	X	X	X	X
332	4/24/2011	−1.40	36.50	2340			X	X	
334	4/24/2011	−4.40	36.10	1228	X	X	X	X	
337	4/25/2011	−5.36	36.00	935	X	X	X	X	X
338	4/25/2011	−5.75	35.95	337				X	

* 75 m

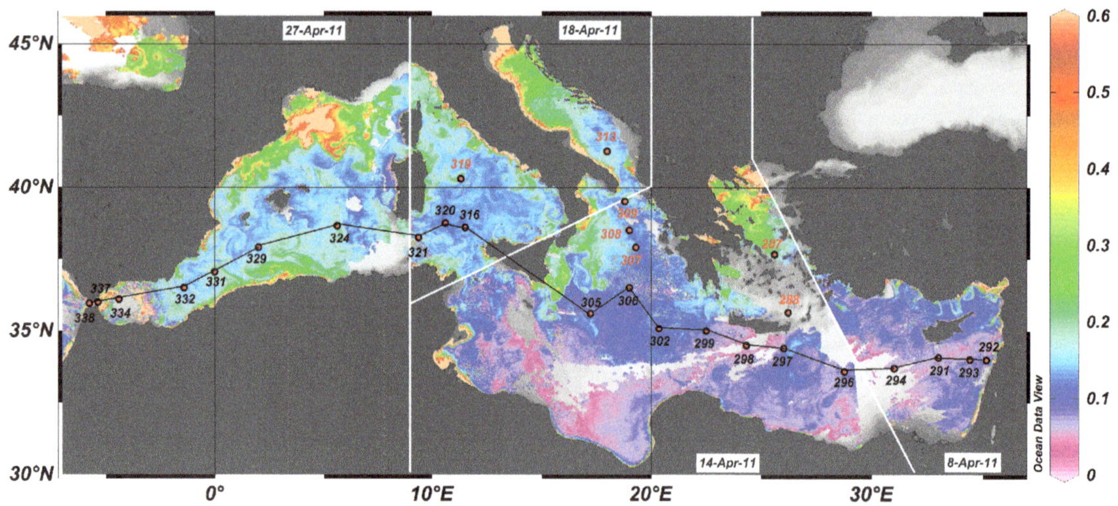

Figure 1. Sampling stations presented in this study collected during the M84-3 research cruise. A composite image is superimposed showing the surface chlorophyll concentration (mg m^{-3}) at an approximate date of the sampling period in the different basins. The transect shown in the following figures includes the stations labeled in black.

relative abundances of diatoms were similar in the two methods. They comprised between 1.7 and 3.3 % (SD = 0.85) of cells counted via optical microscopy and between 0.8 and 4.1 % (SD = 1.6) of cells counted via SEM. Coccolithophore species were identified and their absolute and relative abundances counted. In samples with very few coccospheres a larger filter portion was observed in order to quantify a minimum of 100 cells (a maximum number of 420 cells were counted). Lower (C_L) and upper (C_U) confidence intervals at 95 % significance were estimated following Bollmann et al. (2002). For a 100-cell count these were $C_L = 82$ and $C_U = 102$, and for a 420-cell count: $C_L = 382$ and $C_U = 422$. Cell densities (number of cells per liter of seawater) were calculated. *Emiliania huxleyi* was sub-classified into morphotypes according to Young et al. (2003). For each sample the Shannon–Wiener diversity index (H') was calculated for heterococcolithophores and holococcolithophores. These two groups were treated separately because they represent two different stages of a coccolithophore's life cycle, and taxonomy between the two does not always account for it.

2.3 Environmental parameters

A detailed protocol of all measured environmental variables can be found in Tanhua et al. (2013a). In situ salinity, temperature, and oxygen data were measured with a CTD (described in Sect. 2.1). Overall data accuracies were 0.002 °C for temperature and 0.003 for salinity. Macronutrients (phosphate and nitrate and silicate concentrations) were measured onboard with a QuAAtro autoanalyzer from SEAL Analytical. The following protocols from SEAL Analytical were followed: nitrate (NO_3) (method no. Q-068-05 rev. 4), phosphate (PO_4^{3-}) (method no. Q-031-04 rev. 2), and silicium (Si) (method no. Q-066-05 rev. 3). The nutrient analytical error was determined on 5–7 sample replicates taken at selected stations. The error for nitrate is $0.08\,\mu mol\,kg^{-1}$, for phosphate $0.007\,\mu mol\,kg^{-1}$, and for silicate $0.10\,\mu mol\,kg^{-1}$.

The carbonate system was characterized by measuring dissolved inorganic carbon (DIC), pH, and total alkalinity (TA). DIC content was measured coulometrically using a SOMMA (single-operator multi-metabolic analyzer) system. The precision of the analysis is $\pm 0.6\,\mu mol\,kg^{-1}$ and the accuracy is $2.5\,\mu mol\,kg^{-1}$. The pH was measured by means of double-wavelength spectrophotometry, and it is reported at 25 °C on the total scale. The reproducibility of the pH measurements was 0.0012. TA was analyzed following a double-end-point potentiometric technique. The precision of the TA measurements was $0.1\,\mu mol\,kg^{-1}$. More details about the DIC, TA, and pH measurements and quality control are presented in Álvarez et al. (2014). In situ conditions for other CO_2-related variables were calculated from in situ pH and TA for the first 100 m water column. Calculations were performed using the program CO2Sys (Lewis and Wallace, 1998). Equilibrium constants of Mehrbach et al. (1973) refitted by Dickson and Millero (1987) were chosen (Álvarez et al., 2014).

A characterization of the upper 100 m environmental parameters is shown in Fig. 2; the profiles for the complete water column at a higher spatial horizontal resolution and a full description of the physicochemical setting are presented in Tanhua et al. (2013b) and Álvarez et al. (2014) in this *Ocean Science* special issue.

2.4 Statistical analyses

The E-PRIMER v.5 package was used for the following analyses:

- (1) The BIOENV routine, which computes a rank correlation between the elements of similarity matrices for environmental parameters and biological data, was run to detect the combined changes in environmental parameters and species distribution among stations. The routine examines all possible combinations of environmental variables and gives the "best fit" (higher Rho–Spearman rank correlation) of environmental variables explaining changes in biological communities. This test was performed for all heterococcolithophore and holococcolithophore species contributing > 2 % to the total assemblage of each group. Before running the routine, we checked for mutual correlation among environmental variables and selected a subset of them for this routine. These were salinity, temperature, oxygen, pH, partial pressure of carbon dioxide (pCO_2), and the concentrations of bicarbonate ion (HCO_3^-), carbonate ion (CO_3^{-2}), $NO_3 + NO_2$, and PO_4^{3-}.

- (2) Hierarchical cluster analyses by group average. These were performed for heterococcolithophore and holococcolithophore species. *Emiliania huxleyi* morphotype A was removed of the data set used to run the cluster analyses. This was done to emphasize our results on overall community composition and not on *E. huxleyi* , which largely dominated the assemblages in our samples. When clusters among species were detected, pair-wise Spearman correlations were performed using the software SPSSv18 to assess the environmental parameters influencing changes in each species' abundance.

For the analyses performed using E-PRIMER software, the biological data were transformed in logarithmic scale $\log(1+x)$ to avoid overemphasizing the dominant species. Environmental data were standardized (−mean · STD) to bring data into a comparable scale. Similarity matrices were created for biological and environmental data. For the biological data the Bray–Curtis similarity coefficient was used to examine similarity between each sample's pair. Euclidean distances were used to create the environmental data matrix. Pair-wise Spearman correlations were performed on the basis of non-transformed, non-standardized data.

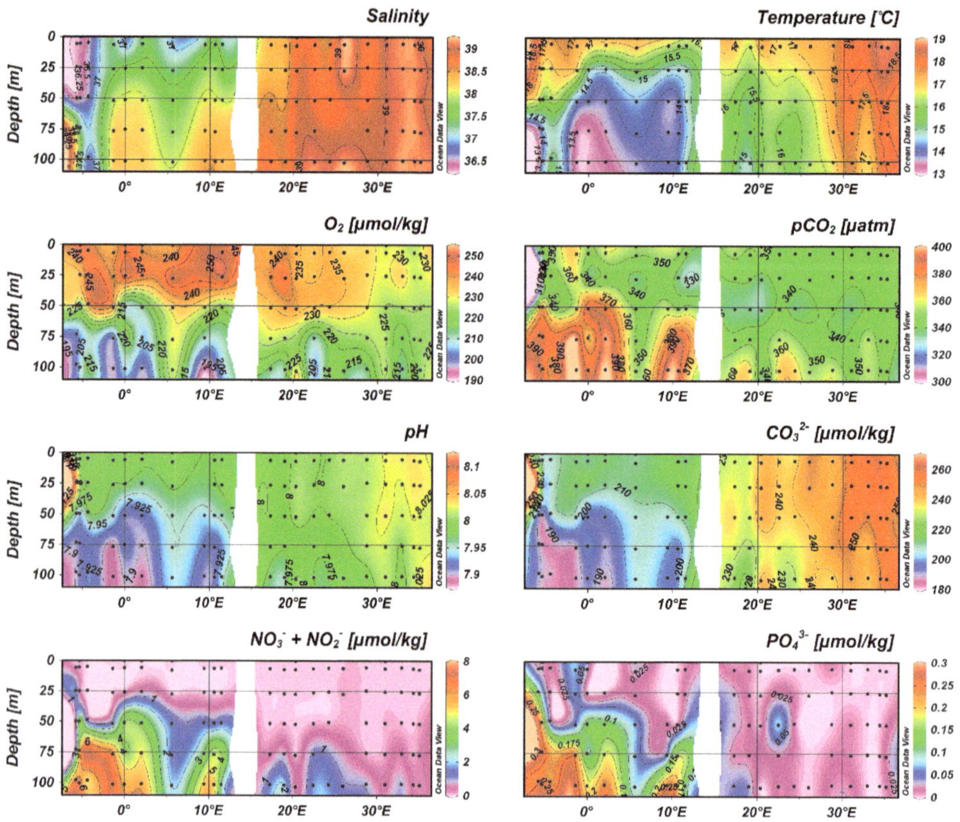

Figure 2. Environmental parameters at sampling stations for surface waters during the M84-3 cruise. The west–east transect along the Mediterranean Sea does not take into account the stations in the Aegean, Adriatic, the northernmost Ionian, or northern Tyrrhenian regions. It includes the stations labeled in black (west–east transect) in Fig. 1.

3 Regional settings

The Mediterranean Sea can be divided into two basins with a general counterclockwise circulation: the western and eastern basins. The Strait of Sicily prevents deep-water exchange between the two. The upper layer (from the surface down to 150–200 m, according to the basin and the season) is occupied by the Atlantic Water (AW) (Ribera d'Alcalà et al., 2003). In the western Mediterranean, AW flows along the eastern part of the Alboran Sea in an anticyclonic fashion, and moves to the western Alboran in a more variable pattern (Robinson et al., 2001). Further east, the AW is transported along the Algerian slope until the Sardinia Channel. Part of the flow then proceeds to the eastern basin along the Tunisian slope, while the other part reaches northern Sicily and circulates along Italy. In the north, the Liguro–Provenco–Catalan Current, is form by AW surrounding Corsica. AW closes its western gyre southeast of Spain, where it encounters the newly entered AW (see Robinson et al., 2001, for more details). In the eastern basin, a jet of Atlantic Water enters through the Strait of Sicily, meanders through the interior of the Ionian Sea and continues to flow through the central Levantine all the way to the shores of Israel (Robinson et al., 2001). Convection occurs when the water approaches the Rhodos gyre. Another part of the AW flow moves to the Ionian and a part of it goes further into the Adriatic Sea. Deep-water formation also occurs in the Adriatic gyre (Pinardi and Masetti, 2000).

During the M84/3 cruise, the surface layer in the western basin was filled by the inflow of relatively low-salinity AW through the Strait of Gibraltar. The salinity minimum, in the surface layer of the strait, shows the entrance of this water into the Mediterranean. Once it reaches the Sicily Channel (close to stations 304/305), the salinity minimum was found at ~ 100 m depth. (Hainbucher et al., 2014). In terms of carbonate chemistry, AW from the surface western Mediterranean was characterized by TA below 2560 μmol Kg^{-1}, DIC near 2250 μmol Kg^{-1}, and pH of 8 (Àlvarez et al., 2014). In the surface layer, AW with salinity lower than 38 was detected in all the Tyrrhenian stations except 319 in the center of the basin, where salinity was higher. There, Álvarez et al. (2014) suggest that the upper ~ 50 dbar was occupied by probably AW affected by evaporation or mixed with the salty intermediate water entering through the Strait of Sicily. During the M84/3 cruise, modified AW and Levantine Surface Water (LSW) dominated the eastern Mediterranean, down to approximately 100 m water depth. The LSW, formed

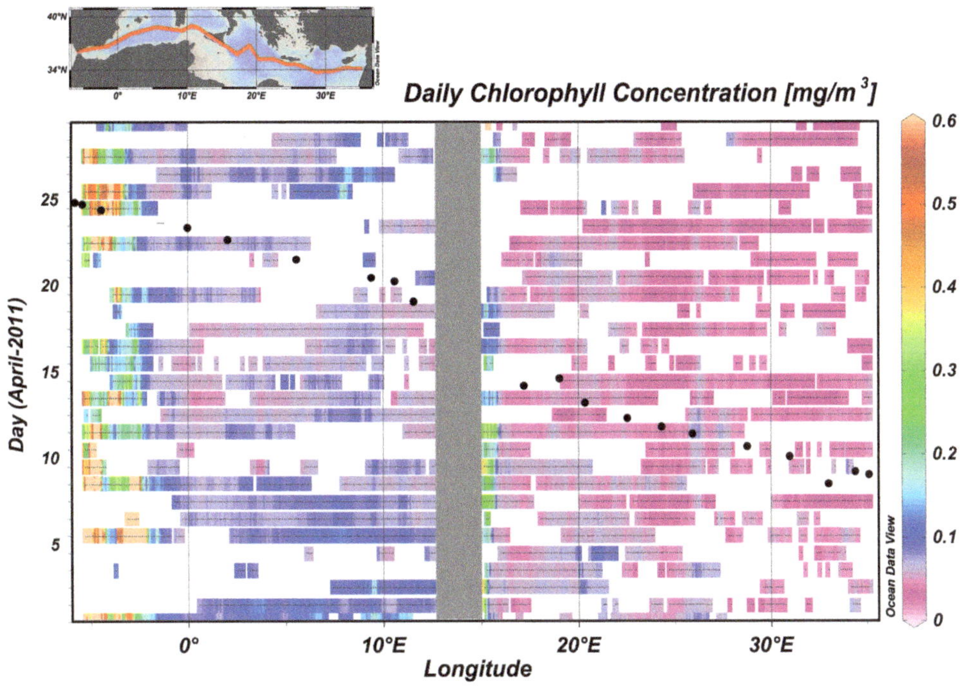

Figure 3. Daily mean chlorophyll *a* concentration from satellite data (Volpe et al., 2012) along the transect of the M84/3 cruise, plotted for the month of April 2011. In this figure, gaps correspond to cloud-covered observations and the temporal–spatial location of the samplings during our study is indicated by black dots. The transect is representative of 18 km in latitude (two pixels at 9 km resolution).

by intensive heating and evaporation, has the largest salinity and temperature of the entire Mediterranean (> 17 °C, > 38.9), and has TA values around 2610 μmol Kg^{-1}, DIC around 2270 μmol Kg^{-1}, and pH25T around 8.03 (Álvarez et al., 2014). Levantine Intermediate Water (LIW) was detected at ∼ 50–100 m in the Levantine Sea, along its path from the northeastern Levantine Sea, and in the Rhodes cyclonic gyre (Hainbucher et al., 2014, and references therein). Close to the M84/3 trajectory, wind-induced upwelling and downwelling zones are found in different seasons (see Bakun and Agostini, 2001): for instance, (1) downwelling along the southern coastal boundary during winter, reversing to coastal upwelling off Libya in spring and off Algeria in summer and fall; (2) strong upwelling in the eastern Aegean Sea throughout the year, becoming remarkably intense in the summer and fall; and (3) strong summer and fall upwelling in the eastern Ionian Sea.

Figure 3 shows daily variations in cloud coverage during the month of April (from satellite observations). It provides an overview of the meteorological situation and possible short-term variability that may mask any relationship with long-term average oceanographic gradients. Approximately half of the satellite observations were cloud covered, with a minimum in cloud coverage (∼ 40 % of observations) at about 25° E (south of Crete). This minimum corresponds to the only segment of the transect directly influenced by northern wind regimes. Such approximation can be compared with the meteorological reports during the M84/3 (data not shown), which shows that ∼ 58 % of the sampling days were cloudy (at the time of recording), with the majority of sunny days from southern Crete towards the Adriatic. Therefore, the sampled days were representative, in terms of cloud coverage, of the conditions during April 2011.

4 Results

4.1 Main phytoplankton community

Overall, phytoplankton cell density was highest in the western Mediterranean Sea, in the Strait of Gibraltar and in the Alboran Sea. For coccolithophores, the pattern was not one of gradual increase towards the west but rather of localized spots of higher cell densities close to Gibraltar and a continuous presence, at lower cell densities, in the rest of the Mediterranean. The satellite-derived chlorophyll *a* concentration during April 2011 (Fig. 3) shows a similar pattern. Coccolithophores were the most abundant phytoplankton group during the sampling, relative to siliceous phytoplankton. They were present in great numbers in all the main Mediterranean basins and accounted for 68 to 99 % of counted phytoplankton. Diatoms, although present in all studied basins, displayed low concentrations in the eastern Mediterranean, increasing towards the west. They were on average 6 % (maximum 25 %) of total phytoplankton. Silicoflagellates (*Dictyocha* spp.) accounted on average for

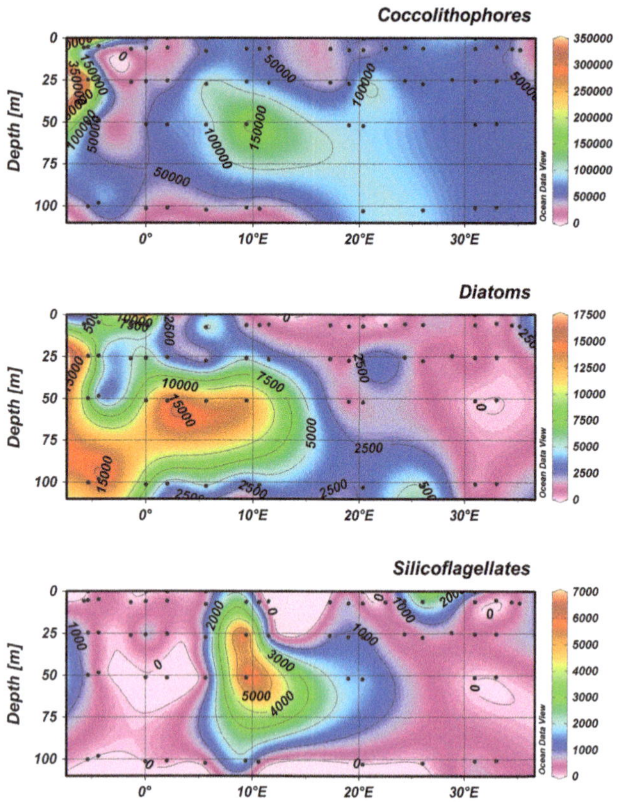

Figure 4. Distribution of coccolithophores (upper panel), diatoms (middle panel), and silicoflagellates (lower panel) in a west–east transect along the Mediterranean Sea (the transect includes the stations labeled in black in Fig. 1).

1 % of phytoplankton (maximum 9 %). Figure 4 shows the distribution of these different phytoplankton groups in the Mediterranean transect. A species present mostly at low cell densities was the xanthophyte *Meringosphaera mediterranea*.

4.2 Environmental factors controlling coccolithophore community distribution

A total of 70 coccolithophore species in heterococcolithophore life stage and 45 in the holococcolithophore stage were recorded (see Appendix A). The majority of the quantified cells were in the heterococcolithophore phase. The species *Emiliania huxleyi* largely dominated the coccolithophore counts in all stations except for station 319 (at the center of the Tyrrhenian Basin), where *Corisphaera gracilis* and *Rabdosphaera clavigera* were the dominant species.

The results from the Spearman's rank-correlation-based routine (BIOENV) suggest that holococcolithophores were preferentially distributed in low-nutrient, high-pH seawaters. ($\rho = 0.328$, Table 2). They were almost absent at 100 m. Heterococcolithophore distribution was best linked to a combination of [CO_3^{2-}], pH, and salinity with $\rho = 0.566$ as well as to [PO_4^{3-}] (Table 3).

4.3 Species assemblages

When clustering all the species within the heterococcolithophore life stage, three groups or assemblages were identified that shared more than 65 % similarities. Figure 5 presents the results of further clustering of the species with high similarities. The distribution of these species along the east–west transect (stations with black labels in Fig. 1) that includes the stations in the Levantine, Ionian (excluding 307–309), Tyrrhenian (excluding 319), Algerian, Alboran, and Gibraltar regions shows that the three groups are distinctively distributed in the Mediterranean Sea (Figs. 5, 6, and 7). The first group comprises species that were more abundant in the eastern stations: *U. tenuis*, *D. tubifera*, *P. vandelii*, *S. pulchra*, *R. clavigera*, and *S. protrudens* (*R. xyphos* was very close to this cluster, with similarity > 50 %). *P. vandelii*, however, was patchily distributed all along the transect. The second group includes *E. huxleyi* Type B/C and the *Gephyrocapsa* species: *G. ericsonii*, *G. muellerae*, and *G. oceanica*. These species were almost exclusive of the western basin. A third group was formed by *Florisphaera profunda* and *Gladiolithus flabellatus*, being closely related to *A. robusta* and *H. carteri* (the last two with similarities between 40 and 60 %). The species in the latter group were almost restricted to depths below 50 m, with higher abundances at 100 m, and were patchily present from the Algerian to the Levantine basins. Clustering analysis for the holococcolith phase did not reveal any pattern in the species composition among the different samples.

Single Spearman correlations for the species that were clustered together reveal that their distribution can be better explained by seawater carbonate chemistry parameters; for instance, species that were mostly abundant at eastern stations thrived in waters with higher [CO_3^{2-}] and pH and in the surface. Among these species, *U. tenuis*, *R. clavigera*, *S. pulchra*, and *S. protrudens* were negatively correlated to phosphate concentrations and only *D. tubifera* showed a high positive correlation with temperature. Finally, *F. profunda* and *G. flabellatus* were correlated to [$NO_3 + NO_2$] and negatively with temperature. Table 4 shows these results.

4.4 Species diversity (H')

Hetero- and holococcolithophore species diversity index (H') changed slightly in the W–E transect. Although correlations between H' and the longitude of the sample sites (°E) in the first 50 m water column were rather weak, the trend was opposite for the two life stages. For instance, heterococcolithophore diversity tended to decrease towards the east ($\rho = -0.419$, $p = 0.000$), while holococcolithophore species diversity tended to increase W–E ($\rho = 0.310$; $p = 0.005$) (Fig. 9). H' index at 100 m was often zero for both

Table 2. Rho values (ρ) for the best Spearman's rank correlations for all possible combinations between the environmental parameters explaining patterns in holococcolithophore assemblages. Only $\rho > 0.2$ are shown. Description as in Table 1. On the left, the number of variables taken into account; on the right, description of the variables and the highest ρ values for the rank correlations at a given number of variables. In the first row, the ρ value for each variable is shown in decreasing order of contribution to explain changes in the biological data. Only $\rho > 0.2$ are shown.

Number of variables	Variables (ρ)
1	$NO_3^- + NO_2^-$ (0.275); PO_4^{3-} (0.236); pH (0.214); pCO_2 (0.204);
2	$NO_3^- + NO_2^-$ pH (0.327)
3	$NO_3^- + NO_2^-$ PO_4^{3-}, pH (0.328)
4	$NO_3^- + NO_2^-$ pH, O_2, Salinity (0.311)
5	$NO_3^- + NO_2^-$ PO_4^{3-}, pH, O_2, salinity (0.311)
6	$NO_3^- + NO_2^-$ PO_4^{3-}, pH, pCO_2, O_2, salinity (0.299)
7	$NO_3^- + NO_2^-$ PO_4^{3-}, pH, pCO_2, O_2, salinity, Temperature (0.283)
8	All (0.275)

Table 3. Rho values (ρ) for the best Spearman's rank correlations for all possible combinations between the environmental parameters explaining patterns in heterococcolithophore assemblages. Description as in Table 2.

Number of variables	Variables (ρ)
1	CO_3^{2-} (0.551); pH (0.498); pCO_2 (0.397); PO_4^{3-} (0.358); $NO_3^- + NO_2^-$ (0.328); salinity (0.310); O_2 (0.226);
2	CO_3^{2-} salinity (0.563)
3	CO_3^{2-} pH, salinity (0.566)
4	CO_3^{2-} pH, PO_4^{3-}, salinity (0.565)
5	CO_3^{2-} pH, pCO_2, PO_4^{3-}, salinity (0.539)
6	CO_3^{2-} pH, pCO_2, PO_4^{3-}, salinity, temperature (0.517)
7	CO_3^{2-} pH, pCO_2, PO_4^{3-}, salinity, O_2, temperature (0.507)
8	All (0.491)

groups, being on average 0.3 for holococcolithophores, which are mostly present at surface, and 1.3 for heterococcolithophores.

5 Discussion

5.1 Main phytoplankton community

Although picoplankton can seasonally dominate phytoplankton assemblages in the Mediterranean Sea (Decembrini et al., 2009; Yacobi et al., 1995), previous studies have often suggested that coccolithophores are one of the most abundant phytoplankton groups in this sea, in both the eastern (e.g., Gotsis-Skretas et al., 1999; Malinverno et al., 2003; Ignatiades et al., 1995, 2009; Rabitti et al., 1994; Ziveri et al., 2000) and western basins (e.g Barlow et al., 1997; Barcena et al., 2004, Ignatiades et al., 2009). Our findings suggest that coccolithophores are a main phytoplankton group in the entire Mediterranean Sea that dominated over siliceous phytoplankton for the period under study (Fig. 4). Silicoflagellates were almost absent at 100 m and more abundant at surface waters of the Tyrrhenian Sea, with cell densities up to 6.7×10^3 cells L^{-1}. Diatoms were preferentially distributed in the western Mediterranean Sea, in waters with lower [CO_3^{2-}] pH and higher PO_4^{3-} at a maximum cell density of 1.6×10^4 cells L^{-1} (Fig. 4). Diatom cell densities in the Mediterranean ranged from a few individuals in the Ionian Sea surface to $\sim 3.5 \times 10^4$ in the Tyrrhenian Sea (June 1999; Ignatiades et al., 2009), from 0.8×10^3 to 2×10^4 (July 2005, Decembrini et al., 2009), from 0.2×10^3 to 5×10^3 (September–October 2004, Lasternas; et al., 2011), and from 0.7×10^3 to 2.6×10^3 cells L^{-1} (December 2005; Decembrini et al., 2009) in the Tyrrhenian Sea. Without taking into account interannual variability, this suggests a pattern of enhanced diatom abundances in the first part of the year and a decrease after July.

In the last decades, many attempts to explain the temporal and spatial phytoplankton distribution based on few key environmental factors have been made. Some of the most discussed are the hypothesis of Sverdrup (1953), stating that changes in phytoplankton abundances are light- and nutrient-controlled features, and the so-called Margalef's mandala (Margalef, 1978), which proposes that phytoplankton succession depends on nutrient concentration and turbulence. Regarding the possible role of light in the relative success of

Table 4. Three highest Spearman's correlation (ρ) results for (1) the dominant species *Emiliania huxleyi*, (2) each of the species belonging to the observed clusters with the environmental parameters, (3) the holo- and hetero-life phases of the same species, and (4) hetero- and holococcolithophore abundances and diversity (H') and abundances of siliceous phytoplankton groups. Preferential depth and distribution along the west–east transect is described. Number of samples taken into account for the analysis is also shown (N). Significance level was 0.05.

Species/group	Variable (r, N)			Pref. depth	Pref. distribution
E. huxleyi Type A	pH (0.468; 73)	CO_2 (-0.448; 73),	CO_3^{2-} (0.444; 73),	0–100 m	W–E
E. huxleyi Type B/C	Salinity (−0.691; 81),	CO_3^{2-} (−0.685;73),	pH (−0.616; 73)	0–100 m	W
G. ericsonii	Salinity (−0.805; 81),	CO_3^{2-} (−0690; 73),	pH (-0.576; 73)	Above 50 m	W
G. oceanica	CO_3^{2-} (−0.833; 73),	Salinity (−0.803; 81),	pH (−0.731; 73)	0–100 m	W
G. muellerae	CO_3^{2-} (−0.835; 73),	Salinity (−0.830; 81),	pH (−0.707; 73)	0–100 m	W
U. tenuis	CO_3^{2-} (0.626; 73),	pH (0.605; 73),	PO_4^{3-} (-0.551; 74)	Above 50 m	E
R. clavigera	CO_3^{2-} (0.727; 73),	pH (0.701; 73),	PO_4^{3-} (-0.546; 74)	Above 50 m	E
D. tubifera	pH (0.600; 73),	Temperature (0.597; 81),	CO_3^{2-} (0.541; 73)	Above 50 m	E
S. pulchra	pH (0.551; 73),	CO_3^{2-} (0.528; 73),	PO_4^{3-} (-0.399; 74)	0–100 m	E
S. protrudens	pH (0.557; 73),	CO_3^{2-} (0.489; 73),	PO_4^{3-} (-0.450; 74)	Above 50 m	E
F. profunda	$NO_3^- + NO_2^-$ (0.637; 81),	Temperature (−0.553; 81),	pH (−0.468; 73)	Below 50 m	W–E
G. flabellatus	$NO_3^- + NO_2^-$ (0.583; 81),	Temperature (−0.522; 81),	O_2 (−0.481; 73)	Below 50 m	W–E
S. pulchra HOL	PO_4^{3-} (−0.308; 74)	CO_2 (−0.261; 73),	O_2 (0.260; 73)	Above 50 m	W–E
C. mediterranea	HCO_3^- (-0.294; 73)	–	–	Above 50 m	W–E
C. mediterranea HOL	O_2 (0.240; 73)	–	–	Above 50 m	W–E
H. carteri	CO_2 (0.341; 73),	pH (−0.339; 73),	–	0–100 m	W–E
H. carteri HOL	$NO_3^- + NO_2^-$ (−0.284; 81),	–	–	Above 50 m	W–E
Hetero-abund.	CO_2 (−0.278; 73),	–	–	0–100 m	W–E
Holo-abund.	Depth (−0.427; 81),	$NO_3^- + NO_2^-$ (−0.391; 81),	CO_2 (0.339; 73)	Above 50 m	E
Hetero-diversity (H')	CO_3^{2-} (−0.486; 73),	Salinity (−0.426; 81),	pH (−0.413; 73)	–	–
Holo-diversity (H')	HCO_3^- (−0.460; 74)	CO_3^{2-} (0.446; 73),	pH (0.388; 73)	–	–
Diatoms	CO_3^{2-} (−0.617; 73),	pH (−0.602; 73),	PO_4^{3-} (0.534; 74)	Below 25 m	W
Silicoflagellates	Si (0.254; 76)	–	–	0–100 m	W–E

coccolithophores over siliceous phytoplankton, we recorded global and UV radiation at the surface, and at the time of sampling at some of the stations, but lack light data at every depth. No significant correlation was observed for any of the phytoplankton groups with radiation data ($p > 0.05$, $N = 21$). Most likely, a response to changes in radiation will not be immediate (specially because the variable that we recorded – cell densities – relates to cell division). In such a case the short temporal extension of our data (by minute during the sampling) could mask a correlation. Large variations in light are also observed on the vertical profile. Holococcolithophores and diatoms correlated with depth (negative and positive correlations, respectively), although for diatoms this correlation was rather weak ($\rho = 0.240$; $p = 0.031$). This could indicate that the two groups are more sensitive to light conditions than heterococcolithophores, but it might also be an artifact of the co-correlation between depth and nutrients. Additionally, light attenuation can differ between stations, and therefore this comparison is only approximate.

In the case of nutrients, it is important to notice that the dominance of coccolithophores over siliceous phytoplankton was clear in all the basins, including the Gibraltar strait, where nitrate and phosphate were available and silicate concentrations were above the half-saturation constant for diatoms (Ks_Si: ca. 3.5 μM (Merico et al., 2006; Leblanc et al., 2003; Sarthou et al., 2005) or 0.8–2.3 μM (Nelson et al., 1976)) and we could have expected the community to be dominated by fast-growing phytoplankton. In the interior of the Mediterranean, [Si] was probably too low (often below 1.0 μmol Kg^{-1}) to support large diatom populations, except for the deeper layers (100 m) of the Ionian and Levantine basins, which ranged from 0.8 to 1.4 μmol Kg^{-1}. Egge and Aksnes (1992) observed that, at Si values lower than 0.6 μM, *Emiliania huxleyi* outcompeted the otherwise dominant *Skeletonema costatum*. For the data set here presented, nutrient variability alone does not explain the dominance of coccolithophores during April 2011, at least not directly. A possible phosphate limitation for other phytoplankton groups (notice the high N / P ratios in both the eastern and western basins as well as in the Strait of Gibraltar) cannot be ruled out for explaining coccolithophore dominance over other groups. Turbulence does not account for it either, since coccolithophores dominated in regions where water density was homogeneous throughout the first 100 m as well as in regions where isoclines can be distinguished (e.g., see salinity and temperature profiles from Fig. 2). Overall, the reason for the dominance of coccolithophores over siliceous phytoplankton remains unclear.

We suggest that the relative success of coccolithophores over siliceous phytoplankton during April 2011 in all Mediterranean Sea basins can be due to a parameter(s) not measured during this study or to a threshold concentration in a resource that limits the growth of the other phytoplankton groups, but that covers the requirements of coccolithophores. The last hypothesis would be better tested experimentally than by correlation analysis.

5.2 Environmental factors controlling coccolithophore distribution

Concerning the coccolithophore community, the study of Knappertsbusch (1993) suggests the occurrence of seasonal variability. His sampling was divided in two periods: late winter (February–March 1988) and late summer (September–October 1986). Knappertsbusch observed higher cell densities during late winter and increasing towards the eastern Mediterranean, reaching 230 000 cells per liter in the Levantine Basin. During September–October this pattern was reversed. Our results resemble those corresponding to September–October in Knappertsbusch work. However, there was no gradual increase towards the west but rather localized higher cell densities close to Gibraltar and a continuous presence, at lower cell densities, in the rest of the Mediterranean (Fig. 4). This is in agreement with the relatively low satellite-derived chlorophyll *a* values in most the surface Mediterranean Sea during the month of April, $\sim 0\,\mathrm{mg\,m^{-3}}$, and the increased values approaching the Strait of Gibraltar, ~ 0.5–$0.6\,\mathrm{mg\,m^{-3}}$ (Fig. 3), and measured data from Rahav et al. (2013) ranging from ~ 0.03 to $\sim 0.1\,\mu\mathrm{g\,L^{-1}}$ in most of the Mediterranean and rising to $0.31\,\mu\mathrm{g\,L^{-1}}$ near the Strait of Gibraltar. It also agrees with the suggested modulation of pH by primary production in the western basin during the cruise (Álvarez et al., 2014). In the study by Ignatiades et al. (2009), the increase in coccolithophore cell densities in the western Mediterranean is mainly due to the significantly higher values at one station (close to the Balearic Islands).

Coccolithophores have a heteromorphic life cycle, with diploid cells producing heterococcoliths and haploid cells producing holococcoliths. Nothing is known about the effects of the ongoing climate change and ocean acidification on coccolithophore life cycle and how this could determine its distribution.

5.2.1 Holococcolithophores

Our findings documented that holococcolithophore distribution during the time of the M84/3 cruise was best explained by changes in nutrients ($[NO_3+NO_2]$ and $[PO_4^{3-}]$) and pH ($\rho = 0.328$, Table 3). They were more abundant in low-nutrient, high-pH seawaters typical of the surface (upper 50 m) Mediterranean waters. Holococcolithophores were, in general, present in all samples collected in Mediterranean surface waters but almost absent at deeper depths (~ 100 m). Higher abundances in very oligotrophic waters of the Mediterranean Sea have been reported before (Kleijne, 1991,1992). Cros and Estrada (2013) reported holococcolithophores being more abundant in the upper photic zone nutrient-depleted waters. They linked the observed segregation to a differentiation of ecological niches. Dimiza et al. (2008) observed that holococcolithophores around Andros Island were more abundant in surface waters together with some heterococcolithophore species such as *Rabdosphaera clavigera*. The lower ρ obtained in the BIOENV analysis for holococcolithophores (Table 2) might be due to the lack, in the statistical analysis, of an important parameter (e.g., irradiance, zooplankton grazing) not measured during this study.

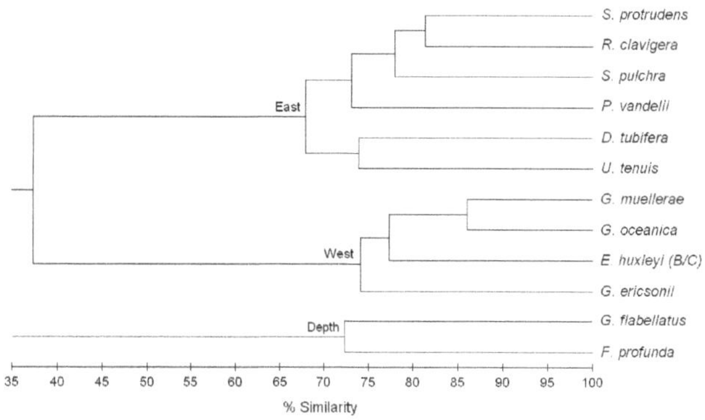

Figure 5. Hierarchical cluster for the heterococcolithophore species that shared more than 65 % similarity in their abundance and distribution patterns.

Interestingly, when both life phases were recorded (i.e., in the species *S. pulchra*, *C. mediterranea* and *H. carteri*), the environmental factors controlling the abundances were considerably different (Table 4). This clearly indicates that the two life phases respond differently to environmental factors which is likely imposing a differential distribution.

5.2.2 Heterococcolithophores

Heterococcolithophore distribution was best linked to a combination of $[CO_3^{2-}]$, pH, and salinity ($\rho = 0.566$; Table 3), and the single abiotic parameter that best grouped heterococcolithophore assemblages in a manner consistent with the sampling locations was $[CO_3^{2-}]$ (Table 3, first row). The ρ value for the combination of four variables – $[CO_3^{2-}]$ pH, salinity, and $[PO_4^{3-}]$ – was not much lower than the "best fit" ($[CO_3^{2-}]$ pH, salinity). This might indicate that, apart from seawater carbonate chemistry, $[PO_4^{3-}]$ is also important in explaining heterococcolithophore distribution. It is worth noting that $[PO_4^{3-}]$ correlated to the heterococcolithophore species diversity (H') (ρ: 0.322, $p < 0.05$) but not to the abundance. Thus, the availability of PO_4^{3-} might co-

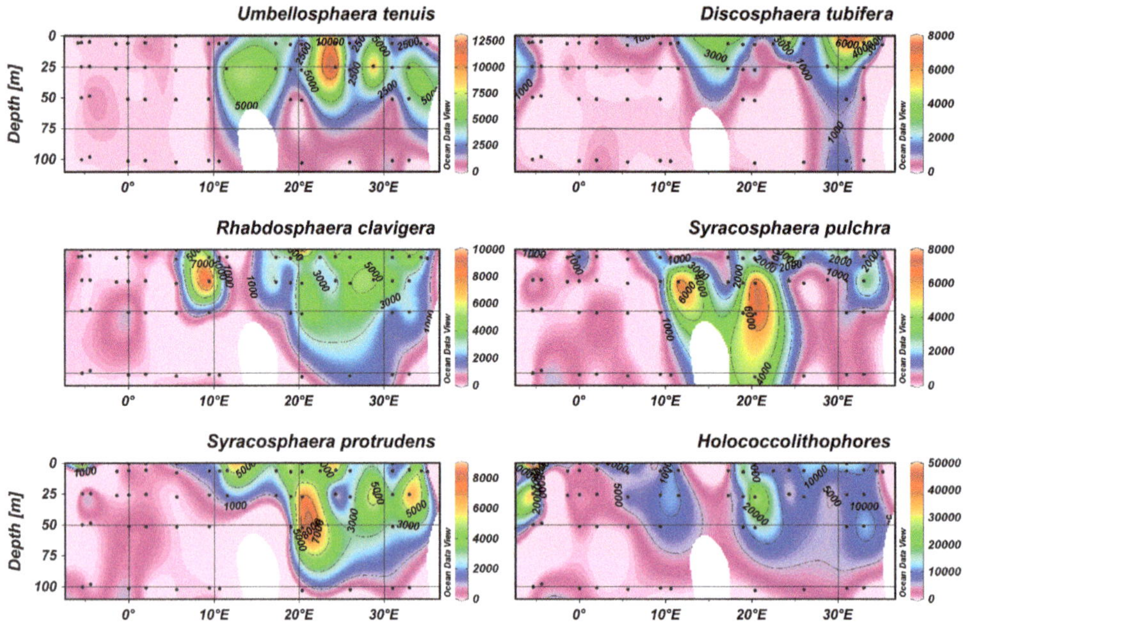

Figure 6. Distribution, along a west–east transect, of the heterococcolithophore species forming the cluster comprised mainly of "eastern Mediterranean species". Holococcolithophores as a group are added in the bottom right panel (the transect includes the stations labeled in black in Fig. 1).

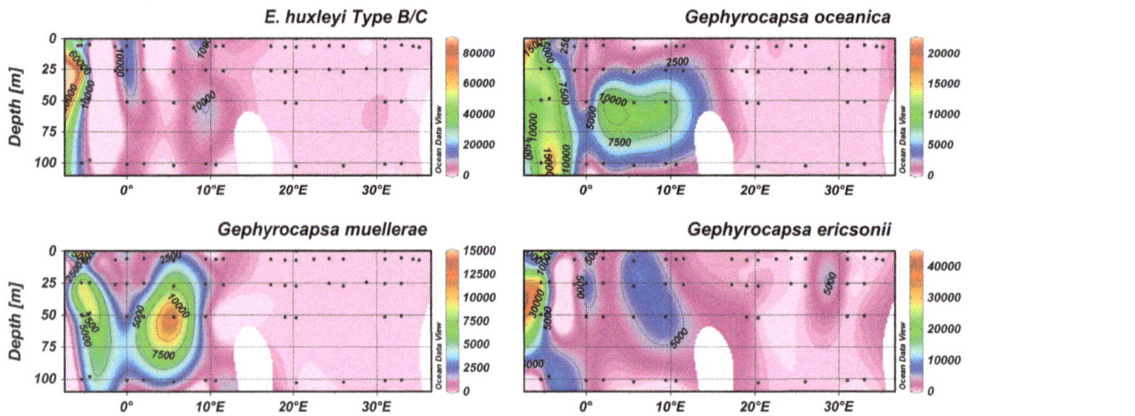

Figure 7. Distribution, along a west–east transect, of the heterococcolithophore species forming the cluster comprised mainly of "western Mediterranean species" (the transect includes the stations labeled in black in Fig. 1).

determine the assemblage composition. Differences in PO_4^{3-} usage have been reported between strains of the same species (Oviedo et al., 2014). It is therefore likely that different species show different sensitivities to $[PO_4^{3-}]$, which could induce differences in community composition upon environmental $[PO_4^{3-}]$. Overall, we suggest $[CO_3^{2-}]$, pH, and probably $[PO_4^{3-}]$ as functionally related important variables in explaining heterococcolithophore distribution in the Mediterranean Sea.

Although heterococcolithophores preferentially use HCO_3^- for their intracellular calcification within a calcifying vesicle (Mackinder et al., 2010) at alkaline pH values, CO_3^{2-} is the major carbon source for $CaCO_3$ and we can assume the uptake of both HCO_3^- and CO_3^{2-} (Ziveri et al., 2012). In the Mediterranean Sea, $[CO_3^{2-}]$ increases gradually towards the east. Thus, it is possible that those species thriving in the eastern basins utilize comparatively more CO_3^{2-} than those that prosper in the western Mediterranean. Using a compilation from worldwide plankton samples and sediments spanning the last 40 kyr, Beaufort et al. (2011) recorded significant correlations between coccolith mass belonging to the family Noelaerhabdaceae (genera *Emiliania*, *Gephyrocapsa*, and *Reticulofenestra*) with $[CO_3^{2-}]$ and $[HCO_3^-]$. They argue that differentially calcified species are distributed in the ocean according to the ocean's carbonate

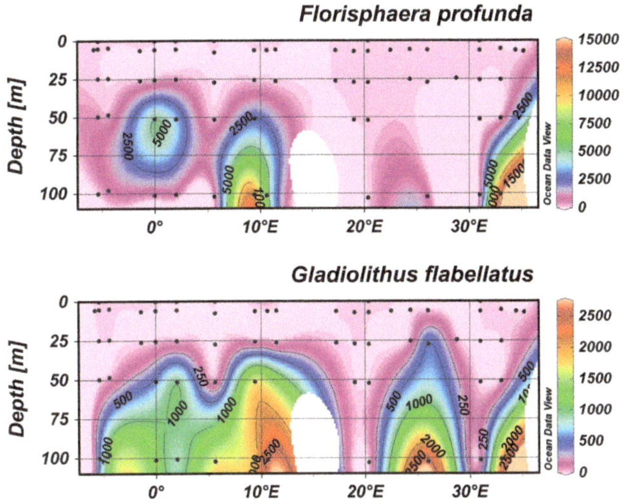

Figure 8. Distribution, along a west–east transect, of the heterococcolithophore species forming the smallest of the observed clusters comprised of two deep photic zone species (the transect includes the stations labeled in black in Fig. 1).

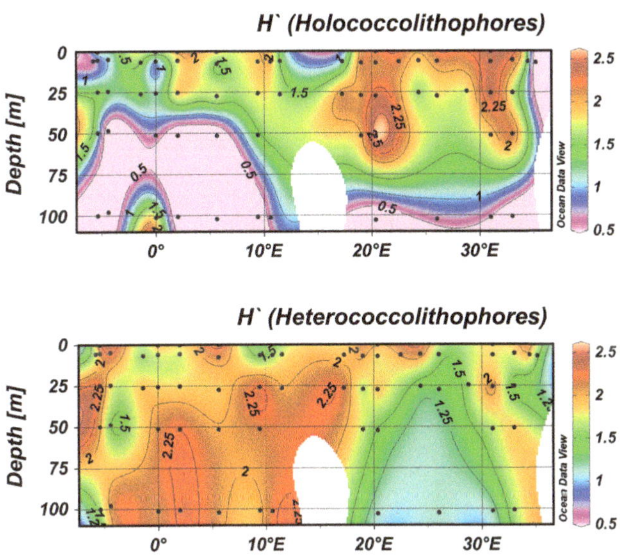

Figure 9. Shannon–Wiener diversity index for holococcolithophores and heterococcolithophores. A higher index, or higher uncertainty in correctly guessing the "next species" that would be sampled, is representative of a more diverse community.

chemistry. Modeling studies also highlighted the importance of $[CO_3^{2-}]$ for coccolithophore distribution, as observed by Merico et al. (2006); Tyrell et al. (2008) observed that *Emiliania huxleyi* blooms in the Baltic Sea coincide with periods of high $[CO_3^{2-}]$.

Changes in pH are concomitant with changes in the ratio between bicarbonate and carbonate ions. This makes difficult to disentangle which parameter of the carbonate system could affect coccolithophore populations. Under laboratory culture conditions, *Emiliania huxleyi* calcification is sensitive to low pH and bicarbonate, while photosynthesis and growth is sensitive to low pCO_2 (Bach et al., 2013) and the coccolith morphogenesis in *Calcidiscus leptoporus* is hampered by pCO_2 and no other parameter of the carbonate chemistry (Langer and Bode, 2011). How will these responses translate to a community scale in the ocean? Charalampopoulou et al. (2011) found that the coccolithophore species distribution between the North Sea and the Atlantic Ocean related to pH and irradiance. In our study, pCO_2, one of the parameters considered to run the BIOENV routine, was not part of the best-fitting variables to explain coccolithophore distribution patterns. This might be an indication that, during the time of the study, there was no evident pCO_2 limitation of photosynthesis in the observed species or that coccolithophore sensitivity to CO_2 among species, even if different (Langer et al., 2006, 2009) do not have major effects on coccolithophore species assemblages in the Mediterranean. Imitating the experimental design of the abovementioned culture experiments on a mesocosm scale could help elucidating how the different effects of carbonate chemistry modification on coccolithophore's physiology and morphology will shape the community.

Salinity, even if one of the environmental variables that optimized the best-fitting combination of variables explaining the biological data, might not be crucial in controlling heterococcolithophore distribution. Experimental evidence (Brand, 1984) indicates strain-specific differences that allow *Emiliania huxleyi* to survive at a wide salinity range. *E. huxleyi* has been found in oceanographic regions characterized by very different salinities (reviewed by Tyrrell et al., 2008). Furthermore, heterococcolithophores isolated from the Mediterranean Sea (CODENET collection) were maintained at salinities of 32–33 with non-observed adverse effects (Probert and Houdan, 2004).

5.3 Species assemblages

Another feature that was different between the coccolithophore life stages is the development of species assemblages. Holococcolithophore species did not form different assemblages along the Mediterranean Sea (no species clustering with high similarities in abundance and distribution). Holococcolithophores seem to behave as a homogeneous group, exploiting a similar ecological niche. However, three heterococcolith clusters were identified whose species shared more than 65 % similarities (Fig. 5). The first group comprises species that were more abundant in the eastern stations: *U. tenuis*, *D. tubifera*, *P. vandelii*, *S. pulchra*, *R. clavigera*, and *S. protrudens*. The second group includes *E. huxleyi* Type B/C and the *Gephyrocapsa* species – *G. ericsonii*, *G. muellerae*, and *G. oceanica* – that were almost exclusive to the western Mediterranean. A third group was formed by *Florisphaera profunda* and *Gladiolithus flabellatus* and found with higher abundances below 50 m. Most of the species in these clusters have been found to share an ecological niche in other studies: *U. tenuis*, *D. tubifera*, *S. pulchra*, and *R. clavigera* have been considered typical in oligotrophic warm waters and/or surface water species (Okada and Honjo 1973; Okada and McIntyre, 1977, 1979; Nishida, 1979; Ziveri et al., 2000; Haidar and Thierstein, 2001; Malinverno et al., 2003; Ziveri et al., 2004, for *Syracosphaera* spp.; Triantaphyllou et al., 2004). The gephyrocapside species have been considered typical of eutrophic areas (Kleijne, 1989; Broerse et al., 2000) with lower density, lower salinity and higher temperature (Takahashi and Okada, 2000; Knappertsbusch, 1993). Finally, *F. profunda* and *G. flabellatus* are widely recognized deep euphotic zone species, often living below 100 m depth (Okada and Honjo, 1973; Boeckel et al., 2006), and are controlled by the dynamics of the nutricline and thermocline (Molfino and McIntyre, 1990; Triantaphyllou et al., 2004). Both species were an important component below the 50 m, with highest abundances around 100m, where fluorescence data (not shown) place the deep chlorophyll maximum. The positive correlation with [NO_3+ NO_2] supports the previous observations. The negative correlation with temperature reflects the fact that nutrients are generally linked to deep mixing and colder waters. For instance, higher abundances of *F. profunda* have been observed in sediments underneath relatively warmer and stratified surface waters with a deep nutricline (Boeckel et al., 2006).

The clustering of coccolithophore species resembles those proposed by Young (1994): an umbelliform group such as *Umbellosphaera tenuis*, found in nutrient-depleted waters; a second group of placolith-bearing cells such as *Emiliania huxleyi* or *Gephyrocapsa* spp. found in coastal or mid-ocean upwelling regions; and a last group, composed of floriform cells, such as *Florisphaera profunda*, associated with deep photic-zone assemblages in low to mid-latitudes. To the first group we would add the rabdolith-bearing species as well as some *Syracosphaera* species widely associated with oligotrophic (Ziveri et al., 2004) surface waters (Malinverno et al., 2003) and cluster them together in our study. As Balch (2004) suggested when referring to this species grouping proposed by Young, "it is likely that the three groups of coccolithophores show differences in their growth strategies which ultimately would relate to their natural abundance".

Heterococcolithophores and holococcolithophores also displayed opposite trends in species diversity (Fig. 9). However the weak correlation obtained between the diversity index H' and the longitude of the sample site should be taken into account. Given the taxonomical problems between the two life stages (see Sect. 2.1) when considering the two as a single group would lead to the overestimation of the number of species, affecting H'. Future work on the topic of discriminating between the two life stages would be necessary in order to clarify the trends suggested here.

Overall, distribution patterns and their relation to environmental parameters are different between the holo- and heterococcolithophore life phases. This is probably the result of a different physiology and reveals that the two phases are exploiting different ecological niches. Differences in the morphology and ecology of the two life phases enable the species to survive under a wider range of environmental conditions and could result in a wider distribution range in space and/or time. For instance, blooms of *Emiliania huxleyi* in heterococcolith phase can end due to viral attacks (Martínez et al., 2007; Vardi et al., 2012) which the haploid phase (i.e., holococcolithophores) can resist (Frada et al., 2008, 2012): Therefore, the occurrence of haploid individuals would serve as new starting point in the case of viral attack in the heterococcolith phase. Another factor that could have affected the results presented here is zooplankton grazing: although generally zooplankton grazing does not cause *E. huxleyi* blooms to end (Nejstgaard et al., 1997), their effect on smaller populations could be more important. Possibly important factors that were not addressed in our study are zooplankton grazing and irradiance. Therefore, their contribution to the control of the observed distribution of holo- and heterococcolith life phases cannot be ruled out.

5.3.1 *Gephyrocapsa oceanica*, *G. muellerae*, and *Emiliania huxleyi morphotype* B/C as tracers of Atlantic water influx

Atlantic Water (AW), with a winter salinity around 36.5 (Rohling, 2009), enter the Mediterranean Sea through the Strait of Gibraltar. In its eastwards pathway following a cyclonic circulation, the surface water increases in salinity, TA, and temperature and decreases in nutrient concentrations. A main part of this AW flows into the Tyrrhenian Sea (Tanhua et al., 2013b, and references therein). Therefore, if there are species that arrive within this water mass (but whose optimal environment is not typical of eastern Mediterranean waters), the Tyrrhenian Sea could still host them.

During April 2011, the distribution of the species *Gephyrocapsa oceanica* and *G. muellerae* and morphotype B/C of the species *Emiliania huxleyi* was highly and negatively correlated with salinity. We have argued that salinity constrain to coccolithophore distribution is not critical. A high correlation with this parameter might just reflect the carry-over of a different species assemblage in a different water mass. *Gephyrocapsa oceanica*, *G. muellerae*, and morphotype B/C of the species *Emiliania huxleyi* were mostly present until $\sim 10°$ E after the Sardinia Channel and in the Tyrrhenian Sea. They are all present in AW (Ziveri et al., 2004; McIntyre and Bé, 1967) and have been reported before for western Mediterranean waters (e.g., Cross, 2002, and Knappertsbusch, 1993; the latter only for *Gephyrocapsa* spp). The morphotype B/C of *E. huxleyi* has been associated with cold (Hagino et al., 2005; Mohan et al., 2008), nutrient-rich (> 10 μmol nitrate kg^{-1}) waters with low calcite saturation states. These characteristics can be found in AW but are lost very soon in the Mediterranean Sea. Knappertsbusch (1993) related *G. oceanica* to surface AW's influence given the highly negative correlation with salinity. Here we propose *Gephyrocapsa oceanica*, *G. muellerae*, and *Emiliania huxleyi* morphotype B/C as tracers for AW influx into the Mediterranean.

6 Conclusions

Our results highlight the importance of seawater carbonate chemistry, especially [CO_3^{2-}], and [PO_4^3-] in unraveling the distribution of heterococcolithophores, the most abundant coccolithophore life stage. Thus, carbonate system parameters might be critical, but overlooked, for solving the coccolithophore distribution patterns. In contrast, holococcolithophore distribution was mainly linked to oligotrophic conditions. This correlation can be due to competitive advantages under such conditions, but this hypothesis remains to be tested.

Environmental parameters that drive the observed patterns in distribution and assemblage composition of the haploid and diploid life phases of coccolithophores (holo- and heterococcolithophore) differ. Our results emphasize the importance of considering holo- and heterococcolithophores separately when analyzing changes in species assemblages and diversity, and the impacts of acidification on coccolithophores. Changes in the biogeographic distribution of hetero- and holococcolithophores can be expected along with the changes in carbonate chemistry and nutrient concentrations that are projected for the next century.

Appendix A

Table A1. Coccolithophore species list. Heterococcolithophores.

[1] *Anoplosolenia brasiliensis* (Lohmann) Deflandre [2] *Calciosolenia murrayi* Gran
Family Coccolithaceae Poche
[3] *Calcidiscus leptoporus* (Murray et Blackman) Loeblich et Tappan [4] *Hayaster perplexus* (Bramlette et Riedel) Bukry [5] *Pleurochrysis carterae* (Braarud et Fagerland) Christensen [6] *Umbilicosphaera sibogae* var. *sibogae* (Weber-Van Bosse) Gaarder [7] *U. foliosa* (Kamptner) Geisen [8] *U. hulburtiana* Gaarder
Family Helicosphaeraceae Black
[9] *Helicosphaera carteri* (Wallich) Kamptner [10] *H. pavimentum* Okada et McIntire
Family Noelaerhabdaceae Jerkovic
[11] *Emiliania huxleyi* (Lohmann) Hay et Mohler [12] *Gephyrocapsa ericsonii* McIntire et Bé [13] *G. ornata* Heimdal [14] *G. oceanica* Kamptner [15] *G. muellerae* Bréhéret [16] *Reticulofenestra parvula* (Okada et McIntyre) Biekart
Family Papposphaeraceae Jordan et Young
[17] *Papposphaera lepida* Tangen
Family Pontosphaeraceae Lemmermann
[18] *Pontosphaera japonica* (Takayama) Nishida [19] *Scyphosphaera apsteinii* Lohmann
Family Rhabdoaphaeraceae Ostenfeld
[20] *Acanthoica biscayensis* Kleijne [21] *A. quattrospina* Lohmann [22] *Algirosphaera cucullata* (Lecal-Schlauder) Young, Probert et Kleijne [23] *Algiropsphaera robusta* (Lohmann) Norris [24] *Anacanthoica acanthos* (Schiller) Deflandre [25] *Cyrtosphaera lecaliae* Kleijne [26] *Discosphaera tubifera* (Murray et Blackman) Ostenfeld [27] *Palusphaera vandeli* Lecal emend. Norris [28] *Rhabdosphaera clavigera* var. *clavigera* Murray et Blackman [29] *R. clavigera var. stylifera* (Lohmann) Kleijne et Jordan [30] *R. xiphos* (Deflandre et Fert) Norris
Family Syracosphaeraceae Lemmermann
[31] *Alisphaera capulata* Heimdal [32] *A. extentata* (Kamptner) Kleijne, Jordan, Heimdal, Samtleben, Chamberlain et Cros [33] *A. gaudi* (Kamptner) Kleijne, Jordan, Heimdal, Samtleben, Chamberlain et Cros [34] *A. unicornis* Okada et McIntire [35] *Calciopappus caudatus* Gaarder et Ramsfjell [36] *C. rigidus* Heimdal

Table A1. Continued.

[37] *Coronosphaera binodata* (Kamptner) Gaarder
[38] *C. mediterranea* (Lohmann) Gaarder
[39] *Michaelsarsia adriaticus* (Schiller) Manton, Bremer et Oates
[40] *M. elegans* Gran, emend. Manton, Bremer et Oates
[41] *Ophiaster formosus* Gran
[42] *Ophiaster hydroideus* Gran
[43] *Syracosphaera ampliora* Okada et McIntire
[44] *S. anthos* (Lohmann) Jordan et Young
[45] *S. bannockii* (Borsetti et Cati) Cros, Kleijne, Zeltner, Billard et Young
[46] *S. borealis* Okada et McIntire
[47] *S. corolla* Lecal
[48] *S. delicata* Cros, Kleijne, Zeltner, Billard et Young
[49] *S.dilatata* Jordan, Kleijne et Heimdal
[50] *S. histrica* Kamptner
[51] *S. lamina* Lecal-Schlauder
[52] *S. marginoporata* Knappertsbusch
[53] *S. molischii* Schiller
[54] *S. nana* (Kamptner) Okada et McIntyre
[55] *S. nodosa* Kamptner
[56] *S. noroitica* Knappertsbusch
[57] *S. ossa* (Lecal) Loeblich et Tappan
[58] *S. pirus* Halldal et Markali
[59] *S. prolongata* Gran, ex Lohmann
[60] *S. protrudens* Okada et McIntyre
[61] *S. pulchra* Lohmann
[62] *S. rotula* Okada et McIntire
[63] *Syracosphaera sp. type D*, sensu Kleijne
[64] *S. tumularis* Sánchez- Suárez
[65] *Syracospahaera* sp.
Sub-Family Umbellosphaeroideae Kleijne
[66] *Umbellosphaera tenuis* (Kamptner) Paasche
[67] *F. profunda* Okada et Honjo
[68] *Gladiolithus flabellatus* (Halldal et Markali) Jordan et Green
[69] *Polycrater galapagensis* Manton et Oates
[70] *Ceratolithus cristatus* Norris

Table A2. Coccolithophore species list. Holococcolithophores.

[1] *Acanthoica quattrospina* HOL = sp. aff *Sphaerocalyptra* of Cros et al. (2000)
[2] *Anthosphaera fragaria* Kamptner emend. Kleijne
[3] *A. lafourcadii* (Lecal) Kleijne
[4] *A. periperforata* Kleijne
[5] *Anthosphaera sp*. Type A of Cros and Fortuño (2002)
[6] *Anthosphaera* sp. Type C of Cros and Fortuño (2002)
[7] *Calcidiscus leptoporus* HOL (Murray et Blackman) Loeblich et Tappan
[8] *C. leptoporus* ssp. *quadriperforatus* HOL (Kamptner) Geisen, Billard, Broerse, Cros, Probert et Young
[9] *Calicasphaera concava* Kleijne
[10] *Calyptrolithina divergens* (Halldal et Markali) Heimdal
[11] *C. divergens var. tuberosa* (Heimdal) Jordan, Kleijne et Heimdal
[12] *C. multipora* (Gaarder) Norris
[13] *Calyptrolithophora papillifera* (Halldal) Heimdal
[14] *Calyptrosphaera cialdii* Borsetti et Cati
[15] *C. dentata* Kleijne
[16*]* *C. heimdalae* Norris
[17] *C. sphaeroidea* Schiller
[18] *Coccolithus pelagicus* ssp. *braarudii* HOL (Gaarder) Geisen, Billard, Broerse, Cros, Probert et Young
[19] *Corisphaera gracilis* Kamptner
[20] *C. strigilis* Gaarder
[21] *C. tyrrheniensis* Kleijne
[22] *Corisphaera* sp. Type A of Kleijne (1991)
[23] *Coronosphaera mediterranea* HOL gracillima-type =*Calyptrolithophora gracillima* (Kamptner) Heimdal
[24] *Coronosphaera mediterranea* HOL hellenica-type =*Zygosphaera hellenica* Kamptner
[25] *Gliskolithus amitakarenae* Norris, orthog. emend., Jordan et Green
[26] *Helicosphaera carteri* HOL =*Syracolithus catilliferus* (Kamptner) Deflandre
[27] *Helladosphaera cornifera* (Schiller) Kamptner
[28] Homozygosphaera arethusae (Kamptner) Kleijne
[29] *H. spinosa* (Kamptner) Deflandre
[30] *H. triarcha* Halldal et Markalii
[31] *Homozygosphaera vercellii* Borsetti et Cati
[32] *Poricalyptra gaarderae* (Borsetti et Cati) Kleijne
[33] *Poritectolithus* sp. 2 of Cros and Fortuño (2002)
[34] *Sphaerocalyptra adenensis* Kleijne
[35] *S. quadridentata* (Schiller) Deflandre
[36] *Sphaerocalyptra* sp. 1 of Cros and Fortuño (2002)
[37] *Sphaerocalyptra* sp. 3 of Cros and Fortuño (2002)
[38] *Sphaerocalyptra* sp. 6 of Cros and Fortuño (2002)
[39] *Syracolithus schilleri* (Kamptner) Loeblich et Tappan
[40] *Syracolithus* sp. type A of Kleijne (1991)
[41] *Syracosphaera anthos* HOL =*Periphyllophora mirabilis* (Schiller) Kamptner
[42] *Syracosphaera bannockii* HOL =*Zygosphaera bannockii* (Borsetti et Cati) Heimdal
[43] *Syracosphaera pulchra* HOL oblonga-type =*Calyptrosphaera oblonga* Lohmann
[44] *Syracosphaera pulchra* HOL pirus-type =*Calyptrosphaera pirus* Kamptner
[45] *Zygosphaera amoena* Kamptner

Acknowledgements. We thank the crew members and scientists onboard the R/V *Meteor* during the M84/3 campaign for making a month of sampling possible and a warm and welcoming time onboard. Thanks to Vincent Taillandier for helping with the chlorophyll data. This work was partially funded by the European Community's Seventh Framework Programme under grant agreement 265103 (project MedSeA) and by project HOTMIX (CTM2011-30010-C02). The authors also acknowledge the "Agencia de Gestión de Ayudas Universitarias y de Investigación" (Becas FI) of the Catalunya Generalitat. We thank Sebastian Meier and the anonymous reviewer for their comments that greatly enriched the manuscript.

Edited by: N. Kress

References

Álvarez, M., Sanleón-Bartolomé, H., Tanhua, T., Mintrop, L., Luchetta, A., Cantoni, C., Schroeder, K., and Civitarese, G.: The CO_2 system in the Mediterranean Sea: a basin wide perspective, Ocean Sci., 10, 69–92, doi:10.5194/os-10-69-2014, 2014.

Bach, L. T., Mackinder, L. C., Schulz, K. G., Wheeler, G., Schroeder, D. C., Brownlee, C. and Riebesell, U.: Dissecting the impact of CO_2 and pH on the mechanisms of photosynthesis and calcification in the coccolithophore Emiliania huxleyi, New Phytol., 199, 121–134, doi:10.1111/nph.12225, 2013.

Balch, W. M.: Re-evaluation of the physiological ecology of coccolithophores, in: Coccolithophores – From Molecular Processes to Global Impact, edited by: Thierstein, H. and Young, J. R., Springer, Berlin, 165–190, 2004.

Bakun, A. and Agostini, V. N.: Seasonal patterns of wind-induced up- welling/downwelling in the Mediterranean Sea, Sciencia Marina, 65, 243–257, 2001.

Barcena, M. A., Flores, J. A., Sierro, F. J., Pérez-Folgado, M., Fabres, J., Calafat, A., and Canals, M.: Planktonic response to main oceanographic changes in the Alboran Sea (Western Mediterranean) as documented in sediment traps and surface sediments, Mar. Micropaleontol., 53, 423–445, 2004.

Barlow, R. G., Mantoura, R. F. C., Cummings, D. G., and Fileman, T. W.: Pigment chemotaxonomic distributions of phytoplankton during summer in the western Mediterranean, Deep-Sea Res. Pt. II, 44, 833–850, doi:10.1016/S0967-0645(96)00089-6, 1997.

Beaufort, L., Couapel, M., Buchet, N., Claustre, H., and Goyet, C.: Calcite production by coccolithophores in the south east Pacific Ocean, Biogeosciences, 5, 1101–1117, doi:10.5194/bg-5-1101-2008, 2008.

Beaufort, L., Probert, I., de Garidel-Thoron, T., Bendif, E. M., Ruiz-Pino, D., Metzl, N., Goyet, C., Buchet, N., Coupel, P., Grelaud, M., Rost, B., Rickaby, R. E. M., and de Vargas, C.: Sensitivity of coccolithophores to carbonate chemistry and ocean acidification, Nature, 476, 80–83, 2011.

Berland, B. R., Benzhitski, A. G., Burlakova, Z. P., Georgieva, L. V., Izmestieva, M. A., Kholodov, V. I., and Maestrini, S. Y.: Conditions hydrologiques estivales en Mediterranee, repartition du phytoplankton et de la matiere organique, Oceanol. Acta, 9, 163–177, 1988.

Berman, T., Townsend, D. W., El Sayed, S. Z., Trees, G. C., and Azov, Y.: Optical transparency, chlorophyll and primary productivity in the Eastern Mediterranean near the Israeli Coast, Oceanol. Acta, 7, 367–372, 1984.

Boeckel, B., Baumann, K. H., Henrich, R., and Kinkel, H.: Coccolith distribution patterns in South Atlantic and Southern Ocean surface sediments in relation to environmental gradients, Deep-Sea Res. Pt. I, 53, 1073–1099, doi:10.1016/j.dsr.2005.11.006, 2006.

Brand, L. E.: The salinity tolerance of forty-six marine phytoplankton isolates, Estuar. Coast. Shelf S., 18, 543–556, 1984.

Brand, L. E.: Physiological ecology of marine coccolithophores, in: Coccolithophores, edited by: Winter, A. and Siesser, W. G., Cambridge University Press, UK, 39–49, 1994.

Broerse, A. T. C., Brummer, G.-J. A., and Van Hinte, J. E.: Coccolithophore export production in response to monsoonal upwelling off Somalia (northwestern Indian Ocean), Deep-Sea Res. Pt. II, 47, 2179–2205, 2000.

Charalampopoulou, A., Poulton, A. J., Tyrrell, T., and Lucas, M. I.: Irradiance and pH affect coccolithophore community composition on a transect between the North Sea and the Arctic Ocean, Mar. Ecol.-Prog. Ser., 431, 25–43, 2011.

Cortès, M. Y., Bollman, J., and Thierstein, H. R.: Coccolithophore ecology at the HOT station Aloha, Hawaii, Deep-Sea Res. Pt. II, 48, 1957–1981, 2001.

Cros, L.: Planktic coccolithophores of the NW Mediterranean. PhD Thesis, Universitat de Barcelona, Barcelona, 181 pp., 2002.

Cros, L. and Estrada, M.: Holo-heterococcolithophore life cycles: ecological implications, Mar. Ecol.-Prog. Ser., 492, 57–68, 2013.

Cros, L. and Fortuno, J.-M.: Atlas of northwestern Mediterranean coccolithophores, Sci. Mar., 66, 7–182, 2002.

Decembrini, F., Caroppo, C., and Azzaro, M.: Size structure and production of phytoplank- ton community and carbon pathways channelling in the Southern Tyrrhenian Sea (Western Mediterranean), Deep-Sea Res. Pt. II, 56, 687–699, 2009.

Dickson, A. G. and Millero, F. J.: A comparison of the equilibrium constants for the dissociation of carbonic acid in seawater media, Deep-Sea Res., 34, 1733–1743, 1987.

Dimiza, M. D., Triantaphyllou, M. V., and Dermitzakis, M. D.: Vertical distribution and ecology of living coccolithophores in the marine ecosystems of Andros Island (Middle Aegean Sea) during late summer 2001, Hellenic J. Geosci., 43, 7–20, 2008.

Dugdale, R. C. and Wilkerson, F. P.: Nutrient sources and primary production in the eastern Mediterranean, Oceanol. Acta, Special Issue, 9, 179–184, 1988.

Egge, J. K. and Aksnes, D. L.: Silicate as regulating nutrient in phytoplankton competition, Mar. Ecol.-Prog. Ser., 83, 281–289, 1992.

Falkowski, P. G.: The role of phytoplankton photosynthesis in global biogeochemical cycles, Photosynth. Res., 39, 235–258, 1994.

Field, C. B., Behrenfeld, M. J., Randerson, J. T., and Falkowski, P. G.: Primary production of the biosphere: integrating terrestrial and oceanic components, Science, 281, 237–240, 1998.

Frada, M., Probert, I., Allen, M. J., Wilson, W. H., and De Vargas, C.: The "Cheshire Cat" escape strategy of the coccolithophore Emiliania huxleyi in response to viral infection, P. Natl. Acad. Sci. USA, 105, 15944–15949, 2008.

Frada, M. J., Bidle, K. D., Probert, I., and de Vargas, C.: In situ survey of life cycle phases of the coccolithophore Emiliania huxleyi (Haptophyta), Environ. Microbiol., 14, 1558–1569, 2012.

Gotsis-Skretas, O., Pagou, K., Moraitou-Apostolopoulou, M., and Ignatiades, L.: Seasonal horizontal and vertical variability in primary production and standing stocks of phytoplankton and zooplankton in the Cretan Sea and the Straits of the Cretan Arc (March 1994–January 1995), Prog. Oceanogr., 44, 625–649, 1999.

Hagino, K., Okada, H., and Matsuoka, H: Spatial dynamics of coccolithophore assemblage in the Euatorial Western-Central Pacific Ocean, Mar. Micropaleontol., 39, 53–72, 2000.

Hagino, K., Okada, H., and Matsuoka, H.: Coccolithophore assemblages and morphotypes of Emiliania huxleyi in the boundary zone between the cold Oyashio and warm Kuroshio currents off the coast of Japan, Mar. Micropaleontol., 55, 19–47, 2005.

Haidar, A. T. and Thierstein, H. R.: Coccolithophore dynamics off Bermuda (N. Atlantic), Deep- Sea Res. Pt. II, 40, 1925–1956, 2001.

Hainbucher, D., Rubino, A., Cardin, V., Tanhua, T., Schroeder, K., and Bensi, M.: Hydrographic situation during cruise M84/3 and P414 (spring 2011) in the Mediterranean Sea, Ocean Sci., 10, 669–682, doi:10.5194/os-10-669-2014, 2014.

Ignatiades, L., Georgopoulos, D., and Karydis, M.: Description of the phytoplanktonic community of the oligotrophic waters of the SE Aegean Sea (Mediterranean), Mar. Ecol.-P. S. Z. N. I., 16, 13–26, 1995.

Ignatiades, L., Gotsis-Skretas, O., Pagou, K., and Krasakopoulou, E.: Diversification of phyto- plankton community structure and related parameters along a large-scale longitudinal east-west transect of the Mediterranean Sea, J. Plankton Res., 31, 411–428, 2009.

Kleijne, A.: Holococcolithophorids from the Indian Ocean, Red Sea, Mediterranean Sea and North Atlantic Ocean, Mar. Micropaleontol., 17, 1–76, 1991.

Kleijne, A.: Extant Rhabdosphaeraceae (coccolithophorids, class Prymnesiophyceae) from the Indian Ocean, Red Sea, Mediterranean Sea and North Atlantic Ocean, ScriptaGeol., 100, 1–63, 1992.

Kleijne, A., Kroon, D., and Zevenboom, W.: Phytoplankton and foraminiferal frequencies in northern Indian Ocean and Red Sea surface waters, Neth. J. Sea Res., 24, 531–539, 1989.

Knappertsbusch, M.: Geographic distribution of living and Holocene coccolithophores in the Mediterranean Sea, Mar. Micropaleontol., 21, 219–247, 1993.

Kroeker, K. J., Kordas, R. L., Crim, R., Hendriks, I. E., Ramajo, L., Singh, G. S., Duarte, C. M., and Gattuso, J. P.: Impacts of ocean acidification on marine organisms: quantifying sensitivities and interaction with warming, Glob. Change Biol., 19, 1884–1896, doi:10.1111/gcb.12179, 2013.

Krom, M. D., Kress, N., Brenner, S., and Gordon, L.: Phosphorous limitation of primary productivity in the Eastern Mediterranean Sea, Limnol. Oceanogr., 36, 424–432, 1991.

Langer, G. and Bode, M.: CO_2 mediation of adverse effects of seawater acidification in Calcidiscus leptoporus, Geochem. Geophy. Geosy., 12, Q05001, doi:10.1029/2010GC003393, 2011.

Langer, G., Geisen, M., Baumann, K.-H., Kläs, J., Riebesell, U., Thoms, S., and Young, J. R.: Species-specific responses of calcifying algae to changing seawater carbonate chemistry, Geochem. Geophy. Geosy., 7, Q09006, doi:10.1029/2005GC00122, 2006.

Langer, G., Nehrke, G., Probert, I., Ly, J., and Ziveri, P.: Strain-specific responses of *Emiliania huxleyi* to changing seawater carbonate chemistry, Biogeosciences, 6, 2637–2646, doi:10.5194/bg-6-2637-2009, 2009.

Lasternas, S., Tunin-Ley, A., Ibañez, F., Andersen, V., Pizay, M.-D., and Lemée, R.: Short-term dynamics of microplankton abundance and diversity in NW Mediterranean Sea during late summer conditions (DYNAPROC 2 cruise; 2004), Biogeosciences, 8, 743–761, doi:10.5194/bg-8-743-2011, 2011.

Leblanc, K., Quéguiner, B., Garcia, N., Rimmelin, P. and Raimbault, P.: Silicon cycle in the NW Mediterranean Sea: seasonal study of a coastal oligotrophic site, Oceanol. Acta, 26, 339–355, 2003.

Lewis, E. and Wallace, D. W. R: Program developed for CO_2 system calculations ORNL/CDIAC-105. Carbon Dioxide Information Analysis CentreOak Ridge National Laboratory, U.S. Department of Energy, Oak Ridge, Tennessee, 1998.

Mackinder, L., Wheeler, G., Schroeder, D., Riebesell, U., and Brownlee, C.: Molecular mechanisms underlying calcification in coccolithophores, Geomicrobiol. J., 27, 585–595, doi:10.1080/01490451003703014, 2010.

Malinverno, E., Ziveri, P., and Corselli, C.: Coccolithophorid distribution in the Ionian Sea and its relationship to eastern Mediterranean circulation during late fall to early winter 1997, J. Geophys. Res., 108, 8115, doi:10.1029/2002JC001346, 2003.

Margalef, R.: Life-forms of phytoplankton as survival alternatives in an unstable environment, Oceanol. Acta, 1, 493–509, 1978.

Martinez, J. M., Schroeder, D. C., Larsen, A., Bratbak, G., and Wilson, W. H.: Molecular dynamics of Emiliania huxleyi and cooccurring viruses during two separate mesocosm studies, Appl. Environ. Microb., 73, 554–562, 2007.

McIntyre, A. and Bé, A. W. H.: Modern coccolithophoridae of the Atlantic Ocean. 1. Placoliths and cyrtoliths, Deep-Sea Res., 14, 561–597, 1967.

Mehrbach, C., Culberson, C. H., Hawley, J. E., and Pytkowitcz, R. M.: Measurements of the apparent dissociation constants of carbonic acid in seawater at atmospheric pressure, Limnol. Oceanogr., 18, 897–907, 1973.

Merico, A., Tyrrell, T., and Cokacar, T.: Is there any relationship between the phytoplankton seasonal dynamics and the carbonate system?, J. Marine Syst., 59, 120–142, 2006.

Mohan, R., Mergulhao, L. P., Guptha, M. V. S. Rajakumar, A., Thamban, M., Anilkumar, N., Sudhakar, M., and Ravindra, R.: Ecology of coccolithophores in the Indian sector of the Southern Ocean, Mar. Micropaleontol., 67, 30–45, 2008.

Molfino, B. and McIntyre, A.: Precession forcing of nutricline dynamics in the equatorial Atlantic, Science, 249, 766–769, 1990.

Nejstgaard, J. C., Gismervik, I., and Solberg, P. T.: Feeding and reproduction by Calanus fin- marchicus, and microzooplankton grazing during mesocosm blooms of diatoms and the coccolithophore Emiliania huxleyi, Mar. Ecol.-Prog. Ser., 147, 197–217, 1997.

Nelson, D. M., Goering, J. J., Kilham, S. S., and Guillard, R. R. L.: Kinetics of silicic acid uptake and rates of silica dissolution in the marine diatom Thalassiosira pseudonana, J. Phycol., 12, 246–252, 1976.

Nishida, S.: Atlas of Pacific nannoplanktons, News Osaka, Micropaleonty, 3, 1–31, 1979.

Okada, H. and McIntyre, A.: Modern coccolithophores of the Pacific and North Atlantic Oceans, Micropaleontology, 23, 1–55, 1977.

Okada, H. and McIntyre, A.: Seasonal distribution of modern coccolithophores in the western North Atlantic Ocean, Mar. Biol., 54, 319–328, 1979.

Okada, H. and Honjo, S.: The distribution of ocean coccolithophorids in the Pacific, Deep-Sea Res., 20, 335–374, 1973.

Oviedo, A. M., Langer, G., and Ziveri, P.: Effect of phosphorus limitation on coccolith morphology and element ratios in Mediterranean strains of the coccolithophore Emiliania huxleyi, J. Exp. Mar. Biol. Ecol., 459, 105–113, 2014.

Pinardi, N. and Masetti, E.: Variability of the large scale general circulation of the Mediterranean Sea from observations and modelling: a review, Palaeogeogr. Palaeoecl., 158, 153–173, 2000.

Probert, I. and Houdan, A.: The Laboratory culture of coccolithophores, in: Coccolithophores: From Molecular Processes to Global Impact, edited by: Thierstein, H. R. and Young, J. R., Springer, Germany, 217–250, 2004.

Psarra, S., Tselepides, A., and Ignatiades L.: Primary productivity in the oligotrophic Cretan Sea (NE Mediterranean): seasonal and interannual variability, Prog. Oceanogr., 46, 187–204, 2000.

Rabitti, S., Bianchi, F., Boldrin, A., Da Ros, L., Socal, G., and Totti, C.: Particulate matter and phytoplankton in the Ionian Sea, Oceanol. Acta, 17, 297–307, 1994.

Rahav, E., Herut, B., Levi, A., Mulholland, M. R., and Berman-Frank, I.: Springtime contribution of dinitrogen fixation to primary production across the Mediterranean Sea, Ocean Sci., 9, 489–498, doi:10.5194/os-9-489-2013, 2013.

Ribera d'Alcalà, M., Civitarese, G., Conversano, F., and Lavezza, R.: Nutrient ratios and fluxes hint an overlooked processes in the Mediterranean sea, J. Geophys. Res., 108, 8106, doi:10.1029/2002JC001650, 2003.

Robinson, A. R., Leslie, W. G., Theocharis, A., and Lascaratos, A.: Mediterranean Sea circulation, in: Encyclopedia of ocean sciences, Academic Press, London, doi:10.1006/rwos.2001.0376, 2001.

Rohling, E. J., Abu-Zied, R., Casford, J. S. L., Hayes, A., and Hoogakker, B. A. A.: The marine environment: present and past, in: The Physical Geography of the Mediterranean, edited by: Woodward, J. C., Oxford University Press, Oxford, ISBN: 978-0-19-926803-0, 33–67, 2009.

Sarthou, G., Timmermans, K. R., Blain, S., and Treguer, P.: Growth physiology and fate of diatoms in the ocean: a review, J. Sea Res., 53, 25–42, 2005.

Sverdrup, H. U.: On conditions for the vernal blooming of phytoplankton, ICES J. Mar. Sci., 18, 287–295, 1953.

Takahashi, K. and Okada, H.: Environmental control on the biogeography of modern coccolithophores in the southeastern Indian Ocean offshore of Western Australia, Mar. Micropaleontol., 39, 73–86, 2000.

Tanhua, T., Hainbucher, D., Cardin, V., Álvarez, M., Civitarese, G., McNichol, A. P., and Key, R. M.: Repeat hydrography in the Mediterranean Sea, data from the *Meteor* cruise 84/3 in 2011, Earth Syst. Sci. Data, 5, 289–294, doi:10.5194/essd-5-289-2013, 2013a.

Tanhua, T., Hainbucher, D., Schroeder, K., Cardin, V., Álvarez, M., and Civitarese, G.: The Mediterranean Sea system: a review and an introduction to the special issue, Ocean Sci., 9, 789–803, doi:10.5194/os-9-789-2013, 2013b.

Triantaphyllou, M., Dimiza, M., Krasakopoulou, E., Malinverno, E., Lianou, V., and Souvermezoglou, E.: Seasonal variation in Emiliania huxleyi coccolith morphology and calcification in the Aegean Sea (Eastern Mediterranean), Geobios., 12th Meeting of the International Nannoplankton Association, Lyon 2008, 43, 99–110, 2010.

Tyrrell, T. and Young, J. R.: Coccolithophores, in: Encyclopedia of Ocean Sciences, 2nd Edn., edited by: Steele, J. H., Turekian, K. K., and Thorpe, S. A., Academic Press, San Diego, US, 3568–3576, 2009.

Tyrrell, T., Schneider, B., Charalampopoulou, A., and Riebesell, U.: Coccolithophores and calcite saturation state in the Baltic and Black Seas, Biogeosciences, 5, 485–494, doi:10.5194/bg-5-485-2008, 2008.

Vardi, A., Haramaty, L., Van Mooy, B. A., Fredricks, H. F., Kimmance, S. A., Larsen, A., and Bidle, K. D.: Host-virus dynamics and subcellular controls of cell fate in a natural coccolithophore population, P. Natl. Acad. Sci. USA, 109, 19327–19332, 2012.

Volpe, G., Colella, S., Forneris, V., Tronconi, C., and Santoleri, R.: The Mediterranean Ocean Colour Observing System – system development and product validation, Ocean Sci., 8, 869–883, 2012

Yacobi, Y., Zohary, T., Kress, N., Hecht, A., Robarts, R., Waiser, M., Wood, A., and Li, W.: Chlorophyll distribution throughout the southern Mediterranean in relation to the physical structure of the water mass, J. Mar. Syst., 6, 179–190, 1995.

Young, J. R.: Functions of coccoliths, in: Coccolithophores, edited by: Winter, A. and Siesser, W. G., Cambridge University Press, Cambridge, UK, 63–82, 1994.

Young, J., Geisen, M., Cros, L., Kleijne, A., Sprengel, C., Probert, I., and Østergaard, J.: A guide to extant coccolithophore taxonomy, J. Nannoplankton Res., 1, 1–125, 2003.

Ziveri, P., Thunell, R. C., and Rio, D.: Seasonal changes in coccolithophore densities in the Southern California Bight during 1991–1992, Deep-Sea Res. Pt. I, 42, 1881–1903, 1995.

Ziveri, P., Rutten, A., de Lange, G., Thomson, J., and Corselli, C.: Present-day coccolith fluxes recorded in central eastern Mediterranean sediment traps and surface sediments, Palaeogeogr. Palaeocl., 158, 175–195, 2000.

Ziveri, P., Baumann, K. H., Böckel, B., Bollman, J., and Young, J. R.: Present day coccolithophore biogeography in the Atlantic Ocean, in: Coccolithophores: From Molecular Processes to Global Impact, edited by: Thierstein, H. R. and Young, J. R., Springer Verlag, Berlin Heidelberg, 2004.

Ziveri, P., Thoms, S., Probert, I., Geisen, M., and Langer, G.: A universal carbonate ion effect on stable oxygen isotope ratios in unicellular planktonic calcifying organisms, Biogeosciences, 9, 1025–1032, doi:10.5194/bg-9-1025-2012, 2012.

10

Using empirical orthogonal functions derived from remote-sensing reflectance for the prediction of phytoplankton pigment concentrations

A. Bracher[1,2], M. H. Taylor[3], B. Taylor[1], T. Dinter[1,2], R. Röttgers[4], and F. Steinmetz[5]

[1]Alfred Wegener Institute Helmholtz Centre for Polar and Marine Research, Bussestraße 24, 27570 Bremerhaven, Germany
[2]Institute of Environmental Physics, University of Bremen, Otto-Hahn-Allee 1, 28359 Bremen, Germany
[3]Leibniz Center for Tropical Marine Ecology, Fahrenheitstraße 6, 28359 Bremen, Germany
[4]Helmholtz Zentrum Geesthacht Center of Materials and Coastal Research, Max-Planck-Str., 21502 Geesthacht, Germany
[5]HYGEOS, Euratechnologies, 165 Avenue de Bretagne, 59000 Lille, France

Correspondence to: A. Bracher (astrid.bracher@awi.de)

Abstract. The composition and abundance of algal pigments provide information on phytoplankton community characteristics such as photoacclimation, overall biomass and taxonomic composition. In particular, pigments play a major role in photoprotection and in the light-driven part of photosynthesis. Most phytoplankton pigments can be measured by high-performance liquid chromatography (HPLC) techniques applied to filtered water samples. This method, as well as other laboratory analyses, is time consuming and therefore limits the number of samples that can be processed in a given time. In order to receive information on phytoplankton pigment composition with a higher temporal and spatial resolution, we have developed a method to assess pigment concentrations from continuous optical measurements. The method applies an empirical orthogonal function (EOF) analysis to remote-sensing reflectance data derived from ship-based hyperspectral underwater radiometry and from multispectral satellite data (using the Medium Resolution Imaging Spectrometer – MERIS – Polymer product developed by Steinmetz et al., 2011) measured in the Atlantic Ocean. Subsequently we developed multiple linear regression models with measured (collocated) pigment concentrations as the response variable and EOF loadings as predictor variables. The model results show that surface concentrations of a suite of pigments and pigment groups can be well predicted from the ship-based reflectance measurements, even when only a multispectral resolution is chosen (i.e., eight bands, similar to those used by MERIS). Based on the MERIS reflectance data, concentrations of total and monovinyl chlorophyll *a* and the groups of photoprotective and photosynthetic carotenoids can be predicted with high quality. As a demonstration of the utility of the approach, the fitted model based on satellite reflectance data as input was applied to 1 month of MERIS Polymer data to predict the concentration of those pigment groups for the whole eastern tropical Atlantic area. Bootstrapping explorations of cross-validation error indicate that the method can produce reliable predictions with relatively small data sets (e.g., < 50 collocated values of reflectance and pigment concentration). The method allows for the derivation of time series from continuous reflectance data of various pigment groups at various regions, which can be used to study variability and change of phytoplankton composition and photophysiology.

1 Introduction

Optical measurements taken from various platforms have been successfully used to determine the total chlorophyll *a* (TChl *a*) concentration (e.g., see the summary by McClain 2009). Those measurements can be taken continuously, thereby allowing for the estimation of TChl *a* concentration at a much higher temporal and spatial resolution than possible from chemical measurements in the laboratory, e.g., by high-performance liquid chromatography (HPLC) analysis of discrete water samples. Chl *a* is the major pigment in all phytoplankton species and is often used as an indicator of phytoplankton biomass. When pigments are measured by HPLC, TChl *a* is defined as the sum of monovinyl Chl *a* (MVChl *a*), divinyl Chl *a* (DVChl *a*) and chlorophyllide *a* (which is mainly formed as an artifact of the former two during the extraction process and therefore included in the calculation). DVChl *a* exists only in the prokaryotic genus *Prochlorococcus*, while MVChl *a* is the Chl *a* pigment for all other phytoplankton (other cyanobacteria and eukaryotes). Besides Chl *a*, there are many other pigments in phytoplankton that are either involved in light harvesting, such as chlorophyll *b* (Chl *b*), chlorophyll *c* (Chl *c*) and photosynthetic carotenoids (PSC), or in protecting Chl *a* and other sensitive pigments from photodamage, such as photoprotective carotenoids (PPC). Some pigments only occur in certain phytoplankton groups and thus are indicator pigments for their identification, e.g., peridinin in dinoflagellates (e.g., Letelier et al., 1993; Vidussi et al., 2001).

When analyzing biogeochemical fluxes in the oceans, however, it is inadequate to consider phytoplankton as a single variable (i.e., TChl *a*) because various groups have different roles in the biogeochemical processes (such as carbon fixation and export, nitrogen fixation, and silicon uptake). TChl *a* is far from being a sole function of phytoplankton biomass and varies, as other phytoplankton pigments do, with taxonomic composition and mean physiological algal assemblage state in response to several factors such as light, temperature and nutrients (Behrenfeld and Boss, 2006). Thus, knowledge of a wider array of phytoplankton pigment concentrations provides insight into phytoplankton composition, overall light absorption and physiological state. Phytoplankton absorption bears the imprints of different types of pigments and can be measured by optical measurements. However, different phytoplankton pigments may correlate in parts of their spectrum, making individual pigment detection difficult.

Several recent studies have investigated the potential of using continuous optical data to derive surface concentrations of pigments other than TChl *a*, with the advantage of being able to supply estimates over larger spatial and temporal scales than obtained with in situ water sampling. Chase et al. (2013) decomposed a large global data set of hyperspectral particulate absorption measurements into Gaussian function components and assessed the magnitude of specific Gaussian functions in relation to the absorption by specific pigments or pigment groups. The method provided robust results for obtaining concentrations of TChl *a*, TChl *b* (sum of different types of Chl *b*), TChl *c* (sum of different types of Chl *c*), PSC, PPC and phycoerythrin (PE). Organelli et al. (2013) used a multivariate approach applied to fourth-derivative spectra of phytoplankton or particulate absorption (a_{ph} and a_p, respectively) data to retrieve TChl *a*, the total concentrations of seven diagnostic pigments and three phytoplankton size classes. However, a_p and a_{ph} are inherent optical properties (IOP) which cannot be directly determined from satellite ocean-color measurements (after successful atmospheric correction), such as the apparent optical properties (AOP). The estimation of IOP from AOP is based on a certain inversion model (e.g., the Quasi-Analytical Algorithm by Lee et al., 2002), which introduces additional uncertainty.

The water-leaving reflectance (ρ_w) is related not only to phytoplankton absorption but also to the scattering and absorption of water and other water constituents and to changes in the radiance distribution in response to environmental conditions such as observation geometry, surface waves and atmospheric conditions.

Pan et al. (2010) developed empirical algorithms based on reflectance ratios to approximate key phytoplankton pigment concentrations. The band-ratio algorithms were developed from underwater radiometric measurements collocated to pigment data taken in northeastern US coastal waters and were successful in deriving the concentration of TChl *a*, TChl *b*, TChl *c* and nine different carotenoids. However, such band-ratio algorithms require a very large database (> 400 collocations with satellite data) from a certain region to derive robust results. Pan et al. (2013) later described that the algorithm had to be adapted by modifying the pigment-specific coefficients based on a regionally specific data set.

Craig et al. (2012) developed local models to estimate TChl *a* and a_{ph} at different wavelengths from hyperspectral in situ measurements of remote-sensing reflectance, $R_{rs}(\lambda)$, in an optically complex water body. The models were based on empirical orthogonal functions (EOF) analysis of normalized $R_{rs}(\lambda)$ spectra and a subsequent linear fitting of measured TChl *a* concentration and $a_{ph}(\lambda)$ as response variables to EOF loadings as predictor variables. Taylor et al. (2013) showed that the method could be used similarly to derive PE concentrations from underwater upwelling radiance spectra, $L_u(\lambda)$, which enabled continuous profile predictions of PE concentrations.

The present study aims to use the spectral information contained in reflectance data to derive the optical signature of different pigments by an automatic and generic technique. The EOF analysis is applied to R_{rs} and to ρ_{wN} (i.e., normalized ρ_w just above surface) data measured in the field and by satellite sensors, respectively, in the Atlantic Ocean. The dominant EOF loadings were subsequently assessed as predictors in a multiple linear regression for the concentration of phytoplankton pigments and pigment groups as response variables. The prediction error of each model is evaluated by a permuted cross-validation routine, which is used to estimate the critical sample sizes necessary for reliable prediction. In addition, we demonstrate the approach's utility in estimating the large-scale distribution and photophysiology of the phytoplankton assemblage.

2 Material and methods

Two sets of optical and pigment data from the Atlantic Ocean were used in the analysis. The first model setup used a data set which included only optical measurements taken in situ (as depth profiles) and collocated surface pigment data collected during three transatlantic RV *Polarstern* cruises in 2008 and 2010. These data enabled us to study the difference in EOF methods between hyper- and multispectral resolution. In the following, we call this data set "field data set". For a second data set, the "satellite-based data set", we considered water reflectance measurements from the satellite sensor Medium Resolution Imaging Spectrometer (MERIS), collocated to pigment data from various researchers in the tropical Atlantic Ocean. These data enabled us to study the generic application of the method.

2.1 Field data set

Samples for the field data set were collected during three RV *Polarstern* cruises: the expeditions *ANTXXIV/4* in April/May 2008 and *ANTXXVI/4* in April/May 2010 followed a south-to-north transect through the Atlantic Ocean from Punta Arenas (Chile) to Bremerhaven (Germany); *ANTXXV/1* in November 2008 followed a north-to-south transect through the eastern Atlantic Ocean from Bremerhaven to Cape Town (South Africa) (see Fig. 1; for more details see Table S1, upper panel in the Supplement). Sampling was generally conducted at 12:00 local time and involved conductivity temperature density (CTD) casts with water samplers, below-water radiance and irradiance measurements and above-water irradiance measurements. Water samples from surface water (< 10 m) for pigment analysis and for PE analysis were filtered on GF/F filters and on 0.4 μm polycarbonate filters, respectively. Filters were immediately shock-frozen in liquid nitrogen and stored at −80 °C until further analysis at the laboratories of the Alfred-Wegener-Institute Helmholtz Centre for Polar and Marine Research (AWI).

2.1.1 Pigment data

The composition of pigments that were soluble in organic solvents was analyzed by HPLC following the method by Barlow et al. (1997) and adjusted to our temperature-controlled instruments (a Waters 600 controller combined with a Waters 2998 photodiode array detector, a Water717plus auto sampler and a LC Microsorb C8 HPLC column) as detailed in Taylor et al. (2011). We determined the list of pigments shown in Table 1 of Taylor et al. (2011) and applied the method by Aiken et al. (2009) for quality control of the pigment data. HPLC data for *ANTXXV/1*, as opposed to the other two cruises, were already published in Taylor et al. (2011) and are available from PANGAEA (doi.pangaea.de/10.1594/PANGAEA.819070). The relative concentration of PE was taken from the data set published

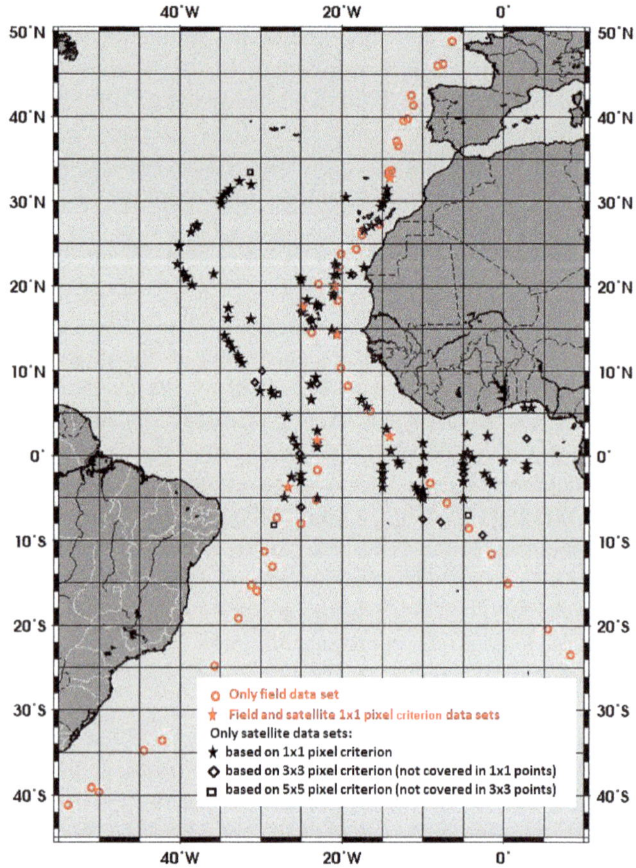

Figure 1. Position of pigment samples used in this study. Red: field data set; black: samples which are collocated to satellite-based but not to field reflectance data; circles: samples which are collocated to field but not to satellite-based reflectance data; stars, diamonds and squares: collocations to MERIS Polymer data based on the 1×1, 3×3 and 5×5 pixel criteria, respectively.

for all three cruises in PANGAEA (doi.pangaea.de/10.1594/PANGAEA.819624) and analyzed in Taylor et al. (2013). As outlined in Taylor et al. (2013), the PE concentration is expressed as a relative value, while all other pigments concentrations are directly measured values.

2.1.2 Reflectance data field data set

For all three cruises as AOP input data, we used $R_{rs}(\lambda)$ data obtained from profiles of radiance and irradiance from 320 to 950 nm, with an optical resolution of 3.3 nm and a spectral accuracy of 0.3 nm, measured with hyperspectral radiometers (RAMSES, TriOS GmbH, Germany) at the same time and place as pigment data of Sect. 2.1.1. R_{rs} data of *ANTXXV/1* were already published in Taylor et al. (2011) and are available from PANGAEA (doi.pangaea.de/10.1594/PANGAEA.819506). For the other two cruises we applied the same technique and instrumentation as in Taylor et al. (2011) to derive the R_{rs} spectrum at each station. To test the influence of spectral range and resolution of AOPs, the hyperspectral field $R_{rs}(\lambda)$ data were used within the range of 350 to 700 nm and 380 to 700 nm and reduced to the multispectral bands (412, 443, 490, 510, 560, 620, 665 and 681 nm) of MERIS by taking the integral over all wavebands within one band (±10 nm around the center wavelength except when 681 nm ± 7.5 nm was used).

2.2 Satellite-based data set

For this data set, pigment concentrations had been determined from the sea surface (< 10 m) with HPLC by several investigators within the area of 35° N–10° S and 42° W–3° E during the MERIS/ENVISAT mission lifetime (2002–2012; for more details on the data set see Supplement Table S1, lower panel). A large part of those data are publicly available from the SEABASS and BODC databases. The remaining pigment data are from the field data set within this area, including additional data from stations where no radiometric measurements had been taken and from four other cruises: pigment data from the RV *Maria S. Merian* cruise *MSM-18/3* were analyzed by AWI as described above in Sect. 2.1.1; data from two RV *Polarstern* cruises (*ANTXXIII/1* and *ANTXXIV/1*) were analyzed by HZG following Zapata et al. (2000); data from the *Bonus Good Hope* (BGH) cruise, conducted by the Laboratoire d'Océanographie de Villefranche, were acquired as outlined in Speich et al. (2008) and analyzed following the method by Ras et al. (2008).

AOP input data is from the MERIS Polymer level 2 $\rho_{wN}(\lambda)$ product given for the same eight wavebands as listed in Sect. 2.1.2. The Polymer algorithm (for details see Steinmetz et al., 2011) provides a powerful atmospheric correction. It is an iterative spectral matching method over the whole available sensor spectrum and uses two decoupled models. First, the water reflectance is modeled using two parameters: the Chl *a* concentration and the particle backscattering coefficient. Second, the reflectance of the atmosphere, including aerosols and contamination by sun glint, is simplified by using an analytical expression that can account for multiple interactions between molecular and aerosol scatterings (and glitter) without referring to a specific aerosol model. Hence, it allows for the retrieval of large amounts of MERIS observations in sun glint, thin clouds or heavy aerosol plumes; these contaminated conditions could not be treated correctly by standard atmospheric correction schemes extrapolating from the near infrared. MERIS Polymer products thus improve the spatial coverage by almost a factor of 2 and have proven successful for retrieving MERIS Ocean Colour products: Polymer was selected as the MERIS processor for atmospheric correction for the Ocean Colour Climate Change Initiative after an extensive validation and intercomparison with other atmospheric correction algorithms in which each algorithm's uncertainty was assessed (Müller and Krasemann, 2012). However, additional uncertainties probably result from the difference in spatial resolution between satellite (1 km by 1 km) and ship-based (20 cm by 20 cm) sampled data.

Matchups between pigment data and MERIS Polymer $\rho_{wN}(\lambda)$ and TChl *a* products were determined according to the MERMAID (MERIS MAtchup In-situ Database) as 1×1 (within the MERIS pixel), 3×3 and 5×5 pixels, respectively, measured on the same day around the field observation (see Barker et al., 2008). For the 3×3 and 5×5 MERIS pixel match-ups, the mean $\rho_{wN}(\lambda)$ and TChl *a* concentrations from the MERIS products were calculated. Then the 1×1, mean 3×3 and mean 5×5 MERIS $\rho_{wN}(\lambda)$ matchup data were used for deriving predicted (modeled) pigment concentrations, as outlined in Sect. 2.3. The mean MERIS Polymer TChl *a* data were validated with the in situ TChl *a* data of the satellite-based data set. The R^2, percent bias (PB), mean percent difference (MPD) and root mean square error (RMSE) between the two collocated data sets were calculated as outlined in Werdell et al. (2013) and used to determine pigment prediction full-fit statistics (see Sect. 2.3.2).

2.3 Statistical methods to retrieve pigment concentrations from reflectance

Figure 1 presents the distribution of collocated pigment and reflectance measurements for both field and satellite-based data sets that were used separately as input for the EOF prediction analysis. The field data set covered 53 collocated reflectance and pigment data points (Fig. 1, red points). We used three setups of the field $R_{rs}(\lambda)$ spectra for the development of pigment-specific models:

1. $R_{rs}(\lambda)$ data in hyperspectral (1 nm resolved, "hyper_R_{rs}") from 350 to 700 nm,
2. "hyper_R_{rs}" from 380 to 700 nm and
3. $R_{rs}(\lambda)$ data in MERIS band resolution ("band_R_{rs}").

The three satellite-based data sets consisted of 139, 155 and 160 collocated reflectance and pigment data points from 2002 to 2012 for the 1×1 (Fig. 1, stars), 3×3 (Fig. 1, diamonds) and 5×5 (Fig. 1, squares) pixel collocation criteria, respectively, covering all months except January, March and December (details on the spatial and temporal distribution of collocations are given in the Supplement Table S1). Eighteen collocations of the field data matched the 1×1 pixel satellite-based data set (Fig. 1, red stars), but no additional field data matched the two other (3×3 and 5×5 pixel) satellite-based data sets.

The following pigments were included in the construction of all different statistical models (further explained below): MVChl *a*, DVChl *a*, chlorophyllide *a*, MVChl *b*, DVChl *b*, Chl $c_{1/2}$ (Chl *c* type 1 and 2), Chl c_3 (Chl *c* type 3), Hex (19-hexanoyl-fucoxanthin), But (19-butanoyl-fucoxanthin), Allo (alloxanthin), Diadino (diadinoxanthin), Diato (diatoxanthin), Fuco (fucoxanthin), Lut (lutein), Peri (peridinin),

Viola (violaxanthin) and Zea (zeaxanthin). In addition, the concentrations of different pigment groups were considered, such as the sum concentration of α- and β-carotene (Caro); MVChl a, DVChl a and chlorophyllide a (TChl a); MVChl b and DVChl b (TChl b); all pheopigments (TPheo); according to Hooker et al. (2005) and Roy et al. (2011), the photosynthetically active carotenoids (PSC: Fuco, But, Hex, Peri), and the photoprotective carotenoids (PPC: Allo, Diadino, Diato, Zea, Caro). PE, although measured by a different technique, was also included since it also had a major impact on the optical $L_u(\lambda)$ data of the field data set (see Taylor et al., 2013). Neo and PE were only considered in the models using the field data set because, for some pigment samples of the data set collocated to the satellite data, these pigments have not been analyzed.

Figure 2 gives an overview describing the various steps of the development and validation of our EOF method to predict various pigments and pigment groups' concentrations, which are described in detail in the following subsections.

2.3.1 Empirical orthogonal function analysis

Following Taylor et al. (2013), the spectral data were subjected to an EOF analysis, also known as a principal component analysis, in order to reduce the high dimensionality of the data and derive the dominant signals ("modes") that best describe variance within the data set. In addition to dimension reduction of spectral data, the use of EOF modes in statistical model building also avoids problems associated with multicollinearity amongst the original predictor variables. All calculations in the following were done with the statistical computing software R (R Development Core Team, 2013).

Spectral data were contained in a data matrix $\mathbf{X}$ with dimensions M, sample rows, by N, reflectance band columns. Spectral samples were collocated to the respective pigment data set $\mathbf{Y}$ with dimensions M by P, pigment columns (pigments and pigment groups included are outlined above). While hyper_R_{rs} data consisted of 350–700 nm ($N = 351$) or 380–700 nm ($N = 321$) bands, band_R_{rs} and the satellite_ρwN data consisted of the eight MERIS visual wavebands ($N = 8$). As in Taylor et al. (2013), spectral data sets $\mathbf{X}$ were standardized for each sample row by first subtracting the mean spectral value (centering) followed by division by the spectral standard deviation (scaling), which focused the analysis on the spectral shape rather than the magnitude. The standardized matrix $\mathbf{X}$ was then subjected to singular value decomposition (SVD) in order to derive EOF modes:

$$\mathbf{X} = \mathbf{U}\Sigma\mathbf{V}^T, \; x_{ij} = \sum_{k=l,N} u_{ik}\sigma_k v_{kj}, \tag{1}$$

where $\mathbf{V}$ is a $N \times N$ matrix containing the EOFs (spectral pattern), $\mathbf{U}$ is an $M \times N$ matrix containing the principal components (PCs), Σ is an $N \times N$ matrix containing the singular values on the diagonal and k is the EOF mode index (length N). Only EOFs $\leq \min(M, N)$ will carry information. This notation differs slightly from that presented in Taylor et al. (2013), where a covariance matrix of the data set was subjected to Eigen decomposition with subsequent projection of data onto EOFs to derive PCs. The results of both approaches are similar except that $\mathbf{U}$ derived via SVD is unitary, and Σ contains standard deviation rather than variance. The SVD method is presented here due to its more straightforward notation: EOFs and PCs are determined in a single step whereas the alternate Eigen decomposition is a three-step calculation (Fig. 2, the upper part of the panel on the left summarizes these steps).

2.3.2 Log transformed general linear model

A general linear model was used to predict log-transformed pigment concentrations of each pigment, y_p, based on a subset of PCs, $\mathbf{U}$, as covariates (Fig. 2, the lower part of the panel on the left summarizes these steps). The linear model uses log-transformed pigment concentrations. Since only positive, non-zero values are permissible with this transformation, a small value was added to all concentrations (0.00001 mg m^{-3}) to allow for the inclusion of samples where pigment concentrations were essentially zero or below the detection limit. A truncated subset of PCs was used as defined by the magnitude of their standard deviation. PCs with standard deviations of ≤ 0.0001 times the standard deviation of the first component were omitted. The resulting multiple regression had the form

$$\log(y_p) = a + b_1 u_1 + b_2 u_2 + \cdots + b_n u_n, \tag{2}$$

where $\log(y_p)$ is the natural log-transformed concentration of pigment p, $u_{1,2,\ldots n}$ are the leading n PC scores from $\mathbf{U}$, a is the intercept and $b_{1,2,\ldots n}$ are the regression coefficients. A bidirectional stepwise routine was used to search for smaller multiple regression models based on fewer predictor terms. Best linear models were selected through minimization of the Akaike information criterion (AIC). Once the best linear model was determined, the relative importance of included terms was defined by the change in AIC (ΔAIC) following each term's removal.

Since the range of concentration varies greatly among the different pigments, we calculated mainly relative error statistics. According to Werdell et al. (2013), the coefficient of determination (R^2), the RMSE, the slope (S) and the intercept (a) of the linear regression are based on the log-scaled predicted ($\log(y_p)$) as opposed to the log-scaled observed ($\log(y_o)$) pigment concentration data, while the MPD, the PB and the median percent difference (MDPD) are based on the non-log-transformed pigment concentrations. The following

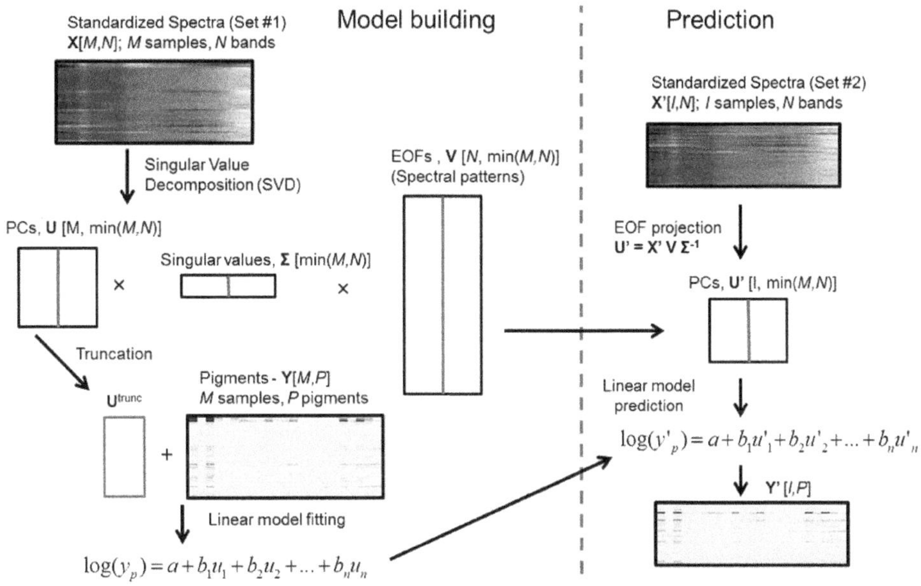

Figure 2. Schematic overview of the steps used in model building and prediction. Multiple linear regression models are fit to log-transformed pigment concentrations, y_p, as the response variable and EOFs derived from a spectral (reflectance) data set, X, as predictor variables. Model building (left) is used for "full-fit" models to all data samples (M) or to a training subset of samples for cross-validation (Sect. 2.3.3). Prediction (right) is used for the assessment of the model error on a validation subset of samples (I) for cross-validation (Sect. 2.3.3) or in the extrapolation of model predictions to an new data set of reflectance spectra, as was done for the larger area of the tropical eastern Atlantic region in this study (Sect. 2.3.4).

equations for these statistics were used:

$$\mathrm{RMSE} = \text{root of } \frac{1}{N}\sum_{i=1}^{N}\left(\log(y_{\mathrm{p}i}) - \log(y_{\mathrm{o}i})\right)^2, \quad (3)$$

$$\mathrm{MPD} = \frac{100}{N}\sum_{i=1}^{N}\left|\left(y_{\mathrm{p}i} - y_{\mathrm{o}i}\right)\right|/\left(y_{\mathrm{o}i}\right) \quad [\%], \quad (4)$$

$$\mathrm{PB} = \frac{100}{N}\sum_{i=1}^{N}\left(y_{\mathrm{p}i} - y_{\mathrm{o}i}\right)/\left(y_{\mathrm{o}i}\right) \quad [\%], \quad (5)$$

$$\mathrm{MDPD} = \text{Median of } \left[\left|\left(y_{\mathrm{p}i} - y_{\mathrm{o}i}\right)\right|/\left(y_{\mathrm{o}i}\right) \times 100\right] \quad i = 1, N \quad [\%]. \quad (6)$$

2.3.3 Model prediction error

In addition to the statistics performed for each pigment linear model (Sect. 2.3.2), we performed a cross-validation of the linear model fitting in order to better test the robustness of the models' prediction error. Data were split into two groups: the first part of the data was used for model fitting (Fig. 2, left panel), while the second part was used for prediction validation (Fig. 2, right panel). According to Craig et al. (2012), we assessed the number of observations required to achieve adequate predictions by the pigment linear models using the variable jack-knife procedure of Wu (1986). So the proportion used for data splitting for the cross-validation procedure was varied as follows, where n is the total number of samples, tp is the number of training points and vp is the number of points used for validation:

$$\mathrm{tp} = n \times d, \text{ with } d = 0.1, 0.15, 0.2, \ldots, 0.9, \quad (7)$$

$$\mathrm{vp} = n\,(1 - d). \quad (8)$$

Since the number of permutations for data splitting affects the overall computing time, the procedure was run for 500 permutations, similar to the recommendation of Craig et al. (2012). Such a high number of permutations rules out the model error being assessed based on a spatially or temporally biased data set.

Each cross-validation procedure was as follows:

1. For 500 permutations, do steps 2–8.
2. Randomly select $n \times d$ of collocated samples to include in training sets $\mathbf{X}^{\mathrm{train}}$ and $\mathbf{Y}^{\mathrm{train}}$ for spectra and pigment data, respectively. Remaining $n(1-d)$ of samples are allocated to the validation sets $\mathbf{X}^{\mathrm{valid}}$ and $\mathbf{Y}^{\mathrm{valid}}$.
3. Standardize $\mathbf{X}^{\mathrm{train}}$ and perform EOF following Eq. (1) to obtain $\mathbf{U}^{\mathrm{train}}$, Σ^{train} and $\mathbf{V}^{\mathrm{train}}$.
4. For each pigment concentration $y_{\mathrm{p}}^{\mathrm{valid}}$ of $\mathbf{Y}^{\mathrm{valid}}$, do steps 5–9.
5. Fit linear model to log-transformed pigment concentrations using selected $\mathbf{U}^{\mathrm{train}}$ as in Eq. (2):

$$\log\left(y_{\mathrm{p}}^{\mathrm{train}}\right) = a + b_1 u_1^{\mathrm{train}} + b_2 u_2^{\mathrm{train}} + \cdots + b_n u_n^{\mathrm{train}}. \quad (9)$$

6. Perform bidirectional stepwise search for smaller linear model.
7. Standardize validation set and project $\mathbf{X}^{\text{valid}}$ onto the EOFs $\mathbf{V}^{\text{train}}$ and the inverse of singular values $\Sigma^{\text{train}^{-1}}$ to derive their PCs $\mathbf{U}^{\text{valid}}$:

$$\mathbf{U}^{\text{valid}} = \mathbf{X}^{\text{valid}} \cdot \mathbf{V}^{\text{train}} \cdot \Sigma^{\text{train}^{-1}}. \tag{10}$$

Use selected PCs of $\mathbf{U}^{\text{valid}}$ as variables in Eq. (10) in order to predict pigment concentrations for the validation data set:

$$\log\left(y_{\text{p}}^{\text{valid}}\right) = a + b_1 u_1^{\text{valid}} + b_2 u_2^{\text{valid}} + \cdots + b_n u_n^{\text{valid}}. \tag{11}$$

8. Record pairs of observed and predicted validation pigment concentrations y_{o} and $y_{\text{p}}^{\text{valid}}$ in a new object for all permutations for later calculation of prediction error.

For each permutation, the R^2 based on the log-scaled predicted ($\log\left(y_{\text{p}}^{\text{valid}}\right)$) versus the log-scaled measured ($\log(y_{\text{p}})$) were derived and finally, over all permutations, the mean value (R^2cv) was calculated. In accordance with statistics in Sect. 2.3.2, the prediction error was described in terms of the absolute squared difference based on log-transformed pigment concentrations, $(\log(y_{\text{p}}^{\text{valid}}) - \log(y_{\text{o}}))^2$, and relative difference based on non-log-transformed pigment concentrations, $(y_{\text{p}}^{\text{valid}} - y_{\text{o}})/y_{\text{o}}$. Mean and median relative difference (MPDcv and MDPDcv, respectively) and the root mean square absolute difference (RMSEcv) over all permutation were determined as follows:

$$\text{MPDcv} = \frac{100}{N}\sum_{i=1}^{N}\left[(y_{\text{p}}^{\text{valid}} - y_{\text{o}})/y_{\text{o}}\right] \quad [\%], \tag{12}$$

$$\text{RMSEcv} = \sqrt{\frac{1}{N}\sum_{i=1}^{N}\left[\log\left(y_{\text{p}}^{\text{valid}}\right) - \log(y_{\text{o}})\right]^2}, \tag{13}$$

$$\text{MDPDcv} = \text{median value of}\left(\frac{y_{\text{i},p}^{\text{pred}} - y_{\text{i},p}^{\text{valid}}}{y_{\text{i},p}^{\text{valid}}} \times 100\right),$$

$$i = 1, N[\%]. \tag{14}$$

2.3.4 Pigment concentration predictions with MERIS reflectance data

In order to predict pigment concentration from MERIS $\rho_{\text{wN}}(\lambda)$ for a whole month of data in November 2008, for which we did not have corresponding pigment measurements, the following method was applied: we projected standardized MERIS $\rho_{\text{wN}}(\lambda)$ data onto the EOF loading (**V**) to derive their principal components (**U**), which were subsequently used for the prediction with the fitted linear model (as in Sect. 2.3.3, step 7, Eq. 11, Fig. 2, right panel), where $b_{1,2,\ldots,n}$ are taken from the EOF model developed with the 1×1 MERIS Polymer $\rho_{\text{wN}}(\lambda)$ matchups (following Fig. 2, left panel).

3 Results and discussion

3.1 Characteristics of input data sets

Figure 3 shows the original and standardized spectra of the field and satellite-based data sets. Considering the conversion of $R_{\text{rs}}(\lambda)$ to $\rho_{\text{wN}}(\lambda)$ data by a factor of π, the magnitude and shape of the original and standardized spectra are similar for the band-resolved data sets, except that the standardized satellite_ρ_{wN} data set contains only one spectrum with maximum reflectance in the green at 560 nm, while the standardized field data set contains four spectra with maxima at 510 nm.

The composition and range of pigments (as detailed with maximum, minimum, mean and standard deviation in Supplement Table S2) show, for all pigments, that the collocations to the field data set contain higher maxima and minima than the collocations to the satellite-based data set (except for Fuco, for which it is equal and for Zea, for which it is inverted). For most pigments, mean values are very similar for both data sets. However, standard deviations for the field data set are 2 to 3 times higher than the mean for all pigments. In the satellite data set, the standard deviation is of a similar magnitude to the mean value. The higher concentration of total pigments in the field data set may explain the small differences in the shape of the reflectance spectra of the two (field versus satellite-based) data sets. However, DVChl b, MVChl b, TChl b, Allo, Diato, Lut, Neo, Peri, Viola and TPheo had values of 0 mg m^{-3} in more than 20 % of all stations in both data sets. Also, Chl c_3 had a concentration of 0 mg m^{-3} in one sample collocated to the field and in over 30 % of samples collocated to the satellite-based data set. Several pigments had concentrations of 0 mg m^{-3} only occasionally (< 10 %) in samples collocated to the satellite-based data set (Caro, Chl $c_{1/2}$, But, Hex, Zea, DVChl a, Diadino and Fuco) and in the field data sets (DVChl a, Diadino and Fuco). All other pigments not listed here had detectable concentrations in all samples.

3.2 EOF analysis – shape of modes and relevance for predictions

Following the EOF truncation criteria outlined in Sect. 2.3.2, the decomposition of the standardized spectra resulted in nine modes (EOF-1 to EOF-9) for the hyper_R_{rs} and seven modes for the band_R_{rs} and satellite_ρ_{wN} data sets (the first four modes are presented in Fig. 4). EOF modes for the three satellite_ρ_{wN} data sets were nearly identical. For simplicity we only show (Fig. 4) and discuss the EOF modes of the 1×1 pixel collocation data set. For all data sets, the first three modes explain over 99.8 % of the variance for all three data sets, with EOF-1 explaining between 94.5 % and 96 % of the variance (Table 1).

The shapes of the first three EOF modes are very similar among all three reflectance data sets. They are nearly iden-

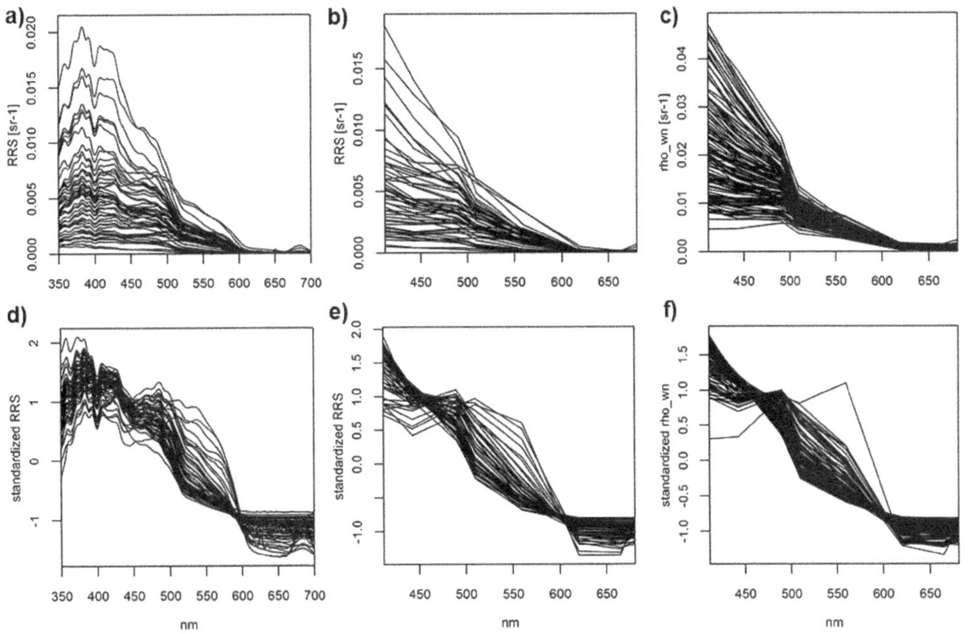

Figure 3. **(a)–(c)** Original (in sr^{-1}) and **(d)–(f)** standardized (subtracted mean and divided by standard deviation) reflectance spectra of hyper_R_{rs} **(a)** and **(d)**, band_R_{rs} **(b)** and **(e)** and satellite_ρ_{wN} (from MERIS Polymer) data within the 1×1 pixel collocation boxes **(c)** and **(f)**.

Table 1. Percent of total variance explained (Expl. variation; upper panel) and cumulative proportion (Cum. proport.; lower panel) by the significant EOFs derived from field R_{rs} spectra in hyperspectral resolution (hyper_R_{rs}) and multispectral resolution (band_R_{rs}) and from satellite_ρ_{wN} (from MERIS Polymer) using the 1×1 pixel collocation criterion.

% Expl. variation	EOF-1	EOF-2	EOF-3	EOF-4	EOF-5	EOF-6	EOF-7	EOF-8	EOF-9
hyper_R_{rs}	95.0	4.1	0.7	0.1	<0.1	<0.1	<0.1	<0.1	<0.1
band_R_{rs}	94.5	5.0	0.4	<0.1	<0.1	<0.1	<0.1		
satellite_ρ_{wN}	95.9	3.9	0.1	<0.1	<0.1	<0.1	<0.1		
% Cum. proport.	**EOF-1**	**EOF-2**	**EOF-3**	**EOF-4**	**EOF-5**	**EOF-6**	**EOF-7**	**EOF-8**	**EOF-9**
hyper_R_{rs}	95.0	99.1	99.8	99.9	99.965	99.981	99.990	99.994	99.996
band_R_{rs}	94.5	99.5	99.9	99.95	99.990	99.999	100.0		
satellite_ρ_{wN}	95.9	99.7	99.9	99.95	99.985	99.996	100.0		

tical for the band_ R_{rs} and the satellite_ρ_{wN} data sets but show smoother shapes and peaks for hyper_R_{rs} for the first two modes. Due to the limited number of wavelengths for the two multispectral data sets, EOFs show evidence of a shift in peak location, starting with EOF-3 (peak at 412 and 443 nm for EOF-3 and EOF-4, respectively), as compared to hyper_R_{rs} (peak at 360 and 410 nm for EOF-3 and EOF-4, respectively). This is likely due to the increased spectral resolution of the hyperspectral data, which allows for more precision in identifying spectral regions of higher variance. For EOF 4, the satellite_ρ_{wN} mode is much flatter beyond 500 nm and shows no trough between 600 and 650 nm compared to the EOF-4 for the other two data sets. Not much similarity is seen among the EOF-5 modes of the different spectra data sets; for EOF-6, the two field data sets are similar in the overall shape, but peak locations are shifted towards longer wavelengths for the satellite data set. EOF-7 and EOF-8 show very similar shapes for hyper_R_{rs} and deviate from EOF-7 in the band data sets, while EOF-9 from hyper_R_{rs} looks much more like the later ones.

The EOF analyses identify dominant modes of variance, which can be interpreted as imprints of changes in the optical properties of water constituents in the water column. For this study, only reflectance spectra taken in high TChl *a* waters with measurable mineral fraction (identified as cluster V for the ANTXXV/1 data in Taylor et al., 2011) show any resemblance to spectral shapes obtained in the case 2 waters of Lubac and Loisel (2007, e.g., class 5) and Craig et al. (2012). The remaining spectra (typical case 1 water) show characteristics not observed in those studies. This difference explains

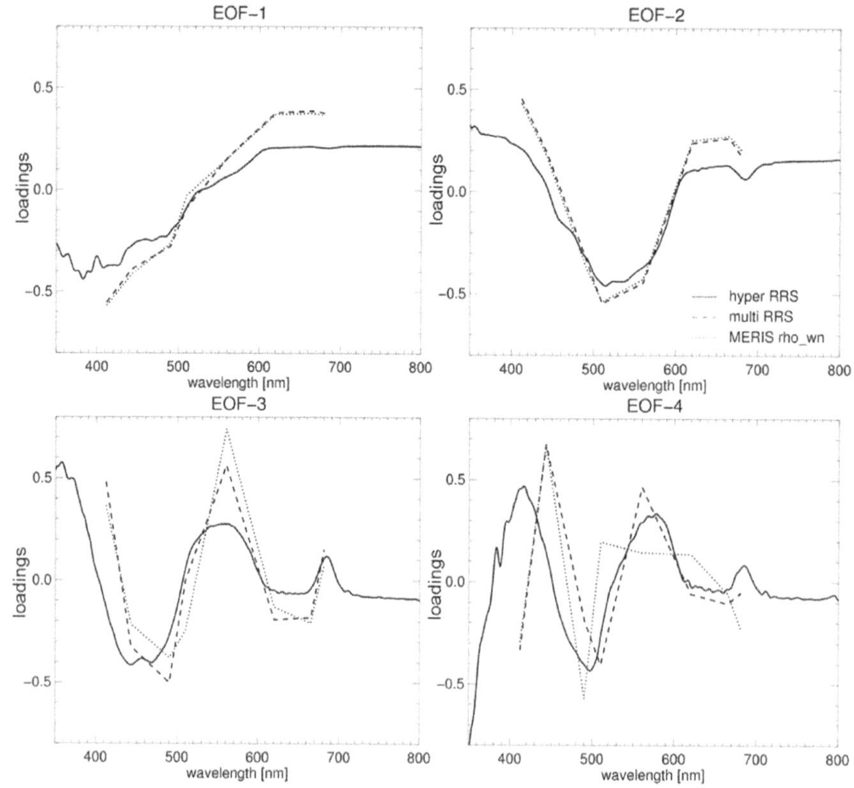

Figure 4. First four EOF modes (EOF-1 to EOF-4) derived from field R_{rs} data set in hyperspectral resolution (hyper_R_{rs}, solid lines) and in multispectral band resolution (band_R_{rs}, dashed lines) and from using satellite_ρ_{wN} (from MERIS Polymer, dotted line) data within the 1×1 pixel collocation box.

the minor variations in the shape and loading of EOFs between their and our data sets. In the following, we focus the discussion on our hyper_R_{rs} data set results with specific comparison to the study by Craig et al. (2012), which was also based on hyperspectral R_{rs} data.

Our first three EOF modes correspond to the ones derived for the hyperspectral case 2 reflectance data set of Craig et al. (2012). As pointed out in their study, EOF-1 is likely the signature of bulk oscillations in phytoplankton biomass concentration (including its effect on backscattering). However, our EOF-1 already explains much more of the variance than in Craig et al. (2012), where it only accounted for 72.4 % and showed much more structure and a weaker exponential decrease from 400 to 550 nm. EOF-2 superficially resembles the overall changes in the total absorption over broad band structures. It strongly decreases from 350 to 510 nm and increases again above 570 nm, which is connected to total pigment and water absorption, respectively. There is a peak around 683 nm which can be linked to MVChl *a* and DVChl *a* fluorescence. While this peak is present in EOF-1 in the Craig et al. (2012) data set, it is not in the EOF-1 of our data set likely because of the lower TChl *a* concentrations.

EOF-3 of our data set as compared to the one of Craig et al. (2012) shows a much steeper decrease with wavelength in the blue spectral range. These changes may reflect concomitant changes of absorption by chlorophyll, colored dissolved organic matter and non-algal particles expected to be co-varying and of much lower concentration in our case 1 waters. Scattering by particles other than phytoplankton was much higher in the case 2 water of Craig et al. (2012), leading to a less steep slope of this EOF mode. EOF-4 appears different in relation to the three peaks. Similar to EOF-2 and EOF-3, these differences are caused by the different composition and overall loading of water constituents of our and their sampled stations.

In summary, in contrast to more coastal waters where measurable mineral fraction can affect R_{rs} properties, the total attenuation is much more affected by total pigment concentration in our open-ocean, case 1 data set. Our data set was largely composed of samples from waters with lower TChl *a* concentration, ranging from 0.005 to 3.553 mg m^{-3}, while in the study of Craig et al. (2012) it ranged from 0.584 to 18.02 mg m^{-3}. EOFs greater than 4 were not presented in Craig et al. (2012) because they were not used to predict TChl *a* from R_{rs} data, as was the case for our TChl *a* (and MVChl *a*) linear model predictions (Sect. 3.3.3). Higher EOF modes probably reflect the influence of specific pigment groups or pigments, as indicated by the results of the ΔAIC values and further discussed in Sect. 3.3.3.

3.3 Pigment prediction by linear models

3.3.1 Field data set linear models

All pigments that were detected in the full set of the field data samples were well predicted by linear models based on hyperspectral (hyper_R_{rs}) or the reduced eight-band (band_R_{rs}) resolution spectra. The correlations between predicted and observed concentrations for these pigments were highly significant ($p < 0.0001$) and cross-validation statistics reached reasonable quality with $R^2\text{cv} \geq 0.5$, MDPDcv ≤ 45 % and MPDcv ≤ 60 % (Table 2a, upper part). For some pigments (TChl a, PSC, MVChl a, Hex, Caro) EOFs based on 380 to 700 nm produced much better linear model results using hyper_R_{rs} data than based on 350 to 700 nm (for all statistical parameters see Supplement Table S3; models based on hyper_R_{rs} (a) at 350 to 700 nm and (b) at 380 to 700 nm and (c) on band_R_{rs}). Lower quality for one statistical parameter for both linear models was reached for Zea (R^2cv 0.35 and 0.28), But (MPDcv 81 and 95 %) and for two parameters for PE (MDPDcv 65 and 67 %, MPDcv 139 and 156 %).

Plots of observed versus predicted values for the full data set of well-predicted pigments TChl a, PSC, PPC, Hex and Zea are shown in Fig. 5. For pigment groups and pigments with a high range of data (TChl a, PSC and Hex), covering about 3 orders of magnitude, the intercept is much lower and the regression closely aligns with the 1 : 1 reference line. The predicted versus observed regression for Zea was of lower quality ($R^2 < 0.6$) likely due to a much lower range of observed concentrations.

For all other pigments, predictions were of low quality (results not shown), demonstrating that the linear model approach does not produce robust predictions for situations where pigment were not detected (i.e., 0 mg m^{-3}) in every sample (see results for all pigment predictions in Supplement Table S3). Even pigments that were only occasionally undetected (e.g., DVChla, TChlb, MVChlb) showed increased error in cross-validation prediction as revealed by MDPDcv and RMSEcv values far above 100 % and 1, respectively. We re-ran the predictions for specific pigments where only a few samples (< 10 %) had concentrations of 0 mg m^{-3}, as was the case for DVChl a, Fuco, Diadino and Chl c_3 (see Supplement Table S2). In those specific linear model runs we only included as input data the data points where the specific pigment concentrations were > 0 mg m^{-3}. The resulting predictions (Table 2a, lower part; for DVChl a see full-fit results in Fig. 5d) from using the adjusted input data for those pigments show robust and significant cross-validation results within the same quality range as for the pigments which were detected in all data. For other pigments, where non-detection occurred more frequently (> 20 % of the samples), the removal of non-detection samples did not result in robust predictions (results not shown).

Cross-validation results of well-predicted pigments (Table 2a) show that, especially regarding the R^2cv and RMSEcv values, hyper_ R_{rs}-based linear models perform either the same (PSC), slightly better (PPC, Chl $c_{1/2}$) or much better (TChl a, MVChl a, But, Hex, Zea, Caro, PE, DVChl a, Chl c_3, Diadino, Fuco) than predictions based on eight wavelengths (band_R_{rs} data set). In particular, RMSEcv is much improved for several pigment predictions where RMSEcv reaches high values (> 0.65 mg m^{-3}), i.e., for PE, Fuco, But, Chl c_3, Diadino and Hex. The benefit was less clear when observing the statistics of MDPDcv and MPDcv in several pigments (MVChl a, Chl $c_{1/2}$, TChl a and PSC predictions). For these pigments the multispectral resolution appears to be sufficient for obtaining similarly robust predictions. TChl a (in line with MVChl a) and PSC dominate the overall phytoplankton pigment composition and absorption. TChl a concentrations have been well retrieved by band-ratio algorithms as a main phytoplankton biomass indicator (e.g., see Brewin et al., 2014). For pigments very similar in spectral range, such as But, Hex and Fuco, the hyperspectral resolution of the linear models provides much more robust pigment predictions (Table 2a). The hyper_R_{rs} linear models also produced better predictions for DVChl a, Zea, Diadino and PPC, where the specific linear models included a much larger set of EOF modes (see Sect. 3.3.3) which may indicate the importance of higher-resolution spectral details not available in the band_R_{rs} data.

3.3.2 Satellite-based data set linear models

Results for the models predicting pigment concentration from the satellite-based data set were very similar when using 1 × 1, 3 × 3 or 5 × 5 collocated MERIS ρ_{wN} data (for all statistical parameters see Supplement Table S3: satellite_ρ_{wN} models based on 1 × 1 (d), 3 × 3 (e) and 5 × 5 (f) collocations). Deviations were within 1 to 3 % for all statistical parameters. R^2cv values were best in all cases for well-predicted pigment concentrations in the 1 × 1 collocations, while MPDcv was best in the 3 × 3 collocations. Results clearly show that even models based on 5 × 5 pixel collocations can produce robust results. For simplicity, in the following we present and discuss the results of the 1 × 1 collocated reflectance data only.

In line with field data linear model results, pigment groups and pigments, which were detected in every sample (MVChl a, TChl a, PSC and PPC; the full-fit linear model results are shown in Fig. 6a–c), are well predicted with similar cross-validation statistic values using the satellite_ρ_{wN} data set (Table 2b, upper part). Also, good predictions for some pigments (DVChl a, Zea, Diadino, Hex, But, Fuco and Chl $c_{1/2}$) could be obtained by re-running the linear model analysis with concentrations of 0 mg m^{-3} excluded (Table 2b, lower part). For example, the full-fit linear model

Table 2. Statistics of linear models using EOF modes based on a) field R_{rs} data in hyperspectral (hyper; normally 350–700 nm; when * then 380–700 nm) resolution and multispectral (band) resolution and (b) the satellite_ρ_{wN} (from MERIS Polymer) using the 1 × 1 pixel collocation criterion data set. Cross-validation results are presented with 500 permutations for data splitting into 80 % of the data used for training and 20 % for validation. Only well-retrieved pigment prediction results, with correlations being highly significant at $p < 0.0001$, are given. Abbreviations of pigments are explained in Sect. 2.3.1. Pigments listed in the upper part of each table show high-quality results using the entire data set. In the lower part of each table (listed under "> 0 mg m^{-3}") models are based only on the data set of collocated R_{rs} samples where the respective pigment reached concentrations above 0 mg m^{-3}. Bold: here band-model performs better than hyper-model. Red signifies only medium quality as specified in the text.

(a)		MDPDcv		MPDcv		RMSEcv		R^2cv	
Band/hyper	N	band	hyper	band	hyper	band	hyper	band	hyper
TChl *a**	53	**28**	32	**42**	43	0.54	0.49	0.72	0.77
PSC*	53	**32**	34	**51**	53	0.62	0.62	0.75	0.75
PPC	53	28	27	51	49	0.57	0.56	0.52	0.52
MVChl *a**	53	**31**	34	45	44	0.54	0.50	0.77	0.79
Chl $c_{1/2}$*	53	**34**	36	**48**	49	0.57	0.56	0.79	0.80
But	53	44	43	95	81	0.90	0.82	0.50	0.55
Hex*	53	37	36	57	53	0.69	0.60	0.64	0.70
Zea	53	31	28	45	42	0.52	0.48	0.28	0.35
Caro*	53	34	33	55	50	0.62	0.54	0.55	0.62
PE*	53	67	65	156	139	1.28	1.16	0.65	0.69
> 0 mg m^{-3}:									
DVChl *a*	49	26	24	45	39	0.49	0.44	0.59	0.67
Chl c_3*	52	40	37	61	58	0.86	0.69	0.71	0.73
Fuco*	52	45	40	92	73	1.02	0.82	0.65	0.66
Diadino	52	37	33	59	53	0.71	0.61	0.57	0.67

(b) satellite_ρ_{wN}	N	MDPDcv	MPDcv	RMSEcv	R^2cv
TChl *a*	139	32	47	0.64	0.67
PSC	139	42	59	0.79	0.70
PPC	139	27	40	0.49	0.50
MVChl *a*	139	40	55	0.67	0.72
> 0 mg m^{-3}:					
DVChl *a*	124	40	59	0.73	0.25
Chl $c_{1/2}$	127	41	89	0.84	0.64
X19_But	126	39	55	0.63	0.74
X19_Hex	132	40	60	0.68	0.68
Fuco	128	40	65	0.85	0.71
Zea	138	29	46	0.51	0.40
Diadino	132	45	91	0.84	0.57

results for DVChl *a*, Hex and Zea are shown in Fig. 6d–f. Nevertheless, some of these pigments show only medium quality for one cross-validation statistical parameter (lower R^2cv for DVChl *a* and Zea, higher MPDcv for Fuco, Chl $c_{1/2}$ and Diadino).

The full-fit results shown in Fig. 6 show that the models based on the satellite data show much poorer predictions (e.g., a, R^2 and RMSE) than the field data models for all pigment or pigment groups (except Zea) even though the satellite data models are based on more samples. This may be caused by the lower quality of water-leaving reflectance data obtained from the satellite as opposed to direct radiometric measurements in the water column. Another explanation may be that the lower standard deviation of the pigments in the satellite-based data set leads to less precision of the EOF-based models. The latter may explain why the full-fit results for predicting Zea concentrations are very similar for the two model types.

Similar to the field data linear models, no robust predictions were obtained for all other pigments that reached

Table 3. ΔAIC for the robust pigment predictions of the pigment groups TChl *a*, PSC and PPC and the pigments MVChl *a*, Zea and DVChl *a* by the EOF models based on field R_{rs} in (a) hyperspectral resolution (hyper_R_{rs}) and (b) multispectral resolution (band_R_{rs}) and (c) the satellite_ρ_{wN} (from MERIS Polymer) using the 1 × 1 pixel collocation criterion. The pigments listed under "no 0 mg m^{-3}" were predicted using a reduced data set where the respective pigment reached concentrations above 0 mg m^{-3}. Bold highlights the EOF mode with the highest ΔAIC.

(a) hyper_R_{rs}	EOF-1	EOF-2	EOF-3	EOF-4	EOF-5	EOF-6	EOF-7	EOF-8	EOF-9
TChl *a*		**80**	16	16					
PSC		**78**	9	8	4				
PPC		**35**	33		0.4		3	2	
MVChl *a*		**88**	16	11					
Zea	5	6	**21**		10	7	1	5	9
no 0 mg m^{-3}:									
DVChl *a*	7	10	21	**21**	1	20	9	4	7
(b) band_R_{rs}	EOF-1	EOF-2	EOF-3	EOF-4	EOF-5	EOF-6	EOF-7		
TChl *a*		**86**	10	15	3	1			
PSC		**80**	5	5	6				
PPC		**38**	20	9		1			
MVChl *a*		**90**	10	9	1				
Zea	2	5	6	**10**	4	7	0.3		
no 0 mg m^{-3}:									
DVChl *a*	12	17		20	**23**				
(c) satellite_ρ_{wN}	EOF-1	EOF-2	EOF-3	EOF-4	EOF-5	EOF-6	EOF-7		
TChl *a*	7	**148**				1			
PSC	5	**146**		0.2		16			
PPC	10	**62**		3		3	4		
MVChl *a*	4	**151**				5			
no 0 mg m^{-3}:									
Zea	38	**63**	17		1	11	6		
DVChl *a*	13		**26**	**35**	2				

> 0 mg m^{-3} in less than 80 % of all samples concentrations, even when only data points with specific pigment concentrations > 0 mg m^{-3} were included (results not shown).

3.3.3 EOF modes relevant for pigment predictions

Table 3 presents the results of EOF significance based on ΔAIC from their removal as model terms. For the hyper_R_{rs} data set, the prediction linear models used EOF-2 and EOF-3 for all pigments. EOF-2 was the most relevant in the respective models for all pigment prediction except for Zea and DVChl *a*, for which EOF-3 was the most important, closely followed by several other EOF modes. For all other well-predicted pigments, EOF-3 followed EOF-2 in importance, except for Chl c_3 (EOF-4) and PE (EOF-1). Besides PE, only EOF-1 was included (with medium importance) for the prediction of But, DVChl *a* and Zea concentrations. Nearly all linear models using the hyper_R_{rs} data set to predict pigment concentrations incorporated the loadings of three to five EOF modes. In contrast, predictive models for DVChl *a*, Zea and PPC incorporated nine, eight and six EOF modes, respectively.

As discussed in Sect. 3.2, EOF-2 reflects the optical imprint of all phytoplankton pigments. The high ΔAIC value of EOF-2 for most pigments' linear models is probably caused by the increase in concentration of these specific pigments and most phytoplankton groups when TChl *a* increases. In contrast to that, cyanobacteria and especially its subgroup *Prochlorococcus*, containing the marker pigments Zea and DVChl *a*, respectively, are the most abundant phytoplankton under low TChl *a* concentrations. This has manifested in the abundance-based algorithms to retrieve picoplankton from TChl *a* data (Uitz et al., 2006; Hirata et al., 2011) and may explain why predictions of those marker pigments by our linear models show lower ΔAIC for EOF-2 and require several different EOF modes in their linear models.

As in Craig et al. (2012), EOF-2 to EOF-4 were relevant for our hyper_R_{rs}-based TChl *a* and MVChl *a* predictions.

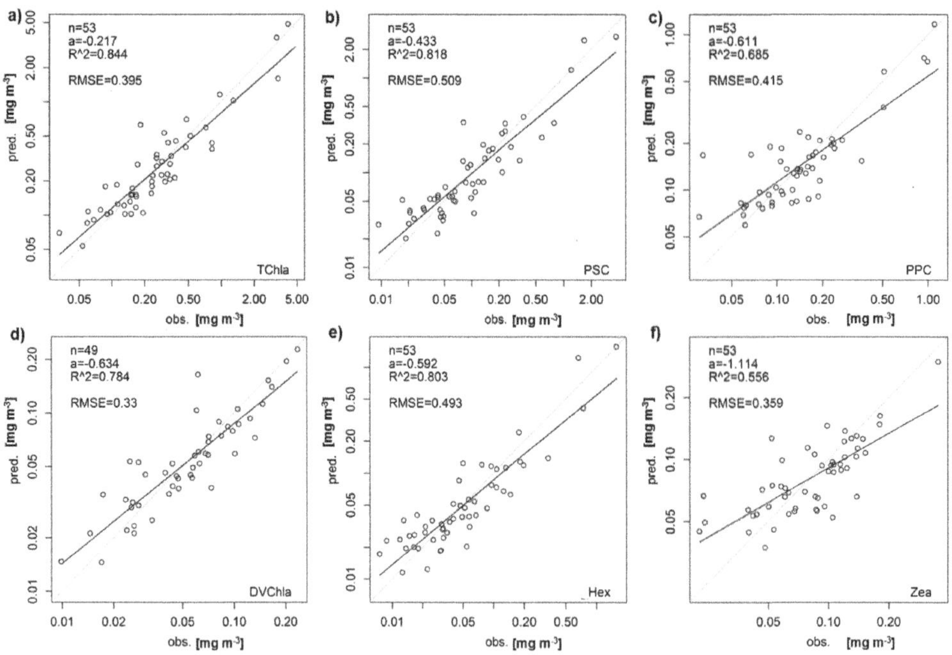

Figure 5. Examples of regressions between observed (obs.) and predicted (pred.) concentrations for pigment groups, **(a)** TChl *a*, **(b)** PSC and **(c)** PPC, and specific pigments, **(d)** DVChl *a*, **(e)** Hex and **(f)** Zea. Observed values have been measured by HPLC (obs.), while predictions are made using a linear model based on EOF modes derived from field R_{rs} data in hyperspectral resolution (hyper_R_{rs}). For DVChl *a*, the model data set was reduced by excluding collocated samples where DVChl *a* had concentrations of 0 mg m^{-3}.

EOF models developed by Taylor et al. (2013) to predict PE concentrations based on L_u data required the first four EOF modes, while our PE prediction based on R_{rs} data required the first three EOFs only. For all other pigments, the higher EOFs were also necessary for robust predictions.

Similarly to the hyper_R_{rs} linear models, the two multispectral linear models also showed EOF-2 to be the most important predictor for specific pigment models except for DVChl *a* (both models) and Zea (only band_R_{rs}).

3.4 Number of data points to construct robust models

Our presented linear models to predict specific pigment or pigment group concentration are calibrated for an ocean-color data set of a specific region with coincidental pigment measurements. Results of the variable jack-knife procedure indicate that the minimal number of training points needed to set up a robust linear model varies among pigments and pigment groups, as revealed by several statistical error measures: the ratio of R^2cv to R^2 (R^2cv / R^2), the ratio of MPDcv to MPD (MPDcv / MPD) and the ratio of RMSE to RMSEcv. Examples for predicting TChl *a*, PSC, PPC and PE are shown in Fig. 7. The ratio R^2cv / R^2 for PPC in all linear models (Fig. 7a, d) drops below 0.8 after a threshold of 50 training data and then decreases exponentially with diminishing data, while other pigments can maintain a high ratio with as few as 30 samples and even 15 samples in the case of the hyper_R_{rs} PE linear model. The threshold where the slope increases in RMSEcv / RMSE (Fig. 7c, f) is for all pigments and linear models probably around 20 to 30 training points. MPDcv/MPD ratios below 1.4, which would indicate robust fits, are obtained for all pigments above 50 training points for the satellite_ ρ_{wN} (Fig. 7e) and above 30 for the hyper_R_{rs} data sets (Fig. 7b). Generally, we observe that band_R_{rs}-based models are more sensitive to training sample size as compared to the hyper_ R_{rs}-based models, especially for TChl *a* and PE. As a general recommendation, a requirement of at least 45 to 50 training data points is advised for most cases, while some pigments (e.g., TChl *a*) may be well predicted with as few as 25 training samples when using models based on hyper_R_{rs} data. Based on these results, we are confident that the models presented in Sect. 3.3 are able to provide robust predictions for both field and the satellite-based data. In the case of PE, the number of samples seems to have been too small, especially for the multispectral resolution, to provide robust PE predictions.

3.5 Comparison to other approaches deriving pigment concentration

Our hyper_R_{rs} TChl *a* linear model results ($R^2 = 0.84$, RMSE = 0.4, R^2cv = 0.77, RMSEcv = 0.49; Fig. 5 and Table 2) are comparable to results by Craig et al. (2012; R^2 = 0.84, RMSE = 0.3, R^2cv = 0.76, RMSEcv = 0.21). Craig et al. (2012) used measurements only from a single location, sampled about weekly over the course of 1 year, while

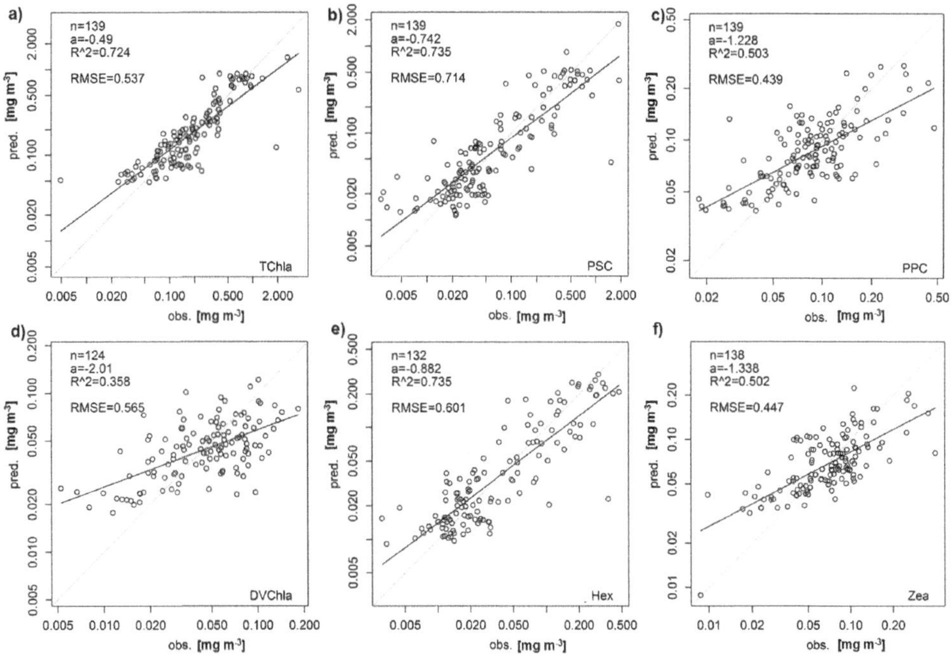

Figure 6. Examples of regressions between observed (obs.) and predicted (pred.) concentrations for pigment groups, **(a)** TChl *a*, **(b)** PSC and **(c)** PPC, and specific pigments, **(d)** DVChl *a*, **(e)** Hex and **(f)** Zea: observed values have been measured by HPLC (obs.), while predictions are made using a linear model based on EOF modes derived from satellite_ρ_{wN} (from MERIS Polymer) data within the 1×1 pixel collocation box. For DVChl *a*, Hex and Zea, the model data set was reduced by excluding collocated samples where DVChl *a* had concentrations of $0\,\mathrm{mg\,m^{-3}}$.

our field data set was from a much larger region (covering 95° in latitude and 85° in longitude) and sampled during two seasons in 2008 and 2010 only. In their study the same linear model setup was used with collocated in situ reflectance and TChl *a* data sampled at Compass Buoy station in the Bedford Basin near Halifax as input data. The number of collocations used for training to obtain robust results for TChl *a* predictions was also similar for both studies, with more than 25 recommended for our hyper_R_{rs} linear model and more than 15 in the Craig et al. (2012) study.

Chase et al. (2013) used Gaussian functions to derive different chlorophyll types, PSC and PPC concentrations from a large global data set of hyperspectral particulate absorption measurements. Their validation results showed MDPD values between predicted and observed concentrations of 30 and 36 %, 40 and 53 %, 49 % and 51 % for TChl *a*, TChl *c*, PSC and PPC, respectively. Our linear models show similar (TChl *a* 27–32 %) or even much better MDPDcv values (Chl $c_{1/2}$: 34–41 %, PSC: 32–43 %, PPC: 24–28 %). We believe that this further indicates the robustness of our approach, especially given that we use a more indirect measure of pigments, AOP (reflectance), as opposed to the IOPs used in their study.

Pan et al. (2010) developed pigment specific band-ratio algorithms with collocated in situ $R_{rs}(\lambda)$ and pigment measurements from the northeastern coast of the United States. Those algorithms are based on deriving pigment-specific coefficients for third-order polynomial functions using the band ratio of either 490–550 nm or 490–670 nm (for SeaWiFS; for MODIS changed accordingly to MODIS bands 488 and 547 nm). Validation of results with collocated satellite (SeaWiFS and MODIS) reflectance data and pigment concentrations showed very good-quality predictions for several pigments (TChl *a*, TChl *c*, Caro, Fuco, Diadino and Zea) using SeaWiFS bands (MPD from 36 to 48 %, RMSE from 0.23 to 0.29, and R^2 from 0.65 to 0.90; similar results were also obtained using MODIS bands). This method was modified to the northern South China Sea using globally derived relationships and locally identified links between pigment concentration and sea surface temperature (Pan et al., 2013) with similar validation results as in Pan et al. (2010). Compared to our linear model results, the quality of pigment concentration prediction is similar: while our results for MPDcv and R^2cv are slightly worse (42–50 % and 0.61–0.80, respectively), our results for RMSEcv (0.48–0.61 $\mathrm{mg\,m^{-3}}$, except Fuco: 0.82 $\mathrm{mg\,m^{-3}}$) are much better.

PE is not well predicted by both our linear models based on the field data set. Still, hyper_R_{rs} linear model cross-validation measures are much better than the PE band_R_{rs} linear model. In Taylor et al. (2013), PE concentrations were predicted from the same underwater light measurements but using L_u instead of R_{rs} data and the model was based on pig-

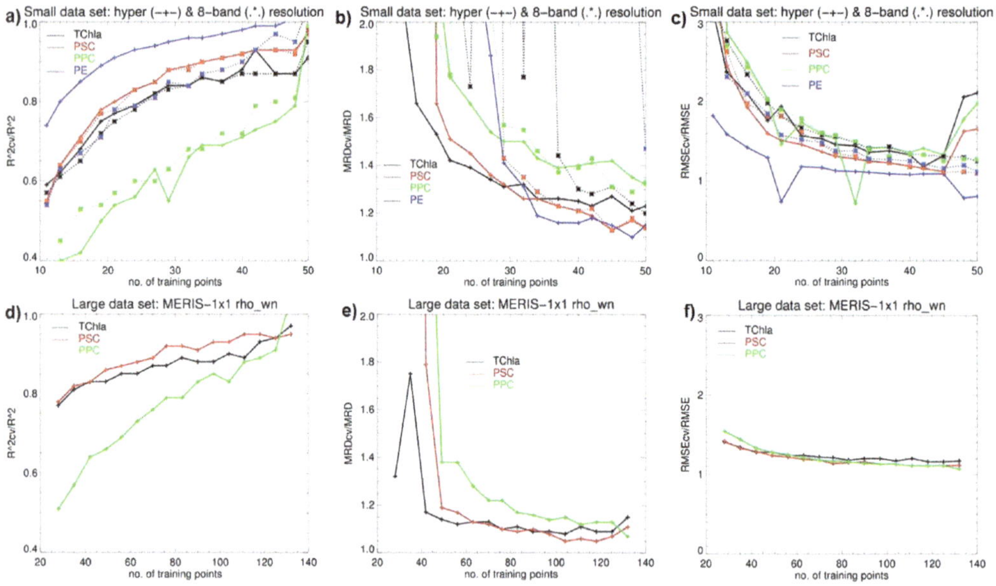

Figure 7. R^2cv / R^2 **(a, d)**, MRDcv / MRD **(b, e)** and RMSEcv / RMSE **(c, f)** as a function of number of training points (tp) for the linear models. Shown are results for specific models for TChl a, PSC, PPC and PE using reflectance data from the field **(a–c)** in hyperspectral (hyper_R_{rs}, solid lines) and multispectral band (band_R_{rs}, dotted lines) resolution and from satellite MERIS Polymer within the 1×1 pixel collocation box (satellite_ρ_{wN}, **(d–f)**). The number of total samples points was $n = 53$ for hyper_R_{rs} and band_ R_{rs} and $n = 139$ for satellite_ρ_{wN}. Cross validation is based on 500 permutations using tp for training and as number of validation points (vp): vp= $n-$tp.

ment concentrations at surface and deeper depths. No cross-validation was performed within their study. Our results for R^2cv (0.69) are even better than their results for using the data from all three cruises for predictions (R^2 of 0.58). The data set of Taylor et al. (2013) was nearly 3 times larger than our field data set and a log-link generalized linear model (GLM) was used instead of a log-transformed linear model. For the latter we tested both settings for our pigment linear models. Cross-validation revealed a similar prediction error for PE using the log-link GLM instead of the log-transformed linear model, but the error increased when GLM was used for other pigment predictions.

As for TChl a predictions from the satellite_ρ_{wN} linear model, validation results of the MERIS Polymer TChl a product collocations with in situ TChl a from the satellite-based data set showed marginal differences for the 1×1, 3×3 or 5×5 pixel collocations (Table 4, upper panel). The TChl a Polymer product obtained 3 % higher *MPD* and similar R^2, RMSE and *PB* values (of about 0.74, 0.51 and 10 % on average, respectively) to the TChl a linear model predictions. In the global validation by Brewin et al. (2015), the OC4V6 (Ocean-Chlorophyll-4 algorithm version 6; O'Reilly et al., 2000) was selected from amongst various TChl a satellite products as the best TChl a algorithm. This algorithm is used to produce the MERIS Polymer TChl a from atmospheric-corrected MERIS Polymer data. Global validation by Brewin et al. (2015), with 1039 collocations and retrievals of TChl a directly from in situ $\rho_{\mathrm{wN}}(\lambda)$ data, showed an R^2 of 0.87 and a RMSE of 0.29 for OC4V6 based on non-log-transformed concentrations (which compares to our RMSE values on log-scale shown in Table 4 of Bracher et al., 2014). We conclude that both MERIS Polymer TChl a products, the level 2 and linear models, show high quality within the eastern Atlantic Ocean although they are retrieved from satellite data and not in situ ρ_{wN} data.

The comparison with other methods of retrieving pigment concentrations from reflectance data shows that our method, based on a linear model using EOFs from reflectance data, gives robust results for pigment groups and pigments that are always present in the region investigated. To test our EOF methods for independent data sets using the method established by a certain testing data set, we have used the cross-validation technique. The technique allows the re-sampling of all data for 500 different subsets (i.e., run by 500 permutations) into testing and validation data sets.

The advantage of our approach is that it allows for the estimation of several pigments and pigment groups using either reflectance data measured directly in the ocean water or obtained from a satellite ocean-color sensor. For the eastern tropical Atlantic Ocean data set, these additional pigments (other than TChl a) include PPC, PSC, DVChl a and MChl a. Additional pigments may also be accurately predicted with this approach; however, the results suggest that the prediction error increases for pigments that are found in lower concentrations or with a high number of samples below the detection limit (i.e., referred to in statistics as "censoring"). This poor performance may be in part due to the fact that pigments found in small concentrations are likely to have a limited ef-

Table 4. TChl *a* validation statistics for MERIS Polymer TChl *a* (left panel) and TChl *a* obtained from the full-fit linear model on satellite_ ρ_{wN} (from MERIS Polymer data, EOF full-fit model, right panel) with different collocation criteria (either 1 × 1 or the mean of 3 × 3 or 5 × 5 pixel values) for the MERIS Polymer data compared to the in situ (from HPLC) value.

MERIS		Polymer				EOF full-fit model			
Collocation	N	R^2	RMSE	PB	MPD	R^2	RMSE	PB	MPD
1 × 1 pixel	139	0.74	0.51	6	39	0.72	0.54	10	37
3 × 3 pixel	155	0.75	0.50	10	39	0.74	0.52	10	36
5 × 5 pixel	160	0.73	0.51	11	40	0.74	0.52	10	36

fect on spectral shape, but further modeling work may also need to focus on better approaches for the treatment of censored values. Generally, we can also see from the field data linear models that using a coherent in situ data set, where all pigments have been measured by the same method and instrumentation, may be better suited for the modeling approach due to the homogeneous error across the range of pigment concentrations. An advantage of our linear method to pigment-specific band algorithms is that we require a much smaller data set for establishing the prediction (about 50 as opposed to several hundreds) of collocated pigment and reflectance data.

3.6 Application of linear model to study large-scale pigment distributions

For demonstrating the application of our linear model, we used the satellite_ρ_{wN} specific pigment's full-fit models for TChl *a*, MVChl *a*, PSC and PPC and ran these specific models using November 2008 MERIS Polymer ρ_{wN} level 2 data to retrieve those pigments for an example time period on a larger spatial scale. By subtracting the MVChl *a* value from TChl *a* we also derived concentrations of DVChl *a*. Figure 8 shows the monthly averages for those various pigment groups and pigments. Also, the MERIS Polymer TChl *a* concentration for the same time and region is shown.

The distributions of TChl *a* from the EOF model prediction or from the Polymer algorithm are very similar, ranging from 0.00003 to 7.52 mg TChl *a* m^{-3}. For this particular month, the total biomass of phytoplankton shows a strong phytoplankton bloom (> 2 mg m^{-3}) at the Mauritanian upwelling spread in two parts, 190–24° N and 14–7° N, and high values (> 0.5 mg m^{-3}) at all coastal areas of the African continent. Enhanced TChl *a* concentrations > 0.3 mg m^{-3} are also spreading into the open ocean especially at 5–20° N and 30–40° W, along the 0° latitude from Africa to South America, and south of this at 3–10° S from 3° E to about 25° W. MVChl *a* follows more or less the TChl *a* distribution, however, only reaching the magnitude indicated by the TChl *a* values at the northern bloom. The deviation between TChl *a* and MVChl *a* is obvious in the distribution of DVChl *a*, which indicates that at the northern part of the Mauritanian upwelling bloom, *Prochlorococcus* (the only phytoplankton genus which contains DVChl *a*) seems to have contributed to this bloom by only a very minor fraction (i.e., a few percent), while elsewhere it presents a substantial background of about 30 % of all phytoplankton.

Our predicted PPC concentrations show values in the same range as TChl *a* at the oligotrophic areas and about 50 % in the enhanced TChl *a* areas and the southern part of the bloom. As for DVChl *a*, in the northern part of the bloom PPC concentrations are significantly lower and only contribute less than 10 % to the total pigment concentrations. PSC concentration in the oligotrophic and enhanced TChl *a* areas are much lower than PPC or even DVChl *a* concentrations but reflect the TChl *a* distribution more or less on the large scale. Within the northern part of the Mauritanian upwelling PSC concentrations reach values even as high as for TChl *a*, while concentrations at the bloom further south contribute to less than 10 % of the total pigment concentrations. In Taylor et al. (2011) the analysis of pigment and additional microscopic data clearly showed very high concentrations of Fuco, a main pigment of PSC, and a high dominance of diatoms within water samples at the northern bloom collected at the same time period.

From our results, we can conclude that the northern phytoplankton bloom at the Mauritanian upwelling seems to have been freshly growing with very high photosynthetic activity, while for most of the other areas a lot of the energy build-up via photosynthesis was used for photoprotection. We have no information on photodegradation since no significant prediction linear model could be developed for phaeopigments. These pigments had only been identified in less than 60 % of all samples collocated to the field and satellite-based data sets, and the results show that this pigment group was not well predicted by the linear model. Based on the biogeography of Longhurst (2006), the oligotrophic areas on our maps fall in the North Atlantic Subtropical Gyre Province East at > 25° N (the border between the two is the subtropical convergence) and the North Atlantic Tropical Gyre Province at 25° N to about 12° N. At the eastern corner towards the coast of these provinces, in the Canary Coastal Province (CNRY), concentrations of all predicted pigments and pigment groups may have been increased due to eddy-driven processes that increase the supply of nutrients. In Taylor et al. (2011), the two blooms analyzed by field samples at CNRY have been

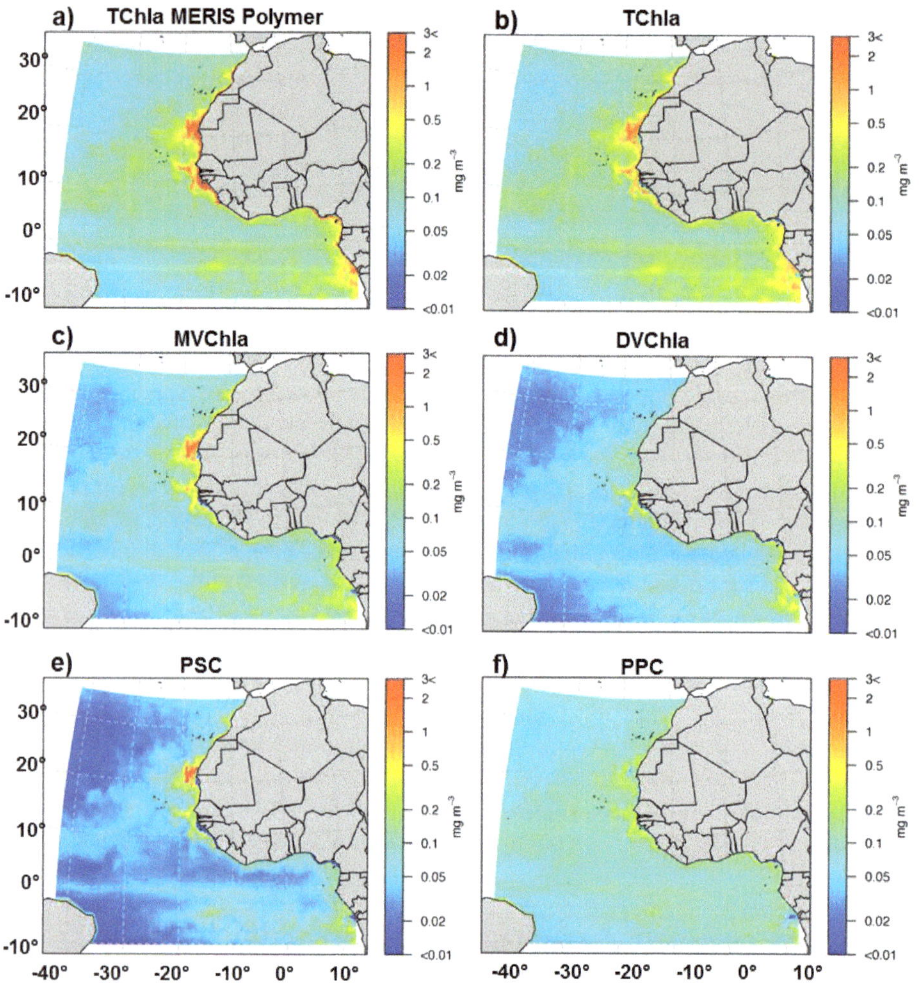

Figure 8. Monthly mean concentrations (in 0.25° grid resolution) for November 2008 of **(a)** TChl *a* of the MERIS Polymer algorithm (TChl *a* MERIS Polymer) and predicted **(b)** TChl *a*, **(c)** MVChl *a*, **(d)** DVChl *a*, **(e)** PSC and **(f)** PPC by the LM, based on the full fit of satellite ρ_{wN} data within the 1 × 1 pixel collocation box and the EOFs of this month's MERIS Polymer ρ_{wN} data.

identified to cluster differently due to their pigment composition. The northern "fresh" bloom with low photoinhibition, high dominance of phytoplankton and strong photosynthetic efficiency was related to a major upwelling focusing in the area south of Cape Blanc (western Sahara) off the coast of Mauritania. DVChl *a* was absent in this bloom, which is in line with our results obtained from the linear model. The southern part of the CNRY bloom was placed within the African dust veil where mineral-rich dust fertilizes the ocean. In the northern bloom stations, the spectral shape and absolute values of particulate non-phytoplankton absorption spectra, presented in Taylor et al. (2011) and coinciding with the pigment data used in our study, clearly indicated that mineral particle absorption was very high.

Comparisons of our predictions to pigment data not used for the development and validation of our EOF model show consistent results: Partensky et al. (1996) measured TChl *a* concentrations of about 1.2 mg m^{-3} in December 1992 (EUMELI 5 cruise) at a station within a phytoplankton bloom at 18° 29′ N and 21° 05′ W, similar to the range of our predicted values at the southern edge of the northern bloom. Barlow et al. (2002) measured, within the area of our predictions, concentrations of TChl *a*, DVChl *a*, PSC and PPC during the AMT-3 cruise in October 1996 at 20° N and 20° W (0.4, 0.05, 0.175 and above 0.09 mg m^{-3}, respectively) and 30° N and 22° W (0.05, 0.01, 0.022 and above 0.04 mg m^{-3}, respectively), similar to our predicted concentrations for the same pigments just east of the northern bloom of the Mauritanian upwelling and at the North Atlantic Subtropical Gyre Province East, respectively.

4 Conclusions

We present robust predictions of concentrations of various pigments and pigment groups from linear models based on fitting empirical orthogonal function on a set of reflectance data to collocated pigment concentrations. Spectral shapes of the reflectance spectra from the eastern Atlantic and of their derived EOF modes reflect typical case 1 water characteristics. In our study, it was shown that EOFs derived from both hyperspectral underwater radiometric measurements and multispectral reflectance data from field or satellite (MERIS Polymer) enable reliable predictions of the concentration of nine different pigments/pigment groups (TChl a, PPC, PSC, MVChl a, Chl $c_{1/2}$, But, Hex, Zea, Caro, PE). A limitation of all predictions was that only those pigments can be predicted that have been identified in every collocated sample; adding a small value (0.0001 mg m^{-3}) to censored samples was not an appropriate solution to this problem.

The method proves for the first time to be applicable for predicting concentrations of not only TChl a and PE but also of other pigments and pigment groups with weaker, but spectrally unique, features on the underwater light field. Statistical resampling used for cross-validation indicates that predictions were robust (R^2cv ≥ 0.5, MDPDcv $\leq 44\,\%$ and MPDcv $\leq 60\,\%$) for all pigments (except for PE, Zea and But, which deviated for one of these measures) and pigment groups. Hyperspectral linear models proved to be already stable with less collocated samples for most pigment or pigment groups used for training ($n > 30$ to 40) than linear models based on multispectral reflectance data ($n > 50$). The linear models using MERIS Polymer reflectance data as input were applied to 1 month of satellite data to predict the concentrations of TChl a, PSC, PPC, MVChl a and DVChl a for the whole eastern tropical Atlantic. For the first time a consistent picture of several phytoplankton pigments indicating group-specific behavior and photophysiology on a larger spatial scale for this area was shown.

Our presented linear models are generic and can be applied to even a small, consistently collocated reflectance and pigment data set to enable various specific pigment predictions from continuous optical measurements. The optical data can be obtained from radiometric measurements based on various platforms (buoys, gliders, floats or satellite). On a global scale, TChl a, PSC and PPC are consistently accurately predicted, while other pigments may be better predicted on smaller spatial scales. Highly temporally resolved time series data, which – depending on the platform – may even provide good spatial coverage, can be used to study variability and change of overall phytoplankton and photophysiological responses to environmental variables. While we established the linear models for prediction of various pigments in typical case 1 waters, the method should be tested in the future for its applicability in case 2 waters as well.

Author contributions. A. Bracher designed and ran the experiments and wrote the manuscript. M. H. Taylor developed the statistical method and wrote the R code; B. Taylor supplied all field and part of the additional pigment data, prepared all input data and designed part of the experiment; T. Dinter contributed to programming; R. Röttgers contributed to R_{rs} data and additional pigment data; and F. Steinmetz supplied MERIS Polymer data and match-ups. All co-authors assisted in writing the manuscript.

Acknowledgements. We thank AWI, Helmholtz Impulse Fond (HGF Young Investigators Group Phytooptics), Total Foundation (project PHYTOSCOPE) and the Leibniz Association (project OCEANET) for funding. We thank Erika Allhusen, Helmke Hepach, Kerstin Heymann, Laila Bentama, Anja Bernhardt and Sonja Wiegmann for work in the laboratory and on-board RVs *Polarstern* and *Merian S. Merian*. We thank the crew, principal investigators and other scientists on *ANTXXIII/1*, *ANTXXIV/1*, *ANTXXIV/4*, *ANTXXV/1*, *ANTXXVI/4* and MSM 18-3 for support on board and fruitful discussions afterwards. We thank Oliver Zielinski and his group at ICBM, University Oldenburg, previously at the University of Applied Sciences, Bremerhaven, for the use of their equipment. We thank ESA for MERIS level 1 data. We thank Amélie Talec for BGH sample collection and Hervé Claustre and Josephine Ras (all LOV) for BGH pigment sample analysis. This study contains data supplied by the Natural Environment Research Council (NERC) via BODC and NASA via SEABASS. We thank all researchers behind these large pigment data sets.

Edited by: O. Zielinski

References

Aiken, J., Pradhan, Y., Barlow, R., Lavender, S., Poulton, A., Holligan, P., and Hardman-Mountford, N.: Phytoplankton pigments and functional types in the Atlantic Ocean: A decadal assessment, 1995–2005, Deep-Sea Res. Pt. II, 56, 899–917, doi:10.1016/j.dsr2.2008.09.017, 2009.

Barker, K., Mazeran, C., Lerebourg, C., Bouvet, M., Antoine, D., Ondrusek, M., Zibordi, G., and Lavender, S.: MERMAID: MERis MAtchup in situ Database. Proceedings of the 2nd (A)ATSR and MERIS Workshop, Frascati, Italy, September 2008, 2008.

Barlow, R. G., Cummings, D. G., and Gibb, S. W.: Improved resolution of mono- anddivinyl chlorophylls a and b and zeaxanthin andlutein in phytoplankton extracts using reverse phase C-8 HPLC, Marine Ecol. Progr. Ser., 161, 303–307, 1997.

Barlow, R. G., Aiken, J., Hooligan, P. M., Cummings, D. G., Maritorena, S., and Hooker, S.: Phytoplankton pigment and absorption characteristics along meridional transects in the Atlantic, Deep-Sea Res. Pt. I, 47, 637–660, 2002.

Behrenfeld, M. I. and Boss, E. B.: Beam attenuation and chlorophyll concentration as alternative optical indices of phytoplankton biomass. J. Marine Res., 64, 431–451, 2006.

Bracher, A., Taylor, M. H., Taylor, B., Dinter, T., Röttgers, R., and Steinmetz, F.: Using empirical orthogonal functions derived from remote sensing reflectance for the prediction of concentrations of phytoplankton pigments, Ocean Sci. Discuss., 11, 2073–2117, doi:10.5194/osd-11-2073-2014, 2014.

Brewin, R. J. W., Sathyendranath, S., Müeller, D., Brockmann, C., Deschamps, P.-Y., Devred, E., Doerffer, R., Fomferra, N., Franz, B. A., Grant, M. G., Groom, S. B., Horseman, A., Hu C., Krasemann, H., Lee, Z., Maritorena, S., Mélin, F., Peters, M., Platt, T., Regner, P., Smyth, T., Steinmetz, F., Swinton, J., Werdell, J., and White III, G. N.: The Ocean Colour Climate Change Initiative: A round-robin comparison on in-water bio-optical algorithms, Remote Sens. Environ., in press, 2015.

Chase, A., Boss, E., Zaneveld, R., Bricaud, A., Claustre, H., Ras, J., Dall'Olmo, G., and Westberry, T. K.: Decomposition of in situ particulate absorption spectra, Methods Oceanogr., 7, 110–124, 2013.

Craig, S. E., Jones, C. T., Li, W. K. W., Lazin, G., Horne, E., Caverhill, C., and Cullen, J. J.: Deriving optical metrics of ecological variability from measurements of coastal ocean colour, Remote Sens. Environ., 119, 72–83, 2012.

Hirata, T., Hardman-Mountford, N. J., Brewin, R. J. W., Aiken, J., Barlow, R., Suzuki, K., Isada, T., Howell, E., Hashioka, T., Noguchi-Aita, M., and Yamanaka, Y.: Synoptic relationships between surface Chlorophyll-*a* and diagnostic pigments specific to phytoplankton functional types, Biogeosciences, 8, 311–327, doi:10.5194/bg-8-311-2011, 2011.

Hooker, S. B., Van Heukelem, L., Thomas, C. S., Claustre, H., Ras, J., and Barlow, R.: The second SeaWiFS HPLC analysis round robin experiment (SeaHARRE-2). NASA TM/2005-212785, NASA Goddard Space Flight Center, Greenbelt, Maryland, 2005.

Lee, Z. P., Carder, K. L, and Arnone, R. A.: Deriving inherent optical properties from water color: A multi-band quasi-analytical algorithm for optically deep waters, Appl. Opt., 41, 5755–5772, 2002.

Letelier, R. M., Bidigare, R. R., Hebel, D. V., Ondrusek, M., Winn, C. D., and Karl, D. M.: Temporal variability of phytoplankton community structure based on pigment analysis. Limnol. Oceanogr., 38, 1420–1437, 1993.

Longhurst, A.: Ecological Geography of the Sea, 2nd Edn., Elsevier Academic press, USA, 2006.

Lubac, B. and Loisel, H.: Variability and classi?cation of remote sensing reflectance spectra in the eastern English Channel and southern North Sea, Remote Sens. Environ., 110, 45–48, 2007.

McClain, C. R.: A Decade of Satellite Ocean Color Observations, Ann. Rev. Marine Sci., 1, 19–42, doi:10.1146/annurev.marine.010908.163650, 2009.

Müller, D. and Krasemann, H.: Product validation and algorithm selection report, part 1 – atmospheric correction. Tech. Rep. AO-1/6207/09/I-LG D2.5, European Space Agency, ESRIN, 2012.

O'Reilly, J. E., Maritorena, S., Siegel, D., and O'Brien, M. C.: Ocean color chlorophyll a algorithms for SeaWiFs, OC2, and OC4: Technical report, edited by: Toole, D., Mitchell, B. G., Kahru, M., Chavez, F. P., Strutton, P., Cota, G., Hooker, S. B., McClain, C. R., Carder, K. L., Muller-Karger, F., Harding, L., Magnuson, A., Phinney, D., Moore, G. F., Aiken, J., Arrigo, K. R., Letelier, R., Culver, M., Hooker, S. B., and Firestone, E. R.: SeaWiFS postlaunch calibration and validation analyses, Part 3. NASA, Goddard Space Flight Center, Greenbelt, Maryland. SeaWiFS Postlaunch Technical Report Series, 11, 9–23, 2000.

Organelli, E., Bricaud, A., Antoine D., and Uitz, J.: Multivariate approach for the retrieval of phytoplankton size structure from measured light absorption spectra in the Mediterranean Sea (BOUSSOLE site), Appl. Opt., 52, 2257–2273, 2013.

Partensky, F., Blanchot, J., Lantoine, F., Neveux, J., and Marie, D.: Vertical structure of picophytoplankton at different trophic sites of the tropical northeastern Atlantic Ocean, Deep-Sea Res. Pt. I , 43, 1191–1213, 1996.

Pan, X., Mannino, A., Russ, M. E., Hooker, S. B., and Harding, L. W.: Remote sensing of phytoplankton pigment distribution in the United States northeast coast, Remote Sens. Environ., 114, 2403–2416, 2010.

Pan, X., Wong, G. T. F., Ho T.-Y., Shiah, F.-K., and Liu H.: Remote sensing of picophytoplankton distribution in the northern South China Sea, Remote Sens. Environ., 128, 162–175, 2013.

R Core Team R: A language and environment for statistical computing. R Foundation for Statistical Computing, Vienna, Austria, available at: http://www.R-project.org/, last access: 14 December 2013.

Ras, J., Claustre, H., and Uitz, J.: Spatial variability of phytoplankton pigment distributions in the Subtropical South Pacific Ocean: comparison between in situ and predicted data, Biogeosciences, 5, 353–369, doi:10.5194/bg-5-353-2008, 2008.

Roy, S., Llewellyn, C. A., and Egeland, E. S.: Phytoplankton pigments: characterization, chemotaxonomy and applications in oceanography, Cambridge University Press, 2011.

Speich, S., Boyé, M., and Dehairs, F.: IPY BONUS-GoodHope cruise Report. May 2008, IPEV-INSU-LEFE-IFREMER-ANR-IRD-IUEM-UBO, 2008.

Steinmetz F., Deschamps P.-Y., and Ramon D.: Atmospheric correction in presence of sun glint: application to MERIS, Opt. Expr., 19, 9783–9800, 2011.

Taylor, B. B., Torrecilla, E., Bernhardt, A., Taylor, M. H., Peeken, I., Röttgers, R., Piera, J., and Bracher, A.: Bio-optical provinces in the eastern Atlantic Ocean and their biogeographical relevance, Biogeosciences, 8, 3609–3629, doi:10.5194/bg-8-3609-2011, 2011.

Taylor, B. B., Taylor, M., Dinter, T., and Bracher, A.: Estimation of relative phycoerythrin concentrations from hyperspectral underwater radiance measurements – a statistical approach, J. Geophys. Res.-Ocean., 118, 2948–2960, doi:10.1002/jgrc.20201, 2013.

Uitz, J., Claustre, H., Morel, A., and Hooker, S.: Vertical distribution of phytoplankton communities in open ocean: an assessment based on surface chlorophyll, J. Geophys. Res., 111, C08005, doi:10.1029/2005JC003207, 2006.

Vidussi, F., Claustre, H., Manca, B. B., Lucheta, A., and Marty, J.-C.: Phytoplankton pigment distribution in relation to upper thermocline circulation in the eastern Mediterranean Sea during winter, J. Geophys. Res., 106, 19939–19956, 2001.

Werdell, P. J., Franz, B. A., Bailey, S. W., Feldman, G. C., Boss, E., Brando, V. E., Dowell, M., Hirata, T., Lavender, S. J., Lee, Z. P., Loisel, H., Maritorena, S., Melin, F., Moore, T. S., Smyth, T. J., Antoine, D., Devred, E., d'Andon, O. H. F., and Mangin, A.:

Generalized ocean color inversion model for retrieving marine inherent optical properties, Appl. Opt., 52, 2019–2037, 2013.

Wu, C. F.: Jackknife, bootstrap and other resampling methods in regression analysis, Annals Stat., 14, 1261–1295, 1986.

Zapata, M., Rodriguez, F., and Garido, J. L.: Separation of chlorophylls and carotenoids from marine phytoplankton: a new HPLC method using a reversed phase C8 column and pyridine-containing mobile phases, Marine Ecol. Progr. Ser., 195, 29–45, 2000.

11

Fate of colloids during estuarine mixing in the Arctic

O. S. Pokrovsky[1,2], **L. S. Shirokova**[1,3], **J. Viers**[1], **V. V. Gordeev**[4], **V. P. Shevchenko**[4], **A. V. Chupakov**[3], **T. Y. Vorobieva**[3], **F. Candaudap**[1], **C. Causserand**[1], **A. Lanzanova**[1], **and C. Zouiten**[1]

[1]Georesources and Environment Toulouse GET, UMR5563, CNRS, Université Paul Sabatier, 14 Avenue Edouard Belin, 31400 Toulouse, France
[2]BIO-GEO-CLIM Laboratory, Tomsk State University, Tomsk, Russia
[3]Institute of Ecological Problems of the Northern Regions, Ural Branch of the Russian Academy of Science, 23 Naberezhnaja Sev. Dviny, Arkhangelsk, Russia
[4]P.P. Shirshov Institute of Oceanology of the Russian Academy of Science, 36 Nakhimovsky Prospekt, 117997 Moscow, Russia

Correspondence to: O. S. Pokrovsky (oleg@get.obs-mip.fr)

Abstract. The estuarine behavior of organic carbon (OC) and trace elements (TE) was studied for the largest European sub-Arctic river, which is the Severnaya Dvina; this river has a deltaic estuary covered in ice during several hydrological seasons: summer (July 2010, 2012) and winter (March 2009) baseflow, and the November–December 2011 ice-free period. Colloidal forms of OC and TE were assessed for three pore size cutoffs (1, 10, and 50 kDa) using an in situ dialysis procedure. Conventionally dissolved (< 0.22 μm) fractions demonstrated clear conservative behavior for Li, B, Na, Mg, K, Ca, Sr, Mo, Rb, Cs, and U during the mixing of freshwater with the White Sea; a significant (up to a factor of 10) concentration increase occurs with increases in salinity. Si and OC also displayed conservative behavior but with a pronounced decrease in concentration seawards. Rather conservative behavior, but with much smaller changes in concentration (variation within ±30 %) over a full range of salinities, was observed for Ti, Ni, Cr, As, Co, Cu, Ga, Y, and heavy REE. Strong non-conservative behavior with coagulation/removal at low salinities (< 5 ‰) was exhibited by Fe, Al, Zr, Hf, and light REE. Finally, certain divalent metals exhibited non-conservative behavior with a concentration gain at low (∼ 2–5 ‰, Ba, Mn) or intermediate (∼ 10–15 ‰, Ba, Zn, Pb, Cd) salinities, which is most likely linked to TE desorption from suspended matter or sediment outflux.

The most important result of this study is the elucidation of the behavior of the "truly" dissolved low molecular weight $LMW_{<1\,kDa}$ fraction containing Fe, OC, and a number of insoluble elements. The concentration of the LMW fraction either remains constant or increases its relative contribution to the overall dissolved (< 0.22 μm) pool as the salinity increases. Similarly, the relative proportion of colloidal (1 kDa–0.22 μm) pool for the OC and insoluble TE bound to ferric colloids systematically decreased seaward, with the largest decrease occurring at low (< 5 ‰) salinities.

Overall, the observed decrease in the colloidal fraction may be related to the coagulation of organo-ferric colloids at the beginning of the mixing zone and therefore the replacement of the $HMW_{1\,kDa–0.22\,\mu m}$ portion by the $LMW_{<1\,kDa}$ fraction. These patterns are highly reproducible across different sampling seasons, suggesting significant enrichment of the mixing zone by the most labile (and potentially bioavailable) fraction of the OC, Fe and insoluble TE. The size fractionation of the colloidal material during estuarine mixing reflects a number of inorganic and biological processes, the relative contribution of which to element speciation varies depending on the hydrological stage and time of year. In particular, $LMW_{<1\,kDa}$ ligand production in the surface horizons of the mixing zone may be linked to heterotrophic mineralization of allochthonous DOM and/or photodestruction. Given the relatively low concentration of particulate versus dissolved load of most trace elements, desorption from the river suspended material was less pronounced than in other rivers in the world. As a result, the majority of dissolved components exhibited either conservative (OC and related elements such as divalent metals) or non-conservative, coagulation-controlled (Fe, Al, and insoluble TE associated with organo-ferric colloids) behavior. The climate warming at high latitudes is likely to intensify the production of $LMW_{<1\,kDa}$ organic ligands and the associated TE; therefore, the delivery of potentially bioavailable trace metal micronutrients from the land to the ocean may increase.

1 Introduction

Over the past several decades, a comprehensive picture of organic carbon and trace element geochemistry in the mixing zone of the river mouths has been completed for many major and "model" rivers around the world (Bianci, 2007; Bianci and Allison, 2009; Gordeev, 2012). Several main processes control the concentration patterns of OC, Fe, and TE in the estuaries: coagulation of organic and organo-mineral colloids (Sholkovitz, 1976; Boyle et al., 1977; Sigleo and Helz, 1981; Benoit et al., 1994; Lisitsyn, 1995; Guieu et al., 1996; Benoit et al., 1994), desorption from river-suspended matter or sediments (Dai et al., 1995; Guieu and Martin, 2002), diffusive flux from sediments (Shiller, 1996; Audry et al., 2006, 2007), and biological uptake (Redfield et al., 1963; Shiller and Boyle, 1991). The relative importance of each of these processes defines the degree of river flux transformation in the mixing zone and net elemental input from the land to the ocean. In this regard, physicochemical and biological mechanisms operating within the mixing zone of the Arctic Ocean are especially important given the following: (1) extremely high contributions of dissolved flux relative to the surface and ocean volume compared to the other oceans (Holmes et al., 2012); (2) dominance of dissolved and colloidal flux over the suspended flux for most major and trace elements of the river water (Gordeev et al., 1996; Pokrovsky et al., 2010); (3) high vulnerability of Arctic river and sea ecosystems to both anthropogenic pollution and ongoing climate change.

The majority of studies of the Arctic estuaries were carried out during summer baseflow periods (MacDonald and Yu, 2006; Gobeil, 2006; Kuzyk et al., 2008, and McClelland et al., 2012, for a review). Although this season in the Arctic is very important due to the biological activity in the water column, it does not allow the other major controlling factors of element transformation in the estuary, such as photodestruction of DOM-related colloidal forms of TE, desorption from river-suspended matter and redox exchange with the sediments, to be tested. The relative importance of the various driving mechanisms fluctuates depending on the season; therefore, systematic studies over several seasons are required. These studies constitute the first objective of the present study, which is aimed at assessing the estuarine profiles of both major and trace elements, including organic carbon, in the largest European Arctic River, the Severnaya Dvina. These profiles will be based on several sampling campaigns performed in the Severnaya Dvina's mouth zone during the main hydrological seasons: summer baseflow (July), beginning of winter and ice formation (November–December) and end of winter baseflow (March).

All boreal rivers of the European Arctic, particularly the Severnaya Dvina River, are major carriers of organic and organo-mineral colloids from the land to the ocean. For example, 20 to 30 % of Mn, Zn, and Cu, 30 to 50 % of Fe and REEs, and up to 80 % of C_{org} and U are transported in colloidal form from the total (dissolved + suspended) flux of this river (Pokrovsky et al., 2010). Despite the importance of the colloidal form in the overall elemental transport from the land to the ocean, the transformation of colloids in the mixing zone is still very poorly studied. The main reason for knowledge gap is the lack of uniform and standardized techniques allowing in situ or on site size fractionation of the dissolved load along the salinity gradient. Until now, tangential cross-flow and frontal ultrafiltration, along with diffusive gradients in thin films, have remained the basic techniques used to assess the colloidal distribution in estuarine zones all over the world (Dai and Martin, 1995; Greenamoyer and Moran, 1997; Ingri et al., 2004; Forsberg et al., 2006; Howell et al., 2006; Waeles et al., 2008; Österlund et al., 2012). The use of these techniques, however, requires significant precautions against experimental artifacts, such as charge separation, filter clogging and polarization, thorough calibration and assumptions on the diffusion efficiency in the gels and, most importantly, the use of a trace-clean procedure during sampling, handling and filtration. These issues are sometimes difficult to control under the field conditions of the Arctic environments; as a result, no systematic studies of colloidal size fractionation over different seasons in Arctic estuaries have been performed so far. The in situ dialysis techniques commonly used in boreal rivers and lakes by our group allow for passive, but adequate separation of colloidal (1 kDa–0.22 μm) and truly dissolved (< 1 kDa) fractions, as well as separation by size distribution within the colloidal pool itself (Pokrovsky et al., 2005, 2011, 2012a; Vasyukova et al., 2010). To date, dialysis has been used to assess the size fractionation in freshwaters (Alfaro-De la Torre et al., 2000; Gimpel et al., 2003; Vasyukova et al., 2012), but has only occasionally been applied under estuarine conditions (cf. Hamilton-Taylor et al., 2002); the second objective of this study was to apply this dialysis technique toward studying the Arctic estuary along the full salinity profile. The specific goals here were as follows: (i) to quantify the distribution of organic carbon and TE among the major pools of colloids and truly dissolved fraction and (ii) to trace the evolution of the colloidal fraction along the salinity gradient. Our working hypothesis states that the coagulation of organic and organo-mineral colloids increases the relative proportion of potentially bioavailable, low molecular weight ($LMW_{<1\,kDa}$) fractions while decreasing the proportions of the total dissolved (< 0.22 μm) and colloidal (1 kDa–0.22 μm) fractions. Testing this scenario during different seasons along the salinity gradient of the Severnaya Dvina River constituted an important objective in this work.

2 Sampling and analyses

The sampling area is located in the northern part of NW Russia within the delta of the Severnaya Dvina River, which belongs to the White Sea basin of the Arctic Ocean (Fig. 1). The field work was performed during several hydrological seasons: winter baseflow in March 2009, summer baseflow in July 2010 and 2012 and the start of ice formation/beginning of winter baseflow in November–December 2011 while the water was still free of ice. Detailed geological, climatic and geographic descriptions of the region are presented elsewhere (Shevchenko et al., 2004, 2010; Pokrovsky et al., 2010; Koukina et al., 2001). Sampling was performed from the surface (0.5 m) horizon using a horizontal water sampler made of polycarbonate/PVC (Aquatic Research Co) hung with nylon cord from the PVC motor boat (July) or from the ice (March). During the ice formation period in December 2011, a coastal pilot ship was used during water sample collection.

In total, 74 samples were collected from the Severnaya Dvina River and its mixing zone using filtration through $0.22\,\mu$m acetate cellulose filters. Approximately 80 % of these samples were processed additionally using in situ dialysis (1 kDa or 1, 15 and 50 kDa membranes). The trace metal sampling, filtration, and dialysis, as well as the chemical analyses, are presented in Sect. 1 in the Supplement. These techniques are very similar to the methods used in our previous studies (Pokrovsky et al., 2006, 2011, 2012a; Shirokova et al., 2013; Vasyukova et al., 2010). In situ dialysis was performed using a large volume (5 to 10 L) of unfiltered estuarine surface water placed in a dark room at the temperature at which the sampled water was collected. Regular agitation of the dialysis vessels was ensured. The equilibrium dialysis was performed at a very high ratio of external solution/internal reservoir without any correction for the external reservoir dilution, which allowed assessing of the proportion of colloidal forms operationally defined as the fraction of 1 kDa–$0.22\,\mu$m. The methodology of dialysis procedure as well as various dialysis artifacts are thoroughly described in Sect. S1 in the Supplement.

3 Results

3.1 Elemental concentration as a function of salinity

Major and trace element concentrations at different salinities and size fraction are listed in Table S1 in the Supplement. Several groups of elements could be distinguished according to their concentration profiles, which were maintained throughout all of the studied seasons:

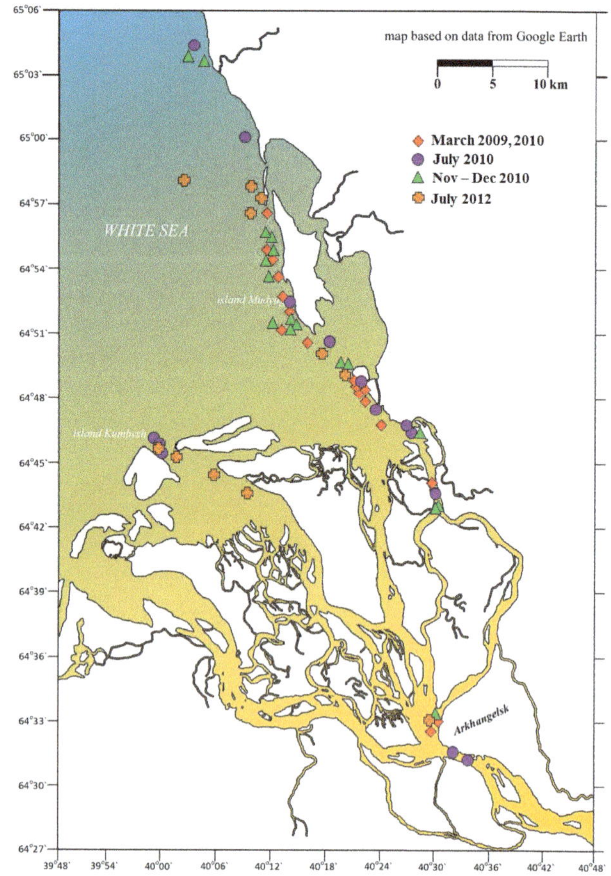

Fig. 1. Schematic map of sampling sites during different seasons. Diamonds, March 2010, 2009; Circles, July 2010, 2012; Triangles, November–December 2011.

1. Elements exhibiting conservative behavior with a strong (up to a factor of 10) linear increase with salinity both in the $LMW_{<1\,kDa}$ and the total dissolved ($<0.22\,\mu$m) fractions: Li, B, Na, Mg, K, Ca, Sr, Rb, Cs, Mo and U, illustrated for B, Mo and U in Fig. 2a–d.

2. Quasi-conservative behavior with a slight or moderate ($\leq$ factor of 3) concentration increase with salinity: V, As, Cr, and Sb, illustrated for the first two elements in Fig. 3a, b.

3. Conservative behavior with strong (> factor of 3) decreases in concentration with the increase in salinity is observed for DOC, Si and Mn (in March), as illustrated in Fig. 4a–c.

4. Conservative behavior with a weak ($\leq$ factor of 2) concentration change or constant level over the full range of salinities, with relative variations within ±30 %: Ni, Cr, As, Co (in March), Cu, Zn, Ga, Ge, Y, heavy REEs (Gd-Yb), as illustrated in Fig. 5a–d for Ni, Zn, Cu and Cr.

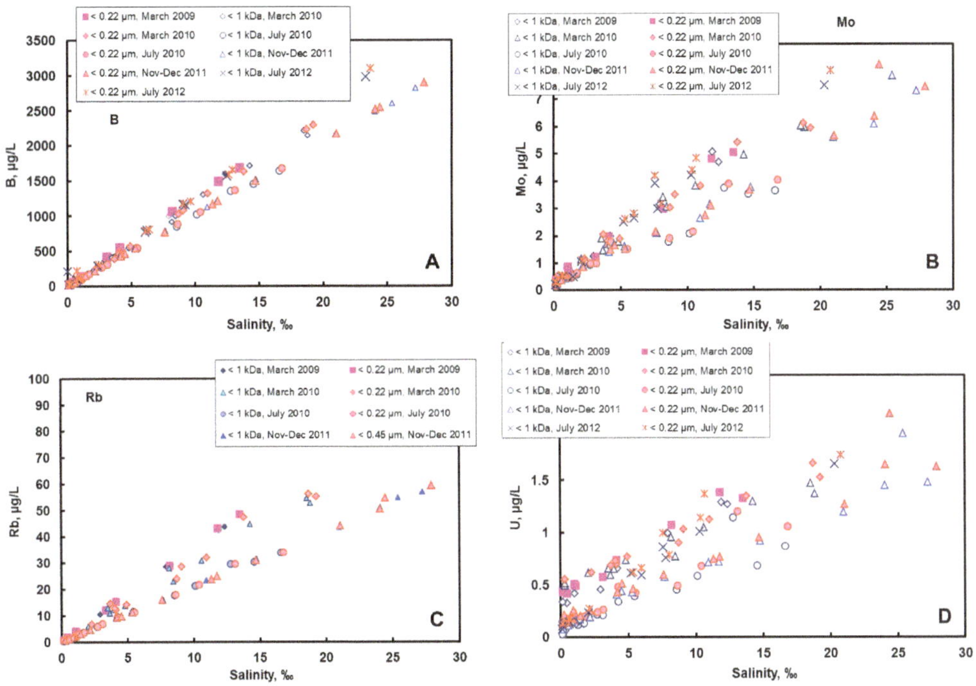

Fig. 2. Boron (**A**), molybdenum (**B**), rubidium (**C**), and uranium (**D**) concentrations in the estuary.

5. Non-conservative behavior with pronounced coagulation and element removal at low (< 5 ‰) salinities: Ti, Fe, Al, Zr, Hf, LREEs (La-Eu), illustrated in Fig. 6a–c for Al, Fe and Zr, respectively.

6. Non-conservative behavior with a concentration gain at low (∼ 2–5 ‰, Ba, Mn) or intermediate (∼ 10–15 ‰, Cd, Zn, Pb) salinities, illustrated for Ba, Cd and Pb in Fig. 7a, b, and c, respectively.

Note that this classification is not exhaustive and the same element, depending on the season, may belong to several groups. For example, V increases linearly with salinity in March but remains constant or only slightly increases during the other seasons; Al increases its concentration in July 2012; REEs that exhibit rather stable concentrations with salinity increase in March; Mn and Ti show both conservative and non-conservative behavior. Note the pronounced fractionation between the light REE and heavy REE in the estuarine mixing zone because it is illustrated by the concentration ratio of La to Yb (Fig. 8): this ratio decreases by almost one order of magnitude from the freshwater to the seawater zone.

3.2 Behavior of colloidal fraction of organic carbon and trace metals along the salinity gradient

Most elements studied in Sect. 3.1 exhibited similar salinity–concentration profiles for < 0.22 µm and < 1 kDa fractions. However, several typical colloid-borne TE and organic carbons demonstrated decoupling in the salinity profiles of the

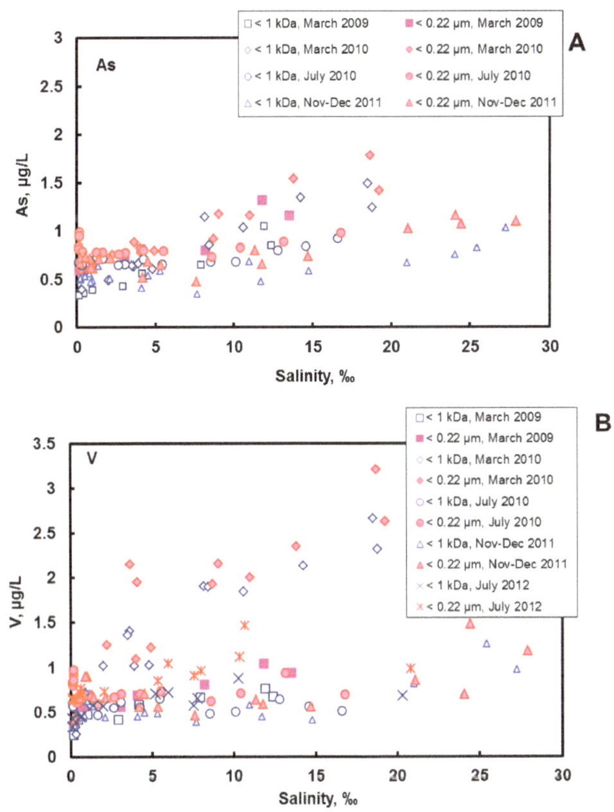

Fig. 3. Arsenic (**A**) and vanadium (**B**) concentrations in the estuary.

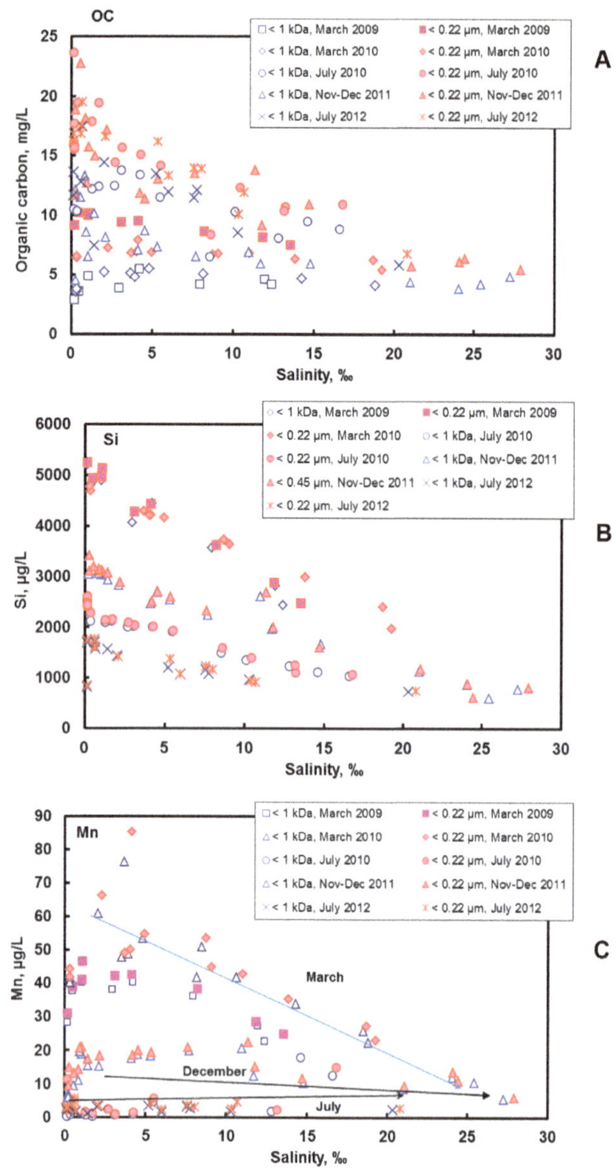

Fig. 4. Organic carbon **(A)**, Si **(B)** and Mn **(C)** concentrations as a function of salinity.

total dissolved and $LMW_{<1\,kDa}$ fractions. The proportion of colloidal organic carbon as a function of the salinity measured over two summers, two winters and one ice formation period is depicted in Fig. 9. The gradual decrease in colloidal OC with increasing salinity occurred during March and November–December, but the concentration of colloids remained rather constant in the estuary during the summer. There was a sharp drop in colloidal carbon within the first 2–3 ‰ in the summer; at higher salinities, the proportion of colloidal OC was the smallest during this period.

The strongest decrease in the colloidal fraction along the salinity gradient was observed for the insoluble trivalent and tetravalent hydrolysates, as illustrated for Al, Fe, Ti and Zr in Fig. 10. This decrease in the colloidal fraction, from ca. 80–90 % of the total dissolved concentration in the freshwater, to 10–20 % at 20–15 ‰ salinity, is the most pronounced for Fe, as recorded through all seasons. Similar to organic carbon, this decrease is the largest in the summer and the smallest during winter-ice formation periods. Large decreases in the colloidal fraction with increasing salinity are also recorded for Mn and Pb; the major changes occurred at low salinities (< 5 ‰), as shown in Fig. 11a and d. In contrast, the decrease was much smaller for Cu and Ni (Fig. 11b and c, respectively), which exhibited a somewhat constant proportion of colloids or only a 10 % decrease over the full range of salinities, respectively. Similarly, oxyanions and neutral molecules linked to colloids, such as V, Cr, As and Sb, demonstrated a general decrease in the % colloids with salinity (Fig. 12), although this decrease was different across the different seasons and most pronounced for As and Cr (Fig. 12b, c). The colloidal fraction of trivalent REEs (Fig. 13) followed the pattern of Fe (for LREEs) and OC (for HREEs). Finally, U strongly decreased its colloidal fraction in the range of low salinities and then remained mostly in dissolved form during all of the studied seasons (Fig. 13). Taken together, the rather uniform distribution of the colloidal fraction versus salinity during the different seasons shown in Figs. 9–13 allows for the size fractionation of major and trace elements to be approximated with an average value calculated for the range of low salinities (0 to 5 ‰, 36 samples) and for the intermediate to high salinities (5 to 30 ‰ , 29 samples), as illustrated in the form of a stack diagram (Fig. 14). Despite the significant uncertainty that stems from the high dispersity of the data points across the different seasons, it is still apparent that both for low- and high-salinity ranges, there are three groups of chemical elements dependent upon their affinity for colloids: (1) < 10–20 % in colloidal fraction: Li, Na, K, Si, Ca, Sr, Rb, Mg, B, Mo, Ba, Cs, P, and U; (2) from 20 to 40 % in colloids: Sb, Mn, Zn, As, V, Y, Ni, Cd, Co, Cr, Ga, OC, Cu, Ti, and Al, and (3) 40–80 % present in colloidal form: Hf, Pb, Zr, all REEs, and Fe. A second important observation is the systematically lower proportion of colloidal material in the 5–30 ‰ saline waters compared to the beginning of the estuary (0–5 ‰). This difference ranges from 5 to 20 % among the different trace elements, with the largest shift being observed for Co and Pb.

3.3 Size fractionation of trace elements among major colloidal pools along the salinity gradient

The use of dialysis membranes with different pore sizes (1, 15, and 50 kDa) during all three studied seasons allowed the quantification of the OC and TE size fractionation along the salinity gradient, from freshwater to seawater as the end-point. Examples of dissolved component size fractionation along the salinity gradient in November–December, July, and March are presented in Figs. 15, 16, and 17, respectively. For this illustration, we have chosen organic carbon, two

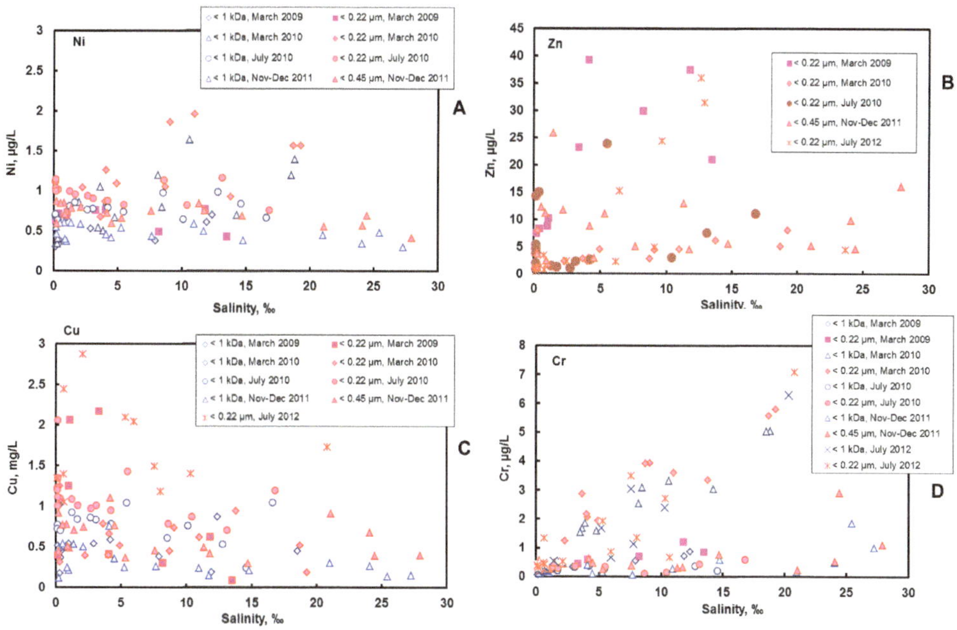

Fig. 5. Ni (**A**), Zn (**B**), Cu (**C**) and Cr (**D**) concentrations as a function of salinity.

insoluble colloidal elements Al and Fe, which also act as carriers of other TE in the form of organo-mineral colloids, U because it changes its speciation from organic colloids to truly dissolved uranyl carbonate complexes, and two metal micronutrients – Cu and Ni.

In November–December, the organic carbon is dominated by $LMW_{<1\,kDa}$ complexes (60 %) and large colloids (50 kDa–0.22 μm) in the freshwater zone. There is a systematic decrease in the large colloids and an increase in the $LMW_{<1\,kDa}$ forms with the appearance of a non-negligible fraction of small colloids (1–15 kDa) in saline environments. In March, the $LMW_{<1\,kDa}$ progressively increases seaward, whereas the $HMW_{50\,kDa–0.22\,\mu m}$ colloidal fraction remains similar between the freshwater and the seawater end point (Fig. 17).

Al, Fe and other tri- and tetravalent insoluble elements mostly appeared as $HMW_{50\,kDa–0.22\,\mu m}$ colloids in the low-salinity zone; this fraction almost disappears (Al) or decreases significantly (Fe) in the high-salinity samples, whereas the $LMW_{<1\,kDa}$ fraction becomes dominant above 9.5 ‰ salinity, which was observed for November–December and July (Figs. 15 and 16). Notably, while the fractions for 1–15 kDa and 15–50 kDa of Fe are minor in the low-salinity range, they constitute 13–19 % of the total dissolved metal at 9.5–24 ‰ salinity.

Uranium exhibits a rapid decrease in the proportion of both medium (1–15 kDa) and high (50 kDa–0.22 μm) molecular weight forms at low-salinity ranges in November–December and in March. During the summer period, this decrease is less pronounced; ca. 10 % of U is still present as 50 kDa–0.22 μm colloids at 10 ‰ salinity.

Divalent metals exhibit a complex pattern of size fractionation as a function of salinity, which resembles neither the organic carbon nor the iron profile. For Ni, the $HMW_{50\,kDa–0.22\,\mu m}$ fraction achieves its maximum in November–December, and in March, the maximum is at 9.0 ± 0.5 ‰; for medium size fractions (15–50 kDa or 1–50 kDa), however, this maximum occurs at 24 ‰ in November–December and at 3.7 ‰ in March. The summer period is characterized by the presence of $\sim$ 40 % of HMW colloids in the freshwater zone and their disappearance with rising salinity at the expense of approximately equal amounts of 1–15 kDa and 15–50 kDa fractions. In contrast to Ni, the HMW of copper disappears above 1.2 ‰ salinity and the medium-size fraction (15–50 kDa) becomes dominant in highly saline water.

4 Discussion

4.1 Conservative versus non-conservative behavior controlled by colloidal coagulation in the estuary

This work demonstrates rather "classic" behavior of major and trace elements in conventionally dissolved (< 0.22 μm) fractions in the mixing zone of the Severnaya Dvina River and the White Sea. The spatial resolution of our measurements does not allow the detection of the flocculation of DOC that is reported to occur within the 2 to 4 ‰ salinity range in the Severnaya Dvina estuary (Artemyev and Romankevich, 1988). A linear decrease in the dissolved organic carbon with salinity increase in the major Arctic estuaries is fairly well known (Kattner et al., 1999; Koehler et al.,

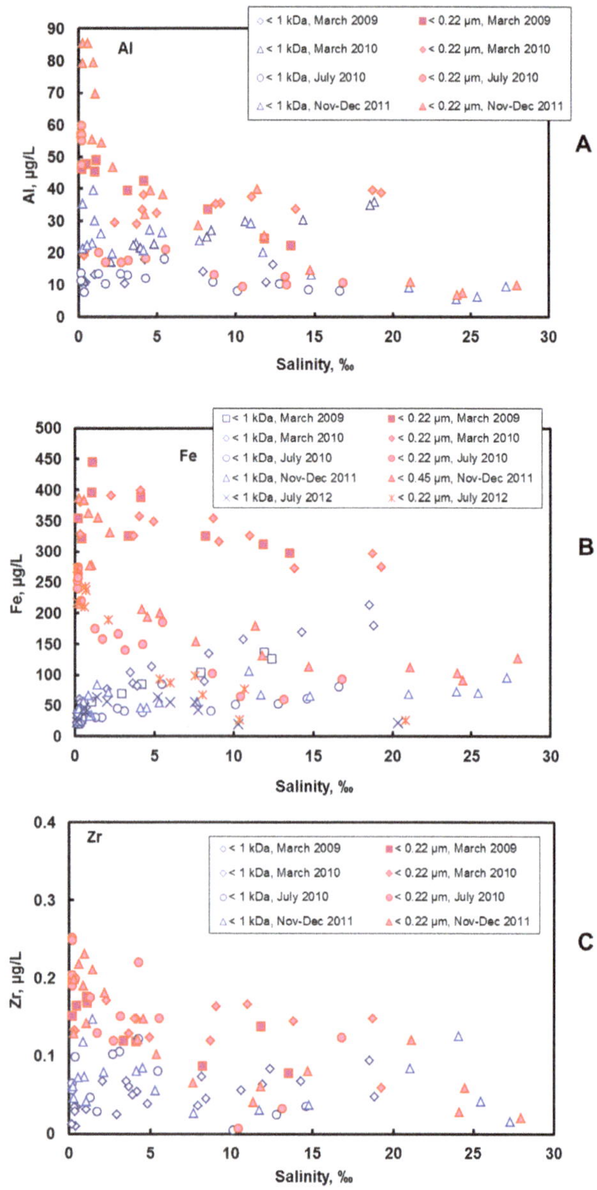

Fig. 6. Al (**A**), Fe (**B**) and Zr (**C**) concentrations as a function of salinity in the estuary.

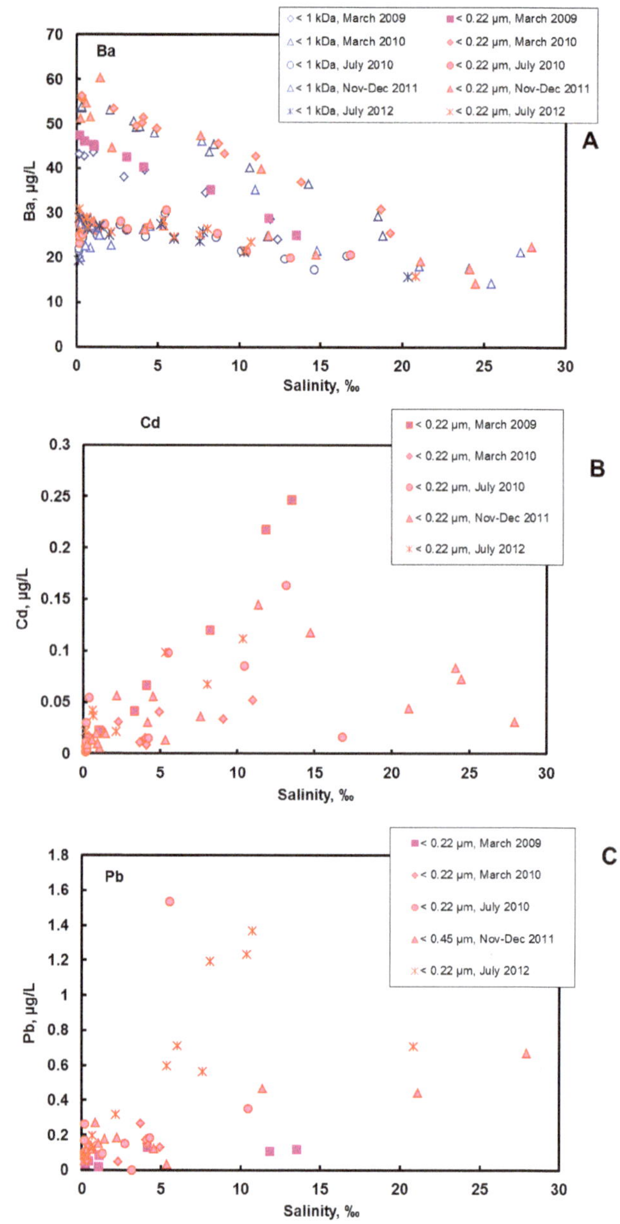

Fig. 7. Ba (**A**), Cd (**B**) and Pb (**C**) concentrations as a function of salinity in the estuary.

2003; Amon et al., 2012). This phenomenon is linked to the rather refractory nature of DOM in the subarctic rivers and the large proportion of $LMW_{<1\,kDa}$ that was not subjected to coagulation compared to large colloids (see Sect. 4.2). In contrast to a productive temperate estuary (i.e., McKenna, 2004), it is very unlikely that the coupled DOC addition and removal may explain the semi-linear dependence of $DOC_{<0.22\,\mu m}$ on salinity. The preservation of this highly conservative behavior takes place during both summer and the glacial season in the Severnaya Dvina estuary (Figs. 4a and 9).

Other examples of the highly conservative behavior observed during the estuarine mixing in the Arctic are boron (Savenko et al., 2003), molybdenum, rubidium, and uranium, the concentrations of which in the seawater are an order of magnitude higher than the content of the freshwater end member. Whereas B, Rb and Mo are unlikely to be affected by colloidal forms (Fig. 2a, b), U does appear in $HMW_{50\,kDa-0.22\,\mu m}$ colloids in the freshwater zone (Figs. 13 and 15–17). However, strong uranyl carbonate complexes start to appear at the beginning of mixing with the seawater, which decreases the affinity of this element for organo-mineral colloid formation and renders it mainly to its truly dissolved form. The removal of dissolved U in the very

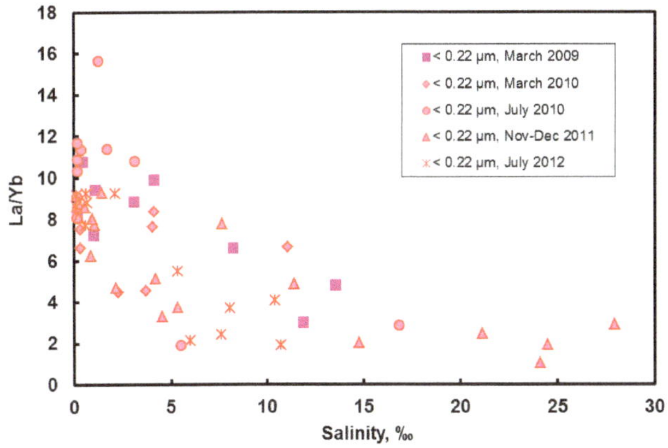

Fig. 8. Plot of the La/Yb 0.22 μm fraction concentration ratio as a function of salinity during different seasons in the Severnaya Dvina estuary.

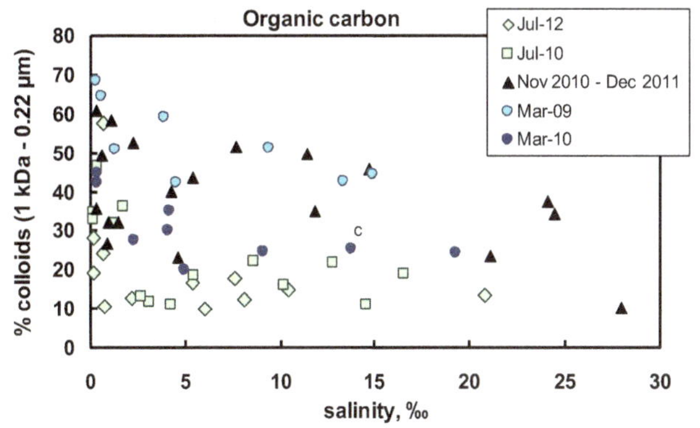

Fig. 9. Relative proportion of colloidal organic carbon as a function of salinity during different seasons in the Severnaya Dvina estuary.

low-salinity zone by the Fe- and organo-rich colloids observed in the Severnaya Dvina River estuary (Figs. 13, 15a, 16) corroborates numerous observations in the Kalix River–Baltic Sea mixing zone (Porcelli et al., 1997; Andersson et al., 1998, 2001), as well as reports for small rivers on the NW German coast (Beck et al., 2012).

Other neutral molecules and oxyanions, such as As, Sb, V, and Cr, demonstrate systematic concentration increases toward the seawater end member (Fig. 3a, b) because these elements are more concentrated in seawater than in the freshwater. This behavior is quite typical for anionic species in the temperate estuaries (i.e., Van der Sloot et al., 1985; Strady et al., 2009). Despite the fact that these elements are bound to 10–30 % in HMW organo-ferric colloids in the Severnaya Dvina River (Pokrovsky et al., 2010), this speciation feature does not influence their concentration pattern, which remains independent for the Fe above ~2 ‰ salinity. Only at the very beginning of the estuarine mixing does a measurable drop (20 to 30 % from the freshwater concentration) occur, as illustrated for As and V in Fig. 3a, b. This drop can be linked to the coagulation of the majority of $HMW_{50\,kDa-0.22\mu m}$ Fe colloids at the beginning of the mixing zone, because the proportion of colloidal forms in total dissolved concentration of As and V achieves 20–25 % (Fig. 14). Chromium is different from other oxyanions, because, as it can be seen in Fig. 5, in addition to the general increase in concentration with increasing salinity, Cr also shows a mid-salinity maximum similar to Ni and Cu. This result is consistent with occurrence of both Cr(III) and Cr(VI) in natural waters (Schroeder and Lee, 1975). The appearance of this maximum may be linked to Cr(III) association with humic substances similar to Ni and Cu or Cr(III) desorption from river-suspended matter similar to other cations.

As expected, Si presents quite conservative behavior during all three studied seasons (Fig. 4b), with the seawater end member being equal to $900 \pm 100\,\mu g\,L^{-1}$ during all seasons and the riverine component ranging from 1800 to $5100\,\mu g\,L^{-1}$. This behavior agrees with numerous previous observations (i.e., Boyle et al., 1974) in the Severnaya Dvina estuary in particular (Savenko and Shevchenko, 2005), and the reported minimal removal of silica in the brackish bay mixing zone of the Baltic Sea (Gustafsson et al., 2000). This result strongly suggests the lack of any active uptake of Si by the diatoms in the mixing zone, as well as a negligible effect on the dissolution of sediments or diagenetic flux. In agreement with the majority of the previous experiments using dialysis and ultrafiltration (Pokrovsky and Schott, 2002; Vasyukova et al., 2010), colloidal Si does not appear in these waters.

The significant coagulation of Al, Fe, and trivalent (REEs) and tetravalent (Ti, Zr, Hf, Th) hydrolysates is pronounced mostly at the very beginning of the estuarine mixing (Figs. 10 and 13) because these elements normally appear as large-size colloids (Figs. 15–17) that are quite unstable in aqueous solution with an elevated ionic strength, as confirmed by numerous observations of the flocculation and removal of colloidal material with increasing salinity in the estuaries (Sholkowitz, 1976, 1978; Sigleo and Helz, 1981). The decrease in the proportion of colloidal Fe with the concomitant increase in salinity is well established from size-fractionation techniques (Kuma et al., 1998; Sañudo-Wilhelmy et al., 1996; Stolpe and Hassellöv, 2007). Insoluble, less mobile elements follow the pattern of Fe tightly, with maximal removal occurring in the low (0–5 ‰) salinity zone. These results are consistent with the dominant status of the trivalent and tetravalent elements in the White Sea watersheds, which occur in the form of large organo-ferric colloids rather than organic complexes capable of controlling the speciation of these elements in the surface boreal waters (Pokrovsky and Schott, 2002; Vasyukova et al., 2010; Pokrovsky et al., 2010, 2012b). The decrease in the La/Yb ratio in the $<0.22\,\mu m$ fraction with rising salinity (Fig. 8) may reflect the removal of the Fe-rich colloids, binding LREE via surface adsorption or

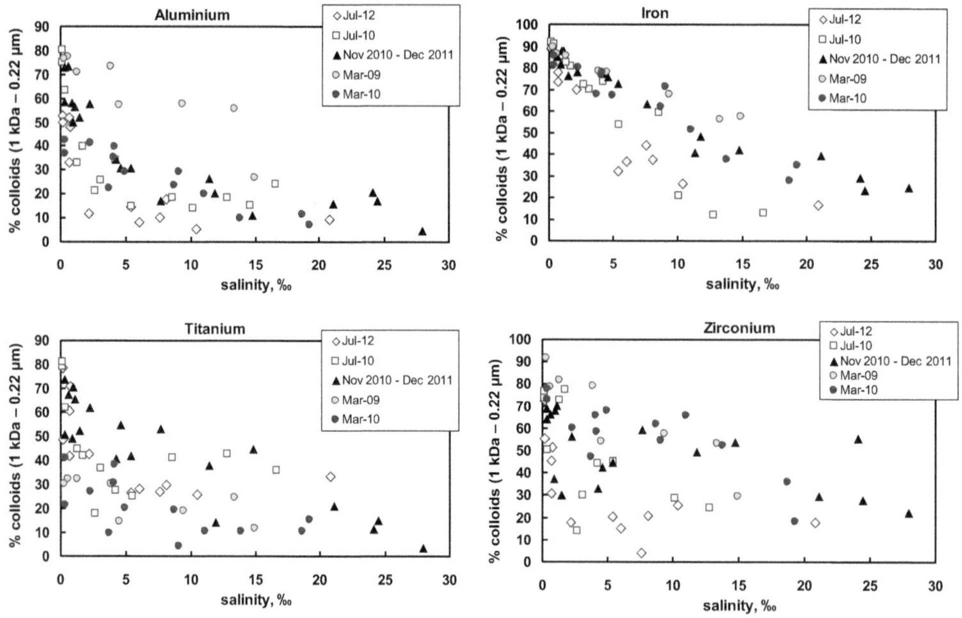

Fig. 10. Proportion of colloidal Al, Fe, Ti and Zr as a function of salinity during different seasons in the Severnaya Dvina estuary.

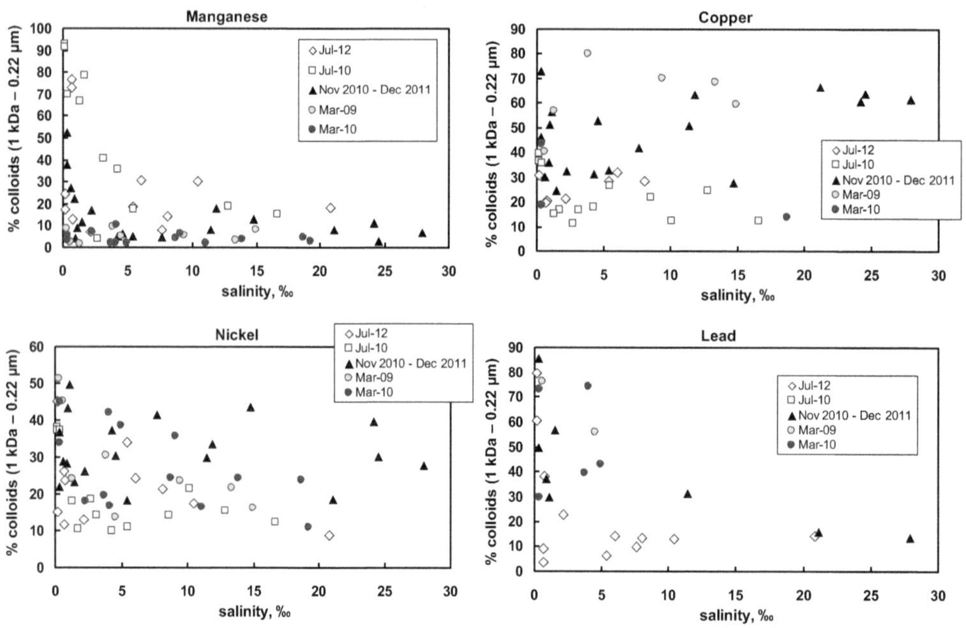

Fig. 11. Proportion of colloidal Mn, Cu, Ni and Pb as a function of salinity during different seasons in the Severnaya Dvina estuary.

incorporation, and the appearance of LMW organic ligands forming strong complexes with HREE. Preferential removal of the light REE was already observed in temperate estuaries (Sholkovitz and Elderfield, 1988; Elderfield et al., 1990) and LREE are known to be removed more efficiently than HREE in small estuaries (Lawrence and Kamber, 2006; Åström et al., 2012). The affinity of light REE for mineral and colloidal surfaces such as HMW organo-ferric colloids, and HREE for soluble organic ligands such as $LMW_{<1\,kDa}$ complexes, is fairly well known both from experimental studies (Bau, 1999) and natural observations in the boreal zone (Pokrovsky et al., 2006). This is also proven from the analysis of filtrates and ultrafiltrates of natural Fe and OC-rich samples (Sholkovitz, 1995, for Hudson river filtrates 0.025 μm, 50 kDa and 5 kDa and Pédrot et al., 2008, for soil solution filtrates 0.22 μm, 30, 5 and 2 kDa).

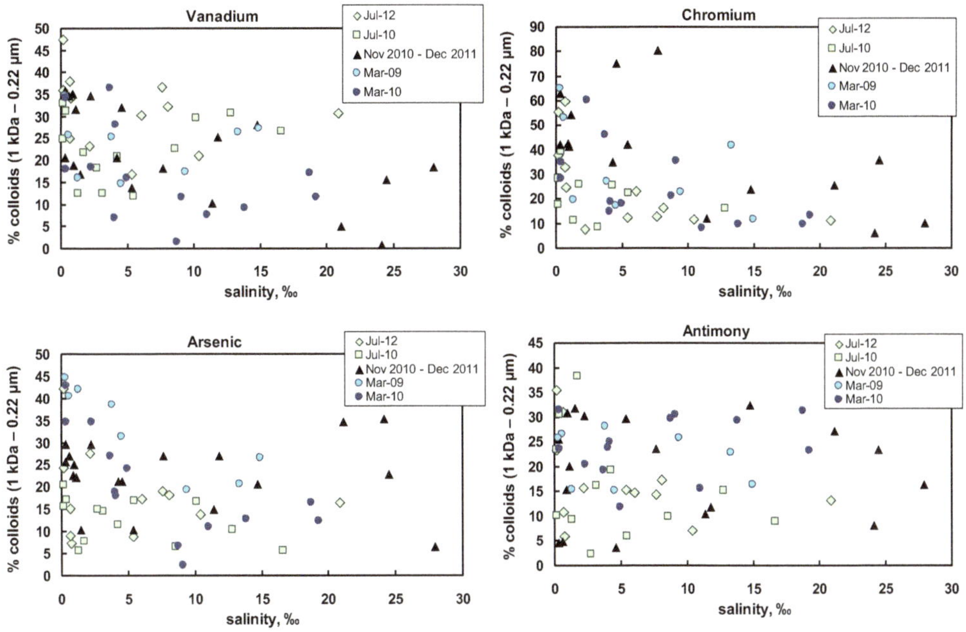

Fig. 12. Proportion of colloidal V, Cr, As and Sb as a function of salinity during different seasons in the Severnaya Dvina estuary.

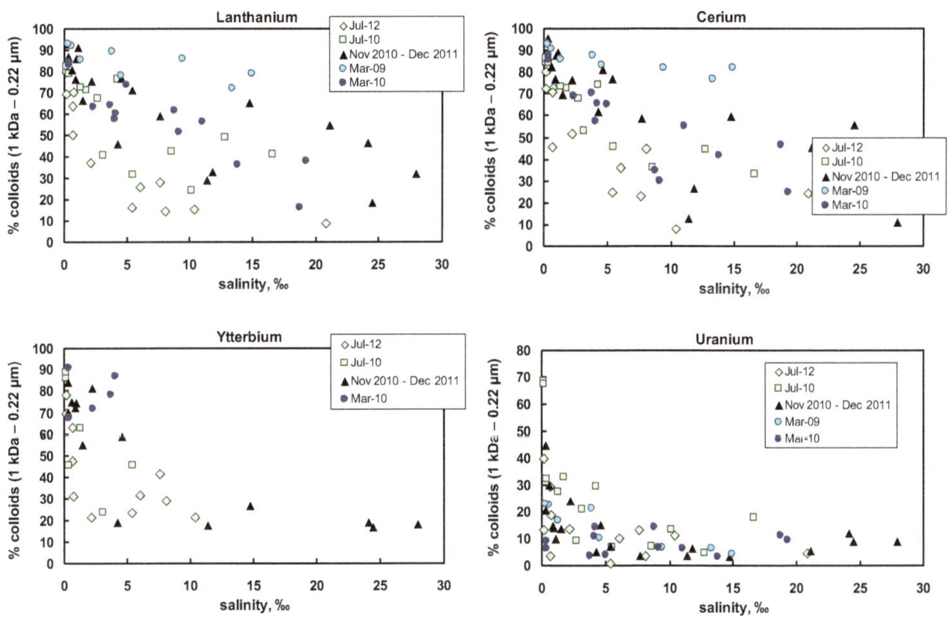

Fig. 13. Proportion of colloidal La, Ce, Yb, and U as a function of salinity during different seasons in the Severnaya Dvina estuary.

The behavior of the divalent transition metals (Ni, Cu, Zn) reflects both seasonal variation in elemental concentration in the freshwater end member, as well as their affinity for organic matter. The concentration–salinity pattern of these elements is quite complex, with generally flat or seaward-decreasing concentration lines with pronounced maxima in the medium-salinity (10–15 ‰) range (Fig. 5a–c). The phytoplankton uptake of these elements is not strongly pronounced in the Severnaya Dvina estuary because the summer-period element concentrations are comparable to the winter time measurement. For the same reason, the desorption of these divalent micronutrients from the river suspended matter (RSM) and bed sediments, which is induced by the inorganic complexation of the metals by seawater anions and also known to occur in the estuaries (Kraepiel et al., 1997; Takata et al., 2012), cannot be proved from our data. Indeed,

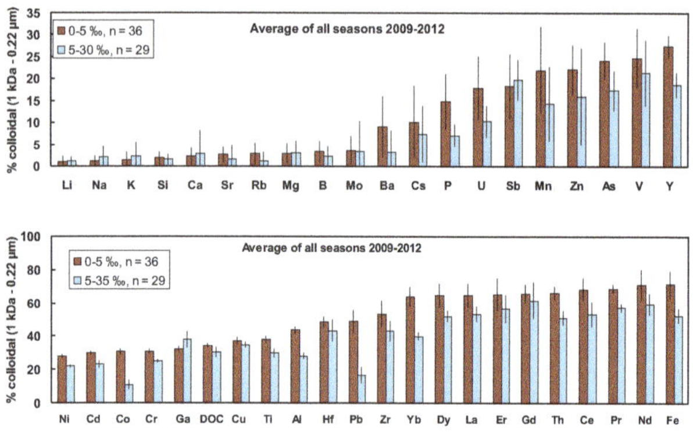

Fig. 14. A stack diagram of colloidal fractions of TE as an average of all seasons in low-salinity and high-salinity zones (dark and light columns, respectively).

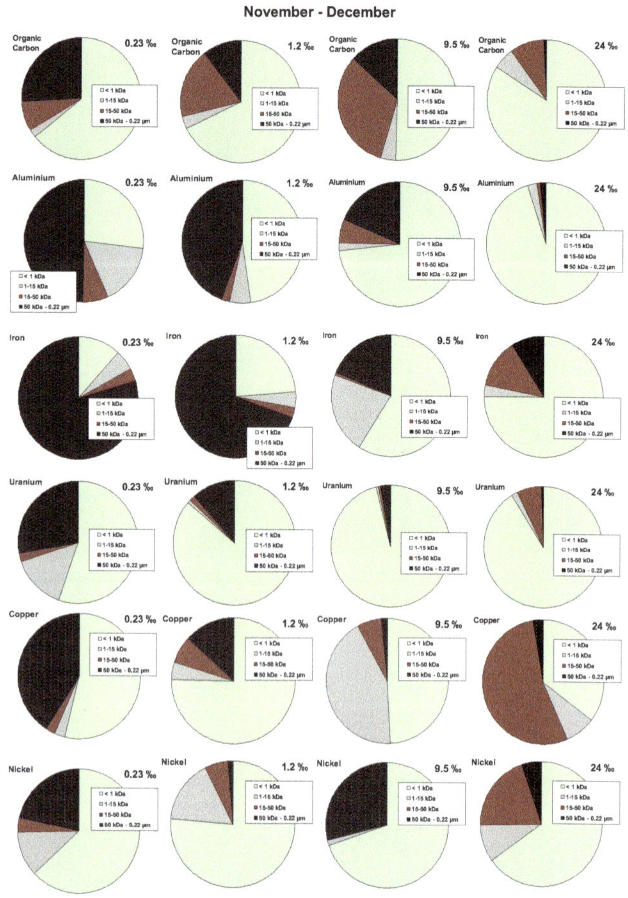

Fig. 15. A diagram of relative fraction of colloidal material in the Severnaya Dvina River estuary during the November–December ice formation period.

the RSM concentration in March is a factor of 2 to 3 lower than that which was observed in the summer (Shevchenko et al., 2004, 2010), yet the concentrations of these three elements exhibit the maximum value in the range of medium salinities during winter baseflow. At the same time, the diffusion flux from the sediments, similar to what is reported for these elements from other world estuaries (Boughriet et al., 1992; Audry et al., 2006, 2007; Point et al., 2007), cannot be excluded because of the existence of several maxima of heavy metal distribution in the sediments of the Severnaya Dvina estuary along the salinity profile (Koukina et al., 2001).

Manganese should be considered separately from other divalent metals because of the following: its concentration (1) strongly varies across the different seasons, reflecting strong accumulation in the winter under the ice because of the enhanced discharge of anoxic groundwater or Mn-rich tributary (Gordeev et al., 2007; Pokrovsky et al., 2010) and (2) depends on the light and oxygen due to the multiple redox states of Mn in boreal waters during the summer (Ingri et al., 2011; Pokrovsky and Shirokova, 2013). The first mechanism is consistent with the peaks of the concentration observed at low (0–5 ‰) salinity ranges that follow the order of March > November–December > July (Fig. 4c). This order also supports the hypothesis that the Mn concentration in the estuary is being controlled by diffusion from the sediments, which is likely to be most pronounced during the winter period. Remobilization of the Mn from the sediments due to strong Mn reductions taking place directly below the sediment–water interface is fairly well established for other Arctic river delta settings (i.e., Nolting et al., 1996) and was also suggested for the Baltic Sea coastal zone (Ingri et al., 2004). The reduction of Mn oxides in the water column itself cannot be excluded either (Klinkhammer and McManus, 2001).

Finally, two indifferent/toxic heavy metals – Cd and Pb – as well as Ba, demonstrated the typical maxima of concentration at moderate (Cd, Pb) and low (Ba) salinities (Fig. 7), which is most likely linked to desorption from the river suspended matter (Shiller and Boyle, 1991; Dai et al., 1995; Elbaz-Poulichet et al., 1996; Shulkin and Bogdanova, 2003), sediments (Stecher III and Kogut, 1999) and/or diffusion flux from the sediments (Point et al., 2007; Du Laing et al., 2009). The particle desorption or the influx from the sediments is known to increase the Cd concentration in the high-salinity zone of temperate estuaries to as much as 360 % of its riverine end member (Waeles et al., 2004) because the seawater Ca^{2+} replaces the Cd^{2+} in the iron/manganese oxide-associated fraction (Waeles et al., 2005a, b). In the case of the Severnaya Dvina estuary, the desorption from the RSM occurring at low salinities is more pronounced in the $LMW_{<1\,kDa}$ fraction (Fig. 7b), demonstrating the importance of the exchangeable (ionic) form of this metal to the estuarine behavior. Despite the high affinity of Pb^{2+} toward the large Fe-rich colloids demonstrated in the boreal waters of the White Sea watershed (Pokrovsky and Schott, 2002; Vasyukova et al., 2010), some temperate rivers (Stolpe et al., 2010) and boreal estuaries (Stolpe and Hassellöv, 2007),

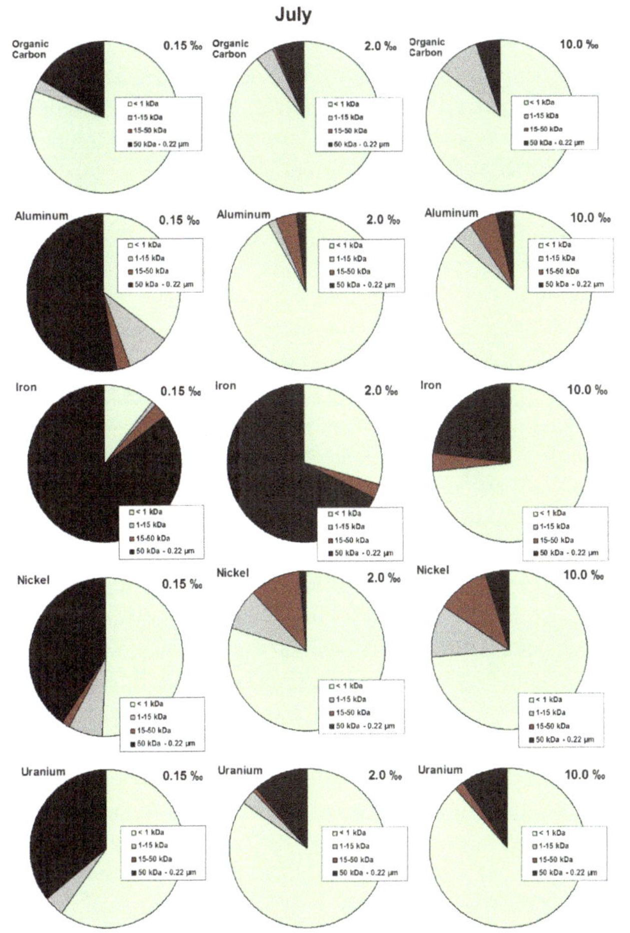

Fig. 16. A diagram of relative fraction of colloidal material in the Severnaya Dvina River estuary during July.

the co-precipitation of Pb with coagulates of organo-ferric colloids at a lower salinity range known for other estuaries (Breuer et al., 1999) is not pronounced in the Severnaya Dvina estuary. Instead we observed an enrichment of the mixing zone in Pb (Fig. 7) likely due to both RSM desorption and sediment mobilization. This effect is most likely due to the low Pb concentration in the riverine end member; however, anthropogenic contamination of estuarine sediments cannot be ruled out, given the highest Pb concentration in clays of the Severnaya Dvina compared to other neighboring subarctic rivers such as the Mezen and Pechora (Kukina et al., 2002).

4.2 Increase in the proportion of $LMW_{<1\,kDa}$ of OC, trace elements and metal micronutrients during estuarine mixing

The first and most significant result is the difference in the degree of conservative or non-conservative behavior between traditional "total" ($<0.22\,\mu$m) and "truly" dissolved $LMW_{<1\,kDa}$ fractions. For the majority of studied metals

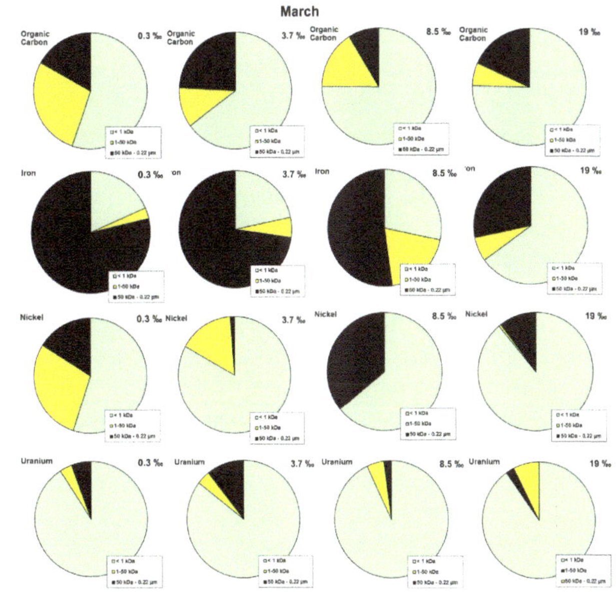

Fig. 17. A diagram of relative fraction of colloidal material in the Severnaya Dvina River estuary during March (winter baseflow).

this result was not counterintuitive because it was demonstrated in earlier studies that the colloidal fraction in the estuary is subjected to the largest degree of coagulation (Dai and Martin, 1995). Additionally, a much smaller decrease in the truly dissolved (< 1 kDa) form of Fe and Cu compared to the $<0.22\,\mu$m fraction when the salinity increases in the coastal zone has been already reported across the range of high salinities (Wells et al., 2000). However, the less expected result is the enrichment of the LMW fraction in many insoluble metals and OC with increased salinity. For example, organic carbon demonstrates an increase in the $LMW_{<1\,kDa}$ fraction in July 2012 and 2010 at approximately 5 ‰ salinity; otherwise, it exhibits a rather flat concentration pattern along the estuary (Fig. 5a). No loss of this fraction occurs; rather, there is some gain in the $LMW_{<1\,kDa}$ fraction at ca. 2 ‰ salinity, presumably due to the autochthonous process of OC transformation/production within the water column. An organic-rich subtropical river in the southeastern US also demonstrated an increase in the LMW fraction of OC at low salinity and relatively flat concentrations towards the seawater end member (Powell et al., 1996); a boreal estuary demonstrated the unchanged concentration of small (0.5–3 nm) fluorescent OM with increasing salt concentrations up to 25 ‰ salinity (Stolpe and Hassellöv, 2007). DOC production during an August phytoplankton bloom in the estuaries is also fairly well known (McKenna, 2004).

Because the increase in the organic carbon $LMW_{<1\,kDa}$ in the Severnaya Dvina River estuary is much stronger and pronounced in July than in winter (Figs. 4a and 9), three main processes may be responsible for it: (i) photochemical oxidation of allochthonous organic matter that produces LMW complexes because it is fairly well known for a number of other boreal and temperate settings (cf. Amon and Benner, 1996; Bertilsson and Tranvik, 2000; Moran et al., 2000; Jonsson et al., 2001; Anesio and Graneli, 2004; Kopáček et al., 2006a, b; Kelton et al., 2007), (ii) heterotrophic mineralization of DOM by aerobic bacterioplankton known for their effect on boreal lakes (cf. Tranvik, 1988; Tranvik and Jørgensen, 1995; Jansson et al., 2000, 2007; Kritzberg et al., 2004; Ask et al., 2008), subtropical marshes (Moran and Hodson, 1990), and temperate estuaries (Raymond and Bauer, 2000), and (iii) the exometabolites of the phytoplankton, whose presence is mostly pronounced at the beginning of the Severnaya Dvina estuarine mixing, in the zone where both marine and freshwater species co-exist (T. Y. Vorobieva, unpublished data); exometabolite production by gram-negative bacteria in the marine zone cannot be ruled out (Wells et al., 1998). The photo degradation of dissolved organic matter is known to produce LMW organic compounds that can significantly stimulate bacterial biomass production (Moran and Zepp, 1997). For example, production of the lower molecular weight species of OC, Fe and metals at higher salinities has been detected in the Trinity Estuary (Galveston Bay, Gulf of Mexico) and suggested to be induced by some photochemical and microbiological reactions (Wen et al., 1999). However, the degree of the increase in the LMW fraction in this tropical organic-poor river ($6\,\mathrm{mg\,L^{-1}}$ DOC in the freshwater zone) was significantly smaller than the organic-rich Severnaya Dvina River estuary.

The most crucial changes in the size fractionation pattern over the estuarine zone occurred with Fe, which is normally present in the freshwater end member of this river as organo-ferric HMW colloids (Pokrovsky et al., 2010), which is consistent with other rivers' speciation studies (Stolpe et al., 2010). It can be observed from Fig. 6b that a clear enrichment of the mixing zone by $\mathrm{LMW}_{<1\,\mathrm{kDa}}$ Fe complexes occurred, with a relative increase between freshwater and 10 ± 5 ‰ salinity as high as a factor of 5. This suggests higher stability of LMW Fe complexes during river water migration through the mixing zone. Indeed, the lifetime of truly dissolved iron in coastal waters is known to be significantly longer than that of colloidal fractions (14 days and < 1 h, respectively, Fujii et al., 2008b). The increase in LMW-reactive Fe fractions at high salinities has been already observed in subtropical, organic-rich estuaries (Powell et al., 1996). On the northwest African coast, LMW ($<0.02\,\mu$m) colloidal Fe decreased its concentration seaward to a smaller degree than that of dissolved ($<0.22\,\mu$m) Fe (Ussher et al., 2010). Experimental studies in the Scheldt estuary demonstrated that, in the range of 0.3 to 10 ‰ salinity, Fe concentration decreased much more sharply for the fraction $<0.2\,\mu$m compared to the fraction < 1 kDa (Gerringa et al., 2007). This analysis demonstrated that, although a smaller degree of coagulation for LMW Fe commonly occurs in the estuary, the increase in this fraction seaward observed in the present study is unprecedented. In the Severnaya Dvina estuary, the maximal increase of $\mathrm{LMW}_{<1\,\mathrm{kDa}}$ Fe fraction was detected during the winter baseflow (Figs. 6b and 15–17). This observation rules out the possibility of superoxide-mediated LMW Fe(II) photochemical formation from organically complexed Fe(III), as it is known for coastal waters from the temperate zone (Kuma et al., 1995; Fujii et al., 2008a). The 50–80 cm ice coverage together with an equal amount of overlaying snow during the Arctic winter period should not allow significant photo-reduction processes. Given that phytoplankton activity during glacial periods is also low, we hypothesize that the process responsible for winter-time increase of LMW compounds is heterotrophic mineralization of allochthonous DOM. Indeed, the efficiency of DOC utilization by bacteria is known to increase with the increase in salinity, as shown by in situ incubation experiments in a temperate estuary (Raymond and Bauer, 2000). Concurrently, the relative decrease in the 1 kDa–$0.22\,\mu$m fraction, i.e., coagulation processes of Fe colloids, is more pronounced during summer, when compared to levels during the winter baseflow (Fig. 10). This result is consistent with the fact that both phytoplankton activity and photo degradation leading to the production of $\mathrm{LMW}_{<1\,\mathrm{kDa}}$ fractions are much more efficient in summer than in winter. Indeed, in the freshwater temperate settings, an increase in LMW carbohydrates during summer time was reported (Wilkinson et al., 1997). Additionally, bacterial consumption of DOC is known to increase with temperature increase in the estuary (Raymond and Bauer, 2000). Production of the LMW (0.3–1 kDa) Fe(III) siderophores by marine prokaryotes is fairly well known (Reid et al., 1993; Barbeau et al., 2001) and strong LMW complexes of this metal need to be reported in sea surface waters (Wen et al., 2006). The effect of redox processes should not be pronounced for dissolved Fe in the Severnaya Dvina estuary because the majority (> 95 %) of this element is present in its trivalent form, as determined from spectrophotometric analysis (Pokrovsky et al., 2010, and references therein). This finding is consistent with the other data, indicating that all dissolved Fe in freshwater appears as Fe-humic complexes or aggregates of Fe hydroxides and humic-Fe (Allard et al., 2004; Lofts et al., 2008; Liu et al., 2013). Another reason for more pronounced total dissolved concentration of OC and Fe decrease in July compared to March is that the river contained a higher concentration of aquatic organic polymers in summer than in winter (i.e., Wilkinson et al., 1997), thus enhancing the aggregation of colloids in July.

The behavior of other insoluble elements linked to organo-ferric colloids (Al, Ti, Zr, REEs, Figs. 10 and 13), as well as As, Cr, Pb and U (Figs. 12 and 13), follows that of Fe with the (i) strong decrease in the colloidal fraction across all seasons, which is linked to coagulation and the mechanical removal of these elements in the form of Fe-rich precipitates; the (ii) intense processes of colloidal transformation during the summer and the appearance of a significant fraction of the $\mathrm{LMW}_{<1\,\mathrm{kDa}}$ forms. This estuarine behavior corroborates with the strong control over the migration of these elements as organo-ferric colloids, which is known for the freshwater settings of the White Sea watershed in general (Pokrovsky and Schott, 2002) and the Severnaya Dvina River in particular (Pokrovsky et al., 2010). At the same time, the chemical nature of strong metal chelators, including LMW forms, which might serve as transports for bio-available iron in the rivers, estuaries and coastal waters have not yet been identified (cf., Krachler et al., 2005; Boye et al., 2010).

Divalent metals (Cu, Ni, Mn, Pb) also exhibit a rapid decrease in their colloidal forms in the summer at the very beginning of the mixing zone and a much more gradual decrease in the colloidal forms during the winter period (Fig. 11). The highly dynamic behavior of the colloidal divalent metals (Cd, Co, Cu, Ni, Pb, Zn) during the summer in the Trinity River estuary has been reported by Wen et al. (1999). The abiotic coagulation of colloids should be similar across the different seasons and thus does not allow explaining of the seasonal variation of colloidal fractions. In contrast, either photo-induced degradation or respiration by bacterioplankton – both being intensified during the ice-free summer period and warmer water temperatures – can be invoked to control the divalent metal speciation in the mixing zone. In addition, the contribution of the LMW phytoplankton exudates to the total dissolved pool of metals can decrease the proportion of the colloidal fraction. This decrease is much smaller for Cu relative to Mn, Pb and Ni (Fig. 11), which is consistent with a very strong complexation of this metal by LMW organic ligands of both terrestrial origin (Shank et al., 2004; Krachler et al., 2005) and from phytoplankton exometabolism (van den Berg et al., 1987; Muller, 1998; Waeles et al., 2005a; Wen et al., 2011). Strong Cu^{2+} binding to humic complexes across the full range of salinity is also known from dialysis mixing experiments (Hamilton-Taylor et al., 2002). In contrast, the similar pattern of % colloids vs. salinity for Mn and Pb (Fig. 11), which are two strongly opposing metals in terms of their biological affinity and the degree of their complexation by DOM (Powell et al., 1996; Kozelka et al., 1997), is not supportive of active biological control of metal complexation due to the LMW phytoplankon exometabolites in the mixing zone.

4.3 Perspective on carbon and trace element delivery to the Arctic Ocean under a climate change scenario

The increase in the concentration of the $LMW_{<1\,kDa}$ fraction in the seaward direction highlighted for the first time in this work may have significant and unexpected consequences for the biogeochemical functioning of the mixing zone. First, this LMW fraction is potentially bioavailable because the pore sizes of cell wall transport channels and pores of the 1 kDa dialysis membranes are comparable (see discussion in Vasyukova et al., 2012). Therefore, a number of metal micronutrients (Fe, Co, Ni, Mn, Zn, V, etc.) should enrich the coastal zone with their potentially bioavailable fraction and thus affect the primary productivity of the Arctic Ocean. Second, the LMW fraction of the trace metals in the river water exhibits isotopic signatures drastically different from those of conventionally dissolved $<0.22\,\mu m$ fraction, as is known for Fe (Ilina et al., 2013a), Sr (Ilina et al., 2013b) and U (Bagard et al., 2011). For example, the $LMW_{<1\,kDa}$ of the river water is 2 to 4 ‰ enriched in ^{57}Fe relative to the $<0.22\,\mu m$ fraction. The unusually high mobility of the $LMW_{<1\,kDa}$ fraction of Fe through the mixing zone is capable of enriching the Arctic Ocean in the heavy isotope of various metals relative to what might be predicted from the flux of the $<0.22\,\mu m$ fraction.

Upon the climate warming, the ongoing increase in the LMW metal and OC concentration occurring in the White Sea watersheds (cf., Shirokova et al., 2013; Pokrovsky et al., 2013) will be further accentuated in the estuaries because this LMW pool passes through the mixing zone without coagulation. This phenomenon might influence the following: (i) the primary productivity of the coastal zone, leading to CO_2 drawdown from the atmosphere, and (ii) heterotrophic respiration of the allochthonous DOC accompanied by CO_2 release into the atmosphere. However, distinguishing quantitatively between these two processes is not yet possible. For organic carbon, the smallest proportion of colloids (1 kDa–0.22 μm) is observed during summer across the entire salinity zone (Fig. 9). The insoluble elements (Al, Fe, Zr, REEs) and divalent metals (Co, Ni, Cu, Pb) follow this scheme, therefore suggesting that the heating of the surface water and extension of the active (vegetative) ice-free period should decrease the proportion of colloids versus "truly" dissolved components of the $LMW_{<1\,kDa}$ fraction.

Finally, the majority of annual colloidal flux in the Severnaya Dvina River occurs during the spring flood period (Pokrovsky et al., 2010). The most important transformation processes of the allochthonous organic matter and related elements should therefore occur in May (i.e., Gustafsson et al., 2000), which is the period not yet investigated using this size fractionation technique along the salinity gradient in Russian Arctic rivers. The particularity of colloidal composition during spring flood on the Severnaya Dvina River has been thoroughly described in Pokrovsky et al. (2010). It consists in the higher proportion of colloidal (1 kDa–0.22 μm) fraction compared to winter and summer baseflow and a factor of 2 to 3 higher concentration of 0.22 μm filtered DOC, Fe and a number of insoluble trivalent and tetravalent hydrolysates. As a result, one may expect significantly more pronounced colloidal coagulation in the estuarine zone during the spring flood in May compared to winter and summer baseflow. On the annual scale, consideration the spring flood period would certainly increase the flux of LMW fraction of OC and TE to the Arctic Ocean. Given highly pronounced Arctic summer insolation, low water temperatures and low phytoplankton activity in May, the relative order of the processes controlling the production of $LMW_{<1\,kDa}$ fraction may be as follows: photodestruction > heterotrophic mineralization > exometabolite production. On the other hand, due to significantly higher discharge in spring, the water residence time in the estuary becomes shorter and thus the time necessary for colloidal coagulation might be insufficient to ensure the non-conservative behavior of insoluble elements. Thus, Fe-rich HMW colloids may travel further seaward, north of Mud'yug Island (Fig. 1) due to significantly higher freshwater zone influence.

5 Conclusions

Systematic sampling of the Severnaya Dvina River mixing zone within the White Sea (European subarctic) revealed several families of major and trace elements, depending on their conservative or non-conservative behavior, which was in accordance with previous studies of both temperate and tropical estuaries. The in situ dialysis of estuarine water performed for the first time with membranes having different pore sizes (1, 15, and 50 kDa) allowed the characterization of the complexity of the colloidal size distribution and allowed us to trace the transformation of HMW into LMW fractions along the estuarine profile during different seasons. The coagulation of $HMW_{50\,kDa–0.22\,\mu m}$ colloids at the beginning (< 5 ‰) of the mixing zone is especially pronounced for the insoluble TE bound to ferric colloids in the river water (Al, Ti, Zr, Hf, Th, REEs, Pb, Mn (in winter), U) and more rarely observed for divalent metals initially present as LMW organic complexes (Cu, Ni) or oxyanions (As, Sb, V, Cr). At the same time, the enrichment of the estuarine surface water in the $LMW_{<1\,kDa}$ organic fraction seaward for most insoluble elements, such as Fe, is a novel and unexpected finding. Therefore, the $LMW_{<1\,kDa}$ fraction of DOC, Fe, Al, and other insoluble trace metals is accumulated in the estuary, despite significant coagulation of the colloidal fraction. The climate warming at high latitudes is likely to accentuate the enrichment of the coastal waters in the $LMW_{<1\,kDa}$ fraction by decreasing the degree of colloidal coagulation in the estuarine mixing zone.

Acknowledgements. This work was supported by the BIO-GEOCLIM mega-grant of the Ministry of Education and Science of the Russian Federation and Tomsk State University (no. 14.B25.31.0001), PICS No 6063, ANR "Arctic Metals", RFBR grants 13-05-00890, 11-05-98810-p, and 12-05-91055_a, RAS Presidium programs (no. 12-5-4-007 Arktika) and UrORAS (no. 12-U-5-1034 and no. 14-05-31533).

Edited by: E. J. M. Delhez

References

Alfaro-De la Torre, M. C., Beaulieu, P. Y., and Tessier, A. T.: In situ measurement of trace metals in lakewater using the dialysis and DGT techniques, Anal. Chim. Acta, 418, 53–68, 2000.

Allard, T., Menguy, N., Salomon, J., Calligaro, T., Weber, T., Calas, G., and Benedetti, M. F.: Revealing forms of iron in river-borne material from major tropical rivers of the Amazon Basin (Brazil), Geochim. Cosmochim. Ac., 68, 3079–3094, 2004.

Amon, R. M. W. and Benner, R.: Photochemical and microbial consumption of dissolved organic carbon and dissolved oxygen in the Amazon River System, Geochim. Cosmochim. Ac., 60, 1783–1792, 1996.

Amon, R. M. W., Rinehart, A. J., Duan, S., Louchouarn, P., Prokushkin, A., Guggenberger, G., Bauch, D., Stedmon, C., Raymond, P. A., Holmes, R. M., McClelland, J. W., Peterson, B. J., Walker, S. A., and Zhulidov, A. V.: Dissolved organic matter sources in large Arctic rivers, Geochim. Cosmochim. Ac., 94, 217–237, 2012.

Andersson, P. S., Porcelli, D., Wasserburg, G. J., and Ingri, J.: Particle transport of $^{234}U/^{238}U$ in the Kalix river and in the Baltic Sea, Geochim. Cosmochim. Ac. 62, 385–392, 1998.

Andersson, P. S., Porcelli, D., Gustafsson, O., Ingri, J., and Wasserburg, G. J.: The importance of colloids for the behavior of uranium isotopes in the low-salinity zone of a stable estuary, Geochim. Cosmochim. Ac., 65, 13–25, 2001.

Anesio, A. M. and Graneli, W.: Photochemical mineralization of dissolved organic carbon in lakes of different pH and humic content, Archiv für Hydrobiologie, 160, 105–116, 2004.

Artemyev, V. E. and Romankevich, E. A.: Seasonal variations in the transport of organic matter in North Dvina Estuary, Mitt. Geol.-Paläont. Inst. Univ. Hamburg, SCOPE/UNEP Sonderband Heft, 66, 177–184, 1988.

Ask, J., Karlsson, J., Persson, L., Ask, P., Byström, P., and Jansson, M.: Whole-lake estimates of carbon flux through algae and bacteria in benthic and pelagic habitats of clear-water lakes, Ecology, 90, 1923–1932, 2008.

Åström, M. E., Österholm, P., Gustafsson, J. P., Nystrand, M., Peltola, P., Nordmyr, L., and Boman, A.: Attenuation of rare earth elements in a boreal estuary, Geochim. Cosmochim. Ac., 96, 105–119, 2012.

Audry, S., Blanc, G., Schäfer, J., Chaillou, G., and Robert, S.: Early diagenesis of trace metals (Cd, Cu, Co, Ni, U, Mo, and V) in the freshwater reaches of a macrotidal estuary, Geochim. Cosmochim. Ac., 70, 2264–2282, 2006.

Audry, S., Blanc, G., Schäfer, J., and Robert, S.: Effect of estuarine sediment resuspension on early diagenesis, sulfide oxidation and dissolved molebdenum and uranium distribution in the Gironde estuary, France, Chem. Geol., 238, 149–167, 2007.

Bagard, M. L., Chabaux, F., Pokrovsky, O. S., Prokushkin, A. S., Viers, J., Dupré, B., and Stille, P.: Seasonal variability of element fluxes in two Central Siberian rivers draining high latitude permafrost dominated areas, Geochim. Cosmochim. Ac., 75, 3335–3357, 2011.

Barbeau, K., Rue, E., Bruland, K., and Butler, A.: Photochemical cycling of iron in the surface ocean mediated by microbial iron (III)-binding ligands, Nature, 413, 409–413, 2001.

Bau, M.: Scavenging of dissolved yttrium and rare earths by precipitating iron oxyhydroxide: Experimental evidence for Ce oxidation, Y-Ho fractionation, and lanthanide tetrad effect, Geochim. Cosmochim. Ac., 63, 67–77, 1999.

Beck, M., Dellwig, O., Fischer, S., Schnetger, B., and Brumsack, H.-J.: Trace metal geochemistry of organic-rich watercourses draining the NW German coast, Estuar. Coast. Shelf S., 104–105, 66–79, 2012.

Benoit, G., Oktay-Marshall, S. D., Cantu II, A., Hood, E. M., Coleman, C. H., Corapcioglu, M. O., and Santschi, P. H.: Partitioning of Cu, Pb, Ag, Zn, Fe, Al, and Mn between filter-retained particles, colloids, and solution in six Texas estuaries, Mar. Chem., 45, 307–336, 1994.

Bertilsson, S. and Tranvik, L. J.: Photochemical transformation of dissolved organic matter in lakes, Limnol. Oceanogr., 45, 753–62, 2000.

Bianci, T. S.: Biogeochemistry of Estuaries, Oxford University Press, Oxford, United Kingdom, 706 pp., 2007.

Bianci, T. S. and Allison, M. A.: Large-river delta-front estuaries as natural "recorders" of global environmental change, P. Natl. Acad. Sci. USA, 106, 8085–8092, 2009.

Boughriet, A., Ouddane, B., Fischer, J. C., Wartel, M., and Leman, G.: Variability of dissolved Mn and Zn in the Seine estuary and chemical speciation of these metals in suspended matter, Water Res., 26, 1359–1378, 1992.

Boye, M., Nishioka, J., Croot, P., Laan, P., Timmermans, K. R., Strass, V. H., Takeda, S., and de Baar, H. J. W.: Significant portion of dissolved organic Fe complexes in fact is Fe colloids, Mar. Chem., 122, 20–27, 2010.

Boyle, E., Collier, R., Dengler, A. T., Edmond, J. M., Ng, A. C., and Stallard, R. F.: On the chemical mass-balance in estuaries, Geochim. Cosmochim. Ac., 38, 1719–1728, 1974.

Boyle, E., Edmond, J. M., and Sholkowitz, E. R.: The mechanism of iron removal in estuaries, Geochim. Cosmochim. Ac., 41, 1313–1324, 1977.

Breuer, E., Sañudo-Wilhelmy, S. A., and Aller, R. C.: Trace metals and dissolved organic carbon in an estuary with restricted river flow and a brown tide bloom, Estuaries, 22, 603–615, 1999.

Dai, M. and Martin, J.: First data on trace metal level and behaviour in two major Arctic river-estuarine systems (Ob and Yenisey) and in the adjacent Kara Sea, Russia, Earth Planet. Sc. Lett., 131, 127–141, 1995.

Dai, M., Martin, J., and Cauwet, G.: The significant role of colloids in the transport and transformation of organic carbon and associated trace metals (Cd, Cu and Ni) in the Rhône delta (France), Mar. Chem., 51, 159–175, 1995.

Du Laing, G., Rinklebe, J., Vandecasteele, B., Meers, E., and Tack, F. M. G.: Trace metal behavior in estuarine and riverine floodplain soils and sediments: A review, Sci. Total Environ., 407, 3972–3985, 2009.

Elbaz-Poulichet, F., Garnier, J.-M., Guan, D. M., Martin, J.-M., and Thomas, A. J.: The conservative behaviour of trace metals (Cd, Cu, Ni and Pb) and As in the surface plume of stratified estuaries: example of the Rhône River (France), Estuar. Coast. Shelf S., 42, 289–310, 1996.

Elderfield, H., Upstill-Goddard, R., and Sholkovitz, E. R.: The rare earth elements in rivers, estuaries, and coastal seas and their significance to the composition of ocean waters, Geochim. Cosmochim. Ac., 54, 971–991, 1990.

Forsberg, J., Dahlqvist, R., Gelting-Nystrom, J., and Ingri, J.: Trace metal speciation in Brackish water using diffusion gradients in thin films and ultrafiltration: comparison of techniques, Environ. Sci. Technol., 40, 3901–3905, 2006.

Fujii, M., Ito, H., Rose, A. L., Waite, D. T., and Omura, T.: Superoxide-mediated Fe(II) formation from organically complexed Fe(III) in coastal waters, Geochim. Cosmochim. Ac., 72, 6079–6089, 2008a.

Fujii, M., Ito, H., Rose, A. L., Waite, D. T., and Omura, T.: Transformation dynamics and reactivity of dissolved and colloidal iron in coastal waters, Marine Chem., 110, 165–175, 2008b.

Gerringa, L. J. A., Rijkenberg, M. J. A., Wolterbeek, H. Th., Verburg, T. G., Boye, M., and de Baar, H. J. W.: Kinetic study reveals weak Fe-binding ligand, which affects the solubility of Fe in the Scheldt estuary, Mar. Chem., 103, 30–45, 2007.

Gimpel, J., Zhang, H., Davison, W., and Edwards, A. C.: In situ metal speciation in lake surface waters using DGT, dialysis, and filtration, Environ. Sci. Technol., 37, 138–146, 2003.

Gobeil, C.: Biogeochemistry and chemical contamination in the St. Lawrence Estuary, The Handbook of Environmental Chemistry, Pt. H, 5, 121–147, doi:10.1007/698_5_023, 2006.

Gordeev, V. V.: Geochemistry of the river-sea system, Moskow, Russia, 452 pp., 2012.

Gordeev, V. V., Martin, J.-M., Sidorov, I. S., and Sidorova, M. V.: A reassessment of the Eurasian river input of water, sediment, major elements, and nutrients to the Arctic Ocean, Am. J. Sci., 296, 664–691, 1996.

Gordeev, V. V., Beeskow, B., and Rachold, V.: Geochemistry of the Ob and Yenisey estuaries: A comparative study, Berichte zur Polar-und Meeresforschung, AWI-Bremerhaven, Germany, 565, 235 pp., 2007.

Greenamoyer, J. M. and Moran, S. B.: Investigation of Cd, Cu, Ni and ^{234}Th in the colloidal size range in the Gulf of Maine, Mar. Chem., 57, 217–226, 1997.

Guieu, C. and Martin, J. M.: The level and fate of metals in the Danube River plume, Estuar. Coast. Shelf S., 54, 501–512, 2002.

Guieu, C., Huang, W. W., Martin, J.-M., and Yong, Y. Y.: Outflow of trace metals into the Laptev Sea by the Lena River, Mar. Chem., 53, 255–267, 1996.

Gustafsson, O., Widerlund, A., Andersson, P. S., Ingri, J., Roos, P., and Ledin, A.: Colloid dynamics and transport of major elements through a boreal river – brackish bay mixing zone, Mar. Chem., 71, 1–21, 2000.

Hamilton-Taylor, J., Postill, A. S., Tipping, E., and Harper, M. P.: Laboratory measurements and modelling of metal-humic interactions under estuarine conditions, Geochim. Cosmochim. Ac., 66, 403–415, 2002.

Holmes, R. M., McClelland, J. W., Peterson, B. J., Tank, S. E., Bulygina, E., Eglinton, T. I., Gordeev, V. V., Gurtovaya, T. Y., Raymond, P. A., Repeta, D. J., Staples, R., Striegl, R. G., Zhulidov, A. V., and Zimov, S. A.: Seasonal and annual fluxes of nutrients and organic matter from large rivers to the Arctic Ocean and surrounding seas, Estuar. Coast., 35, 369–382, doi:10.1007/s12237-011-9386-6, 2012.

Howell, K. A., Achterberg, E. P., Tappin, A. D., and Worsfold, P. J.: Colloidal metals in the Tamar Estuary and their influence on metal fractionation by membrane filtration, Environ. Chem., 3, 199–207, 2006.

Ilina, S. M., Poitrasson, F., Lapitskiy, S. A., Alekhin, Yu. V., Viers, J., and Pokrovsky, O. S.: Extreme iron isotope fractionation between colloids and particles of boreal and temperate organic-rich waters, Geochim. Cosmochim. Ac., 101, 96–111, 2013a.

Ilina, S. M., Viers, J., Mialle, S., Mavromatis, V., Brunet, P., Lapitskiy, S. A., Alekhin, Y. V., and Pokrovsky, O. S.: Stable (Cu, Mg) and radiogenic (Sr, Nd) isotope fractionation in colloids of boreal organic-rich waters, Chem. Geol., 342, 63–75, 2013b.

Ingri, J., Nordling, S., Larsson, J., Rönnegård, J., Nilsson, N., Rodushkin, I., Dahlqvist, R., Andersson, P., and Gustafsson, Ö.: Size distribution of colloidal trace metals and organic carbon during a coastal bloom in the Baltic Sea, Mar. Chem., 91, 117–130, 2004.

Ingri, J., Pekka, L., Dauvalter, V., Rodushkin, I., and Peinerud, E.: Manganese redox cycling in Lake Imandra: impact on nitrogen and the trace metal sediment record, Biogeosciences Discuss., 8, 273–321, doi:10.5194/bgd-8-273-2011, 2011.

Jansson, M., Bergström, A. K., Blomqvist, P., and Drakare, S.: Allochthonous organic carbon and phytoplankton/bacterioplankton production relationship in lakes, Ecology, 81, 3250–3255, 2000.

Jansson, M., Persson, L., De Roos, A. M., Jones, R. I., and Tranvik, L. J.: Terrestrial carbon and intraspecific size-variation shape lake ecosystems, Trends Ecol. Evol., 22, 316–322, 2007.

Jonsson, A., Meili, M., Bergström, A. K., and Jansson, M.: Whole-lake mineralization of allochthonous organic carbon in a large humic lake (Ortrasket, N. Sweden), Limnol. Oceanogr., 46, 1691–700, 2001.

Kattner, G., Lobbes, J., Fritznar, H. P., Engbrodt, R., and Lara, R. J.: Tracing dissolved organic substances and nutrients from the Lena River through Laptev Sea (Arctic), Mar. Chem., 65, 25–39, 1999.

Kelton, N., Molot, L. A., and Dillon, P. J.: Effect of ultraviolet and visible radiation on iron lability in boreal and artificial waters, Aquat. Sci., 69, 86–95, 2007.

Klinkhammer, G. P. and McManus, J.: Dissolved manganese in the Columbia River estuary: Production in the water column, Geochim. Cosmochim. Ac., 65, 2835–2841, 2001.

Koehler, H., Meon, B., Gordeev, V. V., Spitzy, A., and Amon, R. M. W.: Dissolved organic matter (DOM) in the estuaries of Ob and Yenisei and the adjacent Kara-Sea, Russia, Proceed. Marine Sci., 6, 281–309, 2003.

Kopáček, J., Klementová, S., and Norton, S. A.: Photochemical production of ionic and particulate aluminium and iron in lakes, Environ. Sci. Technol., 39, 3656–3662, 2006a.

Kopáček, J., Marešova, M., Norton, S. A., Porcal, P., and Vesely, J.: Photochemical source of metals for sediments, Environ. Sci. Technol., 40, 4455–4459, 2006b.

Koukina, S. E., Calafat-Frau, A., Hummel, H., and Palerud, R.: Trace metals in suspended particulate matter and sediments from teh Severnaya Dvina estuary, Russian Arctic, Polar Res., 37, 249–256, 2001.

Kozelka, P. B., Sanudo-Wilhelmy, S., Flegal, A. R., and Bruland, K. W.: Physico-chemical speciation of lead in South San Francisco Bay, Estuar. Coast. Shelf S., 44, 649–658, 1997.

Krachler, R., Jirsa, F., and Ayromlou, S.: Factors influencing the dissolved iron input by river water to the open ocean, Biogeosciences, 2, 311–315, doi:10.5194/bg-2-311-2005, 2005.

Kraepiel, A. M. L., Chiffoleau, J.-F., Martin, J.-M., and Morel, F. M. M.: Geochemistry of trace metals in the Gironde estuary, Geochim. Cosmochim. Ac., 61, 1421–1436, 1997.

Kritzberg, E. S., Cole, J. J., Pace, M. L., Granéli, W., and Bade, D. L.: Autochthonous versus allochthonous carbon sources of bacteria: results from whole-lake ^{13}C addition experiments, Limnol. Oceanogr., 49, 588–596, 2004.

Kukina, S. E., Sadovnikova, L. K., Kalafat, A., Hummel, H., and Regoli, F.: Distribution of metals in suspended matter and bottom sediments in the estuary of the Northern Dvina, Oceanology, 42, 218–227, 2002.

Kuma, K., Nakabayashi, S., and Matsunaga, K.: Photoreduction of Fe(III) by hydroxycarboxylic acids in seawater, Water Res., 29, 1559–1569, 1995.

Kuma, K., Katsumoto, A., Nishioka, J., and Matsunaga, K.: Size-fractioned iron concentrations and Fe(III) hydroxide solubility in various coastal waters, Estuar. Coast. Shelf S., 47, 275–283, 1998.

Kuzyk, Z. A., MacDonald, R. W., Granskog, M. A., Scharien, R. K., Galley, R. J., Michel, C., Barber, D., and Stern, G.: Sea ice, hydrological, and biological processes in the Churchill River estuary region, Hudson Bay, Estuar. Coast. Shelf S., 77, 369–384, 2008.

Lawrence, M. G. and Kamber, B. S.: The behaviour of the rare earth elements during estuarine mixing-revisited, Mar. Chem., 100, 147–161, 2006.

Lisitsyn, A. P.: The marginal filter of the ocean, Oceanology, 34, 671–682, 1995.

Liu, R., Lead, J. R., and Zhang, H.: Combining cross flow ultrafiltration and diffusion gradients in thin-films approaches to determine trace metal speciation in freshwaters, Geochim. Cosmochim. Ac., 109, 14–26, 2013.

Lofts, S., Tipping, E., and Hamilton-Taylor, J.: The chemical speciation of Fe(III) in freshwaters, Aquat. Geochem., 14, 337–358, 2008.

MacDonald, R. W. and Yu, Y.: The Mackenzie Estuary of the Arctic Ocean, The Handbook of Environmental Chemistry, Pt. H, 5, 91–120, doi:10.1007/698_5_027, 2006.

McClelland, J. W., Holmes, R. M., Dunton, K. H., and MacDonald, R. W.: The Arctic Ocean estuary, Estuar. Coast., 35, 353–368, 2012.

McKenna, J. H.: DOC dynamics in a small temperate estuary: simultaneous addition and removal processes and implications on observed nonconservative behaviour, Estuaries, 27, 604–616, 2004.

Moran, M. A. and Hodson, R. E.: Bacterial production on humic and nonhumic components of dissolved organic carbon, Limnol. Oceanogr., 35, 1744–1757, 1990.

Moran, M. A. and Zepp, R. G.: Role of photoreactions in the formation of biologically labile compounds from dissolved organic matter, Limnol. Oceanogr., 42, 1307–1316, 1997.

Moran, M. A., Sheldon Jr., W. M., and Zepp, R. G.: Carbon loss and optical property changes during long-term photochemical and bi-

ological degradation of estuarine dissolved organic matter, Limnol. Oceanogr., 45, 1254–1264, 2000.
Muller, F. L. L.: Colloid/Solution partitioning of metal-selective organic ligands, and its relevance to Cu, Pb and Cd cycling in the Firth of Clyde, Estuar. Coast. Shelf S., 46, 419–437, 1998.
Nolting, R. F., van Dalen, M., and Helder, W.: Distribution of trace and major elements in sediment and pore waters of the Lena Delta and Laptev Sea, Mar. Chem., 53, 285–299, 1996.
Österlund, H., Gelting, J., Nordblad, F., Baxter, D. C., and Ingri, J.: Copper and nickel in ultrafiltered brackish water: labile or non-labile?, Mar. Chem., 132–133, 34–43, 2012.
Pédrot, M., Dia, A., Davranche, M., Coz, M. B.-L., Henin, O., and Gruau, G.: Insights into colloid-mediated trace element release at the soil/water interface, J. Colloid-Interf. Sci., 325, 187–197, 2008.
Point, D., Monperrus, M., Tessier, E., Amouroux, D., Chauvaud, L., Thouzeau, G., Jean, F., Amice, E., Gall, J., Leynaert, A., Clavier, J., and Donard, O. F. X.: Biological control of trace metal and organometal benthic fluxes in a eutrophic lagoon (Thau Lagoon, Mediterranean Sea, France), Estuar. Coast. Shelf S., 72, 457–471, 2007.
Pokrovsky, O. and Schott, J.: Iron colloids/organic matter associated transport of major and trace elements in small boreal rivers and their estuaries (NW Russia), Chem. Geol., 190, 141–179, 2002.
Pokrovsky, O. S. and Shirokova, L. S.: Diurnal variations of dissolved and colloidal organic carbon and trace metals in a boreal lake during summer bloom, Water Res., 47, 922–932, 2013.
Pokrovsky, O. S., Dupré, B., and Schott, J.: Fe-Al-organic colloids control of trace elements in peat soil solutions, Aquat. Geochem., 11, 241–278, 2005.
Pokrovsky, O. S., Schott, J., and Dupré, B.: Trace element fractionation and transport in boreal rivers and soil porewaters of permafrost-dominated basic terrain in Central Siberia, Geochm. Cosmochim. Ac., 70, 3239–3260, 2006.
Pokrovsky, O. S., Viers, J., Shirokova, L. S., Shevchenko, V. P., Filipov, A. S., and Dupré, B.: Dissolved, suspended, and colloidal fluxes of organic carbon, major and trace elements in Severnaya Dvina River and its tributary, Chem. Geol., 273, 136–149, 2010.
Pokrovsky, O. S., Shirokova, L. S., Kirpotin, S. N., Audry, S., Viers, J., and Dupré, B.: Effect of permafrost thawing on organic carbon and trace element colloidal speciation in the thermokarst lakes of western Siberia, Biogeosciences, 8, 565–583, doi:10.5194/bg-8-565-2011, 2011.
Pokrovsky, O. S., Shirokova, L. S., Zabelina, S. A., Vorobieva, T. Y., Moreva, O. Yu., Chupakov, A., Shorina, N. V., Kokryatskaya, N. M., Andry, S., Viers, J., Zoutien, C., and Freydier, R.: Size fractionation of trace elements in a seasonally stratified boreal lake: control of organic matter and iron colloids, Aquat. Geochem., 18, 115–139, 2012a.
Pokrovsky, O. S., Viers, J., Dupré, B., Chabaux, F., Gaillardetc, J., Audry, S., Prokushkin, A. S., Shirokova, L. S., Kirpotinf, S. N., Lapitsky, S. A., and Shevchenko, V. P.: Biogeochemistry of carbon, major and trace elements in watersheds of northern Eurasia drained to the Arctic Ocean: The change of fluxes, sources and mechanisms under the climate warming prospective, C.R. Geosci., 344, 663–677, 2012b.
Pokrovsky, O. S., Shirokova, L. S., Kirpotin, S. N., Kulizhsky, S. P., and Vorobiev, S. N.: Impact of western Siberia heat wave 2012 on greenhouse gases and trace metal concentration in thaw lakes of discontinuous permafrost zone, Biogeosciences, 10, 5349–5365, doi:10.5194/bg-10-5349-2013, 2013.
Porcelli, D., Andersson, P. S., Wasserburg, G. J., Ingri, J., and Baskaran, M.: The importance of colloids and mires for the transport of uranium isotopes through the Kalix River watershed and Baltic Sea, Geochim. Cosmochim. Ac., 61, 4095–4113, 1997.
Powell, R. T., Landing, W. M., and Bauer, J. E.: Colloidal trace metals, organic carbon and nitrogen in a southeastern U.S. estuary, Mar. Chem., 55, 165–176, 1996.
Raymond, P. A. and Bauer, J. E.: Bacterial consumption of DOC during transport through a temperate estuary, Aquatic Microb. Ecol., 22, 1–12, 2000.
Redfield, A. C., Ketchum, B. H., and Richards, F. A.: The influence of organisms on the composition of seawater, in: The Sea. Ideas and Observation on Progress in the Study of the Seas, 2, edited by: Hill, M. N., John Wiley, London, United Kingdom, 26–77, 1963.
Reid, R., Live, D., Faulkner, D., and Bauer, J.: A siderophore from a marine bacterium with an exceptional ferric ion affinity constant, Nature, 366, 455–458, 1993.
Savenko, A. V. and Shevchenko, V. P.: Seasonal variability of the distribution of dissolved forms of biogenic elements and alkalinity in the Northern Dvina mouth, Water Resour., 32, 417–421, 2005.
Savenko, A. V., Shevchenko, V. P., Novigatskii, A. N., and Efimova, L. V.: Effect of seasonal variability in the chemical composition of the riverine runoff on the distribution of dissolved forms of strontium, calcium, fluorine, and boron in the mouths of the Severnaya Dvina and Onega rivers, Oceanology, 43, 99–107, 2003.
Sañudo-Wilhelmy, S. A., Rivera-Duarte, I., and Flegal, A. R.: Distribution of colloidal trace metals in the San Francisco Bay estuary, Geochim. Cosmochim. Ac., 60, 4933–4944, 1996.
Schroeder, D. C. and Lee, F. G.: Potential transformations of chromium in natural waters, Water Air Soil Pollut., 4, 355–365, 1975.
Shank, G. C., Skrabal, S. A., Whitehead, R. F., and Kieber, R. J.: Strong copper complexation in an organic-rich estuary: the importance of allochthonous dissolved organic matter, Mar. Chem., 88, 21–39, 2004.
Shevchenko, V. P., Lisitzin, A. P., Belyaev, N. A., Filippov, A. S., Golovnina, E. A., Ivanov, A. A., Klyuvitkin, A. A., Malinkovich, S. M., Novigatsky, A. N., Politova, N. V., Rudakova, V. N., Rusakov, V. Yu., and Sherbak, S. S.: Seasonality of suspended particulate matter distribution in the White Sea, Berichte zur Polar- und Meeresforschung, 482, 142–149, 2004.
Shevchenko, V. P., Pokrovsky, O. S., Filippov, A. S., Lisitsyn, A. P., Bobrov, V. A., Bogunov, A. Yu., Zavernina, N. N., Zolotyh, E. O., Isaeva, A. B., Kokryatskaya, N. M., Korobov, V. B., Kravchishina, M. D., Novigatsky, A. N., and Politova, N. V.: On the elemental composition of suspended matter of the Severnaya Dvina river (White Sea region), Doklady Earth Sciences, 430, 228–234, 2010.
Shiller, A. M.: The effect of recycling traps and upwelling on estuarine chemical flux estimates, Geochim. Cosmochim. Ac., 60, 3177–3185, 1996.
Shiller, A. M. and Boyle, E. A.: Trace elements in the Mississippi River Delta outflow region: Behavior at high discharge, Geochim. Cosmochim. Ac., 55, 3241–3251, 1991.

Shirokova, L. S., Pokrovsky, O. S., Moreva, O. Y., Chupakov, A. V., Zabelina, S. A., Klimov, S. I., Shorina, N. V., and Vorobieva, T. Ya.: Decrease of concentration and colloidal fraction of organic carbon and trace elements in response to the anomalously hot summer 2010 in a humic boreal lake, Sci. Total Environ., 463–464, 78–90, 2013.

Sholkovitz, E. R.: Flocculation of dissolved organic and inorganic matter during the mixing of river water and seawater, Geochim. Cosmochim. Ac., 40, 831–845, 1976.

Sholkovitz, E. R.: The flocculation of dissolved Fe, Mn, Al, Cu, Ni, Co and Cd during estuarine mixing, Earth Planet. Sc. Lett., 41, 77–86, 1978.

Sholkovitz, E. R.: The aquatic chemistry of rare earth elements in rivers and estuaries, Aquat. Geochem., 1, 1–34, 1995.

Sholkovitz, E. R. and Elderfield, H.: The cycling of dissolved rare-earth elements in Chesapeake Bay, Global Biogeochem. Cy., 2, 157–176, 1988.

Shulkin, V. M. and Bogdanova, N. N.: Mobilization of metals from riverine suspended matter in seawater, Mar. Chem., 83, 157–167, 2003.

Sigleo, A. C. and Helz, G. R.: Composition of estuarine colloidal material: major and trace elements, Geochim. Cosmochim. Ac., 45, 2501–2509, 1981.

Stecher III, H. A. and Kogut, M. B.: Rapid barium removal in the Delaware estuary, Geochim. Cosmochim. Ac., 63, 1003–1012, 1999.

Stolpe, B. and Hassellöv, M.: Changes in size distribution of fresh water nanoscale colloidal matter and associated elements on mixing with seawater, Geochim. Cosmochim. Ac., 71, 3292–3301, 2007.

Stolpe, B., Guo, L., Shiller, A. M., and Hassellöv, M.: Size and composition of colloidal organic matter and trace elements in the Mississippi River, Pearl River and the northern Gulf of Mexico, as characterized by flow field-flow fractionation, Mar. Chem., 118, 119–128, 2010.

Strady, E., Blanc, G., Schäfer, J., Coynel, A., and Dabrin, A.: Dissolved uranium, vanadium and molybdenum behaviours during contrasting freshwater discharges in the Gironde Estuary (SW France), Estuar. Coast. Shelf S., 83, 550–560, 2009.

Takata, H., Aono, T., Tagami, K., and Uchida, S.: Influence of dissolved organic matter on particle-water interactions for Co, Cu and Cd under estuarine conditions, Estuar. Coast. Shelf S., 111, 75–83, 2012.

Tranvik, L.: Availability of dissolved organic carbon for planktonic bacteria in oligotrophic lakes of differing humic content, Microb. Ecol., 16, 311–322, 1988.

Tranvik, L. J. and Jørgensen, N. O. G.: Colloidal and dissolved organic matter in lake water: Carbohydrate and amino acid composition, and ability to support bacterial growth, Biogeochemistry, 30, 77–97, 1995.

Ussher, S. J., Achterberg, E. P., Sarthou, G., Laan, P., de Baar, H. J. W., and Worsfold, P. J.: Distribution of size fractionated dissolved iron in the Canary Basin, Marine Environ. Res., 70, 46–55, 2010.

Van den Berg, C., Merks, A., and Duursma, E.: Organic complexation and its control of the dissolved concentrations of copper and zinc in the Scheldt estuary, Estuar. Coast. Shelf S., 24, 785–797, 1987.

Van der Sloot, H. A., Hoede, D., Wijkstra, J., Duinker, J. C., and Nolting, R. F.: Anionic species of V, As, Se, Mo, Sb, Te and W in the Scheldt and Rhine estuaries and the Southern Bight (North Sea), Estuar. Coast. Shelf S., 21, 633–651, 1985.

Vasyukova, E. V., Pokrovsky, O. S., Viers, J., Oliva, P., Dupre, B., Martin, F., and Candaudap, F.: Trace elements in organic- and iron-rich surficial fluids of the boreal zone: Assessing colloidal forms via dialysis and ultrafiltration, Geochim. Cosmochim. Ac., 74, 449–468, 2010.

Vasyukova, E. V., Pokrovsky, O. S., Viers, J., and Dupré, B.: New operational method of testing colloid complexation with metals in natural waters, Appl. Geochem., 27, 1226–1237, 2012.

Waeles, M., Riso, R. D., Maguer, J.-F., and Le Corre, P.: Distribution and chemical speciation of dissolved cadmium and copper in the Loire estuary and North Biscay continental shelf, France, Estuar. Coast. Shelf S., 59, 49–57, 2004.

Waeles, M., Riso, R. D., and Le Corre, P.: Seasonal variations of dissolved and particulate copper species in estuarine waters, Estuar. Coast. Shelf S., 62, 313–323, 2005a.

Waeles, M., Riso, R. D., and Le Corre, P.: Seasonal variations of cadmium speciation in the Penzé estuary, NW France, Estuar. Coast. Shelf S., 65, 143–152, 2005b.

Waeles, M., Tanguy, V., Lespes, G., and Riso, R. D.: Behaviour of colloidal trace metals (Cu, Pb and Cd) in estuarine waters: An approach using frontal ultrafiltration (UF) and stripping chronopotentiometric methods (SCP), Estuar. Coast. Shelf S., 80, 538–544, 2008.

Wells, M. L., Kozelka, P. B., and Bruland, K. W.: The complexation of 'dissolved' Cu, Zn, Cd and Pb by soluble and colloidal organic matter in Narragansett Bay, RI, Mar. Chem., 62, 203–217, 1998.

Wells, M. L., Smith, G. J., and Bruland, K. W. The distribution of colloidal and particulate bioactive metals in Narragansett Bay, RI, Mar. Chem., 71, 143–163, 2000.

Wen, L.-S., Santschi, P., Gill, G., and Paternostro, C.: Estuarine trace metal distributions in Galverston Bay: importance of colloidal forms in the speciation of the dissolved phase, Mar. Chem., 63, 185–212, 1999.

Wen, L.-S., Jiann, K.-T., and Santschi, P. H.: Physicochemical speciation of bioactive trace metals (Cd, Cu, Fe, Ni) in the oligotrophic South China Sea, Mar. Chem., 101, 104–129, 2006.

Wen, L.-S., Santschi, P. H., Warnken, K. W., Davison, W., Zhang, H., Li, H.-P., and Jiann, K.-T.: Molecular weight and chemical reactivity of dissolved trace metals (Cd, Cu, Ni) in surface waters from the Mississippi River to Gulf of Mexico, Estuar. Coast. Shelf S., 92, 649–658, 2011.

Wilkinson, K. J., Joz-Roland, A., and Buffle, J.: Different roles of pedogenic fulvic acids and aquagenic biopolymers on colloid aggregation and stability in freshwaters, Limnol. Oceanogr., 42, 1714–1724, 1997.

12

Turbulent dispersion properties from a model simulation of the western Mediterranean Sea

H. Nefzi[1], D. Elhmaidi[1], and X. Carton[2]

[1]Faculté des Sciences de Tunis, Université Tunis ElManar, Tunisia
[2]Laboratoire de Physique des Océans-Université Occidentale de Bretagne Brest, France

Correspondence to: D. Elhmaidi (dalila.elhmaidi@fst.rnu.tn)

Abstract. Using a high-resolution primitive equation model of the western Mediterranean Sea, we analyzed the dispersion properties of a set of homogeneously distributed, passive particle pairs. These particles were initially separated by different distances D_0 ($D_0 = 5.55$, 11.1 and 16.65 km), and were seeded in the model at initial depths of 44 and 500 m.

This realistic ocean model, which reproduces the main features of the regional circulation, puts into evidence the three well-known regimes of relative dispersion.

The first regime due to the chaotic advection at small scales lasts only a few days (3 days at 44 m depth, a duration comparable with the integral timescale), and the relative dispersion is then exponential. In the second regime, extending from 3 to 20 days, the relative dispersion has a power law t^{α} where α tends to 3 as D_0 becomes small. In the third regime, a linear growth of the relative dispersion is observed starting from the twentieth day. For the relative diffusivity, the D^2 growth is followed by the Richardson regime $D^{4/3}$. At large scales, where particle velocities are decorrelated, the relative diffusivity is constant.

At 500 m depth, the integral timescale increases (> 4 days) and the intermediate regime becomes narrower than that at 44 m depth due to the weaker effect of vortices (this effect decreases with depth). The turbulent properties become less intermittent and more homogeneous and the Richardson law takes place.

1 Introduction

Satellite observations and numerical models of the western Mediterranean Basin have revealed the presence of coherent vortices with different scales (between ten and two hundred kilometers). They are generated by the instability of the inflowing Atlantic water through the Gibraltar strait. Such vortices can survive for long durations (three years), carrying waters characteristic of their formation site and playing an essential role in transport and mixing processes (Millot, 1999; Millot and Taupier-Letage, 2005). In the ocean, mesoscale horizontal velocities are much larger than vertical velocities and the motion is quasi-two-dimensional (McWilliams, 1984).

Numerical simulations of two-dimensional turbulence have shown that the presence of vortices induces an anomalous intermediate dispersion regime between the ballistic regime and the Brownian one. The study of the absolute dispersion of neutral floats in two-dimensional flows and in the western Mediterranean Basin has revealed that the absolute dispersion anomalous regime is characterized by a power law t^r where r varies between 5/3 and 5/4 (Elhmaidi et al., 1993, 2010). The corresponding intermediate regime for the relative dispersion is known as the Richardson regime (Richardson, 1926; Babiano et al., 1990).

Many analyses of oceanic data have shown the existence of the Richardson regime. LaCasce and Ohlmann (2003) have used the data from the SCULP (Surface Current and Lagrangian Drift Program) experiment to study the relative dispersion in the Gulf of Mexico. They found an exponential growth in a range of scales from 5 to 50 km with a duration of about 10 days. At longer times, the time dependence was a power law with an exponent lying between 2 and 3. Ollitrault et al. (2005) analyzed subsurface float pairs in the western and eastern North Atlantic from the TOPOGULF experiment. They found an exponential regime for scales smaller than the Rossby radius of deformation ($D < 25$ km) and for times $t < 6$ days, followed by a Richardson's law from scales 40 to 300 km, and a linear regime was then observed. Lumpkin and Elipot (2010) analyzed the relative dispersion of a surface pair in CLIMODE (the CLIVAR Mode Water Dynamic Experiment) in the Gulf Stream region. They showed the existence of a Richardson regime at scales from 1–3 km to 300–500 km. At larger scales, the spreading of particles followed a random walk described by a constant diffusivity. The small-scale exponential regime was not observed.

Nevertheless, though large, the number of float or drifter data in the world ocean remains small compared to the ocean surface (or volume), so that their average separation is relatively large and that they do not provide uniform coverage of the ocean. The use of synthetic drifters in numerical simulations can palliate this deficiency and help test the accordance between numerical and experimental data.

Often, particle tracking in numerical models was used to follow the path or mixing of water masses. Here we aim to put into evidence the theoretical regimes of turbulent dispersion in the oceanographic context using a numerical ocean model and particle tracking.

In the present work, we study the dispersion of passive particle pairs, near the surface, in the western Mediterranean Sea. We use a high-resolution model able to reproduce a realistic level of turbulence (mesoscale eddies). In Sect. 2, we recall the numerical conditions of the study (the model and its parameters). In Sect. 3, we briefly recall the theoretical results of turbulent dispersion in quasi-2-D flows. Section 4 is devoted to the analysis of the model results and to the calculation of dispersion and of relative diffusivity. Finally conclusions are drawn.

2 Material and methods: the ROMS model and particle tracking

The Regional Ocean Modeling System (ROMS) is a primitive equation model able to simulate both coastal and oceanic dynamics (Shchepetkin and McWilliams, 2005). It is based on the Boussinesq approximation and on the hydrostatic balance. ROMS is a split-explicit, free surface oceanic model discretized in coastline- and terrain-following curvilinear coordinates. Short time steps are used to solve the barotropic momentum equations and longer ones to advance the baroclinic momentum equations. A third-order, upstream-biased advection scheme implemented in ROMS allows the generation of steep gradients, enhancing the effective resolution of the solution for a given grid size (Penven et al., 2008). The explicit lateral viscosity is zero everywhere in the model domain except in the sponge layers near the boundaries where the viscosity increases smoothly toward the lateral open boundaries.

The computational domain for this study extends from 33° to 44.5° N and from 7° W to 15° E. The model grid, forcing, and initial and boundary conditions are built using the ROMSTOOLS package (Shchepetkin and McWilliams, 2005; Penven et al., 2008). The bottom topography is derived from the ETOPO2 2′ resolution database (Penven and Tan, 2007). A slope parameter $r = 2.5$ is used to prevent errors in the computation of the pressure gradient (Smith and Sandwell, 1997). The ROMS configuration used here has a 1/8° horizontal resolution and 20 vertical levels with stretched s coordinates, using surface and bottom stretching parameters ($\theta_s = 0$, $\theta_b = 6$) (Haidvogel and al., 2000). The temporal resolution is 15 min. Mean monthly temperature and salinity data are obtained from the World Ocean Atlas (WOA2005) database (monthly climatology at $1° \times 1°$ resolution) (Conkright et al., 2001; Song and Haidvogel, 1994); these data are used to initialize the model and restore conditions in the sponge layers.

Figure 1 represents the model grid and its bathymetry. At the two open boundaries (west and east), an active, implicit, upstream-biased radiation condition connects the model solution to the surrounding sea or ocean (Marchesiello et al., 2001). The model is started from rest and is forced with winds, heat and salinity fluxes from the Comprehensive Ocean-Atmosphere Data Set (COADS05), which is a monthly climatology giving data with a spatial resolution of 0.5° (Da Silva et al., 1994). The width of the sponge layers is 150 km and the maximum viscosity in these layers is set to $1000\,\mathrm{m^2\,s^{-1}}$.

The ROMS model is integrated until a statistical equilibrium is obtained. Figure 2a shows the simulated surface currents in a subdomain (2° W to 10° E and 35.5° to 40° N) of the western Mediterranean Basin after the stationary state has been reached. The flow is dominated by coherent vortices with different scales that propagate to the east in accordance with available in situ and satellite data sets (Beranger et al., 2005). A vertical cross section of temperature, salinity and perpendicular horizontal velocity (respectively Fig. 2b, c, and d) through vortex V_1 indicates that the thermohaline anomalies associated with this vortex are essentially intensified in the surface layer (0–200 m), while the dynamical signal extends further down (in particular down to 500 m depth), though it is weaker at depth than near the surface.

The spatial distribution of the relative vorticity $\varsigma = \frac{\partial v}{\partial x} - \frac{\partial u}{\partial y}$ where (u, v) is the horizontal velocity (Fig. 2e) in a subdomain in the western Mediterranean Basin shows that the flow is dominated by cyclonic and anticyclonic vortices surrounded by filaments with small scales. The horizontal kinetic energy spectrum calculated in this subdomain (figure not shown) indicates the presence of the two inertial ranges with the spectral slopes of two-dimensional turbulence ($-5/3$ and -3, respectively).

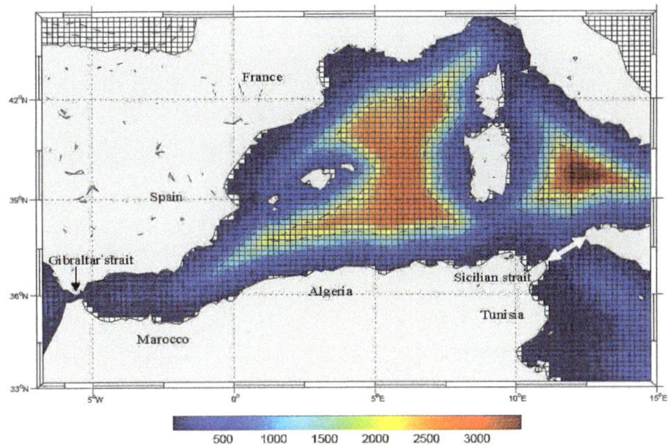

Fig. 1. The model domain and its bathymetry (m).

To study the dispersion processes, we performed two sets of numerical simulations with the ROMS model from the statistical equilibrium state. We seeded neutral, passive particle pairs slightly below the surface (at depth $h = -44$ m) and at a much larger depth ($h = -500$ m). They were advected with the model horizontal velocity. The Lagrangian velocities were calculated using a centered scheme; the interpolation scheme for the particle trajectories was a fourth-order accurate Adams–Bashford–Moulton predictor–corrector scheme. Transport of particles was due to advection only, as no diffusion term was introduced.

3 Relative dispersion

To study particle dispersion, relative dispersion and relative diffusivity are commonly used (Richardson, 1926; Babiano et al., 1990). They are defined respectively by

$$D^2(t, D_0) = \sum_{i \neq j} < x_i(t, D_0) - x_j(t, D_0) >^2, \qquad (1)$$

$$Y(t) = \frac{1}{2}\frac{\mathrm{d}}{\mathrm{d}t} D^2(t, D_0), \qquad (2)$$

where $x_i(t, D_0)$ and $x_j(t, D_0)$ are the vector positions of a pair separated initially by D_0 and the average is over all particle pairs.

Relative dispersion of particles in a turbulent field is governed by the distribution of velocities with spatial scales. Thus, dispersion is related to the spatial spectrum of kinetic energy. In two-dimensional turbulence, this spectrum has two ranges, called inertial ranges, where nonlinear effects (due to advection) dominate the dynamics. These inertial ranges are, respectively, the range of inverse energy cascade (at larger scales) where the energy cascades upscale, and the range of direct enstrophy cascade (at smaller scales), where the enstrophy cascades downscale. Enstrophy is the square of relative vorticity. In these two ranges, the spectral density of kinetic energy is, respectively,

$$E(k) \approx k^{-3} \text{ and } E(k) \approx k^{-5/3},$$

where k is the horizontal, isotropic wavenumber.

The corresponding Richardson and Lin laws for the relative dispersion of two-dimensional turbulence are a cubic power for the evolution with time in the inverse energy cascade range, and an exponential evolution in the enstrophy cascade range (Babiano et al., 1990).

At intermediate times (in the inverse energy range), the mean square relative velocity defined by $< \delta u^2(t, D_0) > = < \frac{\mathrm{d}D}{\mathrm{d}t} \cdot \frac{\mathrm{d}D}{\mathrm{d}t} >$ (D is the separation vector of a pair of particles) has linear behavior (Babiano et al., 1990).

After a sufficiently long time, the separation distance of the particle pair reaches the scale of the energetic eddies and the velocities of the two particles decorrelate. Due to homogeneity, the mean-square relative velocity becomes just four times the eddy kinetic energy (Ollitrault et al., 2005; LaCasce and Bower, 2000):

$$< \left(\frac{\mathrm{d}D}{\mathrm{d}t}\right)^2 > = < \delta u^2(t, D_0) >$$
$$= < u_i^2 > + < u_j^2 > -2 < u_i u_j > \approx 2 < u_i^2 >,$$

where u_i and u_j are the individual velocities of the particle pair.

The relative diffusivity has two types of variations: $Y(t) \approx D^2(t)$ in the enstrophy cascade range, and $Y(t) \approx D^{4/3}(t)$ in the inverse energy range.

At larger separation distances, relative dispersion is a random walk and relative diffusivity is constant.

At early times, the individual particle velocities are correlated and the mean-square relative velocity tends toward ZD_0^2, where Z is total enstrophy defined by $Z = \frac{1}{2} < \varsigma^2 >_x$ ($< . >_x$ is the spatial average of the relative velocity over the entire domain) (Babiano et al., 1990; Ollitrault et al., 2005).

The enstrophy can be calculated through the relative dispersion characteristic time defined by

$$\tau(D) = \frac{D^2}{Y(t)},$$

which is constant, and tends toward $\frac{2}{Z^{1/2}}$ when D tends to zero, and has a power law $\tau(D) \approx D^{2/3}$ in the inverse energy cascade range (Babiano et al., 1990).

4 Results

2726 passive particles pairs were homogeneously seeded at the subsurface ($h = -44$ m) of the model with different separation distances D_0. Note that the results presented hereafter were independent of this initial depth as long as the particles were seeded in the turbulent surface layer (from 0 to -100 m). The particle initial positions, for the different

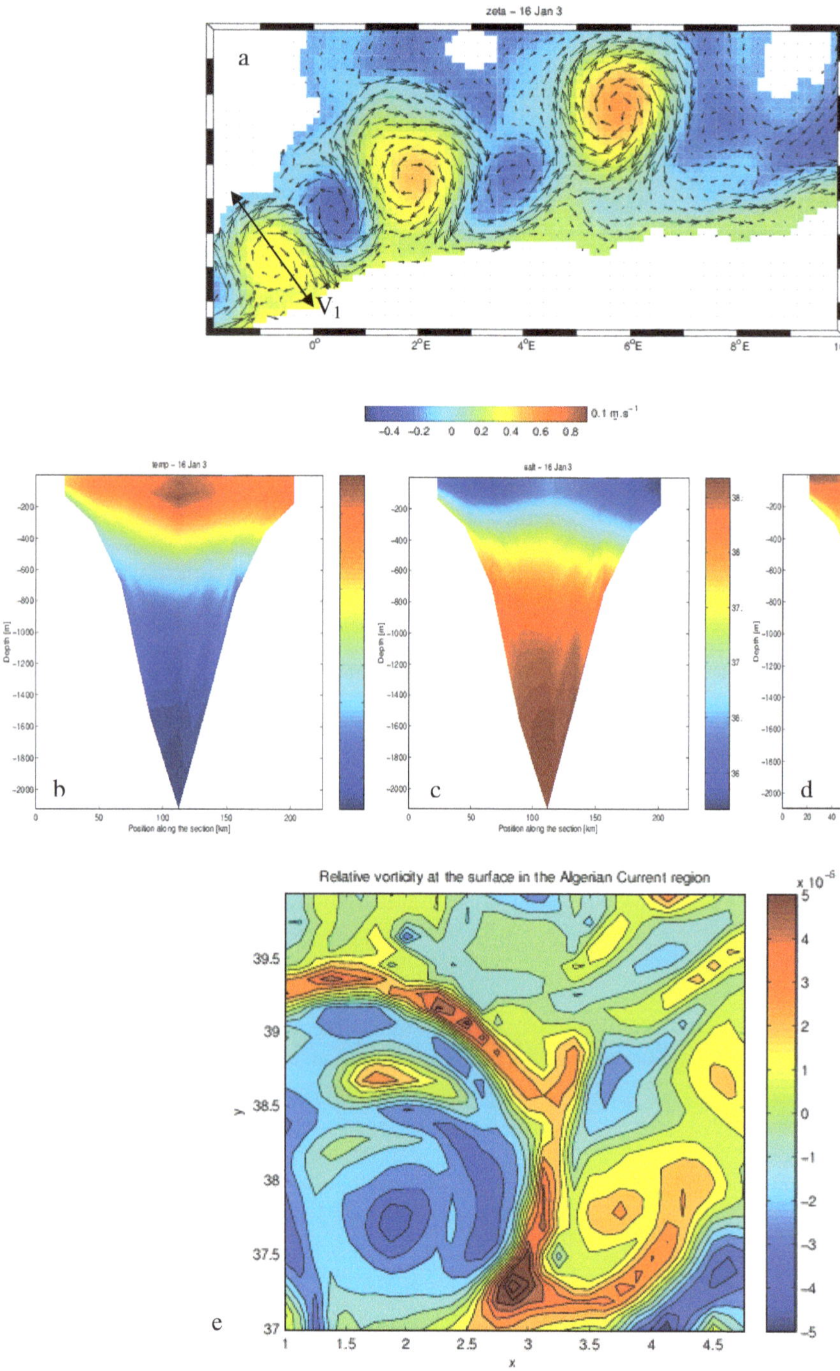

Fig. 2. (**a**) Simulated surface currents superimposed on the surface elevation in a subdomain of the western Mediterranean Basin. Vertical section of (**b**) temperature, (**c**) salinity and (**d**) perpendicular velocity through vortex V_1. (**e**) Relative vorticity in a subdomain of the western Mediterranean Basin.

simulations, were located between 36° and 39° N in latitude and 0° and 12° E in longitude. They were advected during two months. Then the particles remained at the same depth for the whole duration of the simulations. The float positions, for $D_0 = 5.55$ km, are shown in Fig. 3 at the end of the simulation. In the inset we show the initial particle positions. The particles travel east in accordance with the main circulation. They disperse in the whole domain and some of them aggregate around vortices in energetic domains. This particle evolution is similar to that in two-dimensional turbulence where vortices constitute a barrier transport, trap

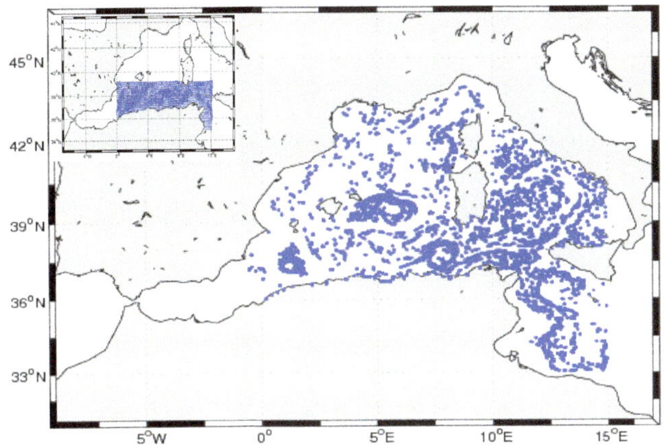

Fig. 3. The particle positions after two months of integration. In the inset we show the initial particle positions.

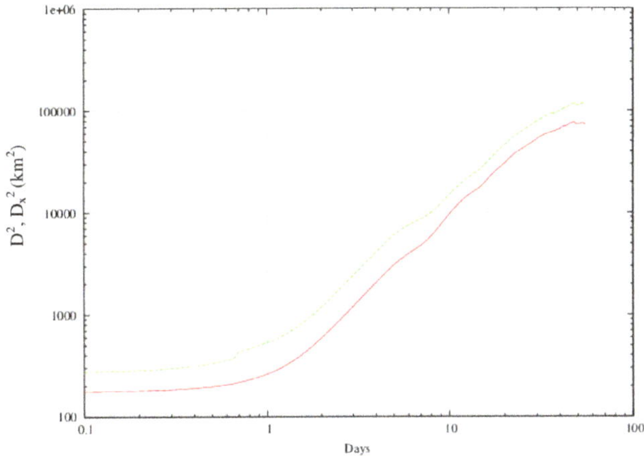

Fig. 4a. Relative dispersion (green curve), zonal relative dispersion (red curve) for $D_0 = 16.65$ km.

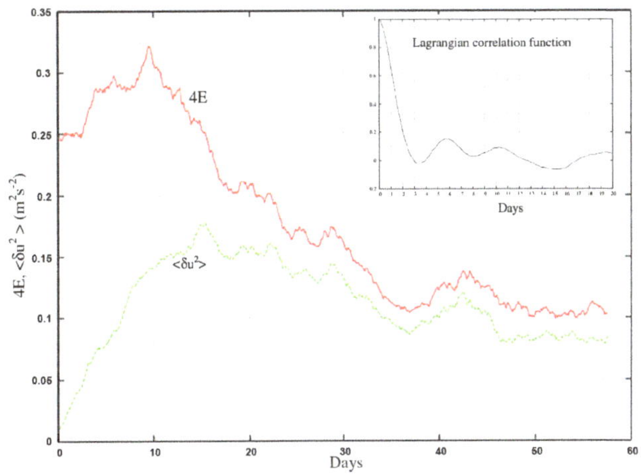

Fig. 4b. Temporal evolution of four times the kinetic energy (4E) (red curve) and of the mean square relative velocity $<\delta u^2>$ (green curve) of the particle pairs separated initially by $D_0 = 11.1$ km. In the inset we show the temporal evolution of the Lagrangian correlation function at 44 m depth.

passive floats and induce a far field effect (Elhmaidi et al., 1993, 2005).

In the western Mediterranean Basin, the zonal motion is dominant. We show in Fig. 4a that the behavior of the total relative dispersion and the zonal one ($D_x^2(t, D_0) = \sum_{i \neq j} < x_i(t, D_0) - x_j(t, D_0) >^2$, $x_i(t, D_0)$ and $x_j(t, D_0)$ are the zonal particle positions of a pair separated initially by D_0) are similar. Due to this similarity, we consider in the following statistics only the zonal components of the particles.

The temporal evolution of the mean square relative velocity $< \delta u^2 >$ and four times the kinetic energy (4E) are plotted for the value $D_0 = 11.1$ km in Fig. 4b. This evolution can be described in three different phases. The duration of these phases has been determined by fitting the phenomenological laws and by minimizing the error between the numerical curve and the theoretical laws.

In the first one (lasting during the first three days), the energy remains constant, as the Lagrangian velocities are correlated. This time interval is roughly equal to the Lagrangian integral timescale defined by

$$T_L = \int_0^\infty R(\tau)\mathrm{d}\tau,$$

where

$$R(\tau) = \frac{< u(t).u(t+\tau) >}{< (u(t).u(t) >}$$

is the Lagrangian velocity autocorrelation function displayed in the inset of Fig. 4b. This Lagrangian integral timescale is about 3 days, as shown by the first root of this function.

In the second phase (lasting from days 3 to 20), the Lagrangian energy initially increases because the particles are trapped in the intense circulation cells around vortices. Then, after day 10, a decrease in energy can be observed given that the majority of passive particles reach the background field where the energy is low (Elhmaidi et al., 1993). During this phase, the mean square relative velocity has a quasi-linear evolution.

In the final phase ($t > 20$ days), the evolutions of 4E and $< \delta u^2 >$ are correlated and the mean square relative velocity roughly equilibrates to four times the kinetic energy.

The zonal mean square relative velocity, divided by t, is displayed in Fig. 4c for the different separation distances. Between days 3 and 12, $< \delta u_x^2 > /t$ tends to a plateau especially for the smallest value of D_0. During this time interval, $< \delta u_x^2 >$ has a power law t^b where b is roughly equal to 1, corresponding to the theory (Babiano et al., 1990). For the smallest time and for the different values of $D_0 =$ 5.55, 11.1, 16.65 km, we can compute $< \delta u_x^2 > \approx 25$, 52, 80 km^2 days^{-2}, respectively (see the inset of Fig. 4c). As

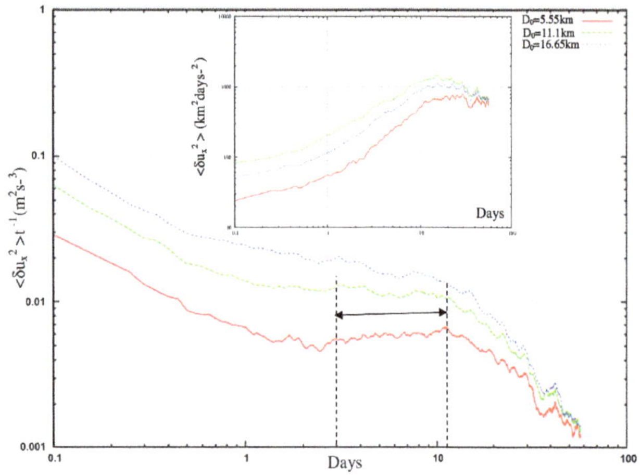

Fig. 4c. Temporal evolution of the mean square relative velocity divided by t for different initial separation distances. The plateau indicates the linear $< \delta u_x^2 >$ law. In the inset, the temporal evolution of $< \delta u_x^2 >$ in km^2 $days^{-2}$ is shown.

$< \delta u_x^2 >$ tends to $Z D_0^2$ for short times, we can deduce that the enstrophy Z fluctuates between 0.3 and 0.81 $days^{-2}$. Using these values of Z and given that $Z = \frac{1}{2} < \varsigma^2 >_x$, the magnitude of the relative vorticity is in accordance with that of Fig. 2e.

In Fig. 5a, the temporal evolution of the zonal relative dispersion is plotted for different separation distances ($D_0 =$ 5.55, 11.1 and 16.6 km). This evolution can be described in three phases.

The first one lasts three days. To analyze the temporal evolution of the zonal relative dispersion in this phase, D_x^2 is displayed in a log-linear plot in Fig. 5b. A linear evolution is observed for the different values of D_0. This is an indication of the exponential growth, while the dispersion is dominated by eddies with scales larger than the pair separations. Then D_x^2 evolves as $\exp(kt)$ where k varies between 0.74 and 0.82. This exponential regime was found previously for separation distances smaller than the Rossby radius of deformation (LaCasce and Bower, 2000; LaCasce and Ohlmann, 2003; Ollitrault et al., 2005).

In the second phase ($3 < t < 20$ days), D_x^2 grows like t^a where a varies between 1.66 and 2.3. We show in Fig. 5c the temporal evolution of D_x^2 in this range for $D_0 = 16.5$ km and 5.55 km. These two curves are respectively fitted by the linear laws 340 $t^{1.66}$ and 13 $t^{2.3}$ with a relative error not exceeding 10 %.

In the third phase, a linear growth of the relative dispersion is observed starting from the twentieth day.

The zonal relative dispersion coefficient $Y_x(D)$ was calculated by finite time differencing $D_x^2(t, D_0)$. Its evolution versus the distance D is shown in Fig. 6a for several initial distances D_0 ($D_0 = 5.55, 11.1$ and 16.6 km). The relative diffusivity is fitted by the curve 0.4 D^2 for $D < 20$ km, and for $D > 20$ km the relative diffusivity is fitted by the curve 4 $D^{4/3}$

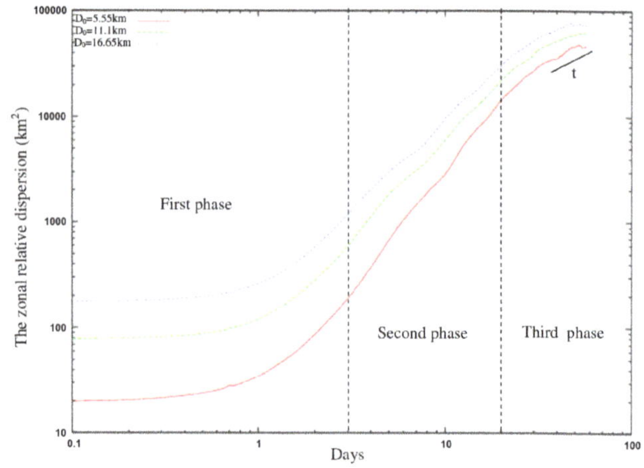

Fig. 5a. Temporal evolution of the zonal relative dispersion D_x^2. The solid black line indicates the linear law.

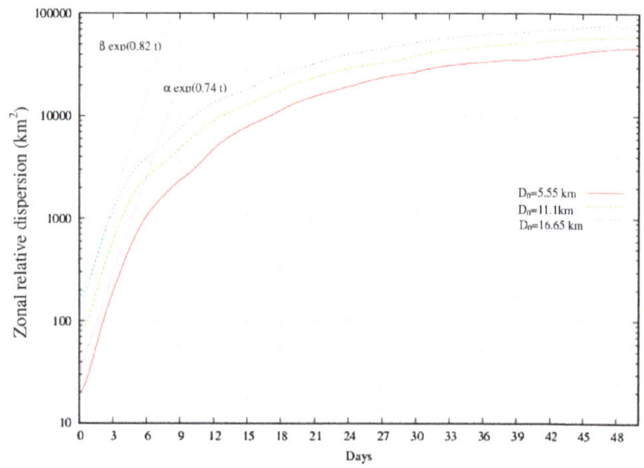

Fig. 5b. The zonal relative dispersion versus time and D_0. The $\alpha \exp(0.74t)$ and $\beta \exp(0.82t)$ laws are given with $\alpha = 30$ and $\beta =$ 120.

and $Y(D)$ is constant for distances larger than 100 km. The D^2 regime confirms the presence of an enstrophy cascade range only for small scales. For large scales, the Richardson $D^{4/3}$ is the signature of the inverse energy cascade.

The enstrophy Z of the flow can be calculated via the evolution of the zonal relative characteristic dispersion time $\tau_x(D)(\tau_x(D) = D_x^2/Y_x)$ reported in Fig. 6b. For small D, the minimum characteristic time varies between two and three days. Then we deduce that the enstrophy fluctuates between 0.44 and 1 day^{-2}. This range of enstrophy values is in agreement with that calculated with the minimum of the square relative velocity. For $D > 20$ km, $\tau_x(D)$ is fitted by the curve 0.3 $D^{2/3}$ showing the presence of the Richardson regime in the inverse energy cascade range.

In a previous work we have shown that, at large depths, the intermediate absolute dispersion follows a power law over a period of time shorter than near the surface due to the weaker

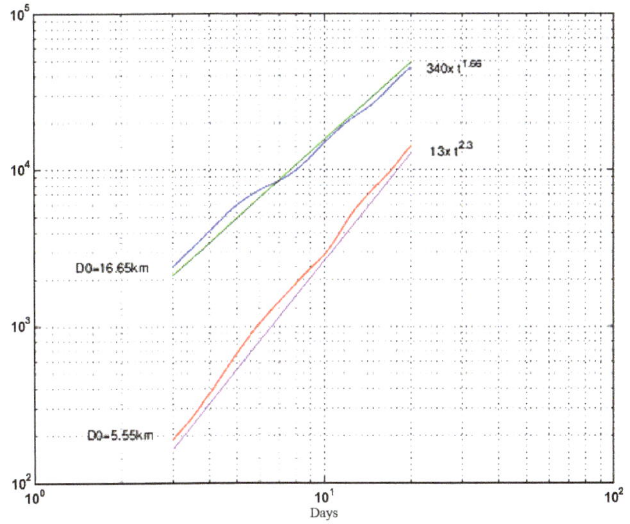

Fig. 5c. Zonal relative dispersion D_x^2 in the second phase ($3 < t < 20$ days) for $D_0 = 5.55$ km (red curve) and $D_0 = 16.65$ km (blue curve). The two curves are fitted respectively by $13\,t^{2.3}$ (purple curve) and $340\,t^{1.66}$ (green curve).

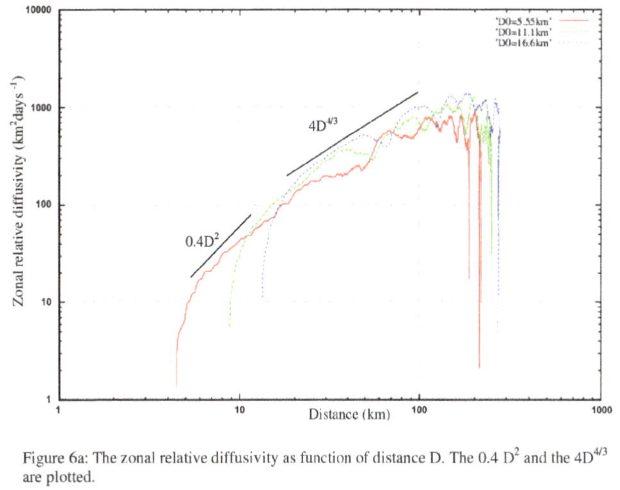

Fig. 6a. The zonal relative diffusivity as a function of distance D. $0.4\,D^2$ and $4\,D^{4/3}$ are plotted.

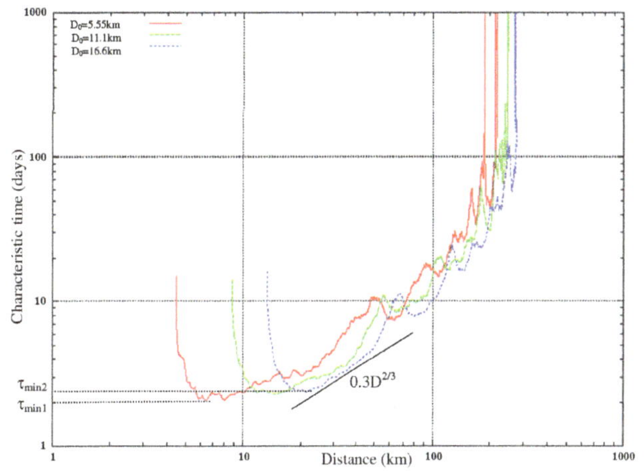

Fig. 6b. The zonal characteristic dispersion time $\tau_x(D) = D_x^2 / Y_x$ as a function of D and D_0. $\tau_{\min 1} = 2$ days for $D_0 = 5.55$ km and $\tau_{\min 2} = 3$ days. The enstrophy varies between 0.44 and 1 days^{-2}. The $0.3\,D^{2/3}$ curve is given.

effect of vortices (this effect decreases with depth, Elhmaidi et al., 2010).

To verify this result for the relative dispersion, 833 pairs of particles were seeded at the 500 m depth initially separated by the values of D_0 already used. These pairs were released in a rectangular area (37.5° to 39.1° N in latitude and 2° to 5° E in longitude).

The temporal evolution of the surface and deep Lagrangian correlation functions for $D_0 = 5.55$ km (Fig. 7) shows that the surface particle velocities decorrelate more rapidly than the deep one. Thus, the duration of the depth first phase is longer than the surface one. We show in Fig. 8 the surface and the 500 m depth relative zonal dispersion divided by t^3 for the smallest value of D_0. A plateau is well observed for the 500 m depth relative dispersion from the 6th to the 16th day, which is the signature of the Richardson t^3 law.

Given the weaker effect of vortices at depth, the turbulent properties become less intermittent and more homogenous and the Richardson law takes place.

Our results are consistent with the analysis of the subsurface drifters released respectively in the Gulf Stream and in the North Atlantic Current at 700 m depth (LaCasce and Ohlmann, 2003; Ollitrault et al., 2005; Lumpkin, and Elipot, 2010).

5 Conclusions

In this work, we have used a high-resolution primitive equation model to study the relative dispersion of neutral passive particle pairs, separated by different initial distances D_0 ($D_0 = 5.55$, 11.1 and 16.65 km), deployed in the western Mediterranean Sea at initial depths 44 and 500 m. The particles were advected with the ROMS horizontal model velocity interpolated over the particle trajectories.

The analysis of the model output has shown that the relative dispersion laws are similar to those of 2-D turbulence with the presence of the three well-known phases.

The first phase characterizes the chaotic advection at small scales, lasts only a few days (3 days, a duration equivalent to the integral timescale), and the relative dispersion is exponential.

In the second phase, extending from 3 to 20 days, the relative dispersion follows a power law t^α where α tends to 3 for small D_0, according to the Richardson regime.

In the third phase, a linear growth of the relative dispersion is observed starting from the twentieth day.

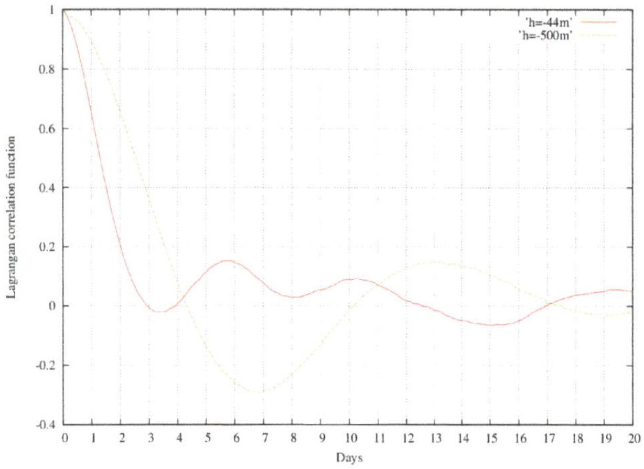

Fig. 7. Temporal evolution of the Lagrangian correlation function at depths 44 and 500 m.

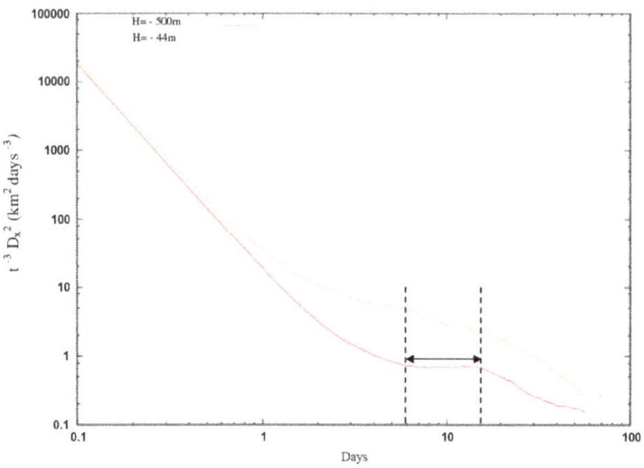

Fig. 8. The surface (green curve) and the 500 m depth (red curve) relative zonal dispersion divided by t^3 for $D_0 = 5.55$ km. The plateau of the 500 m depth extends from the sixth to the sixteenth day.

The relative diffusivity also has three phases, following those of the relative dispersion. It grows respectively as D^2 and $D^{4/3}$ for distances smaller or larger than roughly 20 km.

At 500 m depth, the integral timescale increases and the intermediate regime becomes narrower than the surface one due to the weaker effect of vortices (which decreases with depth). The turbulence becomes more homogeneous and the Richardson law takes place.

Though the western Mediterranean Sea is a complex environment where density stratification and Earth rotation act, we observe a similarity of its coherent structures to those of two-dimensional turbulence and the presence of two inertial ranges in the energy spectrum, and the laws of relative dispersion derived in the context of two-dimensional turbulence hold well near the surface in the western Mediterranean Basin. Indeed, the model was able to reproduce with reasonable accuracy these dispersion properties in this region.

The complementary analysis at 500 m depth shows that the effect of the mesoscale eddies, produced in particular by the baroclinic instability of boundary currents, is most intense at the level of their thermohaline core (that is, above 400 m depth), though it is known that many eddies have a dynamical influence below this core, as shown by the vertical section of horizontal velocity.

Further work will investigate the submesoscale dynamics. Indeed, the intense coherent vortices form filaments that play a significant role for marine ecosystems in the western Mediterranean Basin. The use of a biogeochemical model coupled with the physical circulation model will allow us to assess the impact of these eddies on biological processes.

Acknowledgements. The authors thank the editor of the special issue and the referees for helping them to improve this article. They are grateful to the University of Tunis El Manar-Tunisia and the University of Brest-France for support during the course of this study. This research was performed in partial fulfillment of the requirements for the PhD by Halima Nefzi. This work was supported by the Ministry of Higher Education, Scientific Research and Technology of Information and Communication in the frame work of the Tuniso-Fresh collaboration (DGRS/CNRS) [grant number 14R1001].

Edited by: T. Tanhua

References

Babiano, A., Basdevant, C., Leroy, P., and Sadourny, R.: Relative dispersion in two- dimensional turbulence, J. Fluid Mech., 214, 535–557, 1990.

Béranger, K., Mortier, L., and Crépon, M.: Seasonal variability of water transports through the Straits of Gibraltar, Sicily and Corsica, derived from a high resolution model of the Mediterranean circulation, Prog. Oceanogr., 66, 341–364, 2005.

Conkright, M. E., Locarnini, R. A., Garcia, H. E., O Brien, T. D., Boyer, T. P., Stephens, C., and Antonov, J. I.: World Ocean Atlas 2001: Objective Analyses, Data Statistics, and Figures, CD-ROM Documentation 2001, National Oceanographic Data Center, Silver Spring, MD, Report no. 17, 2002.

Da Silva, A. M., Young, C. C., and Levitus, S.: Atlas of surface marine data, algorithms and procedures. Technical Report U.S. Department of commerce, Vol. 1, NOAA, 1994.

Elhmaidi, D., Provenzale, A., and Babiano, A.: Elementary topology of the two-dimensional turbulence from a Lagrangian viewpoint and single particle dispersion, J. Fluid Mech., 257, 533–558, 1993.

Elhmaidi, D., Hardenberg, J. V., and Provenzale, A.: Large-scale dissipation and filament instability in two-dimensional turbulence, Phys. Rev. Lett., 95, 1–4, doi:10.1103/PhysRevLett.95.014503, 2005.

Elhmaidi, D., Nefzi, H., Carton, X., and Lili, T.: Particle dispersion in the Western Mediterranean basin, The Open Oceanography Journal, 4, 137–143, 2010.

Haidvogel, D. B., Arango, H. G., and Hedstrom, K.: Model evaluation experiments in the north atlantic basin: simulations in nonlinear terrain-following coordinates, Dynam. Atmos. Ocean, 32, 239–281, 2000.

LaCasce, J. H. and Bower, A.: Relative dispersion in the subsurface North Atlantic, J. Mar. Res., 58, 863–894, 2000.

LaCasce, J. H. and Ohlmann, J. C.: Relative dispersion at the surface of the Gulf of Mexico, J. Mar. Res., 61, 285–312, 2003.

Lumpkin, R. and Elipot, S.: Surface drifter pair spreading in the North Atlantic, J. Geophys. Res., 115, C12017, doi:10.1029/2010JC006338, 2010.

Marchesiello, P., McWilliams, J. C., and Shchepetkin, A.: Open boundary condition for long-term integration of regional oceanic models, Ocean Model., 3, 1–21, 2001.

McWilliams, J.: The emergence of isolated coherent vortices in turbulent flow, J. Fluid Mech., 146, 21–43, 1984.

Millot, C.: Circulation in the western Mediterranean Sea, J. Mar. Syst., 20, 423-442, 1999.

Millot, C. and Taupier-Letage, I.: Additional evidence of LIW entrainment across the Algerian Basin by mesoscale eddies and not by a permanent westward flow, Prog. Oceanogr., 66, 231–250, 2005.

Ollitrault, M., Gabillet, C., and Colin De Verdière, A.: Open ocean regimes of relative dispersion, J. Fluid Mech., 533, 381–407, 2005.

Penven, P., Cambon, G., Tan, T., Marchesiello, P., and Debreu, L.: Roms_agrif/romstools user's guide – roms_agrif v2.1 – romstools v2.1, available at: http://www.romsagrif.org/index.php/documentation (last access: 7 March 2013), 2010.

Penven, P., Marchesiello, P., Debreu, L., and Lefèvre, J.: Software tools for pre- and post processing of oceanic regional simulations, Environ. Model. Softw., 23, 660–662, 2008.

Richardson, L. F.: Atmospheric diffusion shown on a distance-neighbor graph, P. Roy. Soc. London A, 110, 709–737, 1926.

Shchepetkin, A. F. and McWilliams, J. C.: The regional oceanic modeling system (ROMS): a split-explicit, free-surface, topography following – coordinate oceanic model, Ocean Model., 9, 347–404, 2005.

Smith, W. H. F. and Sandwell, D. T.: Global sea floor topography from satellite altimetry and ship depth soundings, Science, 227, 1956–1962, 1997.

Song, Y. and Haidvogel, D. B.: A semi-implicit ocean circulation model using a generalized topography-following coordinate system, J. Comput. Phys., 115, 228–244, 1994.

13

Secchi depth in the Oslofjord–Skagerrak area: theory, experiments and relationships to other quantities

E. Aas[1], J. Høkedal[2], and K. Sørensen[3]

[1]Department of Geosciences, University of Oslo, Gaustadalleen 21, 0349 Oslo, Norway
[2]Narvik University College, Lodve Langesgt. 2, 8508 Narvik, Norway
[3]Norwegian Institute for Water Research, Gaustadalleen 21, 0349 Oslo, Norway

Correspondence to: E. Aas (eyvind.aas@geo.uio.no)

Abstract. The Secchi depth and its relationships to other properties of the sea water in the Oslofjord–Skagerrak area have been investigated. White and black disks of different sizes have been applied, and the Secchi depth has been observed with the naked eye, through colour filters and with a water telescope. Spectral luminances and illuminances have been calculated from recordings of radiance and irradiance, and attenuation coefficients have been determined. A theoretical expression for the Secchi depth based on luminances has been tested against field observations, and it is found that the field results for the product of Secchi depth and attenuation coefficients are on average only 4 % less than the predicted value for the white disk. For the Secchi depths observed through colour filters or for the black disk, the average field results are more than 30 % smaller than the theoretical estimates. The reduction in the disk diameter from 30 to 10 cm should theoretically reduce the Secchi depths by 13–22 %, while the field observations show an average reduction of 10–20 %. Similarly we find from theory that the removal of sun glitter should increase the Secchi depth by 12 %, while the observed increase is 14 % on average for the white disk. Our overall conclusion is that the theoretical expression works well for the white disk, but less so for the colour filter observations and the black disk.

Statistical relationships between Secchi depths and attenuation coefficients have been determined, and it is found that the root-mean-square errors relative to the mean value are smaller for the beam attenuation coefficients (12–24 %, white disk) than for the vertical attenuation coefficients (16–65 %, white disk). The depth of the 1 % level of surface quanta irradiance (PAR) can be estimated with a relative root-mean-square error of 23 % from observations of the white Secchi depth. Similar estimates of chlorophyll *a* and total suspended material will have rms errors in the range 40–90 %. Our conclusion becomes that the Secchi depth observation is a very useful tool for checking the value and order of magnitude of other related quantities in the Oslofjord–Skagerrak area.

1 Introduction

1.1 Motivation for the present study

The threshold depth of observation for the Secchi disk is a direct measure of the vertical visibility in water, and it is one of several parameters used by environmental authorities to describe water quality. In some branches of aquatic science it is termed transparency. The depth is determined by the optical properties of the water and can therefore be related to these properties. Observations of the Secchi depth can never be satisfactory substitutes for direct recordings of the other optical properties, but they can serve as independent checks of these properties.

The Norwegian Institute for Water Research and the University of Oslo have collected numerous observations of the Secchi depth as well as recordings of other optical quantities. We have analysed parts of this database to check a theoretical expression for the Secchi depth, and we have checked some of the assumptions on which the Secchi depth theory is based, as well as some results that can be deduced from the theory. Different disks and a water telescope have been applied, and threshold depths have been determined by the naked eye as well as with coloured glass filters in front of the eye. The results are based on observations from the Oslofjord–Skagerrak area. The test of the theoretical expression, as well as the statistical relationships found between Secchi depth and other optical properties, has been the main motivation for this study.

In Sect. 2 we present what is essentially Tyler's (1968) theory of the Secchi depth, and Sect. 3 describes some of the regional environmental characteristics and the applied data sets and instruments. Section 4 estimates the different optical properties appearing in the theoretical expression for the Secchi depth, and the expression as well as an assumption about the applied attenuation coefficients are compared to observations. Empirical relationships between the Secchi depth and other optical properties are discussed in Sect. 5. The results are summed up in Sect. 6.

1.2 History

The visibility of the Pacific Ocean was studied by the Russian naval officer Otto von Kotzebue as early as 1817 by means of a red piece of cloth being lowered into the sea, and on one occasion by a white plate, according to Krümmel (1886). This is probably the first known scientific investigation of the optical properties of the ocean. Other early transparency measurements and occasional observations are mentioned by Boguslawski (1884), a later work by Krümmel (1907), and by Wernand (2010).

In 1866, almost fifty years after von Kotzebue's measurements, Alessandro Cialdi, Commander of the Papal Navy, published a report containing a section by Frater Pietro Angelo Secchi, where the factors influencing the visibility in the sea of submerged disks of different sizes and colourings were discussed (Secchi, 1866). In the years to come the white version of this device became a standard instrument in marine investigations. In some scientific communities the disk is referred to as "the white disk", and in others as "the Secchi disk", although the suggestion of using a white disk came from Cialdi.

Today the marine standard method of measurement is to lower a white disk, with a diameter of approximately 30 cm, supported on a cord and with its plane horizontal, from the ship rail and into the sea to a depth where the disk cannot be seen any longer. The disk is then hauled upwards to a depth where it can be recognised once again. The mean value of the two threshold depths is termed "the Secchi depth". In limnology some communities prefer using a 20 cm disk, painted in black and white quarters, but this device will not be discussed in this paper.

However, during the first century of Secchi depth measurements a satisfactory theory describing the relationship between the threshold depth and the optical properties of the sea was missing. The factors influencing the depth were known (Krümmel, 1889), and Sauberer and Ruttner (1941) had set up the equations governing the contrast and the upward directed light, but the final step connecting the Secchi depth and the attenuation coefficients was not taken. According to Shifrin (1988) Gershun had solved the problem in 1940, but his results were published in Russian and therefore not well known outside the Soviet Union. In the western world the breakthrough came when Tyler (1968) applied a contrast formula, presented sixteen years earlier by Duntley (1952), to derive an expression for the Secchi depth. Holmes (1970) tested the constant of this expression by field measurements. A different contrast formula, including the halo of scattered light around the disk, was suggested by Levin (1980). Preisendorfer (1986) discussed the assumptions and limitations of the Secchi depth theory and procedure, using attenuation coefficients of photopic quantities.

Krümmel (1889) referred to observations made by Austrian oceanographers in the Adriatic and Ionian seas in 1880 with disks of different metals and paintings. Lisitzin (1938) observed the Secchi disk through coloured glass filters in the Baltic Sea as early as in the 1920s, and Takenouti (1950) performed similar measurements in Japanese lakes. Højerslev (1977, 1978) succeeded in relating such glass filter observations to other marine-optical properties. Levin (1980) mentioned recordings with glass filters in the Black Sea. An interesting study of the black disk was made in 1988 by Davies-Colley. Haltrin (1998) reported observations with disks that were painted blue, green and red.

For more than two decades remote sensing of water colour has been used to estimate the Secchi depth (http://www.globcolour.info/data_access_demo.html), which is one of the ESA GlobColour products.

2 Theory of the Secchi depth

An important factor that enters the theory of the Secchi depth is the properties of the human eye as a contrast sensor. Studies in air (Blackwell, 1946) have demonstrated that the human eye is able to distinguish a target from its background down to a lower limit or threshold value of the contrast between the target and its background. In our case the target is the Secchi disk, and the definition of the contrast C becomes $C = (L_D - L)/L$, where L_D is the luminance from the disk and L the luminance from the background. The eye integrates the total spectrum of radiances within the direction to the disk and the background, and weights these radiances by the eye's spectral sensitivity. During daylight conditions this is represented by the photopic sensitivity of the eye, and we have expressed it by the CIE 1924 curve for luminous efficiency (e.g. Walsh, 1958). This curve, which is also termed the photopic luminosity function, has its maximum at 555 nm, half-peak values at 510 and 610 nm, and 1 % of the maximum at 438 and 687 nm. The resulting integrals become the luminances L_D and L. If the eye of the observer is at the depth $z = 0$, which in this paper will mean just beneath the surface, or if a water telescope is being used, that is a tube through the water surface with a glass window at the bottom, then the observed contrast $C(0)$ at this depth between the nadir luminances $L_D(0)$ and $L(0)$ from the white Secchi disk and the background, respectively, will be

$$C_0 = \frac{L_D(0) - L(0)}{L(0)}. \tag{1}$$

When the Secchi disk is lowered, $L(0)$ remains constant while $L_D(0)$ is reduced until the disk reaches the depth where it disappears from sight, namely the Secchi depth. Then $C(0)$ has been reduced to the threshold value C_t. According to Blackwell (1946) C_t is not a constant, but depends on the angle α subtended by the observed target, the luminance $L(0)$ of the background, the probability of detection and the exposure time. C_t decreases with decreasing detection probability and with increasing α, $L(0)$ and exposure time.

Photometric luminance is not a practical quantity if we want to use the *equation of radiative energy transfer*, because the luminance represents a spectral integral, and therefore its attenuation coefficients are not constant in space, even in optically homogeneous waters. However, when we determine the Secchi depth, we notice that the colour of the disk is the same as that of the surrounding waters. This colour corresponds to the wavelength region where the water has its maximum beam transmittance, which is also where the upward scattered radiance obtains its spectral maximum. We can therefore assume as a first approximation that the observed luminances in Eq. (1) correspond to the radiances at the wavelength of the spectral transmittance maximum. In the present discussion we have let the symbol L represent these monochromatic radiances as well as the spectrally narrow-banded photopic luminances. The magnitude of the errors introduced by the monochromatic assumption will be discussed in Sect. 4.8.

The equation of radiative energy transfer for the nadir radiance L_D from the Secchi disk sounds

$$-\frac{dL_D(z)}{dz} \approx -c\,L_D(z) + L_{*D}(z), \tag{2}$$

where z is the vertical coordinate with zero at the surface and positive downwards, c is the beam attenuation coefficient at the wavelength of maximum transmittance, and L_{*D} is the path function along the path from the disk to the observer.

The same equation for the nadir radiance L from the background becomes

$$-\frac{dL(z)}{dz} \approx -cL(z) + L_*(z). \tag{3}$$

Here L_* is the path function along an upward directed path outside the disk.

Tyler (1968) applied Duntley's (1952) contrast formula and assumed that at each depth z the path function $L_{*D}(z)$ above the Secchi disk is practically uninfluenced by the disk, and therefore approximately equal to the path function $L_*(z)$ of the background. This assumption may be questioned, and especially at positions close to the disk. However, let us assume that it is still valid for most of the path between the disk and the surface. By making the approximation $L_{*D}(z) \approx L_*(z)$ and subtracting Eq. (3) from Eq. (2) we obtain

$$\frac{d[L_D(z) - L(z)]}{dz} \approx c[L_D(z) - L(z)]. \tag{4}$$

Provided c is constant with depth the integration of this equation between the surface and the Secchi depth $z = Z_D$ results in

$$[L_D(0) - L(0)] = [L_D(Z_D) - L(Z_D)]\,e^{-cZ_D}. \tag{5}$$

We recognise the left side of this equation as the numerator of Eq. (1). On the right-hand side we have the radiance $L_D(Z_D)$ which is the reflected part of the downward irradiance $E_d(Z_D)$ incident at the disk. The reflectance of this radiance may be termed ρ_{DL}, defined by

$$\rho_{DL} = \frac{L_D(Z_D)}{E_d(Z_D)}. \tag{6}$$

At the same level outside the disk the upward radiance $L(Z_D)$ will be related to the downward irradiance by the radiance reflectance R_L:

$$R_L(Z_D) = \frac{L(Z_D)}{E_d(Z_D)}. \tag{7}$$

In Eqs. (6)–(7) it is assumed that the downward irradiance at the Secchi depth is practically unaffected by the presence of the disk, and has the same value above as outside the disk. By using Eqs. (6)–(7) the radiances on the right side of Eq. (5) may be written as functions of $E_d(Z_D)$, and Eq. (5) becomes

$$[L_D(0) - L(0)] = [\rho_{DL}E_d(Z_D) - R_L(Z_D)E_d(Z_D)]e^{-cZ_D}. \tag{8}$$

At this point we see that it would be very convenient if we could also express the denominator $L(0)$ on the right-hand side of Eq. (1) as a function of $E_d(Z_D)$. $L(0)$ can be transformed in several ways, but the simplest expression is obtained by using $E_d(z)$ and $R_L(z)$ and writing $L(z)$ as an exponential function of z:

$$L(0) = L(Z_D)e^{K_L Z_D} = R_L(Z_D)E_d(Z_D)e^{K_L Z_D}, \tag{9}$$

where K_L is the average vertical attenuation coefficient of $L(z)$ in the depth range 0–Z_D.

If we insert for the numerator in Eq. (1) from Eq. (8) and for the denominator from Eq. (9), we obtain:

$$C(0) = \frac{L_D(0) - L(0)}{L(0)} = \frac{[\rho_{DL} - R_L(Z_D)]e^{-cZ_D}}{R_L(Z_D)e^{K_L Z_D}} = \frac{[\rho_{DL} - R_L(Z_D)]}{R_L(Z_D)}e^{-(c+K_L)Z_D}, \tag{10}$$

which can be more conveniently written as

$$Z_D(c+K_L) = \ln\left(\frac{\frac{\rho_{DL}}{R_L(Z_D)}-1}{C(0)}\right). \tag{11}$$

The contrast $C(0)$ is equal to the threshold value C_t, provided the eye of the observer is below the surface. Tyler (1968) presented a version of Eq. (11), and Hou et al. (2007) obtained a similar expression by using modulation transfer theory.

If the observer's eye is above the surface, the radiance from the direction of the disk will consist of two terms. The first is $L_D(0)\ \tau/n^2$, where τ is the radiance transmittance for a ray of normal incidence at the water–air interface, and n is the index of refraction for water. This is the transmitted part of the radiance $L_D(0)$. The other term is the radiance L_r from sun and sky reflected at the surface towards the observer. The sum of the terms becomes $L_D(0)\ \tau/n^2 + L_r$. Similarly the radiance from the background will now be $L(0)\ \tau/n^2 + L_r$. It should be noted that the first part of the latter sum is the quantity that is usually termed the water-leaving radiance L_w

$$L_w = L(0)\frac{\tau}{n^2}. \tag{12}$$

The contrast of the Secchi disk observed in air through a flat sea surface becomes

$$\begin{aligned} C_{air} &= \frac{\left(L_D(0)\frac{\tau}{n^2}+L_r\right)-\left(L(0)\frac{\tau}{n^2}+L_r\right)}{\left(L(0)\frac{\tau}{n^2}+L_r\right)} \\ &= \frac{L_D(0)-L(0)}{L(0)\left(1+\frac{L_r}{L_w}\right)} = \frac{C(0)}{1+\frac{L_r}{L_w}}. \end{aligned} \tag{13}$$

If capillary waves are present, they will have a blurring effect on the image of the disk, and as a result the apparent contrast will be reduced. Preisendorfer (1986) expressed the transmittance W of $C(0)$ at the water–air interface as a function of the angular subtense of the Secchi disk as seen from just below the surface, and the variance of the slopes of the capillary waves. In our notation his expression can be written

$$W = 1-\exp\left(-\frac{D^2k}{Z_D^2U}\right) \le 1, \tag{14}$$

where the symbols k and U represent the constant 787 m s^{-1} and the wind speed, respectively, and D is the diameter of the disk. When W is included in Eq. (13), the equation becomes

$$C_{air} = \frac{C(0)}{1+\frac{L_r}{L_w}}W = C_t. \tag{15}$$

In this case it is C_{air} that represents the threshold value C_t. Equation (15) can also be written as

$$C(0) = \frac{C_t}{W}\left(1+\frac{L_r}{L_w}\right). \tag{16}$$

The contrast $C(0)$ has to be greater than C_t, and in inverse proportion to the blurring effect W. $C(0)$ will also be a linear function of the ratio L_r/L_w between the reflected and water-leaving radiances. The greater the surface reflection is, the greater $C(0)$ has to be. It will be convenient for our purposes to express L_r and L_w as functions of the same input above the surface, namely the downward irradiance E_{air}. L_r can be written as

$$L_r = \bar{\rho}_{L,air}E_{air}, \tag{17}$$

where $\bar{\rho}_{L,air} = L_r/E_{air}$ is the Fresnel-reflected radiance towards the zenith, depending on the slope distribution of the surface and the angular distribution of radiance from sky and sun. L_w can be transformed by using the quantity $\mathfrak{R}$ introduced by Morel and Gentili (1996):

$$\mathfrak{R} = \frac{L_w}{L(0)}\frac{E_d(0)}{E_{air}} = \frac{L_w}{E_{air}}\frac{1}{R_L(0)}. \tag{18}$$

By inserting for L_r and L_w from Eqs. (17) and (18) into Eq. (16) we obtain

$$C(0) = \frac{C_t}{W}\left(1+\frac{L_r}{L_w}\right) = \frac{C_t}{W}\left(1+\frac{\bar{\rho}_{L,air}}{\mathfrak{R}R_L(0)}\right). \tag{19}$$

Equation (11) can then be written, by substituting for $C(0)$ from Eq. (19), as

$$Z_D(c+K_L) = \ln\left(\frac{\frac{\rho_{DL}}{R_L(Z_D)}-1}{1+\frac{\bar{\rho}_{L,air}}{\mathfrak{R}R_L(0)}}\frac{W}{C_t}\right) = \ln(A), \tag{20}$$

where A is defined as the product of the two fractions inside the greater parentheses. This is our basic equation for the threshold depth of the white Secchi disk observed from above the surface. The numerator of the first fraction is a function of the reflective properties ρ_{DL} of the disk and the radiance reflectance $R_L(Z_D)$ of the background at the Secchi depth, the denominator is a function of the ratio between the surface-reflected and water-leaving radiances, W represents the influence of capillary waves on the transmittance of $C(0)$ through the surface, while C_t is the threshold contrast of the human eye.

Observation of the disk through coloured glass filters will change the threshold depths, and the coefficients c, K_L, ρ_{DL}, R_L, $\bar{\rho}_{L,air}$ and $\mathfrak{R}$ will refer to other wavelengths than the spectral region of maximum transmittance. Otherwise Eq. (20) remains the same.

Equation (20) contains 10 variables, and if we had had precise values of these at a large number of stations, it would have been very interesting to test the equation. However, the usefulness of an expression needing the input of nine variables to predict the one remaining would be limited. It would have been far more convenient if the right-hand side of Eq. (20) had been a constant. The Secchi depth Z_D would then become linearly proportional to $1/(c+K_L)$, or $(c+K_L)$

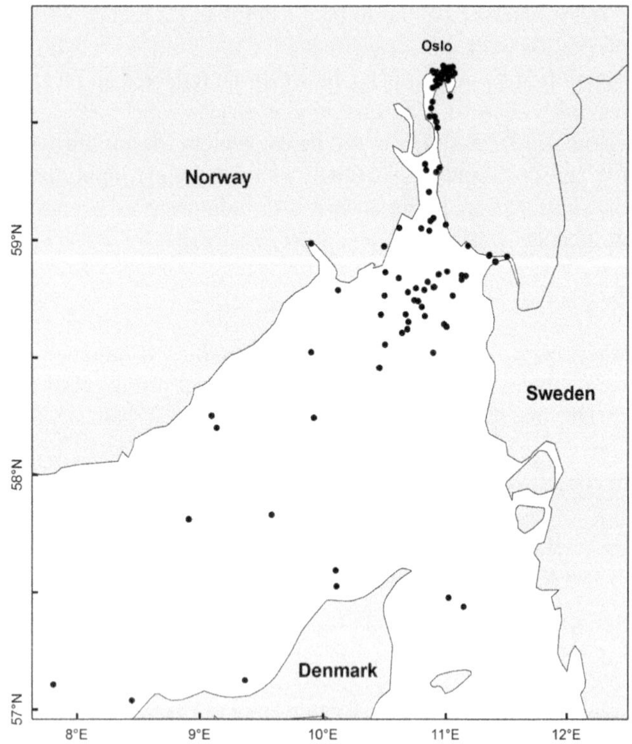

Fig. 1. Locations of the stations.

would be proportional to $1/Z_\mathrm{D}$. Unfortunately, R_L, $\bar{\rho}_{L,\mathrm{air}}$ and $\mathfrak{R}$ are variables determined by the optical properties of the sea water and sea surface, C_t is a function of the Secchi depth and the background luminance, while W varies with the wind speed and the Secchi depth. Thus the variables on the right-hand side of Eq. (20) are neither independent nor constant, but the logarithmic function will reduce the variation of the expression inside the parentheses, which is why the equation may still provide a useful support for other marine-optical observations.

A more practical form of Eq. (20) for our purposes is

$$Z_\mathrm{D}(c+K_L) = \ln\left(\frac{\rho_\mathrm{DL}}{R_L(Z_\mathrm{D})} - 1\right) - \ln\left(1 + \frac{\bar{\rho}_{L,\mathrm{air}}}{\mathfrak{R}R_L(0)}\right) + \ln(W) - \ln(C_\mathrm{t}) = \ln(A_1) - \ln(A_2) + \ln(W) - \ln(C_\mathrm{t}) = \ln(A), \quad (21)$$

where A_1 and A_2 are defined by the expressions inside the parentheses. In order to test Eq. (21) we will calculate the mean value of the expression on the left-hand side of the equation, as well as the mean values of the different logarithmic functions on the right-hand side.

Some special cases can be pointed out:

- For a perfectly black disk, that is a disk where $\rho_\mathrm{DL} = 0$ (Eq. 6), the quantity A_1 of Eq. (21) obtains the value -1, while C_t will be a negative number. The two negative signs will cancel each other inside the parenthesis of $\ln(A)$.
- If $L_\mathrm{r}/L_\mathrm{w} \approx 0$ (Eqs. 16–17), the quantity A_2 becomes 1 and $\ln(A_2) = 0$.
- When Z_D is observed by using a water telescope, the effects of waves and surface-reflected sky radiance are eliminated, so that $W = 1$ in addition to $\bar{\rho}_{L,\mathrm{air}} \approx 0$, and $\ln(A_2) = \ln(W) = 0$.

The different steps leading up to Eq. (21) make it possible to discuss the different factors that influence the Secchi depth. We will determine the attenuation coefficients c and K_L from calculated luminances in Sect. 4.1, estimate mean values of the quantities $\ln(A_1)$, $\ln(A_2)$, $\ln(W)$ and $\ln(C_\mathrm{t})$ in Sects. 4.2–4.3, and then see how the mean value of $(c+K_L)_\mathrm{phot} Z_\mathrm{D,white}$ relates to the combined effects of these quantities according to Eq. (21) in Sect. 4.4. In Sects. 4.5–4.7 we will look at the effects of colour filters, disk size, sun glitter, ship shadow and waves. Section 4.8 will check how the assumption made early in Sect. 2, about attenuation coefficients of luminances in photopic units corresponding to monochromatic coefficients at the wavelength of maximum transmittance, agrees with our results.

3 Environment, data sets and methods

In this section we describe briefly the environmental conditions of the investigated area, the data sets and the applied instruments and methods.

3.1 Environmental conditions

Our main area of investigation has been the Oslofjord, with additional data from the Skagerrak and two stations from the Kattegat (Fig. 1). The waters of the Oslofjord are in general eutrophic due to a supply of nutrients from the surrounding settlements (Ibrekk and Holtan, 1988). There is an estuarine circulation in the fjord, but rather weak in the inner part of the fjord. There is usually an upper and a lower layer separated by a transition layer, the pycnocline. The surface layer may sometimes be well-mixed, but often there will be a gradual change of properties from the upper 1–2 m down to the pycnocline (Aure et al., 1996; Staalstrøm et al., 2012). Salinities in the surface layer are typically in the range 20–29, and in deep waters up to 34 (Gade, 1963, 1967; Staalstrøm et al., 2012). Such values are of interest because a low salinity means a high content of fresh water, and this may indicate a high content of yellow substance, which is a significant optical component in these areas. The Secchi depth is found above the lower layer in the fjord. Maps of the Secchi depth, observed in 1988, have been presented by Aas et al. (1989), and Andresen (1993) has studied seasonal and annual changes of the Secchi depth in the period from 1936 to 1992.

The Skagerrak serves as a transition zone between the North Sea and the Baltic (Aarup et al., 1996a, b; Højerslev et al., 1996), and it also supplies the more saline waters to the Oslofjord. A very thorough analysis of the Secchi depths in the North Sea–Baltic Sea region has been made by Aarup (2002). Surface salinities in the northeastern part of the Skagerrak are in the range 25–32, and below the surface layer the waters become more Atlantic, with salinities up to 35 (Højerslev et al., 1996; Aarup et al. 1996a).

The environmental parameters salinity, wind speed, cloudiness and Secchi depth are presented for the different parts of our investigated area in Table 1. The salinity ranges are based on observations from different years. The Secchi data for the Inner and Outer Fjord are based on observations from 1982–1992, tabulated by Andresen (1993), while the data for the Skagerrak are observations from 1900–1999, based on a figure presented by Aarup (2002). The Norwegian Meteorological Institute has calculated mean values and standard deviations of wind speed and cloudiness at two locations: Fornebu, representing the Inner Fjord, and the Færder Lighthouse at 59.03° N, representing the border between the Outer Fjord and the Skagerrak. The input data are diurnal means from the entire 30 yr period 1961–1990. It should be pointed out, however, that cloudiness in these regions does not follow a Gaussian distribution.

Average values for the inherent optical properties of the Skagerrak–Outer Oslofjord area were found by Sørensen et al. (2007). At 442 nm the absorption coefficient of yellow substance was $a_y(442) = 0.62\,\mathrm{m}^{-1}$, the bleached particle absorption $a_{bp}(442) = 0.065\,\mathrm{m}^{-1}$, and the particle scattering $b_p(442) = 0.645\,\mathrm{m}^{-1}$. The approximate spectral variations of these coefficients within the range 400–550 nm were $a_y(\lambda) = a_y(442)\,e^{-(0.0105\,\mathrm{nm}^{-1})(\lambda-442\,\mathrm{nm})}$, $a_{bp}(\lambda) = a_{bp}(442)\,e^{-(0.0089\,\mathrm{nm}^{-1})(\lambda-442\,\mathrm{nm})}$ and $b_p(\lambda) = b_p(442)\left[\frac{442\,\mathrm{nm}}{\lambda}\right]^{0.376}$ where λ is the wavelength, implying that the contributions from yellow substance and particles to the attenuation coefficient c are of the same order of magnitude in the blue part of the spectrum, while particles will tend to dominate in the red part.

During 2002–2003 spectra of upward radiance just beneath the surface in the Oslofjord and Skagerrak were recorded for calibration and validation purposes (Sørensen et al., 2003, 2004, 2007), related to the ESA committee MAVT. The peak values of the upward radiance spectra usually occurred in a wavelength range from 480 to 570 nm, with the mean wavelength around 525 nm. The bluish maxima were only observed at stations that were strongly influenced by Atlantic waters (salinity close to 35). The half-peak bandwidths of these spectra were typically 150 nm. Figure 2 shows the mean value of the spectra at the MERIS channels, normalised at 555 nm, as well as the standard deviation. An example of the rather rare Atlantic spectrum is included for comparison. When the radiance spectra are multiplied by the CIE photopic efficiency function described in Sect. 2, the shapes of the resulting spectral upward luminances are significantly narrowed. The half-peak bandwidths will now be typically around 80 nm, and the peak values will be situated between 550 and 570 nm.

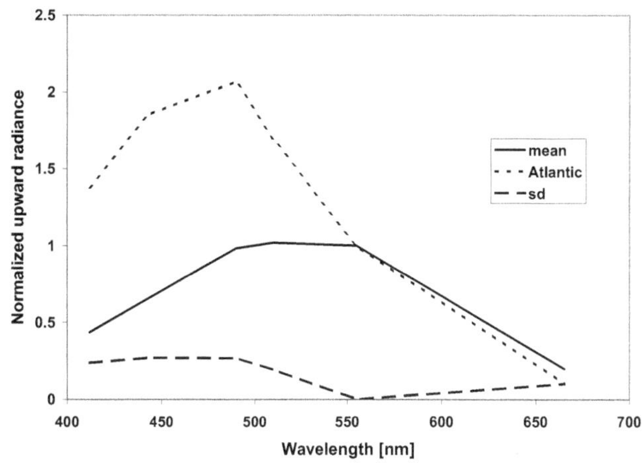

Fig. 2. Mean value of the upward radiance spectra just beneath the surface, normalised at 555 nm and based on 48 stations, and the standard deviation of the normalised spectra. For comparison a station of typical Atlantic water is presented.

3.2 First data set and methods

The first data set consists of data from 26 stations, collected in 1992 between May and December, in the Inner Oslofjord. The set contains the threshold depths of different types of Secchi disks observed by the open eye as well as with coloured glass filters. Two sizes of white and black disks were used, the standard size with a diameter of 30 cm, and a smaller one with a diameter of 10 cm. It is difficult to obtain a perfectly black disk, defined as a disk with no reflectance at all. Disks painted black or made from black materials may still have a radiance reflectance ρ_{DL} that is greater than the corresponding reflectance R_L of the background water. The closest approximation to the non-reflecting disk that we have used has been a bowl-shaped lamp shade of brass, painted black for the scientific purpose, with diameter 30 cm. This device was originally acquired by the University of Copenhagen, but later on kindly donated to the University of Oslo by N. K. Højerslev. The instrument is supposed to work in principle very much like the light trap termed Rayleigh's Horn. Measurements were also taken with a 3.0 m-long water telescope (donated by the University of Copenhagen) that reached from the ship rail into the sea. The properties of the photopic sensitivity of the eye alone and in combination with blue, green and red glass filters (Mikaelsen and Aas, 1990) are shown in Table 2. The filters were produced by Schott and termed BG12, VG9 and RG1. The latter filter corresponds to OG590 in the latest Schott catalogue. The combined effect of

Table 1. Environmental conditions in the Oslofjord–Skagerrak area. S is salinity range between the surface and the Secchi depth, U is mean $\pm$ standard deviation of wind speed in m s^{-1}, C is mean $\pm$ sd of cloudiness in octas, and $Z_{D,white}$ is mean $\pm$ sd of Secchi depth in m.

Area	Inner Fjord north of 59.67° N	Outer Fjord 59.00°–59.67° N	Skagerrak 57.00°–59.00° N
S (all year)[a,b]	15–28	16–30	20–32
U (summer)[c]	2.8 ± 0.5	6.7 ± 1.1	
U (winter)[c]	2.0 ± 0.7	7.9 ± 1.6	
C (summer)[c]	5.0 ± 0.8	4.8 ± 0.7	
C (winter)[c]	5.5 ± 0.9	5.6 ± 0.9	
$Z_{D,white}$ (summer)[d]	4.4 ± 1.7	4.3 ± 1.8	
$Z_{D,white}$ (winter)[d]	10.5 ± 2.8	11.1 ± 3.2	
$Z_{D,white}$ (all year)[d,e]	7.4 ± 2.3	7.7 ± 2.6	8.3 ± 2.8

References: [a] Gade (1963, 1967); Aure et al. (1996); Staalstrøm et al. (2012); [b] Aarup et al. (1996a); Højerslev et al. (1996); [c] Norwegian Meteorological Institute (personal communication, 2014); [d] Andresen (1993); [e] Aarup (2002).

eye sensitivity, filter transmittance and a typical spectrum of upward luminances $L(0)$ is also shown, and we see that the half-peak bandwidth of the naked eye is reduced from 100 to 78 nm, otherwise the bandwidths and peak wavelengths remain the same. The Secchi depth was observed on both the sunlit and shaded sides of the ship.

The radiance reflectance ρ_{DL} of the white disk was determined in 1992 from laboratory recordings of the spectral downward irradiance E_d and the corresponding reflected upward radiance L_D from the white disk, as expressed by Eq. (6). Both were measured in water just above the submerged disk. In order to avoid the shadow of the radiance meter on the disk, the instrument could not be held directly above it, but at an angle of 30°–40° away from the axis of the disk. Still, we think that the obtained values of ρ_{DL} were close to the correct ones, being 0.25–0.29–0.30 at the wavelengths 450–520–550 nm, respectively. These values will be discussed in Sect. 4.2.

The accuracy of a Secchi depth observation depends mainly on the wind conditions. Winds will produce waves and may make the ship drift, with a sloping line from the observer down to the disk as a result. It will then become more difficult for the observer to estimate the correct depth below the waves. Other effects of waves are discussed in Sects. 4.2 and 4.3, and values of observed wind speed are presented for the different investigated areas in Table 1. Based on experience we estimate the possible error of $Z_{D,white}$ to be in the range 0.2–0.5 m, depending on wind speed. It may be mentioned that the observations in this and the other data sets were made by experienced oceanographers.

In addition to the Secchi observations quanta irradiance was recorded in the sea with the Underwater Quantum Sensor LI-192SB from LI-Cor, inc., Lincoln, Nebraska, and with the LI-190SB Quantum Sensor on deck as a reference.

3.3 Second data set and methods

A second data set consists of 143 stations from the Inner and Outer Oslofjord, collected between 1973 and 2008 by the University of Oslo during courses and project excursions, with recordings of quanta irradiance and Secchi depths. The period of observation included all months from February to December. The Secchi depth was determined with the 30 cm white disk on the sunlit side of the ship, and quanta irradiance was recorded with the Li-Cor instrument mentioned above.

3.4 Third data set and methods

The third data set consists of observations at 79 stations. Secchi depths, spectral irradiances, radiances, and absorption and attenuation coefficients were measured in the Oslofjord–Skagerrak area during the years 2002–2003, in the summer period from early May to early September.

The Secchi depth was observed with the 30 cm white disk, by the naked eye and with colour filters, often on both the sunlit and shaded sides of the ship. Vertical sub-surface profiles of the downward and upward irradiances and the upward radiance were measured with the PRR-600 from Biospherical Instruments, San Diego, California, with the deck instrument PRR-610 used as a reference. The spectral channels were 412, 443, 490, 510, 555 and 665 nm, corresponding to the channels of the satellite sensors SeaWiFS and MERIS. Vertical profiles of the upward radiance L and the downward and upward irradiances E_d and E_u were recorded. According to the radiance model of Aas and Højerslev (1999), the quantity $Q = L/E_u$ should obtain values in the interval from π to 2π only, and a few stations where Q lay outside this range have been omitted. Immersion coefficients determined by the manufacturer and self-shading effects (Gordon and Ding, 1992; Zibordi and Ferrari, 1995; Aas and Korsbø, 1997) were taken into account. Radiances and irradiances were plotted in semi-logarithmic diagrams and extrapolated

Table 2. Wavelength λ_p of peak value and half-peak bandwidth $\Delta\lambda$ of the eye's photopic sensitivity, alone and in combination with the upward spectral luminance L_u; for the open eye and with filters for the white disk (Mikaelsen and Aas, 1990).

	Open (white and black disk)	Blue filter	Green filter	Red filter
Eye's sensitivity				
λ_p [nm]	555	460	540	620
$\Delta\lambda$ [nm]	100	48	56	38
Eye's sensitivity and L_u				
λ_p [nm]	555	470	540	620
$\Delta\lambda$ [nm]	78	47	55	34

up to the surface. The uncertainty of the resulting surface values was estimated to be $\pm 10\,\%$. A few series of spectral upward radiance just beneath the surface, recorded with the hyperspectral Ramses-ARC radiance sensor from TriOS, have been included to examine the wavelength of the spectral peak.

Vertical profiles of the spectral absorption and scattering coefficients were recorded with an ac-9 from WET Labs, Philomath, Oregon. The applied instrument records at 412, 440, 488, 510, 532, 555, 650, 676, and 715 nm. The data were cleaned for obvious noise, and unrealistic spikes were avoided by using a median filter, resulting in an estimated uncertainty of $\pm 10\,\%$. The recordings with the Biospherical instrument and the ac-9 were analysed and sent to the ESA for calibration and validation purposes (Sørensen et al., 2003, 2004, 2007).

At 19 stations from the 2002–2003 data set chlorophyll *a* (Chl) and total suspended material (TSM) were sampled on glass fibre filters. The concentrations of Chl were determined by the high-performance liquid chromatographic (HPLC) method, and the TSM by a gravimetric method (Sørensen et al., 2007, and references therein).

Figure 1 shows the locations of the stations for the three data sets. The Inner Fjord contains data from all three sets (168 stations), the Outer Fjord data from the second and third sets (48 stations), while the Skagerrak only contains data from the third set (32 stations).

Tables 3, 5, 6, 7, and 9 are based on data solely from the third set, Table 4 on data from the first and third sets, while Table 8 has applied data from all three sets.

4 Test of Eq. (21) in photopic units

In Sects. 4.1–4.3 we will obtain mean values for the different terms in Eq. (21), and in Sect. 4.4 we will test the equation with these values. Other results of our Secchi depth experiments, related to Eq. (21), will be discussed in Sects. 4.5–4.8.

4.1 Values of c_{phot} and $K_{L,\mathrm{phot}}$

In Sect. 2 it was pointed out that the luminance is not a practical quantity in our marine-optical research, because the corresponding coefficient of beam attenuation will depend on the spectral shape of the luminance at the point where the attenuation starts and the distance along the beam. The coefficient c of Eq. (4) describes the beam attenuation of two different luminances: (1) the upward luminance from the disk and (2) the corresponding upward luminance from the surrounding waters. The radiance reflectance ρ_{DL} of the white Secchi disk is defined by Eq. (6). If the albedo of the disk had been 1 and the disk had acted like a perfect Lambert diffuser, the reflected radiance would have been constant for all directions with a radiance reflectance equal to $1/\pi = 0.32$. A real non-perfect Secchi disk will not have this value, but a number of the same order of magnitude. Tyler (1968) applied an estimate of ρ_{DL} equal to $0.82/\pi = 0.26$. Haltrin (1998) presented albedos for white and coloured disks, and provided the disks had acted like perfect diffusers, the values of $\rho_{\mathrm{DL}}$for the white disk would have been 0.23–0.26–0.26 at 450–520–550 nm. Our earlier determinations of ρ_{DL}, described in Sect. 3.2, are very close to the values presented above, being 0.25–0.29–0.30 at the wavelengths 450–520–550 nm, respectively. The latter values have been applied here.

From the recorded spectral downward irradiances $E_{\mathrm{d}}(Z_{\mathrm{D}})$ at the Secchi depth in the Oslofjord and Skagerrak the reflected upward radiances $L_{\mathrm{D}}(Z_{\mathrm{D}})$ from the disk were then estimated by Eq. (6). The beam attenuations of the spectral radiances from this depth and up to the surface were calculated by applying recorded values of spectral c, and the resulting spectrum of radiances just beneath the surface was determined. The radiance spectra at the Secchi depth and at the surface were then integrated spectrally by using the photopic efficiency function, and from the resulting two luminances the efficient attenuation coefficient $c_{\mathrm{phot,disk}}$ could be obtained. The attenuation coefficient c_{phot} of the upward background luminance $L(z)$ between the Secchi depth and the surface was calculated in a similar way. All values of the ratio $c_{\mathrm{phot,disk}}/c_{\mathrm{phot}}$ have been found to lie within the range 1.00 ± 0.01. Thus the two attenuation coefficients are practically equal.

The upward luminances at the surface and the Secchi depth, $L(0)$ and $L(Z_D)$, were found by integrating the spectral upward luminances at the two depths, and then $K_{L,\mathrm{phot}}$ could be determined by using Eq. (9). Because the observed values of Z_D for the black disk were less than for the white disk (Table 3), the spectral luminance $L(Z_D)$ at this smaller depth differed from the former luminance, and accordingly c_{phot} and $K_{L,\mathrm{phot}}$ for the black disk also became slightly different (Table 3). The attenuation coefficients c_{phot} and $K_{L,\mathrm{phot}}$ for the luminances observed through colour filters were calculated in the same way as for the open-eye coefficients.

4.2 Values of ρ_{DL}, R_L, $\mathfrak{R}$, $\bar{\rho}_{L,\mathrm{air}}$, $\ln(A_1)$ and $\ln(A_2)$

It was explained in Sect. 4.1 how the downward illuminance and the reflected luminance from the disk were determined, and which spectral values of ρ_{DL} we have applied. From the latter values ρ_{DL} was estimated equal to 0.30 for the white disk observed through the red filter, and by definition equal to 0 for the black disk. The values of ρ_{DL} for the white disk observed with the open eye and through the blue and green filters were determined by using the spectral recordings of downward illuminance, and the results are 0.29, 0.27 and 0.29, respectively (Table 3).

The mean value of $R_L(0)$, defined as the ratio between upward luminance and downward illuminance just beneath the surface (Eq. 7), has been calculated from 32 stations as 0.56 % for the open eye. This value is valid for both the white and black disks (Table 3). However, the mean value of $R_L(Z_D)$ at the Secchi depth becomes different for the white and black disks, being 0.83 % and 0.61 %, respectively, and the reason for this is that Z_D differs for these disks. The estimated values of R_L for the colour filters are shown in Table 3.

It was found by Aas et al. (2009) that an average value for the quantity $\mathfrak{R}$ in the Oslofjord–Skagerrak area was 0.506 ± 0.045 for solar altitudes between 15° and 60°, which is close to similar results found by Morel and Gentili (1996) and Mobley (1999). $\mathfrak{R}$ was assumed to be independent of wavelength, and the found value has been applied for both the white and black disks (Table 3).

The radiance reflected towards the zenith at the surface of the sea is described by the radiance reflectance $\bar{\rho}_{L,\mathrm{air}}$. Its value for a clear sky depends on the angular distribution of sky radiance, the solar altitude, the ratio between diffuse sky irradiance and direct solar irradiance, and the statistical distribution of surface slopes. Polynomials for calculating $\bar{\rho}_{L,\mathrm{air}}$ were obtained during a previous work (Aas, 2010). The problem was simplified by looking at average values for all possible angles between wind direction and solar azimuth, using the Cox and Munk model with a one-dimensional Gaussian distribution for the surface slopes (Cox and Munk, 1954a, b). The input was spectral sky and solar radiance data from the Oslo region (Høkedal and Aas, 1998; Aas and Høkedal, 1999). However, there is the problem that while the wave height and wave spectrum are functions of wind speed, duration and fetch, only the wind speed appears in the Cox–Munk model. A comparison between wave heights and wind speeds at those of our stations where both quantities were observed, shows that in many cases either the wind duration or the fetch must have had a limiting effect on the wave height. Still, the mean values ± the standard errors of wind speed and wave height, $5.5 \pm 0.4\,\mathrm{m\,s^{-1}}$ and 0.7 ± 0.1 m, respectively, are consistent with the conditions for a fully developed sea shown in the diagram by Gröen and Dorrestein (1976, also shown by WMO, 1998). Accordingly we have tentatively chosen the wind speed $5.5\,\mathrm{m\,s^{-1}}$ to represent the average conditions. The mean values and standard deviations of $\bar{\rho}_{L,\mathrm{air}}$ shown in Table 3 are based on 10 different cases of atmospheric radiance distribution. The values of $\bar{\rho}_{L,\mathrm{air}}$ for the open eye (the white and black disks) and for the blue, green and red filters were obtained from atmospheric radiance distributions at 550, 470, 540 and 620 nm, respectively (Table 3).

The estimated mean value and standard deviation of $\ln(A_1)$ were obtained by using the value of ρ_{DL} shown in Table 3, and varying values of $R_L(Z_D)$ for each station. Similarly the values of $\ln(A_2)$ were found from the values of $\bar{\rho}_{L,\mathrm{air}}$ and $\mathfrak{R}$ suggested by Table 3 and varying values of $R_L(0)$. The results for $\ln(A_1)$ and $\ln(A_2)$ are presented in Table 3.

The size distributions of the optical quantities estimated here are often highly asymmetric, implying that the standard deviation does not always provide a satisfactory description of the range of variation, like when the standard deviation is greater than the mean value.

4.3 Value of C_t and W

It was mentioned in Sect. 2 that according to Blackwell (1946) the contrast threshold C_t will depend on the angle α subtended by the observed target, the luminance $L(0)$ of the background, the probability of detection, and the exposure time. In Part I of Blackwell's investigation circular targets, brighter than the background, had an exposure time of six seconds, and the contrast thresholds were those that corresponded to a 50 % detection probability. In Part II the targets were darker than the background, and the conclusion was that in most cases negative stimuli are equivalent to positive stimuli of the same area and contrast. The results in Part III were obtained when an indefinitely long exposure time was used. A comparison between Blackwell's experiments and Secchi disk observations is not straightforward, because Blackwell's target images were constant during the time of exposure, while at sea the image of the Secchi disk is usually varying. We have chosen an indefinitely long exposure time and the highest detection probability, 100 %.

The apparent angle α from the observer's eye across the Secchi disk is a function of the diameter D of the disk, the height H of the observer's eye above the surface of the sea, the Secchi depth Z_D and the refractive index n. Because α is a small angle, $\tan(\alpha/2)$ can be approximated by $\alpha/2$, and

Table 3. Mean value and standard deviation of the different quantities and terms of Eqs. (19)–(20) in the Oslofjord and Skagerrak, for the 30 cm white and black disks. Colour filters are for the white disk. Recordings of radiance and irradiance at 32 stations have been applied.

	White	Blue filter	Green filter	Red filter	Black
Z_D [m]	7.8 ± 2.7	4.1 ± 1.4	5.7 ± 2.0	4.2 ± 1.5	2.0 ± 0.7
$(c + K_L)_{phot}$ [m^{-1}]	1.09 ± 0.67	1.41 ± 0.99	1.08 ± 0.67	1.48 ± 0.56	1.11 ± 0.68
c_{phot} [m^{-1}]	0.83 ± 0.51	0.97 ± 0.65	0.83 ± 0.52	1.03 ± 0.41	0.84 ± 0.51
$K_{L,phot}$ [m^{-1}]	0.26 ± 0.19	0.44 ± 0.36	0.25 ± 0.18	0.45 ± 0.18	0.27 ± 0.19
ρ_{DL} [%]	29 ± 0	27 ± 1	29 ± 0	30	0
$R_L(0)$ [%]	0.56 ± 0.37	0.47 ± 0.38	0.70 ± 0.47	0.17 ± 0.12	0.56 ± 0.37
$R_L(Z_D)$ [%]	0.83 ± 1.38	0.66 ± 1.33	0.80 ± 1.09	0.28 ± 0.13	0.61 ± 0.51
$\bar{\rho}_{L,air}$ [%]	0.12 ± 0.08	0.20 ± 0.10	0.14 ± 0.09	0.10 ± 0.08	0.12 ± 0.08
$\Re$ [%]	51	51	51	51	51
C_t [%]	0.56 ± 0.02	0.67 ± 0.12	0.56 ± 0.03	0.59 ± 0.10	-0.54 ± 0.01
W [%]	33 ± 23	66 ± 25	49 ± 27	64 ± 25	94 ± 8
$\ln(A_1)$	3.9 ± 0.7	4.1 ± 0.7	3.8 ± 0.6	4.7 ± 0.4	0
$\ln(A_2)$	0.4 ± 0.1	0.7 ± 0.2	0.4 ± 0.1	0.9 ± 0.3	0.4 ± 0.1
$\ln(W)$	-1.3 ± 0.7	-0.5 ± 0.4	-0.9 ± 0.6	-0.5 ± 0.4	-0.1 ± 0.1
$\ln(C_t)$	-5.2 ± 0.0	-5.0 ± 0.2	-5.2 ± 0.1	-5.1 ± 0.1	-5.2 ± 0.0
$\ln(A)$	7.3 ± 0.9	7.9 ± 0.7	7.8 ± 0.8	8.4 ± 0.7	4.7 ± 0.1
$Z_D(c + K_L)_{phot}$	7.0 ± 1.3	4.8 ± 1.2	5.1 ± 0.9	5.5 ± 0.7	1.9 ± 0.3

$\tan(j)$, where j is the corresponding angle of refraction in water, can be approximated by $\alpha/(2n)$, using Snell's Law. The angle α in radians becomes

$$\alpha = \frac{D}{H + (Z_D/n)}. \tag{22}$$

If we use Eq. (22) with $D = 0.3$ m, $H = 3$ m, and $n = 1.33$, and if Z_D lies in the interval from 0.5 to 14 m, then the angle α across the Secchi disk will lie between 1.3° and 5.1°. This is within the range in Blackwell's study where the angles of α extended from 0.01° to 6°. Our recordings showed that the background luminances $L(0)$ in the Oslofjord–Skagerrak area were in the range 0.4–1600 cd m^{-2} for the observations with open eye and colour filters. Blackwell's luminances extended from 3×10^{-6} to 3400 cd m^{-2}. Thus our environmental conditions were within the ranges described by Blackwell's experiments. We have selected those of Blackwell's data that are closest to our observed background luminances and angles of subtense, and interpolated these data to fit our ranges. The estimated mean values of C_t and $\ln(C_t)$ for the 30 cm disk, observed with the open eye and through colour filters, are presented in Table 3.

Rather than calculating C_t for each single case and then finding the mean value and standard deviation from the resulting data set, we have found it necessary to restrict the calculations of C_t to the mean, maximum and minimum conditions. The mean conditions were assumed to produce the mean value of C_t, and a crude estimate of the standard deviation was obtained by the expression $(C_{t,max} - C_{t,min})/4$. For a normal distribution, 95 % of the observations will fall within a range of ±2 standard deviations from the mean value.

In Table 3 the mean value of C_t is 0.56 % for the white disk, and the mean value of $\ln(C_t)$ becomes −5.2. Krümmel (1889) referred to Helmholtz for the threshold contrast $1/133 = 0.75$ % which produces $\ln(C_t) = -4.9$. The relative difference between the last number and our estimate is only 6 %. Tyler (1968) applied $C_t \approx 0.66$ %, leading to $\ln(C_t) = -5.0$, which is even closer to our estimate. Gordon and Wouters (1978) applied C_t values from 0.15 to 0.6 % or $\ln(C_t)$ from −5.1 to −6.5 in their model studies. Højerslev (1986) deduced from his Baltic recordings a threshold contrast slightly greater than our estimate, $C_t = (0.70 \pm 0.03)$ %, resulting in $\ln(C_t) = -(4.96 \pm 0.04)$. Considering that C_t is not supposed to be a constant, these values are surprisingly similar.

It is noteworthy that Blackwell's experiments involved both the colour and night visions of the human eye. It is well known that night vision requires some time to be activated and adapted, and that a sudden bright glint may temporarily change the vision from night to colour mode. The range where both colour and night visions are active, termed mesopic vision, is 0.001–10 cd m^{-2} according to the CIE. Thus some of our colour filter observations are within this range, but to what extent the varying light conditions have influenced the resulting Secchi depths, we cannot say.

Since our Secchi depths usually are from coastal areas where the surrounding land masses reduce the influence of the wind on the ordinary sea waves, the use of Eq. (14) may perhaps also be overestimating the effect of the wind on the capillary waves. Still, we have tentatively estimated W by using the observed wind speeds and Secchi depths. The average values of W and $\ln(W)$, based on the 32 stations constituting

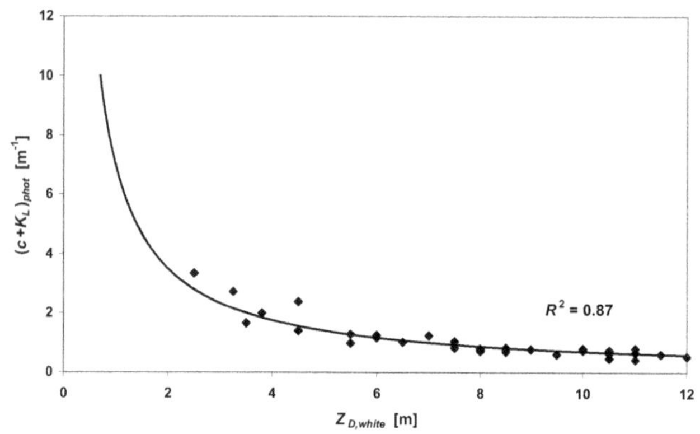

Fig. 3. $(c + K_L)_{\text{phot}}$ as a function of $Z_{\text{D,white}}$. The line is the function $(c + K_L)_{\text{phot}} = 7.0/Z_{\text{D,white}}$. See text in Sect. 4.4 for discussion. R^2 is the coefficient of determination for this line.

the data set where we have complete observations of c, K_L and R_L, are shown in Table 3.

4.4 Value of ln(A) and comparison to observations of $Z_{\text{D,white}}\,(c + K_L)_{\text{phot}}$

There are different ways to test Eq. (21) and to estimate a value for ln(A), and the result will depend on the chosen method. By using the estimated mean values of ln(A_1), ln(A_2), ln(W) and ln(C_t) from Table 3, ln(A) in Eq. (21) obtains the value 7.3 for the 30 cm white disk. The mean value of $Z_{\text{D,white}}(c + K_L)_{\text{phot}}$, found from observations, is 7.0 (Table 3), which is 4 % less than the value predicted by ln(A). Early in the analysis it was discovered that whenever $Z_{\text{D,white}}$ was in the range 1–2 m, the values of $Z_{\text{D,white}}(c + K_L)_{\text{phot}}$ tended to increase to 10–14, and consequently the few stations where $Z_{\text{D,white}} < 2$ m have been omitted. The reason for the discrepancy between theory and observations for small Secchi depths may be that one or more of the assumptions on which Eq. (21) is based becomes invalid. For instance, the assumption $L_{*\text{D}}(z) \approx L_*(z)$ made for Eq. (4) may be less good when Z_{D} is small. Figure 3 presents $(c + K_L)_{\text{phot}}$ as a function of $Z_{\text{D,white}}$ at the 32 stations where we have sufficient observations down to the Secchi depth, with the curve

$$(c + K_L)_{\text{phot}} = \frac{7.0}{Z_{\text{D,white}}} \tag{23}$$

added. The curve represents the mean value $(c + K_L)_{\text{phot}} Z_{\text{D,white}} = 7.0$, according to Table 3. It has been extended to values of $Z_{\text{D,white}} < 2$ m to illustrate that it will significantly underestimate $(c + K_L)_{\text{phot}}$ in this range. The symbol $R^2 = 0.87$ in Fig. 3 is the coefficient of determination. The general definition of this quantity for a data set (x, y) and a chosen trendline $f(x)$ is $R^2 = 1 - (\varepsilon^2/s_y^2)$, where ε is the root mean square of the residuals y–$f(x)$, and s_y is the standard deviation of y. It should be pointed out that while the quantity R^2 depends on the chosen trendline, the coefficient of correlation r is solely a function of the data set (x, y), since r is defined by $r = S_{x,y}/(s_x s_y)$, where $S_{x,y}$ is the covariance of x and y, and s_x is the standard deviation of x.

Another way to test the mean value ln(A) = 7.3 in Eq. (21) is to correlate $(c + K_L)_{\text{phot}}$ with $1/Z_{\text{D,white}}$. The best-fit line through the origin obtains the slope 7.5, and if $(c + K_L)_{\text{phot}}$ is estimated by $7.5/Z_{\text{D,white}}$, the rms of the error $[(c + K_L)_{\text{phot}} - 7.5/Z_{\text{D,white}}]$ becomes 0.23 m^{-1}. The mean value of $(c + K_L)_{\text{phot}}$ is 1.09 m^{-1} (Table 3), and the rms error relative to this value represents 21 %. It is interesting that if we estimate $(c + K_L)_{\text{phot}}$ by using Eq. (23), the rms error only changes to 0.24 m^{-1} or 22 %. The coefficient of determination in this case becomes $R^2 = 0.88$.

If the observer's eye is below the surface, $\bar{\rho}_{L,\text{air}} = 0$ and $W = 1$, and the value of ln(A) in Eq. (21) becomes 8.7. Tyler (1968) estimated the number 8.69 for this case, while Holmes (1970) found for sub-surface observations in turbid coastal waters that the average value of $Z_{\text{D,white}}(c + K_L)_{\text{phot}}$ was 9.4. Højerslev (1977) obtained the same value for the sub-surface case. Thus our theoretical estimate of ln(A) for the sub-surface case agrees with Tyler, but deviates 8 % from the estimates by Holmes and Højerslev. Højerslev (1977) found that for observations above the surface, the product $Z_{\text{D,white}}(c + K_L)_{\text{phot}}$ should be in the range 7.9–9.4. This is 8–29 % above our estimate in Table 3.

4.5 Effect of colour filters and the black disk

When the white Secchi disk is observed through the coloured glass filters, the optical coefficients and ln(A) change values, as shown by Table 3. The table demonstrates that in this case the observed mean values of $Z_{\text{D}}(c + K_L)_{\text{phot}}$ are more than 30 % smaller than the estimated mean values of ln(A).

The best-fit correlation lines $y = A + Bx$ and $y = B_0 x$ for the observations of $Z_{\text{D,blue}}$, $Z_{\text{D,green}}$ and $Z_{\text{D,red}}$ as functions of $Z_{\text{D,white}}$ have been determined, and the found constants A, B and B_0 as well as the mean values of the ratios $(y/x) = Z_{\text{D,filter}}/Z_{\text{D,white}}$ are presented in Table 4. We see that the colour filters reduce the Secchi depths to 50–70 % of the depths for the open eye. The result that $Z_{\text{D,green}}$ on average is reduced to almost 70 % of $Z_{\text{D,white}}$, although the wavelengths of peak photopic sensitivity for $Z_{\text{D,green}}$ and $Z_{\text{D,white}}$ are very close (Table 2), may be due to the half-peak bandwidth of the green filter, which is only half of that for the open eye.

Table 4 also shows the correlation lines $y = A + Bx$ with the constant term A. When the values of B_0, B and $(y/x)_{\text{mean}}$ are close to each other, it means that the observations lie close to a straight line through the origin.

Mikaelsen and Aas (1990) observed the Secchi depth in the Inner Oslofjord during 1986–1987. Based on their data from 11 stations the mean values of the ratio $Z_{\text{D,filter}}/Z_{\text{D,white}}$ become 0.61, 0.81 and 0.62 for the blue, green and red filter,

Table 4. Linear relationships of the forms $y = A + Bx$ and $y = B_0 x$ obtained by correlation analysis of Z_D (white disk) observed by the open eye and with blue, green and red glass filters, and $Z_{D,\text{black}}$ (disk) and $Z_{B,\text{black}}$ (bowl) observed by the open eye. Mean values of y/x and y and their standard deviations are included. The size of the disk is 30 cm if not otherwise indicated. r is the correlation coefficient, ε is the root-mean-square error of the deviations $(y–A–Bx)$, and ε_0 is the rms of $(y–B_0x)$. N is the number of data pairs (y, x).

y [m]	x [m]	r	A [m]	B	B_0	$(y/x)_{\text{mean}\pm\text{sd}}$	$y_{\text{mean}\pm\text{sd}}$ [m]	ε [m]	ε_0 [m]	N
$Z_{D,\text{blue}}$	$Z_{D,\text{white}}$	0.89	0.2	0.49	0.52	0.53 ± 0.10	4.0 ± 1.6	0.7	0.7	25
$Z_{D,\text{green}}$	$Z_{D,\text{white}}$	0.96	−0.5	0.79	0.73	0.71 ± 0.09	5.5 ± 2.4	0.7	0.7	25
$Z_{D,\text{red}}$	$Z_{D,\text{white}}$	0.94	0.5	0.48	0.54	0.56 ± 0.08	4.2 ± 1.5	0.5	0.5	25
$Z_{B,\text{black}}$	$Z_{D,\text{black}}$	0.63	0.8	0.77	1.20	1.29 ± 0.44	2.2 ± 0.7	0.5	0.5	19
$Z_{B,\text{black}}$	$Z_{D,\text{white}}$	0.11	1.9	0.03	0.23	0.26 ± 0.11	2.2 ± 0.7	0.7	0.9	19
$Z_{D,\text{black}}$	$Z_{D,\text{white}}$	0.54	0.8	0.17	0.26	0.29 ± 0.10	2.2 ± 0.9	0.8	0.8	50
$Z_{D,\text{white},10}$	$Z_{D,\text{white},30}$	0.99	−0.4	0.85	0.81	0.80 ± 0.06	6.9 ± 2.8	0.5	0.5	21
$Z_{D,\text{blue},10}$	$Z_{D,\text{blue},30}$	0.98	−0.1	0.91	0.90	0.88 ± 0.08	3.2 ± 1.5	0.3	0.3	13
$Z_{D,\text{green},10}$	$Z_{D,\text{green},30}$	0.94	1.0	0.70	0.83	0.87 ± 0.12	5.3 ± 2.5	0.8	0.9	13
$Z_{D,\text{red},10}$	$Z_{D,\text{red},30}$	0.95	0.1	0.88	0.90	0.91 ± 0.10	3.9 ± 1.5	0.4	0.4	13
$Z_{D,\text{all},10}$	$Z_{D,\text{all},30}$	0.98	0.4	0.78	0.83	0.86 ± 0.10	5.1 ± 2.7	0.6	0.6	60
$Z_{D,\text{white,shade}}$	$Z_{D,\text{white,sun}}$	0.96	−0.2	0.99	0.97	0.97 ± 0.10	8.3 ± 3.3	0.9	0.9	34
$Z_{D,\text{blue,shade}}$	$Z_{D,\text{blue,sun}}$	0.91	0.5	0.74	0.83	0.87 ± 0.15	4.1 ± 1.8	0.7	0.8	29
$Z_{D,\text{green,shade}}$	$Z_{D,\text{green,sun}}$	0.94	0.1	0.90	0.91	0.91 ± 0.12	5.6 ± 2.6	0.9	0.9	29
$Z_{D,\text{red,shade}}$	$Z_{D,\text{red,sun}}$	0.85	0.4	0.82	0.89	0.91 ± 0.16	4.0 ± 1.4	0.7	0.8	29
$Z_{D,\text{all,shade}}$	$Z_{D,\text{all,sun}}$	0.95	−0.2	0.96	0.93	0.91 ± 0.14	5.6 ± 3.0	0.9	0.9	121
$Z_{D,\text{white},10,\text{tel}}$	$Z_{D,\text{white},10}$	0.96	−1.0	1.37	1.22	1.14 ± 0.20	7.2 ± 3.8	1.1	1.2	15
$Z_{D,\text{blue},10,\text{tel}}$	$Z_{D,\text{blue},10}$	0.93	−0.1	1.18	1.15	1.14 ± 0.18	3.3 ± 1.8	0.6	0.6	11
$Z_{D,\text{green},10,\text{tel}}$	$Z_{D,\text{green},10}$	0.96	−0.7	1.27	1.15	1.09 ± 0.15	5.2 ± 2.9	0.7	0.8	11
$Z_{D,\text{red},10,\text{tel}}$	$Z_{D,\text{red},10}$	0.92	0.1	1.14	1.15	1.16 ± 0.18	4.2 ± 1.8	0.7	0.7	11
$Z_{D,\text{all},10,\text{tel}}$	$Z_{D,\text{all},10}$	0.96	−0.5	1.28	1.19	1.14 ± 0.18	5.2 ± 3.1	0.9	0.9	48

respectively, which is 11–15 % greater than our results of 0.53, 0.73 and 0.54 in Table 4. Both the first and the latter set of values resemble Lisitzin's results from the Baltic Sea (1938) when $Z_{D,\text{white}} < 10$ m. In the clear and blue waters of the Florida Shelf where $Z_{D,\text{white}}$ was 21–26 m, Højerslev's colour filter observations (1985) yielded mean values of the ratios equal to 0.90, 0.91 and 0.29 for the blue, green and red filters. The difference from our results in greenish coastal waters is clearly a result of the difference in water colour.

Table 3 predicts that the average value of $Z_{D,\text{black}}(c + K_L)_{\text{phot}}$ should be $\ln(A) = 4.7$ for the 30 cm black disk, but the observed value is less than half of this: 1.9. This could be because an ordinary disk of black plastic was used in these observations, and not the bowl-shaped device. The surface of the plastic disk seemed to be entirely black on deck, but looked dark grey in the sea, and brighter than its background. In 1992 a series of measurements at 19 stations was taken with the black-painted bowl together with the black and white disks. The depth $Z_{B,\text{black}}$ of the black bowl was on average 20–30 % greater than the depth $Z_{D,\text{black}}$ of the black disk (Table 4). Thus the bowl produces a smaller reflection and a greater contrast than the disk. Since the estimated attenuation $(c + K_L)_{\text{phot}}$ is approximately the same for the black and white disks as shown by Table 3 (the minor difference being due to $Z_{D,\text{black}}$ being smaller than $Z_{D,\text{white}}$), the values of $\ln(A)$ in Table 3 imply that the ratio $Z_{B,\text{black}}/Z_{D,\text{white}}$ should be approximately $4.7/7.3 = 0.64$. However, the observed ratio in Table 4 is only 0.23–0.26. This could mean that our black bowl is not perfectly black, but it could also be that the use of colour filters and the black disk or bowl introduces effects that are not included in Eq. (21).

4.6 Effect of size

The observations with the 10 cm disks were made in 1992, when instruments for spectral recordings of c and K_L were not available. Consequently the two sides of Eq. (21) cannot be tested directly for the 10 cm disk, but other experiments have been made. The angle α of the total field of view across the 10 cm Secchi disk can be calculated from Eq. (22) with $D = 0.1$ m, $H = 3$ m and $n = 1.33$. Because the observed values of Z_D for this smaller disk lie in the interval from 0.5 to 12 m, the angle α across the Secchi disk becomes 0.5°–1.7°. The threshold contrast for the smaller disk becomes greater than for the 30 cm disk, while the value of W will be reduced according to Eq. (14), because D^2 is reduced more than Z_D^2. The combined effect is that the values of $\ln(A)$ in Table 3 for the 30 cm white and black disks will be reduced by 22 and 13 %, respectively. For the blue, green and red colour filters the decreased size produces similar reductions of 18, 21, and 17 %. This can be tested directly by comparing the values of Z_D for disk diameters of 10 and

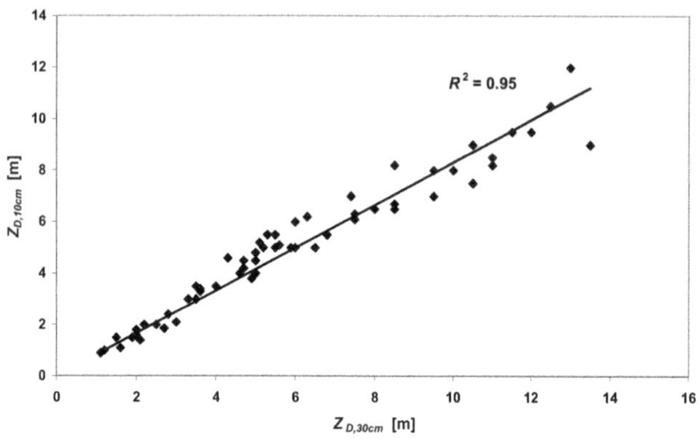

Fig. 4. All naked eye and colour filter observations of Z_D with a 10 cm disk as a function of the observations with a 30 cm disk. The best-fit line through the origin is the function $Z_{D,10\,cm} = 0.83\, Z_{D,30\,cm}$.

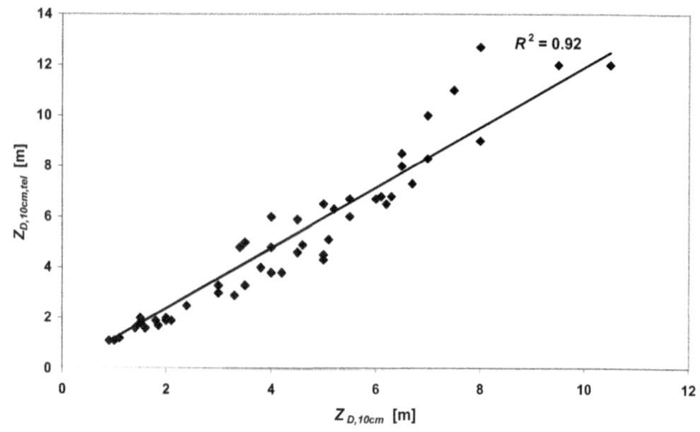

Fig. 5. All naked eye and colour filter observations of $Z_{D,10\,cm,tel}$ with a telescope as a function of $Z_{D,10\,cm}$ without a telescope. The best-fit line through the origin is the function $Z_{D,10\,cm,tel} = 1.19 Z_{D,10\,cm}$.

30 cm. The results in Table 4 show that the relative influence of size is greater for the open eye than for the colour filters, and that the decrease in diameter from 30 to 10 cm reduces the Secchi depth by an average of 10–20 %, in agreement with our estimates. Figure 4 presents all open-eye and colour filter observations for 10 and 30 cm disks put together, and the best fit line through the origin obtains a slope of 0.83, indicating an average Secchi depth reduction of 17 %.

4.7 Effects of sun glitter, water telescope and ship shadow

The effect of a water telescope is to eliminate sun glitter and skylight reflection at the surface, and to reduce the blurring effect of capillary waves. If we insert a surface reflectance $\bar{\rho}_{L,\mathrm{air}} = 0$ and a wave factor $W = 1$ into Eq. (21), we find that $\ln(A)$ should increase by 12 % for the 30 cm white disk, based on the estimated quantities in Table 3. For the colour filters and black disk the effect is in the range ± 6 %. Mikaelsen and Aas (1990) tested the effect at four stations in 1987 with a 30 cm disk. According to their observations $Z_{D,\mathrm{white}}$ increased by 11 % on average using the water telescope, and by including the observations with the colour filters the increase became 15 %. New experiments were done in 1992, and a 10 cm disk was used in order to see it properly within the field of view of the telescope. The results, listed in the last five rows of Table 4, show that the telescope increased $Z_{D,\mathrm{white},10}$ by an average of 14 %, and the effect on $Z_{D,\mathrm{filter},10}$ was of the same magnitude. If all open-eye and filter observations are put together (Table 4), we see that the average increase in the Secchi depth by using the telescope is the same, 14 %. The best-fit line through the origin obtains the slope 1.19 (Fig. 5), indicating an average increase of 19 %.

Holmes' use of a water telescope (1970) only increased the observed depth by 2–4 %, which is significantly less than our result. Højerslev (1986) found for Baltic waters that the ratio between the depth $Z_{D,\mathrm{tel}}$ observed with a water telescope and the ordinary Z_D could be approximated by a function of the wave height H in units of metres: $Z_{D,\mathrm{rel}}/Z_D = 1 + 0.4\, H$. This expression underestimates the effect at our 11 water telescope stations where H was < 0.1 m, but we have not tested it in more rough seas. Sandén and Håkansson (1996) investigated the effect of wind on Z_D, and it seems like the presence of wind tended to reduce the depth by ~ 10 %.

If we observe the Secchi disk on the sunlit side of the ship, sun glitter from the sea will reduce its threshold depth. On the shadow side there may be less glitter, but the ship shadow will also reduce the luminances from both the background and the disk, and this may require a greater contrast, and thus lead to a smaller depth. The balance between gains and losses related to the absence or presence of direct sunlight is demonstrated by Fig. 6, which presents the Secchi depths observed on both sides of the ship with the white disk. The best-fit line through the origin obtains the slope 0.97, which is close to 1. If we do the same experiment for all depths observed with colour filters and the open eye (Table 4), we see that the slope becomes 0.93, that is Z_D is now on average reduced by 7 % on the shadow side.

4.8 The monochromatic assumption

In Sect. 2 it was assumed as a first approximation that the beam attenuations of the upward luminances resembled the attenuations of the radiances at the wavelength of maximum spectral transmittance, and Sect. 4.1 described how the efficient attenuation coefficients $c_{\mathrm{phot,disk}}$ and c_{phot} of upward luminance from the disk and background could be estimated. In this section we will investigate the relationships between the photopic and monochromatic coefficients as well as the errors introduced by the monochromatic assumption.

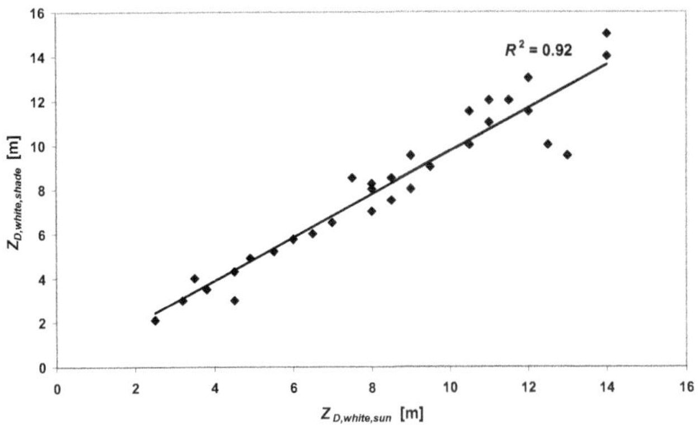

Fig. 6. All naked eye and colour filter observations of Z_D, on the shaded versus the sunlit side of the ship. The best-fit line through the origin is the function $Z_{D,shade} = 0.93 Z_{D,sun}$.

The coefficients $c_{phot,disk}$ and c_{phot} have been correlated with the monochromatic beam attenuation coefficient c_{555} at 555 nm, and the results are presented in Table 5. Similarly, the vertical attenuation coefficient $K_{L,phot}$ has been correlated with $K_{L,555}$. The correlations are described by best-fit lines on the forms $y = A+Bx$ and $y = B_0x$. Table 5 also displays values of $(y/x)_{mean}$ and y_{mean}. We see that the values of B, B_0 and $(y/x)_{mean}$ are close to each other, indicating that the points (x, y) lie close to the lines. This is confirmed by the very small errors ε and ε_0 introduced by using the correlation lines ($\approx 0.01\,m^{-1}$). ε and ε_0 are defined as the root mean squares of the deviations $(y–A–Bx)$ and $(y–B_0x)$, respectively. The coefficient c_{phot}, for instance, is almost identical to c_{555}, with an average ratio of 1.02. However, Table 5 also demonstrates that the deviations between the photic and monochromatic coefficients are greatest for $K_{L,phot}$ versus $K_{L,555}$, which displays a slope of 1.12. The sum $(c+K_L)_{phot}$ versus $(c+K_L)_{555}$ obtains the slope 1.03.

It follows from the descriptions of c_{phot} and $K_{L,phot}$ that these coefficients are apparent optical properties, depending on the ambient light field. It could not be stated a priori that they would be so strongly correlated with the monochromatic coefficients at 555 nm, with a coefficient of correlation equal to 1.00. We have not investigated how the found relationships will work in more clear and bluish sea waters, but we assume that the correlations may be weaker there. Holmes (1970) equipped his irradiance and beam attenuation meters with filters where the spectral shape of the transmittance resembled the spectral sensitivity of the eye, but the attenuation meter then recorded a lamp spectrum that probably was different from the natural spectrum within the sea.

The correlation results for attenuation coefficients derived from luminances observed through the colour filters (Table 5) versus monochromatic coefficients show that the correlation coefficients are all very close to 1.0. The slope B_0 for the blue and green filters obtains values between 0.95 and 1.01, while the red filter deviates more from 1.0 with the slopes 1.06 and 1.21. The slopes B_0 for the sum $(c+K_L)$ are in the range 0.97–1.10. None of these deviations from the slope 1.0 explain the discrepancies between theory and observations displayed by Table 3 for the colour filters. Our overall conclusion becomes that the monochromatic assumption works satisfactorily, and that the mentioned discrepancies must have other causes. In Blackwell's experiments (1946) the targets were projections of "white" light from electrical lamps onto a white screen. The different parts of the visible spectrum were not investigated separately, and it could perhaps be that C_t also has a spectral dependence.

It remains to test the ability of Eq. (21) to describe the relationships between Z_D and monochromatic values of c, K_L and $(c+K_L)$. This will be done in the next section.

5 Relationships between the Secchi depth and other quantities

5.1 Estimates of monochromatic coefficients

In Sect. 2 we assumed that the Secchi depth observed with the open eye might be determined by the attenuation coefficients at the wavelength of maximum water transmittance, that is around 555 nm. Section 4.4 demonstrated that the numerical values of the constants in the relationships between Z_D and $(c+K_L)$ would depend on the chosen method. If our intention is to estimate $(c+K_L)_{555}$ from observed $Z_{D,white}$, the obvious choice, based on the form of Eq. (21), is to perform a linear correlation analysis between $(c+K_L)_{555}$ and $1/Z_{D,white}$. The slope of the line through the origin obtains the value 7.4, which is practically the same as the slope 7.5 found in Sect. 4.4 for $(c+K_L)_{phot}$. However, usually in marine-optical research the singular values of c and K_L are more interesting than their sum $(c+K_L)$, and therefore c_{555} and $K_{L,555}$ have been separately correlated with $1/Z_{D,white}$, as shown by Table 6. While the correlation coefficient is 0.95 for c_{555}, which is almost as perfect as it can be expected to be, it is reduced to 0.72 for $K_{L,555}$. Similar results have been obtained for the channels at 412, 443, 490, 510, 620 and 665 nm (Table 6). The root-mean-square errors of K_L and c, estimated from the correlation lines, are in the range $0.1–0.6\,m^{-1}$, but because K_L is smaller than c the relative errors become greater for K_L than for c. The result that the correlation coefficients are greater for c than for K_L is as expected since c contributes more than K_L to the sum $(c+K_L)$ which determines $Z_{D,white}$.

Often the vertical attenuation coefficient of downward irradiance, K_d, will be a more useful quantity than K_L. The correlation results for K_d at the MERIS channels as a function of $1/Z_{D,white}$ are presented in Table 6. Some of the values are remarkably similar to those obtained for K_L, and the reason for this is that the mean value of the ratio K_d/K_L usually is close to 1. In order to complete the investigation, the coefficient K_u of upward irradiance has been included in

Table 5. Linear relationships of the forms $y = A + Bx$ and $y = B_0x$ obtained by correlation analysis of attenuation coefficients of spectrally integrated luminance for the open eye and with blue, green and red glass filters and coefficients of monochromatic radiance at 470, 540, 555 and 620 nm, and mean values of y/x and y. The attenuation coefficients are averages over the depth range 0–$Z_{D,white}$. r is the correlation coefficient. The error ε is the root mean square of the deviations (y–A–Bx), and ε_0 is the rms of $(y - B_0x)$. The analysis is based on 32 stations.

y [m^{-1}]	x [m^{-1}]	r	A [m^{-1}]	B	B_0	$(y/x)_{mean\pm sd}$	$y_{mean\pm sd}$ [m^{-1}]	ε [m^{-1}]	ε_0 [m^{-1}]
$c_{phot,disk}$	c_{555}	1.00	0.02	0.99	1.01	1.02 ± 0.01	0.83 ± 0.50	0.01	0.02
c_{phot}	c_{555}	1.00	0.01	1.00	1.01	1.02 ± 0.01	0.83 ± 0.51	0.01	0.01
$K_{L,phot}$	$K_{L,555}$	1.00	0.00	1.12	1.12	1.11 ± 0.05	0.26 ± 0.19	0.01	0.01
$(c + K_L)_{phot}$	$(c + K_L)_{555}$	1.00	0.00	1.03	1.03	1.03 ± 0.01	1.09 ± 0.67	0.01	0.01
$c_{phot,blue}$	c_{470}	1.00	0.06	0.90	0.95	0.98 ± 0.04	0.97 ± 0.65	0.04	0.05
$c_{phot,green}$	c_{540}	1.00	0.02	0.97	0.98	1.00 ± 0.01	0.83 ± 0.52	0.01	0.02
$c_{phot,red}$	c_{620}	1.00	0.18	0.91	1.06	1.12 ± 0.07	1.03 ± 0.41	0.03	0.08
$K_{L,phot,blue}$	$K_{L,470}$	0.99	0.05	0.95	1.01	1.10 ± 0.05	0.44 ± 0.36	0.04	0.05
$K_{L,phot,green}$	$K_{L,540}$	1.00	0.01	0.96	0.98	1.00 ± 0.01	0.25 ± 0.18	0.00	0.01
$K_{L,phot,red}$	$K_{L,620}$	0.99	0.09	1.05	1.21	1.27 ± 0.08	0.45 ± 0.18	0.02	0.04
$(c + K_L)_{phot,blue}$	$(c + K_L)_{470}$	1.00	0.11	0.92	0.97	1.02 ± 0.04	1.41 ± 0.99	0.07	0.09
$(c + K_L)_{phot,green}$	$(c + K_L)_{540}$	1.00	0.03	0.96	0.98	1.00 ± 0.01	1.08 ± 0.67	0.01	0.02
$(c + K_L)_{phot,red}$	$(c + K_L)_{620}$	1.00	0.25	0.94	1.10	1.16 ± 0.07	1.48 ± 0.56	0.04	0.11

Table 6. Relationships of the forms $y = A + Bx$ and $y = B_0x$, where A, B and B_0 are coefficients from the correlation analysis between y and $x = 1/Z_{D,white}$. The error ε is the root mean square of the deviations (y–A–Bx), and ε_0 is the rms of (y–B_0x). r is the correlation coefficient and N is the number of stations.

y [m^{-1}]	r	A [m^{-1}]	B	B_0	$(y/x)_{mean\pm sd}$	$y_{mean\pm sd}$ [m^{-1}]	ε [m^{-1}]	ε_0 [m^{-1}]	N
c_{412}	0.86	−0.24	10.6	9.6	8.8 ± 2.3	1.67 ± 1.20	0.58	0.60	79
c_{443}	0.91	−0.23	9.0	8.0	7.2 ± 1.7	1.38 ± 0.97	0.37	0.39	79
c_{490}	0.94	−0.22	7.6	6.6	6.0 ± 1.2	1.14 ± 0.79	0.22	0.28	79
c_{510}	0.95	−0.21	7.2	6.4	5.7 ± 1.1	1.10 ± 0.75	0.20	0.26	79
c_{555}	0.95	−0.17	6.6	5.9	5.4 ± 1.0	1.02 ± 0.68	0.18	0.20	79
c_{620}	0.95	0.03	6.1	6.3	6.3 ± 0.9	1.14 ± 0.63	0.18	0.18	79
c_{665}	0.93	0.18	5.8	6.5	6.9 ± 1.1	1.21 ± 0.60	0.19	0.21	79
$K_{L,412}$	0.66	0.01	4.6	4.6	4.5 ± 1.8	0.91 ± 0.77	0.57	0.57	53
$K_{L,443}$	0.64	−0.05	4.2	4.0	3.7 ± 2.0	0.78 ± 0.72	0.55	0.55	53
$K_{L,490}$	0.72	−0.02	2.4	2.3	2.2 ± 0.9	0.45 ± 0.36	0.25	0.25	53
$K_{L,510}$	0.72	0.00	2.0	2.0	1.9 ± 0.7	0.39 ± 0.30	0.21	0.21	53
$K_{L,555}$	0.72	0.05	1.3	1.5	1.6 ± 0.5	0.31 ± 0.20	0.14	0.14	53
$K_{L,620}$	0.77	0.15	1.5	2.1	2.4 ± 0.7	0.45 ± 0.22	0.14	0.15	53
$K_{L,665}$	0.78	0.22	1.7	2.5	3.0 ± 0.8	0.55 ± 0.24	0.15	0.18	53
$K_{d,412}$	0.59	−0.02	5.9	5.8	5.4 ± 3.3	1.14 ± 1.11	0.89	0.89	53
$K_{d,443}$	0.64	−0.05	4.2	4.0	3.7 ± 2.0	0.78 ± 0.72	0.55	0.55	53
$K_{d,490}$	0.70	−0.03	2.5	2.4	2.2 ± 1.0	0.46 ± 0.39	0.28	0.28	53
$K_{d,510}$	0.71	−0.02	2.1	2.0	1.9 ± 0.8	0.40 ± 0.32	0.22	0.22	53
$K_{d,555}$	0.74	0.02	1.4	1.5	1.5 ± 0.5	0.30 ± 0.21	0.14	0.14	53
$K_{d,620}$	0.82	0.25	1.2	2.2	2.8 ± 0.8	0.49 ± 0.16	0.09	0.15	53
$K_{d,665}$	0.86	0.41	1.1	2.7	3.7 ± 1.2	0.62 ± 0.13	0.07	0.21	53
$K_{u,412}$	0.43	0.18	6.5	7.2	7.0 ± 6.5	1.42 ± 1.61	1.43	1.43	51
$K_{u,443}$	0.52	0.05	5.2	5.4	5.1 ± 3.5	1.05 ± 1.08	0.92	0.92	51
$K_{u,490}$	0.67	−0.03	3.1	3.0	2.8 ± 1.4	0.58 ± 0.51	0.38	0.38	53
$K_{u,510}$	0.68	0.00	2.5	2.5	2.4 ± 1.1	0.50 ± 0.41	0.30	0.30	53
$K_{u,555}$	0.64	0.06	1.6	1.9	2.0 ± 0.8	0.38 ± 0.28	0.21	0.22	53
$K_{u,620}$	0.71	0.20	1.9	2.7	3.1 ± 1.1	0.58 ± 0.29	0.21	0.23	53
$K_{u,665}$	0.72	0.30	2.1	3.3	3.9 ± 1.3	0.71 ± 0.32	0.22	0.26	53

Table 7. Relationships of the forms $y = A + Bx$ and $y = B_0x$ obtained by correlation analysis of attenuation coefficients of monochromatic radiance and irradiance and Secchi depths observed with the open eye and with blue, green and red glass filters. The attenuation coefficients are averages over the depth range 0–$Z_{D,white}$, and r is the correlation coefficient. The error ε is the root mean square of the deviations $(y–A–Bx)$, and ε_0 is the rms of $(y–B_0x)$. The analysis is based on 25 stations.

y [m^{-1}]	x [m^{-1}]	r	A [m^{-1}]	B	B_0	$(y/x)_{mean \pm sd}$	$y_{mean \pm sd}$ [m^{-1}]	ε [m^{-1}]	ε_0 [m^{-1}]
c_{470}	$1/Z_{D,blue}$	0.84	−0.23	4.3	3.7	3.4 ± 1.1	1.09 ± 0.79	0.42	0.43
c_{470}	$1/Z_{D,white}$	0.94	−0.34	9.0	7.3	6.4 ± 1.6	1.09 ± 0.79	0.26	0.30
c_{540}	$1/Z_{D,green}$	0.94	−0.12	4.5	4.1	3.9 ± 0.9	0.91 ± 0.59	0.19	0.20
c_{540}	$1/Z_{D,white}$	0.96	−0.18	6.9	5.9	5.5 ± 1.1	0.91 ± 0.59	0.15	0.17
c_{620}	$1/Z_{D,red}$	0.89	0.24	2.6	3.3	3.6 ± 0.7	1.00 ± 0.49	0.22	0.25
c_{620}	$1/Z_{D,white}$	0.97	0.09	5.7	6.2	6.4 ± 0.9	1.00 ± 0.49	0.12	0.13
$K_{L,470}$	$1/Z_{D,blue}$	0.74	−0.17	2.0	1.6	1.4 ± 0.7	0.45 ± 0.42	0.28	0.29
$K_{L,470}$	$1/Z_{D,white}$	0.80	−0.20	4.1	3.1	2.6 ± 1.2	0.45 ± 0.42	0.25	0.26
$K_{L,540}$	$1/Z_{D,green}$	0.79	−0.04	1.3	1.2	1.2 ± 0.4	0.27 ± 0.21	0.13	0.13
$K_{L,540}$	$1/Z_{D,white}$	0.80	−0.06	2.0	1.7	1.6 ± 0.6	0.27 ± 0.21	0.12	0.13
$K_{L,620}$	$1/Z_{D,red}$	0.80	0.11	0.9	1.2	1.4 ± 0.4	0.38 ± 0.19	0.11	0.13
$K_{L,620}$	$1/Z_{D,white}$	0.83	0.07	1.9	2.3	2.5 ± 0.6	0.38 ± 0.19	0.10	0.11
$K_{d,470}$	$1/Z_{D,blue}$	0.73	−0.17	2.0	1.6	1.4 ± 0.7	0.44 ± 0.42	0.29	0.29
$K_{d,470}$	$1/Z_{D,white}$	0.82	−0.21	4.1	3.0	2.6 ± 1.2	0.44 ± 0.42	0.24	0.26
$K_{d,540}$	$1/Z_{D,green}$	0.80	−0.05	1.3	1.2	1.1 ± 0.4	0.26 ± 0.20	0.12	0.12
$K_{d,540}$	$1/Z_{D,white}$	0.81	−0.06	2.0	1.7	1.6 ± 0.5	0.26 ± 0.20	0.12	0.12
$K_{d,620}$	$1/Z_{D,red}$	0.80	0.26	0.62	1.3	1.7 ± 0.4	0.44 ± 0.13	0.07	0.15
$K_{d,620}$	$1/Z_{D,white}$	0.83	0.23	1.3	2.4	3.1 ± 0.8	0.44 ± 0.13	0.07	0.13
$K_{u,470}$	$1/Z_{D,blue}$	0.72	−0.34	3.1	2.2	1.8 ± 1.0	0.60 ± 0.65	0.44	0.47
$K_{u,470}$	$1/Z_{D,white}$	0.78	−0.38	6.2	4.3	3.4 ± 2.0	0.60 ± 0.65	0.39	0.43
$K_{u,540}$	$1/Z_{D,green}$	0.80	−0.06	1.6	1.4	1.4 ± 0.5	0.31 ± 0.25	0.15	0.15
$K_{u,540}$	$1/Z_{D,white}$	0.80	−0.07	2.4	2.0	1.9 ± 0.7	0.31 ± 0.25	0.15	0.15
$K_{u,620}$	$1/Z_{D,red}$	0.77	0.17	1.0	1.5	1.7 ± 0.5	0.47 ± 0.22	0.14	0.16
$K_{u,620}$	$1/Z_{D,white}$	0.82	0.12	2.2	2.8	3.1 ± 1.0	0.47 ± 0.22	0.13	0.14

Table 6. The results indicate that the errors of the estimated vertical coefficients are greatest for K_u and smallest for K_L, and less in the green and red part of the spectrum than in the blue part. We suspect that the rms errors at 412 and 443 nm of 90–100 % are due to varying amounts of yellow substance which will influence the coefficients more than the Secchi depth. In addition weak signals compared to noise and detection level at these wavelengths create errors.

Gordon and Wouters (1978) found in their model study that in relatively turbid water, defined by the authors as water where the backscattering probability is constant, the product $cZ_{D,white}$ would be approximately constant, while $(c + K_d)Z_{D,white}$ would vary more. This result is in agreement with the results in Table 6.

In Table 6 the errors of the estimates produced by the best-fit lines $y = A + Bx$ and $y = B_0x$, where y is an attenuation coefficient and $x = 1/Z_{D,white}$, are shown. Usually the line $y = B_0x$ through the origin works satisfactorily, but at some wavelengths the addition of a constant term A reduces the error of the estimates significantly. Figure 7 shows observations of K_d at 555 and 665 nm versus $x = 1/Z_{D,white}$, together with the best-fit lines $y = A + Bx$. The value of A is 0.02 m^{-1} at 555 nm, and 0.41 m^{-1} at 665 nm (Table 6). It is interesting that Jerlov's K_d values for the clearest ocean water type I are of the same order of magnitude: 0.07 m^{-1} at 555 nm and 0.40 m^{-1} at 665 nm (e.g. Jerlov, 1976). Figure 7 indicates that in the green part of the spectrum the constant term can be omitted without any practical consequences, while in the red part a better fit is obtained if A is included. The inclusion of the constant term A reduces the error at 665 nm from 0.21 m^{-1} to 0.07 m^{-1} according to Table 6. For the vertical attenuation coefficients K_L and K_u of upward directed light the improvements by adding a constant term A to the correlation lines are less.

Mikaelsen and Aas (1990) analysed five stations in the Oslofjord with recordings of c, and 11 stations with K_d. Z_D was observed by the naked eye and with colour filters. Most of their results overlap with the present findings. Sørensen et al. (1993) obtained from 189 stations in the Oslofjord–Skagerrak area the mean value and standard deviation $c_{green}Z_{D,white} = 6.2 \pm 1.3$, which fall within the corresponding range of Table 6.

The relationships between the Secchi depths observed through the glass filters and the coefficients c, K_L, K_d and K_u at 470, 540 and 620 nm are presented in Table 7. The results, based on 25 stations, show that the errors of the estimated

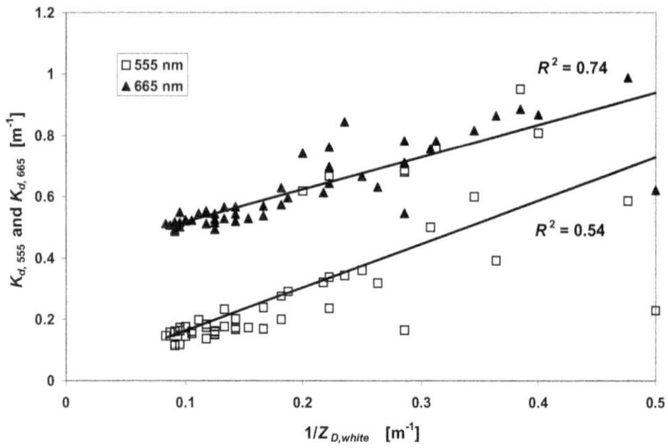

Fig. 7. K_d at 555 nm and 665 nm as a function of $1/Z_{D,\text{white}}$. The upper best-fit line is the function $K_{d,665} = [0.41\,\text{m}^{-1}] + 1.1/Z_{D,\text{white}}$ and the lower line is $K_{d,555} = [0.02\,\text{m}^{-1}] + 1.4/Z_{D,\text{white}}$.

coefficients are in the range 0.07–0.47 m^{-1}, or 10–80 % of the corresponding mean values. It is noteworthy that compared to the observations with the open eye, the colour filter observations represent no improvement in the accuracy of the estimated vertical attenuation coefficients. Especially for the coefficients of beam attenuation it is clear that $Z_{D,\text{white}}$ produces better estimates than $Z_{D,\text{filter}}$. We think this result is valid within our area of investigation, but not necessarily in other sea regions.

It may be noted that Table 7 contains analyses of attenuation coefficients at 620 nm that are parallel to those in Table 6, and that the results are different. The reason for this may be that Table 7 is based on stations with colour filter observations, and accordingly the number of applicable stations will be reduced to less than half of those in Table 6. Evidently the number of stations influences the statistical results.

Højerslev (1977) suggested the approximate mean values 3, 6 and 9 for the products $K_d Z_D$, cZ_D and $(c + K_d)Z_D$, respectively, where K_d, c and Z_D are observed through colour filters, but these findings are not confirmed by Table 7.

Because remote sensing of ocean colour is used to estimate coefficients like K_d, the technique can also be used to estimate the Secchi depth. Within our area of investigation the relationship between $1/Z_{D,\text{white}}$ and the water-leaving radiance L_w recorded in the red part of the spectrum (630–690 nm) by the TM3 sensor at the Landsat satellite was found to be $1/Z_{D,\text{white}} = [0.203\,\text{m}^{-1}] + [0.072\,\text{W}^{-1}\,\text{m sr}\,\mu\text{m}]L_w$, while the coefficient of correlation was 0.92 (Sørensen and Aas, 1994). For the TM1, TM2 and TM4 sensors in the blue-green (450–520 nm), green (520–600 nm) and near-infrared (760–900 nm) parts of the spectrum the coefficients of correlation became 0.69, 0.87 and 0.65, respectively. Kratzer et al. (2003) and Zhang et al. (2003) have discussed estimates of $1/Z_{D,\text{white}}$ based on satellite observation of the Baltic Sea, and Morel et al. (2007) have examined the Secchi depth estimates from various ocean colour sensors for the open ocean case. References to numerous investigations in lake waters can be found on the internet.

5.2 Quanta irradiance – PAR

The spectrally integrated quanta irradiance (400–750 nm), also termed PAR (photosynthetically available radiation), is one of several factors determining the primary production in the sea. If the transmittance of this irradiance between the surface and the Secchi depth is denoted T_D, then the average vertical attenuation coefficient K_q of the irradiance over the depth range $0 - Z_{D,\text{white}}$ is related to $Z_{D,\text{white}}$ by

$$T_D = e^{-K_q Z_{D,\text{white}}}. \tag{24}$$

The observed mean value ± the standard deviation of T_D, obtained from our third data set of 143 stations from the Oslofjord inside the Færder Light House at the Skagerrak border, is (9 ± 4) %. Based on the total data set of 205 stations, including stations from the nearby Skagerrak and Kattegat, the range becomes slightly greater: (9 ± 6) %.

Equation (24) can also be written as

$$K_q Z_{D,\text{white}} = -\ln(T_D), \tag{25}$$

and by using $T_D \approx 0.09 \pm 0.06$, the range of the product $K_q Z_{D,\text{white}}$ becomes 1.9–3.5. This is consistent with the mean value and standard deviation of the product obtained from observed pairs of K_q and $Z_{D,\text{white}}$: 2.5 ± 0.5 (Table 8). By linear correlation K_q may be estimated from $1/Z_{D,\text{white}}$ with an rms error that is less than 20 % of the mean value of K_q.

The depth of the euphotic zone, defined as the surface layer where there is a net positive production from photosynthesis, is often estimated as the depth $Z_q(1\,\%)$ where the quanta irradiance is reduced to 1 % of its surface value. If K_q is approximately constant with depth, $Z_q(1\,\%)$ can be determined from the equation

$$0.01 = e^{-K_q Z_q(1\,\%)} \tag{26}$$

which can be transformed to

$$Z_q(1\,\%) = \frac{4.61}{K_q}. \tag{27}$$

By inserting for K_q from Eq. (25), this expression becomes

$$Z_q(1\,\%) = \frac{4.61}{-\ln(T_D)} Z_{D,\text{white}}, \tag{28}$$

indicating a linear proportionality between $Z_q(1\,\%)$ and $Z_{D,\text{white}}$, provided $\ln(T_D)$ is approximately constant. Similar expressions can be deduced for $Z_q(3\,\%)$, $Z_q(10\,\%)$ and $Z_q(30\,\%)$.The correlation results for $Z_q(p\,\%)$ as a function of $Z_{D,\text{white}}$ are presented in Table 8, together with other related statistics.

Table 8. Linear relationships of the forms $y = A + Bx$ and $y = B_0x$, obtained by correlation analysis of the vertical attenuation coefficient K_q of downward quanta irradiance, the depths $Z_q(p\,\%)$ where the quanta irradiance is reduced to p percent of the surface value, and the Secchi disk depth $Z_{D,white}$. The error ε is the root mean square of the deviations $(y–A–Bx)$, ε_0 is the rms of $(y–B_0x)$, and r is the correlation coefficient. The analysis is based on 205 stations.

y [m^{-1}]	x [m^{-1}]	r	A [m^{-1}]	B	B_0	$(y/x)_{mean\pm sd}$	$y_{mean\pm sd}$ [m^{-1}]	ε [m^{-1}]	ε_0 [m^{-1}]
K_q	$1/Z_{D,white}$	0.82	0.18	1.6	2.2	2.6 ± 0.7	0.52 ± 0.20	0.11	0.14
y [m]	x [m]	r	A [m]	B	B_0	$(y/x)_{mean\pm sd}$	$y_{mean\pm sd}$ [m]	ε [m]	ε_0 [m]
$Z_q(30\,\%)$	$Z_{D,white}$	0.72	1.0	0.22	0.36	0.44 ± 0.16	2.3 ± 0.8	0.6	0.7
$Z_q(10\,\%)$	$Z_{D,white}$	0.78	2.1	0.47	0.77	0.92 ± 0.30	4.9 ± 1.6	1.0	1.4
$Z_q(3\,\%)$	$Z_{D,white}$	0.79	3.5	0.79	1.3	1.5 ± 0.5	8.1 ± 2.7	1.7	2.2
$Z_q(1\,\%)$	$Z_{D,white}$	0.74	5.0	1.1	1.8	2.2 ± 0.7	11.5 ± 4.1	2.8	3.5
$Z_q(30\,\%)$	$Z_q(1\,\%)$	0.80	0.4	0.16	0.20	0.20 ± 0.04	2.3 ± 0.8	0.5	0.5
$Z_q(10\,\%)$	$Z_q(1\,\%)$	0.91	0.7	0.36	0.41	0.43 ± 0.06	4.9 ± 1.6	0.7	0.7
$Z_q(3\,\%)$	$Z_q(1\,\%)$	0.97	0.6	0.65	0.69	0.70 ± 0.05	8.1 ± 2.7	0.6	0.6

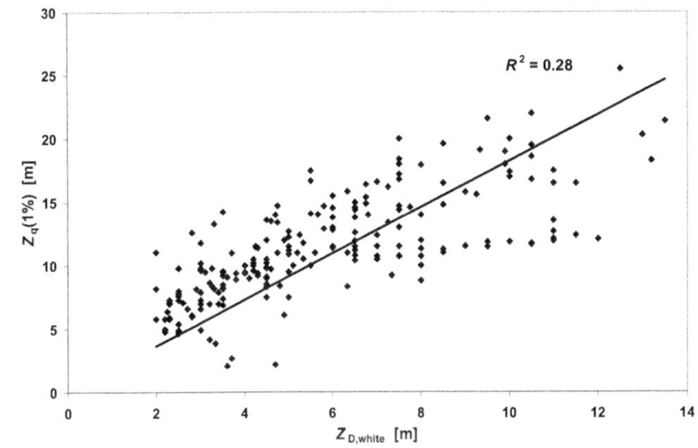

Fig. 8. The depth of the euphotic zone, estimated by the 1 % level $Z_q(1\,\%)$ of the surface quanta irradiance, as a function of the Secchi depth $Z_{D,white}$. The best-fit line through the origin is $Z_q(1\,\%) = 1.8\,Z_{D,white}$.

An old rule of thumb for the Oslofjord says that $Z_q(10\,\%)$ corresponds roughly to $Z_{D,white}$ (also suggested by Paulson and Simpson, 1977), and $Z_q(1\,\%)$ to twice this depth. We have already found that the average percentage of the quanta irradiance at $Z_{D,white}$ was 9 %, and when we calculate the average ratios $Z_q(10\,\%)/Z_{D,white}$ and $Z_q(1\,\%)/Z_{D,white}$, they become 0.9 and 2.2, respectively (Table 8). Figure 8 illustrates the significant scattering of points around the correlation line through the origin for $Z_q(1\,\%)$ as a function of $Z_{D,white}$. If we apply the relationship $y = A + Bx$ instead of the line through the origin, a slightly better fit is obtained, and the rms error for the estimates of $Z_q(1\,\%)$ as a function of $Z_{D,white}$ is reduced from 3.1 to 2.7 m (Table 8). Compared to the mean value 11.5 m of $Z_q(1\,\%)$, these numbers represent relative errors of 27 and 23 %, respectively.

Table 8 shows that estimates of Z_q from observed $Z_{D,white}$ are likely to have relative mean errors in the range 20–30 %. This is because $Z_{D,white}$ is primarily a function of $c + K_L$ (Eq. 20), while Z_q is a function of K_q only (Eq. 27). The vertical attenuation coefficients K_L and K_q are mainly functions of the absorption coefficient a, while the beam attenuation coefficient c consists of $a + b$, where b is the scattering coefficient. Thus Z_q is less influenced by particle scattering than Z_D, a property which reduces the correlation between the two quantities. It is our experience that an increased particle content in the sea may have a strong reducing effect on Z_D, while the influence on Z_q is much smaller. Still, the relationships in Table 8 will provide very useful checks of our irradiance measurements.

Not surprisingly the inter-correlations between $Z_q(1\,\%)$, $Z_q(3\,\%)$, $Z_q(10\,\%)$ and $Z_q(30\,\%)$ are stronger than between Z_q and $Z_{D,white}$ (Fig. 9). The statistical relationships are shown in Table 8.

Mikaelsen and Aas (1990) found $K_q Z_{D,white} = 2.7 \pm 0.6$, $Z_q(10\,\%) = 0.66\,Z_{D,white}$ and $Z_q(1\,\%) = 1.7\,Z_{D,white}$, while Sørensen et al. (1993) obtained $K_q Z_{D,white} = 2.3 \pm 0.4$. All of these results are close to the values in Table 8. Kratzer et al. (2003) found for Baltic waters that $K_q \approx 1.7/Z_{D,white}$, implying that for the same K_q the Secchi depth $Z_{D,white}$ tends to be greater in the Oslofjord than in the Baltic.

5.3 Chlorophyll *a* and total suspended material

The chlorophyll content is perhaps the most used concept when the amount of algae in the sea shall be described. Unfortunately the concentration of chlorophyll *a* (Chl) is not an optical property, although it influences the absorption and scattering coefficients of sea water. As a result our estimates obtained by correlation analysis of Chl as a function of $1/Z_D$, based on 41 stations north of 57.0° N, have average errors of

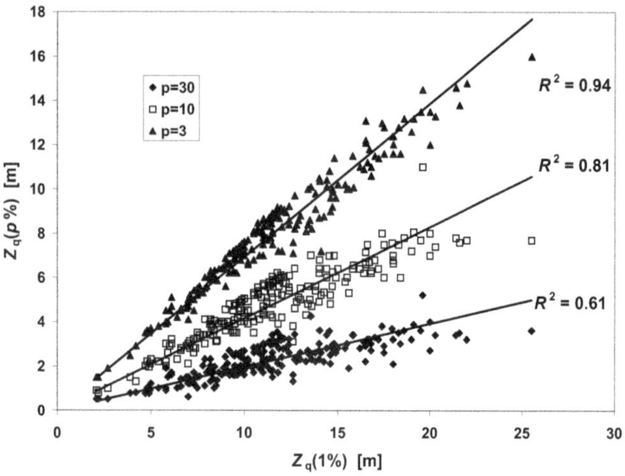

Fig. 9. The depth of the $p\,\%$ levels $Z_q(p\,\%)$ of the surface quanta irradiance, as a function of $Z_q(1\,\%)$ for p equal to 3, 10 and 30. The best-fit lines through the origin are $Z_q(3\,\%) = 0.69 Z_q(1\,\%)$, $Z_q(10\,\%) = 0.41 Z_q(1\,\%)$, and $Z_q(30\,\%) = 0.20 Z_q(1\,\%)$.

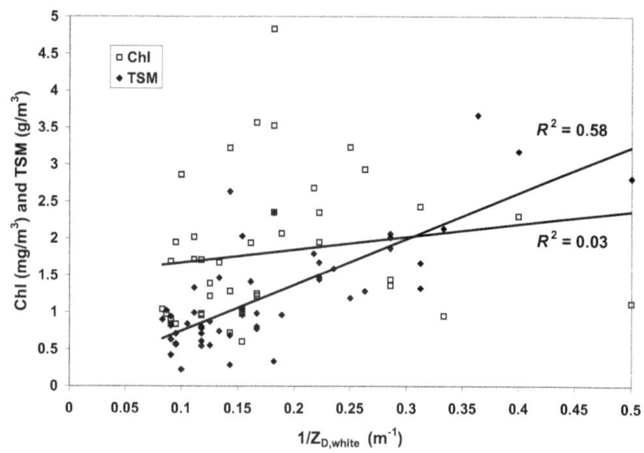

Fig. 10. Content of Chl and TSM versus $1/Z_{D,white}$ at stations north of 57° N. The best-fit lines are $Chl = [1.50\,mg\,m^{-3}] + [1.73\,mg\,m^{-2}]/Z_{D,white}$ and $TSM = [0.12\,g\,m^{-3}] + [6.22\,g\,m^{-2}]/Z_{D,white}$.

50–90 % relative to the mean value of chlorophyll content (Table 9, Fig. 10). The estimates of total suspended material TSM (51 stations) for the same area have average errors of 40–70 % (Table 9, Fig. 10). Still, such relationships provide useful checks because they quantify the order of magnitude of the concentrations, within the area of investigation. Better correlations may sometimes be found for smaller regions. In general the relationships discussed in Sect. 5 are likely to show regional variations.

In the earlier investigation by Sørensen et al. (1993) the products $Z_{D,white}$Chl (249 stations) and $Z_{D,white}$ TSM (275 stations) obtained the mean values and standard deviations $(19 \pm 16)\,mg\,m^{-2}$ and $(7.8 \pm 3.8)\,g\,m^{-2}$, respectively, in agreement with the corresponding ranges of Table 9. An estimate of the product $Z_{D,white}$Chl in the Baltic Sea, based on observations by Fleming-Lehtinen and Laamanen (2012), is $\sim 25\,mg\,m^{-2}$.

The turbidity Turb, expressed in nephelometric formazine units (NFU), has not been included in the present study, because the accuracy of the corresponding turbidity data was not satisfactory. However, in the mentioned investigation by Sørensen et al. (1993) the mean value and standard deviation of the product $Z_{D,white}$ Turb (308 stations) was (3.5 ± 1.4) m NFU^{-1}.

6 Summary and conclusions

We have analysed the Secchi depth Z_D and its relationship to other properties of the sea water in the Oslofjord–Skagerrak area. White and black disks of different sizes have been applied, and the Secchi depth has been observed with the naked eye, through colour filters and with a water telescope. A few stations where $Z_{D,white}$ was less than 2 m or the Q factor was outside the range $\pi - 2\pi$ have been omitted. The upward luminance will typically have its spectral maximum in the green part of the spectrum, close to 555 nm.

An extended version of Tyler's equation for Z_D, as expressed by Eq. (21), has been tested. The right-hand side of the equation, $\ln(A)$, is a function of the reflectance $\bar{\rho}_{L,air}$ at the air–sea interface, the reflectances $R_L(0)$ and $R_L(Z_D)$ of the sea water just below the surface and at the Secchi depth, the reflectance ρ_{DL} of the disk, the contrast threshold C_t of the human eye, the contrast transmittance W through the surface and the factor $\mathfrak{R}$. With the estimated mean values for these quantities (Table 3), $\ln(A)$ obtains the mean value ± the standard deviation 7.3 ± 0.9 for the 30 cm white disk observed with the naked eye. The left-hand side of the equation is the product of the observed Secchi depth $Z_{D,white}$ and the attenuation coefficients $(c + K_L)_{phot}$, and its mean value becomes 7.0 ± 1.3 (Table 3, Fig. 3). Thus on average the observed value is only 4 % less than the one predicted for the white disk. The slope of the best-fit line through the origin for $(c + K_L)_{phot}$ as a function of $1/Z_{D,white}$ becomes 7.5. The deviation between the observed $(c + K_L)_{phot}$ and its estimate $7.5/Z_{D,white}$ obtains an rms value of $0.23\,m^{-1}$, which represents 21 % of the mean value of $(c + K_L)_{phot}$ (Table 3).

The agreement between $\ln(A)$ of Eq. (21) and the observed $Z_D(c + K_L)_{phot}$ that we find for the white disk is reduced when the Secchi disk is observed through coloured glass filters. In this case the observed mean values of $Z_D(c + K_L)_{phot}$ are more than 30 % smaller than the estimated mean values of $\ln(A)$ (Table 3). We speculate if there may be a spectral dependence of C_t.

The observed depths $Z_{D,black}$ of the black disk and the related products $Z_{D,black}(c + K_L)_{phot}$ were also smaller than predicted. According to Table 3 the average value of

Table 9. Relationships of the forms $y = A+Bx$ and $y = B_0x$ obtained by correlation analysis of chlorophyll *a* (Chl), total suspended material (TSM) and inverse Secchi depths ($1/Z_D$) observed with the open eye and with blue, green and red glass filters. The concentrations of Chl and TSM are averages over the depth range 0–$Z_{D,white}$, and r is the correlation coefficient. The error ε is the root mean square of the deviations $(y-A-Bx)$, and ε_0 is the rms of $(y-B_0x)$. The analysis is based on 41 (Chl) and 51 (TSM) stations north of 57.0° N.

y [mg m^{-3}]	x [m^{-1}]	r	A [mg m^{-3}]	B [mg m^{-2}]	B_0 [mg m^{-2}]	$(y/x)_{mean\pm sd}$ [mg m^{-2}]	$y_{mean\pm sd}$ [mg m^{-3}]	ε [mg m^{-3}]	ε_0 [mg m^{-3}]
Chl	$1/Z_{D,white}$	0.16	1.50	1.7	8.5	11.5 ± 6.2	1.8 ± 1.0	0.9	1.1
Chl	$1/Z_{D,blue}$	0.16	1.50	1.0	5.0	5.8 ± 3.1	1.8 ± 1.0	1.0	1.7
Chl	$1/Z_{D,green}$	0.09	1.71	0.3	3.0	8.0 ± 4.1	1.8 ± 1.0	0.9	1.4
Chl	$1/Z_{D,red}$	0.24	1.33	1.5	4.9	6.2 ± 3.1	1.8 ± 1.0	0.9	1.1
y [g m^{-3}]	x [m^{-1}]	r	A [g m^{-3}]	B [g m^{-2}]	B_0 [g m^{-2}]	$(y/x)_{mean\pm sd}$ [g m^{-2}]	$y_{mean\pm sd}$ [g m^{-3}]	ε [g m^{-3}]	ε_0 [g m^{-3}]
TSM	$1/Z_{D,white}$	0.76	0.12	6.2	6.8	7.1 ± 3.0	1.2 ± 0.8	0.5	0.5
TSM	$1/Z_{D,blue}$	0.54	0.71	1.4	2.5	3.6 ± 1.9	1.2 ± 0.8	0.6	0.8
TSM	$1/Z_{D,green}$	0.68	0.27	3.8	4.6	5.1 ± 2.4	1.2 ± 0.8	0.5	0.6
TSM	$1/Z_{D,red}$	0.73	0.08	3.6	3.8	4.0 ± 1.8	1.2 ± 0.8	0.5	0.5

$Z_{D,black}(c+K_L)_{phot}$ should be equal to $\ln(A) = 4.7$ for the 30 cm black disk, but the observed mean value is only 40 % of this. Observations with a black bowl, $Z_{B,black}$, resulted in 20–30 % greater depths than with the black disk, but the product $Z_{B,black}(c+K_L)_{phot}$ is still less than half of the predicted value. One explanation for the deviation could be that the black bowl is not perfectly black. It could also be that the use of colour filters and a black disk or bowl introduces effects that we have not included in $\ln(A)$.

If the diameters of the white and black disks are reduced from 30 to 10 cm, the values of C_t and W will change, and the corresponding Secchi depths including colour filter observations are predicted by the right-hand side of Eq. (21) to be reduced within the range 13–22 %. The observations show that the decrease in size reduces the Secchi depth on average by 10–20 %, in agreement with this theoretical estimate (Table 4). The best-fit line through the origin for all depths of the 10 cm disk as a function of the depths of the 30 cm disk obtains a slope of 0.83 (Fig. 4), indicating an average Secchi depth reduction of 17 % for the 10 cm disk.

The use of a telescope changes $\bar{\rho}_{L,air}$ to 0 and W to 1, and according to Eq. (21) the depth of the 30 cm white disk should then increase by 12 %. A few earlier observations confirm this estimate. Experiments made with a 10 cm disk (Table 4) show that the telescope increases $Z_{D,white,10}$ on average by 14 %, and that the effect on $Z_{D,filter,10}$ is of the same order of magnitude. If all open-eye and filter observations are put together (Fig. 5), the best-fit line through the origin obtains the slope 1.19. Thus we may state, based on Table 4, that the use of a telescope increases Z_D within the range 10–20 %.

There is practically no difference between observations of the white Secchi disk on the sunlit and shadow sides of the ship (Fig. 6), while the depths observed with colour filters may be reduced by up to 17 % on the shadow side (Table 4).

The mean value ± the standard deviation of the ratio $(c+K_L)_{phot}/(c+K_L)_{555}$ becomes 1.03 ± 0.01, supporting our assumption in Sect. 2 about the monochromatic character of the photopic coefficients. Similarly, the ratios c_{phot}/c_{555} and $c_{phot,disk}/c_{555}$ are both 1.02 ± 0.01, while the mean value and standard deviation of $K_{L,phot}/K_{L,555}$ are greater, 1.11 ± 0.05 (Table 5). The attenuation coefficients $c_{phot,filter}$ and $K_{L,filter}$ derived from luminances observed through colour filters are strongly correlated with the corresponding monochromatic coefficients with correlation coefficients very close to 1.0 (Table 5). The slope of the best-fit line through the origin for the blue and green filters obtains values between 0.95 and 1.01, while the red filter deviates more from 1.0 with the slopes 1.06 and 1.21. Thus our conclusion is that the monochromatic assumption works satisfactorily.

The correlation coefficient r is 0.95 for the linear relationship between c_{555} and $1/Z_{D,white}$, while r is reduced to 0.72 for $K_{L,555}$ as a function of $1/Z_{D,white}$ (Table 6). Similar results are found for the other MERIS channels at 412, 443, 490, 510, 620 and 665 nm. While r is in the range 0.86–0.95 for c, it is reduced to the range 0.66–0.78 for K_L. This result is reasonable since c contributes more than K_L to the sum $(c+K_L)$ which determines $Z_{D,white}$. The correlation coefficients for K_d as a function of $1/Z_{D,white}$ at the MERIS channels are in the range 0.59–0.86, and for K_u in the range 0.43–0.72 (Table 6). The errors of the estimated vertical coefficients are greatest for K_u and smallest for K_L. In the blue and green parts of the spectrum there are no significant differences between the best-fit line through the origin and the line with a constant term. At 665 nm, however, the introduction of a constant term reduces the rms error from 0.21 m^{-1} to 0.07 m^{-1} (Fig. 7).

Linear relationships between the Secchi depths observed through glass filters and the coefficients c, K_L, K_d and K_u at the wavelengths of peak visual sensitivity at 470, 540 and

Table 10. Nomenclature.

Symbol	Description	Introduced
A	Dimensionless function	Eq. (20)
A	Constant of correlation analysis	Sect. 4.5
A_1	Dimensionless function	Eq. (21)
A_2	Dimensionless function	Eq. (21)
a_{bp}	Absorption coefficient of bleached particles	Sect. 3.1
a_{y}	Absorption coefficient of yellow substance	Sect. 3.1
B	Constant of correlation analysis	Sect. 4.5
B_0	Constant of correlation analysis	Sect. 4.5
b_{p}	Scattering coefficient of particles	Sect. 3.1
C	Visual contrast between Secchi disk and background	Eq. (1)
C_{air}	Contrast of the Secchi disk observed in air	Eq. (13)
Chl	Concentration of chlorophyll *a*	Sect. 3.4
C_{t}	Visual threshold contrast	Eq. (1)
c	Beam attenuation coefficient	Eq. (2)
c_{phot}	Beam attenuation coefficient for nadir luminance	Sect. 4.1
D	Diameter of the Secchi disk	Eq. (14)
E_{air}	Downward illuminance or irradiance in air	Eq. (17)
E_{d}	Downward illuminance or irradiance in the sea	Eq. (6)
E_{u}	Upward illuminance or irradiance in the sea	Sect. 3.4
H	Height of the observer's eye above the surface of the sea	Eq. (22)
j	Angle of refraction in water	Eq. (22)
K_{d}	Average vertical attenuation coefficient of downward irradiance	Sect. 5.1
K_L	Average vertical attenuation coefficient of $L(z)$	Eq. (9)
$K_{L,\mathrm{phot}}$	Average vertical attenuation coefficient of luminance from nadir	Sect. 4.1
K_{q}	Average vertical attenuation coefficient of quanta irradiance (PAR)	Eq. (24)
K_{u}	Average vertical attenuation coefficient of upward irradiance	Sect. 5.1
k	$787\,\mathrm{m\,s^{-1}}$	Eq. (14)
L	Nadir radiance and luminance from the background in the sea	Eq. (1)
L_{D}	Nadir radiance and luminance from the Secchi disk	Eq. (1)
L_{r}	Radiance from sun and sky reflected at the surface	Eq. (12)
L_{W}	Water-leaving radiance	Eq. (12)
L_*	Path function along path outside the disk	Eq. (3)
$L_{*\mathrm{D}}$	Path function along path from disk to observer	Eq. (2)
n	Index of refraction for water	Eq. (12)
Q	Radiance/irradiance ratio	Sect. 3.4
R^2	Coefficient of determination	Sect. 4.4
R_L	Sub-surface radiance reflectance of the sea	Eq. (7)
r	Correlation coefficient	Sect. 4.4
$\mathfrak{R}$	Radiance/irradiance ratio	Eq. (18)
s_{d}	Standard deviation	Table 4
s_x	Standard deviation of x	Sect. 4.4
s_y	Standard deviation of y	Sect. 4.4
$S_{x,y}$	Covariance of x and y	Sect. 4.4
T_{D}	Transmittance of quanta irradiance between the surface and the Secchi depth	Eq. (24)
TSM	Concentration of total suspended material	Sect. 3.4
Turb	Turbidity	Sect. 5.3
U	Wind speed	Eq. (14)
W	Contrast transmittance at the water–air interface	Eq. (14)
x	Independent variable	Sect. 4.4
y	Dependent variable	Sect. 4.4
z	Vertical coordinate, positive downwards, zero at surface	Eq. (2)
$z_{\mathrm{B,black}}$	Secchi depth of the black bowl	Sect. 4.5
$z_{\mathrm{D,black}}$	Secchi depth of the black disk	Sect. 4.5
$z_{\mathrm{D,blue}}$	Secchi depth of the white disk observed with blue filter	Sect. 4.5
$z_{\mathrm{D,green}}$	Secchi depth of the white disk observed with green filter	Sect. 4.5
$z_{\mathrm{D,red}}$	Secchi depth of the white disk observed with red filter	Sect. 4.5
$z_{\mathrm{D,white}}$	Secchi depth of the white disk observed with naked eye	Eq. (5)
$z_{\mathrm{q}}(p\,\%)$	Depth where the quanta irradiance is reduced to $p\,\%$ of the surface value	Sect. 5.2
α	Apparent angle from the observer's eye across the Secchi disk	Eq. (22)
ε	Root-mean-square error	Sect. 4.8
ε_0	Root-mean-square error	Sect. 4.8
λ	Wavelength	Sect. 3.1
ρ_{DL}	Luminance or radiance reflectance of the Secchi disk	Eq. (6)
$\bar{\rho}_{L,\mathrm{air}}$	Fresnel-reflected luminance or radiance towards the zenith	Eq. (17)
τ	Luminance or radiance transmittance for a ray of normal incidence at the water–air interface	Eq. (12)

620 nm (Table 2) have also been calculated (Table 7). The rms errors of the estimated coefficients are in the range 0.07–0.47 m^{-1}, or 10–80 % of the corresponding mean values. The accuracy of the estimates based on the white disk and naked eye is equal to or better than the estimates from colour filter observations.

The mean vertical attenuation coefficient of quanta irradiance or PAR between the surface and the Secchi depth, K_q, may be estimated from $1/Z_{D,white}$ with an rms error less than 22 % of the mean value of K_q (Table 8). The mean values of the ratios $Z_q(10\,\%)/Z_{D,white}$ and $Z_q(1\,\%)/Z_{D,white}$ are 0.9 and 2.2, respectively, or very close to 1 and 2. Table 8 shows that estimates of Z_q from observed $Z_{D,white}$ will have relative errors in the range 20–30 %. The depth of the euphotic zone may be defined as $Z_q(1\,\%)$ (Fig. 8), and according to Table 8 this depth can be estimated by an rms error of 24 %. Consequently these relationships will provide very useful checks of our irradiance measurements. Relationships for $Z_q(3\,\%)$, $Z_q(10\,\%)$ and $Z_q(30\,\%)$ as functions of $Z_q(1\,\%)$ are presented in Fig. 9 and Table 8.

The Secchi depth may also be used to estimate the concentrations of chlorophyll *a* (Chl) and total suspended material (TSM) (Fig. 10). The estimates of Chl and TSM as functions of $1/Z_D$ have average errors of 50–90 % and 40–70 %, respectively (Table 9, Fig. 10). Still, these estimates will provide the order of magnitude of the concentrations.

Our overall conclusion becomes that Eq. (21) quantifies well the relationships between the white disk and the optical attenuation coefficients, but less so for the colour filters and the black disk. The Secchi depths provide very useful checks of the monochromatic attenuation coefficients and of chlorophyll *a* and total suspended material. We assume that there may be regional differences for the found relationships, and they should be tested out for different oceanic regions and water types.

Acknowledgements. The in-water spectral irradiance and radiance data from the Oslofjord–Skagerrak area were collected in 2002–2003, as parts of two validation projects: "Validation of MERIS Data Products" (VAMP) funded by ESA, the Norwegian Space Centre and the Norwegian Institute for Water Research (PRODEX contract no. 14849/00/NL/Sfe(IC)), and "Regional Validation of MERIS Chlorophyll Products in North Sea Coastal Waters" (REVAMP), funded by an FP5 research contract from the European Commission (contract no. EVG1–CT–2001–00049).

Edited by: M. Hoppema

References

Aarup, T.: Transparency of the North Sea and Baltic Sea – a Secchi depth data mining study, Oceanologia, 44, 323–337, 2002.

Aarup, T., Holt, N., and Højerslev, N. K.:, Optical measurements in the North Sea-Baltic Sea transition zone, II. Water mass classification along the Jutland west coast from salinity and spectral irradiance measurements, Cont. Shelf Res., 16, 1343–1353, 1996a.

Aarup, T., Holt, N., and Højerslev, N. K.: Optical measurements in the North Sea-Baltic Sea transition zone, III. Statistical analysis of bio-optical data from the Eastern North Sea, the Skagerrak and the Kattegat, Cont. Shelf Res., 16, 1355–1377, 1996b.

Aas, E.: Estimates of radiance reflected towards the zenith at the surface of the sea, Ocean Sci., 6, 861–876, doi:10.5194/os-6-861-2010, 2010.

Aas, E. and Højerslev, N. K.: Analysis of underwater radiance observations: Apparent optical properties and analytic functions describing the angular radiance distribution. J. Geophys. Res., 104, 8015–8024, 1999.

Aas, E. and Høkedal, J.: Reflection of spectral sky irradiance on the surface of the sea and related properties, Remote Sens. Environ., 70, 181–190, 1999.

Aas, E. and Korsbø, B.: Self-shading effect by radiance meters on upward radiance observed in coastal waters, Limnol. Oceanogr., 42, 968–974, 1997.

Aas, E., Andresen, T., Løyning, T., and Sørgård, E.: Eutrofisituasjonen i Ytre Oslofjord. Delprosjekt 3.7: Optiske observasjoner – Overflatevannets kvalitet sett ut fra observasjoner i overflatelaget. SFT (Norw. Environ. Agency) Report 388/90, 50 pp., 1989.

Aas, E., Højerslev, N. K., and Høkedal, J.: Conversion of subsurface reflectances to above-surface MERIS reflectance, Int. J. Remote Sens., 30, 5767–5791, 2009.

Andresen, A.: Siktedyputviklingen i Oslofjorden 1936–92, Master thesis, Dept. Geophys., Univ. Oslo, 228 pp., 1993.

Aure, J., Molvær, J., and Stigebrandt, A.: Observations of inshore water exchange forced by a fluctuating offshore density field, Mar. Poll. Bull., 33, 112–119, 1996.

Blackwell, H. R.: Contrast thresholds of the human eye, J. Opt. Soc. Am., 36, 624–632, 1946.

Boguslawski, G. V.: Handbuch der Ozeanographie, Band I. Räumliche, physikalische und chemische Beschaffenheit der Ozeane, Verlag von J. Engelhorn, Stuttgart, 400 pp., 1884.

Cox, C. and Munk, W.: Statistics of the sea surface derived from sun glitter, J. Mar. Res., 13, 198–227, 1954a.

Cox, C. and Munk, W.: The measurements of the roughness of the sea surface from photographs of the sun's glitter, J. Opt. Soc. Am., 44, 838–850, 1954b.

Davies-Colley, R. J.: Measuring water clarity with a black disc, Limnol. Oceanogr., 33, 616–623, 1988.

Duntley, S. Q.: The visibility of submerged objects. Final Rep., Visibility Lab., Mass. Inst. Tech., 74 pp., 1952.

Fleming-Lehtinen, V. and Laamanen, M.: Long-term changes in Secchi depth and the role of phytoplankton in explaining light attenuation in the Baltic Sea, Est. Coast. Shelf Sci., 102–103, 1–10, 2012.

Gade, H. G.: Some hydrographic observations of the Inner Oslofjord during 1959, Hvalrådets skrifter, Nr. 46, Universitetsforlaget, Oslo, 62 pp., 1963.

Gade, H. G.: Delrapport nr. 2, Hydrografi. Oslofjorden og dens forurensningsproblemer, I. Undersøkelsen 1962–1965, Norsk institutt for vannforskning, Oslo, 164 pp., 1967.

Gordon, H. R. and Ding, K.: Self-shading of in-water optical instruments. Limnol. Oceanogr., 37, 491–500, 1992.

Gordon, H. R. and Wouters, A. W.: Some relationships between Secchi depth and inherent optical properties of natural waters, Appl. Opt., 17, 3341–3343, 1978.
Gröen, P. and Dorrestein, R.: Zeegolven. KNMI Opstellen op Oceanografisch en Maritiem Meteorologisch Gebied, 11, 124 pp., 1976.
Haltrin, V. I.: Spectral relative clarity of Black and Aegan Seas. Geoscience and Remote Sens. Symp. Proc., IGARSS '98. 1998 IEEE International, 2, 913–915, 1998.
Holmes, R. W.: The Secchi disk in turbid coastal waters, Limnol. Oceanogr., 15, 688–694, 1970.
Hou, W., Lee, Z., and Weidemann, A. D.: Why does the Secchi disk disappear? An imaging perspective, Optics Expr., 15, 2791–2802, 2007.
Højerslev, N. K.: Spectral daylight irradiance and light transmittance in natural waters measured by means of a Secchi Disc only, Int. Council Explor. Sea, C. M. 1977/C:42, 19 pp., 1977.
Højerslev, N. K.: Daylight measurements appropriate for photosynthetic studies in natural sea waters, J. Cons. Int. Explor. Mer., 38, 131–146, 1978.
Højerslev, N. K.: Bio-optical measurements in the Southwest Florida Shelf ecosystem, J. Cons. Int. Explor. Mer, 42, 65–82, 1985.
Højerslev, N. K.: Visibility of the sea with special reference to the Secchi disc, Ocean Optics VIII, SPIE Vol. 637, 294–305, 1986.
Højerslev, N. K., Holt, N., and Aarup, T.: Optical measurements in the North Sea-Baltic Sea transition zone, I. On the origin of the deep water in the Kattegat, Cont. Shelf Res., 16, 1329–1342, 1996.
Høkedal, J. and Aas, E.: Observations of spectral sky radiance and solar irradiance. Rep. 103, Dept. Geophys., Univ. Oslo, Oslo, 73 pp., 1998.
Ibrekk, H. O. and Holtan, G.: Eutrofisituasjonen i Ytre Oslofjord, Delprosjekt 3.1: Forurensningstilførsler til Ytre Oslofjord, NIVA-Rapport 352/88, Oslo, 44 pp., 1988.
Jerlov, N. G.: Marine Optics. Elsevier, Amsterdam, 231 pp., 1976.
Kratzer, S., Håkansson, B., and Sahlin, C.: Assessing Secchi and photic zone depth in the Baltic Sea from satellite data, Ambio, 32, 577–585, 2003.
Krümmel, O.: Der Ozean, Eine Einführung in die allgemeine Meereskunde, G. Freytag, Leipzig, F. Tempsky, Prag, 242 pp., 1886.
Krümmel, O.: Bemerkungen über die Durchsichtigkeit des Meerwassers, Ann. d. Hydr. mar. Met., 2, 62–78, 1889.
Krümmel, O.: Handbuch der Ozeanographie, Band I. Die räumlichen, chemischen und physikalischen Verhältnisse des Meeres, Verlag von J. Engelhorn, Stuttgart, 526 pp., 1907.
Levin, I. M.: Theory of the white disk, Izvest. Atmos. Ocean. Phys., 16, 678–683, 1980.
Lisitzin, E.: Über die Durchsichtigkeit des Wassers im nördlichen Teil des Baltischen Meeres, Fennia, 65, 1–22, 1938.
Mikaelsen, B. and Aas, E.: Secchi disk depths and related quantities in the Oslofjord 1986–87, Rep. 77, Dept. Geophys., Univ. Oslo, Oslo, 55 pp., 1990.
Mobley, C. D.: Estimation of the remote-sensing reflectance from above-surface measurements, Appl. Opt., 38, 7442–7455, 1999.
Morel, A. and Gentili, B.: Diffuse reflection of oceanic waters, III. Implication of bi-directionality for the remote-sensing problem, Appl. Opt., 35, 4850–4862, 1996.
Morel, A., Huot, Y., Gentili, B., Werdell, P. J., Hooker, S. B., and Franz, B. A., Examining the consistency of products derived in open ocean (Case 1) waters in the perspective of a multi-sensor approach. Remote Sens. Environ., 111, 69–88, 2007.
Paulson, C. A. and Simpson, J. J.: Irradiance measurements in the upper ocean, J. Phys. Oceanogr., 7, 952–956, 1977.
Preisendorfer, R. W.: Secchi disk science: Visual optics of natural waters, Limnol. Oceanogr., 31, 909–926, 1986.
Sandén, P. and Håkansson, B.: Long-term trends in Secchi depth in the Baltic Sea, Limnol. Oceanogr., 41, 346–351, 1996.
Sauberer, F. and Ruttner, F.: Die Strahlungsverhältnisse der Binnengewässer, Akademische Verlagsgesellschaft, Leipzig, 240 pp., 1941.
Secchi, A.: Esperimente per determinare la transparenza del mare, in: Sul moto ondoso del mare e su le correnti di esso specialmente su quelle littorali, edited by: Cialdi, A., Rome, 258–288, 1866.
Shifrin, K. S.: Physical optics of ocean water, Am. Inst. Phys., New York, 285 pp., 1988.
Staalstrøm, A., Aas, E., and Liljebladh, B.: Propagation and dissipation of internal tides in the Oslofjord, Ocean Sci., 8, 525–543, doi:10.5194/os-8-525-2012, 2012.
Sørensen, K. and Aas, E.: Remote sensing of water quality, Proc. SPIE vol. 2258, Ocean Optics XII, edited by: Jaffe, J. S., 332–341, 1994.
Sørensen, K., Aas, E., Faafeng, B., and Lindell, T.: Fjernmåling av vannkvalitet – Videreutvikling av optisk satellittfjernmåling som metode for overvåking av vannkvalitet, NIVA Rep., ISBN 82-577-2262-6, 115 pp., 1993.
Sørensen, K., Høkedal, J., Aas, E., Doerffer, R., and Dahl, E.: Early results for validation of MERIS water products in the Skagerrak. Proceedings of Envisat Validation Workshop, 9–13 December 2002, Frascati, Italy, ESA SP-531, March 2003, 2003.
Sørensen, K., Aas, E., Høkedal, J., Severinsen, G., Doerffer, R., and Dahl, E.: Validation of MERIS water products in the Skagerrak. MAVT Workshop. Proceedings of Envisat Validation Workshop, 20–24 October 2003, Frascati, Italy, ESA CD-ROM, 2004.
Sørensen, K., Aas, E., and Høkedal, J.: Validation of MERIS water products and bio-optical relationships in the Skagerrak, Int. J. Remote Sens., 28, 555–568, 2007.
Takenouti, Y.: On the transparency of sea water, Oceanogr. Mag., Tokyo, 4, 129–135, 1950.
Tyler, J. E.: The Secchi Disc, Limnol. Oceanogr., 13, 1–6, 1968.
Walsh, J. W.: Photometry. Dover, New York, 544 pp., 1958.
Wernand, M. R.: On the history of the Secchi disk, J. Eur. Opt. Soc., Rap. Publ., 5, 100135-1-6, 2010.
WMO: Guide to wave analysis and forecasting, 2nd Edn., WMO-No. 702, 159 pp., 1998.
Zhang, Y., Pulliainen, J., Koponen, S., and Hallikainen, M.: Empirical algorithms for Secchi disk depth using optical and microwave remote sensing data from the Gulf of Finland and the Archipelago Sea, Boreal Environ. Res., 8, 251–261 (ISSN 1239-6095), 2003.
Zibordi, G. and Ferrari, G. M.: Instrument self-shading in underwater optical measurements: experimental data, Appl. Opt., 34, 2750–2754, 1995.

14

Adapting to life: ocean biogeochemical modelling and adaptive remeshing

J. Hill[1], E. E. Popova[2], D. A. Ham[3], M. D. Piggott[1,4], and M. Srokosz[2]

[1]Applied Modelling and Computation Group, Department of Earth Science and Engineering, Imperial College London, SW7 2AZ, UK
[2]National Oceanography Centre, Southampton, University of Southampton Waterfront Campus, European Way, Southampton, SO14 3ZH, UK
[3]Applied Mathematics and Mathematical Physics, Department of Mathematics, Imperial College London, SW7 2AZ, UK
[4]Grantham Institute for Climate Change, Imperial College London, SW7 2AZ, UK

Correspondence to: J. Hill (jon.hill@imperial.ac.uk)

Abstract. An outstanding problem in biogeochemical modelling of the ocean is that many of the key processes occur intermittently at small scales, such as the sub-mesoscale, that are not well represented in global ocean models. This is partly due to their failure to resolve sub-mesoscale phenomena, which play a significant role in vertical nutrient supply. Simply increasing the resolution of the models may be an inefficient computational solution to this problem. An approach based on recent advances in adaptive mesh computational techniques may offer an alternative. Here the first steps in such an approach are described, using the example of a simple vertical column (quasi-1-D) ocean biogeochemical model.

We present a novel method of simulating ocean biogeochemical behaviour on a vertically adaptive computational mesh, where the mesh changes in response to the biogeochemical and physical state of the system throughout the simulation. We show that the model reproduces the general physical and biological behaviour at three ocean stations (India, Papa and Bermuda) as compared to a high-resolution fixed mesh simulation and to observations. The use of an adaptive mesh does not increase the computational error, but reduces the number of mesh elements by a factor of 2–3. Unlike previous work the adaptivity metric used is flexible and we show that capturing the physical behaviour of the model is paramount to achieving a reasonable solution. Adding biological quantities to the adaptivity metric further refines the solution. We then show the potential of this method in two case studies where we change the adaptivity metric used to determine the varying mesh sizes in order to capture the dynamics of chlorophyll at Bermuda and sinking detritus at Papa. We therefore demonstrate that adaptive meshes may provide a suitable numerical technique for simulating seasonal or transient biogeochemical behaviour at high vertical resolution whilst minimising the number of elements in the mesh. More work is required to move this to fully 3-D simulations.

1 Introduction

Biogeochemical processes in the ocean are a key component of Earth's climate system and, in particular, of the carbon cycle. Understanding of the latter is crucial for projections of future climate change. However, modelling the biogeochemistry of the oceans presents a formidable challenge as many of the key processes occur at scales (both horizontal and vertical) not resolved by state-of-the-art climate models. Both mesoscale and sub-mesoscale processes are thought to be important – for example, in the supply of nutrients to the surface water of the oligotrophic gyres that cover one-third of the earth's surface (Oschlies, 2002; McGillicuddy et al., 2003; Lévy et al., 2012). It is generally considered that the representation of the physical processes is a key part in correctly representing ocean biogeochemistry (e.g. Friedrichs et al., 2006), including the representation of the light spectra

(Kettle and Merchant, 2008). Furthermore, many of the processes affecting biogeochemistry at the mesoscale and submesoscale have significant vertical structure (Lévy et al., 2012), meaning that vertical resolution is also important. In addition, the surface fluxes that drive mixed-layer depth (MLD) behaviour can greatly affect the vertical nutrient fluxes (Berline et al., 2007), highlighting the importance of the representation of vertical processes. Thus there is a need to have sufficiently high vertical resolution to correctly represent vertical advection together with mixed-layer deepening and shallowing. Current ocean models use decreasing resolution with increasing depth, concentrating resolution in the upper layers (e.g. Popova et al., 2006). Multiscale resolution is possible using an unstructured mesh where horizontal resolution can vary spatially by orders of magnitude, but the same method can be applied in the vertical also. A number of coastal and regional models use such an approach to model complex coastlines and bathymetries (e.g. Ji et al., 2008; Luo et al., 2012). These models have been successfully coupled to biogeochemical models to study nutrient cycling and plankton blooms (Khangaonkar et al., 2012; Ji et al., 2008), and water quality (Menendez et al., 2013). In addition to multiscale resolution which alters resolution spatially, it is also possible to alter the resolution temporally – mesh adaptivity, which aims to alter resolution only when and where it is required (e.g. Hiester et al., 2011). This approach aims to reduce the number of elements required whilst maintaining some measure of error. Here, as a first step, the suitability of mesh adaptivity for providing appropriate vertical resolution is tested using a simple vertical-column coupled physics and ecosystem model. We neglect vertical advection terms and focus on mixed-layer dynamics only.

The behaviour of ocean ecosystems, and the associated biogeochemistry, is driven largely by physical processes (stirring and mixing). These vary depending on location; for example, differing between the subpolar and subtropical gyres. Therefore, simulations at different locations in the ocean may require different resolution structure (meshes) in the vertical. Adaptivity should allow the best mesh structure to be chosen for each location. By carefully selecting the adaptivity metric and parameters controlling the mesh, computational cost can in principle be minimised by reducing the number of degrees of freedom (Hiester et al., 2011; Hill et al., 2012). There is also a need to conserve biogeochemical quantities, so interpolation between meshes during adaptation can therefore be key in ensuring conservation. Adaptivity has been used previously in ocean-type settings. Hill et al. (2012) showed that adaptivity can reduce the number of elements required to model the mixed layer, using Fluidity, the model also used here. Adaptive techniques have been shown to reduce levels of numerical mixing in a number of idealised examples (Hofmeister et al., 2010). Other models have shown effective use of adaptive grids to improve the representation of vertical mixing processes (Burchard and Beckers, 2004). This paper represents the first study to assess the effect of adaptive meshes on ocean ecosystem model numerical accuracy.

The adaptive mesh technique used in Fluidity differs from previous implementations of adaptive mesh techniques used in similar models in that the number of elements (or in the case of finite-difference models, grid points) can change throughout the simulation. For example, in both Burchard and Beckers (2004) and Hanert et al. (2006) the number of grid points remains fixed: the adaptive mesh moves them to locations to minimise an error metric; in essence a mesh movement algorithm. The techniques of Burchard and Beckers (2004) have been extended to 3-D by allowing each horizontal location to have a different vertical mesh (Hofmeister et al., 2010). Again, the number of grid points is fixed. In addition to the number of elements being able to vary throughout the simulation, the model presented here also allows the adaptivity metric weights to be user defined, giving a great deal of flexibility on the adaptivity metric composition. This allows a comparison of using only biological tracers in the adaptivity metric, only physical variables, or a combination of both.

In order to examine a wide range of ocean conditions, three ocean stations (Fig. 1) were chosen to test the performance of mesh adaptivity in conjunction with ocean biogeochemistry models. These were Ocean Weather Station Papa, Ocean Weather Station India and the Bermuda Atlantic Time-series. These stations show very different mixed-layer and biological behaviours and so test a model's ability to accurately simulate a range of physical and biological behaviours. Whilst Papa is ideal for carrying out 1-D studies due to the lack of significant horizontal advection (Denman and Miyake, 1973; Gaspar et al., 1990; Burchard and Bolding, 2001), India and Bermuda both experience significant horizontal advection. Previous attempts to model Bermuda in one dimension have resorted to ad hoc "fixes" (Anderson and Pondaven, 2003; Weber et al., 2007) in order to simulate the physical and biological behaviour here. However, the aims of these previous studies were to understand the processes occurring in more detail. In this study we are concerned with how well adaptive remeshing can replicate the results of a fixed-mesh simulation whilst minimising the number of elements used. We therefore do not expect a perfect match to observed data for these two stations, but the simulations must replicate the general observed behaviour at all three stations. In addition, we do not necessarily expect model skill to increase when using adaptive remeshing. We are instead examining the numerical response of the model. Without numerical confidence in the model, altering biological parameters to tune the model may "lead to misconceptions in the interpretation of ecosystem dynamics" (Oschlies and Garçon, 1999).

In Sect. 2 the numerical model used in this study is described, including the biogeochemical model used, turbulence parameterisation and mesh adaptivity algorithm. This model is then evaluated in Sect. 3, before the results from fixed-mesh simulations are described. Section 4 then presents

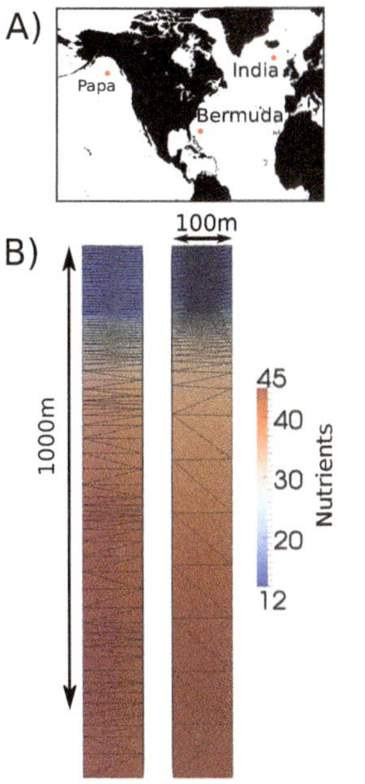

Fig. 1. Map of station locations **(A)** and 2-D view of the model domain at showing two different meshes produced by the adaptivity algorithm **(B)**.

results from the adaptive mesh simulation. Finally, two experiments are described in which the mesh is adapted to concentrate resolution not only in critical regions such as at the MLD, but also to track sinking detritus at Station Papa to well below the mixed layer depth and the subsurface chlorophyll maximum at Bermuda, which occurs below the mixed layer. The paper then assesses the merits of the adaptive algorithm presented and draws some conclusions.

2 Methods

Here, the non-hydrostatic Boussinesq equation system is considered in the context of Fluidity (Ford et al., 2004; Pain et al., 2005; Piggott et al., 2008b), a highly flexible finite element/control volume modelling framework which allows for the numerical solution of the following set of equations:

$$\frac{\partial \boldsymbol{u}}{\partial t} + \boldsymbol{u}\cdot\nabla\boldsymbol{u} + f\underline{k}\times\boldsymbol{u} = -\nabla\left(\frac{p}{\rho_0}\right) - \frac{\rho}{\rho_0}g\underline{k} + \nabla\cdot(\boldsymbol{\nu}\nabla\boldsymbol{u}), \quad (1)$$

$$\nabla\cdot\boldsymbol{u} = 0, \quad (2)$$

$$\frac{\partial T}{\partial t} + \boldsymbol{u}\cdot\nabla T = \nabla\cdot(\boldsymbol{\kappa}_T\nabla T), \quad (3)$$

$$\frac{\partial S}{\partial t} + \boldsymbol{u}\cdot\nabla S = \nabla\cdot(\boldsymbol{\kappa}_S\nabla S), \quad (4)$$

$$\rho \equiv \rho(T,S), \quad (5)$$

where $\boldsymbol{u}$ is the 3-D velocity vector, t represents time, p is the pressure, g is the acceleration due to gravity acting in the $\underline{k} = (0,0,1)^T$ direction, T is temperature and S is salinity. ρ is the density which is given in terms of an equation of state function with temperature and salinity as input arguments, and ρ_0 is a constant background value for density. $\boldsymbol{\nu}$ is the tensor of kinematic viscosities and $\boldsymbol{\kappa}_T$, $\boldsymbol{\kappa}_S$ are the thermal and saline diffusivity tensors respectively. f is the Coriolis parameter which in this work is assumed constant. We also assume, for simplicity, a Cartesian coordinate system with $\underline{k}$ pointing in the direction of gravity.

The above equations were discretised on an unstructured mesh of tetrahedral elements using the finite-element method. The form of the discretisation is determined by the order of the polynomials used for the different solution variables and whether or not they are continuous or discontinuous across element faces. A constant time step of 360 s is used with Crank–Nicolson temporal discretisation.

Here, we use a linear continuous Galerkin method for velocity and pressure, with a control volume formulation used for all tracer fields, including turbulence and biological tracers. Discontinuous linear Galerkin discretisations have also been tested and work successfully. For further details see Piggott et al. (2008b, 2009).

2.1 Boundary conditions

The domain used is quasi-1-D, 100 m square in the horizontal with depths of either 1000 m for Station Papa and Bermuda, or 2000 m for Station India. This ensures that the maximum MLD is well above the lower boundary at all stations. The lateral boundaries have a Dirichlet condition applied to the velocity such that the vertical component is zero. The top and bottom surfaces also have this condition applied. Boundary conditions for the turbulent quantities are as described in Hill et al. (2012) and are Neumann conditions for both turbulent equations. The top surface cell is subjected to heat, momentum and salinity fluxes. These are derived via the Large and Yeager (2004) bulk formulae, with atmospheric data supplied from ERA40 (Uppala et al., 2005). Both Station Papa and Station India use atmospheric forcing from 1970 onwards as this is when most observation data from those stations are available. Bermuda uses atmospheric forcing from 1980 onwards, again as most observational data was available during this period. Briefly, the three surface kinematic fluxes calculated – heat, $\langle w\theta\rangle$; salt, $\langle ws\rangle$; and momentum, $\langle wu\rangle$ and $\langle wv\rangle$ – can be related to the surface fluxes of heat Q, the freshwater F and the momentum $\boldsymbol{\tau} = (\tau_u, \tau_v)$ via

$$\langle w\theta\rangle = Q\left(\rho C_p\right)^{-1} \quad (6)$$

$$\langle ws\rangle = F\left(\rho^{-1}S_0\right) \quad (7)$$

$$(\langle wu\rangle, \langle wv\rangle) = \boldsymbol{\tau}\rho^{-1} = (\tau_u, \tau_v)\rho^{-1}, \quad (8)$$

where ρ is the ocean density, C_p is the heat capacity ($4000\,\mathrm{J\,kg^{-1}\,K^{-1}}$) and S_0 is a reference ocean salinity, which is the current sea surface salinity. These fluxes are then applied as upper-surface Neumann boundary conditions on the appropriate fields.

2.2 Biogeochemical model

The model used here is a six-component model similar to the globally applicable model of Popova et al. (2006). Heuristically, the model consists of nutrients (ammonium and nitrate) which are fixed by phytoplankton in the presence of sunlight (photosynthetic active radiation – PAR). Phytoplankton shading is included, reducing the amount of PAR down the water column. Zooplankton grazes on phytoplankton and detritus partially recycling them back into inorganic nutrients and partially converting into detritus. Phytoplankton and zooplankton mortality also produce detritus which is gradually converted back to nutrients as it sinks through the water column.

Nutrients (ammonium and nitrate), detritus, phytoplankton and zooplankton are solved prognostically, whilst chlorophyll is a diagnostic variable, derived from phytoplankton. Initial conditions for phytoplankton, detritus and zooplankton concentrations were the same for all stations and are 0.1 mmol m^{-3} above 100 m and 0.005 below 100 m. Ammonia was set to an initial value of 0.01 mmol m^{-3}. Nitrate profiles were taken from Kleypas and Doney (2001) for Station Papa and Bermuda and Popova et al. (2006) for station India. For more details of this model see Appendix A1, which includes all parameters used. Note the parameters were tuned to give a good fit to all stations presented here, but were not altered between any model runs.

2.3 Vertical turbulence model

The generic length scale (GLS) turbulence parameterisation simulates vertical turbulence at a finer than that of the mesh. As the GLS model is a RANS parameterisation there is no dependency on the mesh resolution, provided the advective model simulates no turbulent processes, so is ideal for adaptive ocean-scale problems. GLS has the additional advantage that it can be set up to behave as a number of classical turbulence models: k–ϵ, k–kl, k–ω, and an additional model based on Umlauf and Burchard (2003), the *gen* model. The GLS model has been implemented within Fluidity and shown to work well with adaptive remeshing (Hill et al., 2012). Here, we use it in k–ϵ mode as detailed in Hill et al. (2012), which is a two-equation model. The first equation deals with the turbulent kinetic energy, k:

$$\frac{\partial k}{\partial t} + \boldsymbol{u}_i \frac{\partial k}{\partial x_i} = \frac{\partial}{\partial z}\left(\frac{\nu_M}{\sigma_k}\frac{\partial k}{\partial z}\right) + P + B - \epsilon, \tag{9}$$

where σ_k is the turbulence Schmidt number for k, and P and B represent production by shear and buoyancy which are defined as

$$P = -\overline{u'w'}\frac{\partial u}{\partial z} - \overline{v'w'}\frac{\partial v}{\partial z} = \nu_M M^2$$

$$M^2 = \left(\frac{\partial u}{\partial z}\right)^2 + \left(\frac{\partial v}{\partial z}\right)^2, \tag{10}$$

$$B = -\frac{g}{\rho_0}\overline{\rho' w'} = -\nu_H N^2$$

$$N^2 = -\frac{g}{\rho_0}\frac{\partial \rho}{\partial z}. \tag{11}$$

Here, N is the buoyancy frequency, ν_M is the kinematic eddy viscosity and ν_H is the kinematic eddy diffusivity, given by:

$$\nu_M = \sqrt{k} l S_M + \nu_M^0, \quad \nu_H = \sqrt{k} l S_H + \nu_H^0. \tag{12}$$

ν_H^0 is the background diffusivity (set to 1×10^{-6} m^2 s^{-1}), ν_M^0 is the background viscosity (1×10^{-6} m^2 s^{-1}), S_M and S_H are often referred to as stability functions, k is the turbulent kinetic energy, and l is a length scale. When using GLS the values of ν_M and ν_H become the vertical components of the tensors $\boldsymbol{\nu}$ and $\boldsymbol{\kappa}_T$ in Eqs. (1) and (3) respectively. Other tracer fields, such as salinity use the same diffusivity as temperature, i.e. $\boldsymbol{\kappa}_T = \boldsymbol{\kappa}_S$.

There is also the option to add an extra term to account for additional oceanic physics, such an internal waves breaking. This is based on the NEMO ocean model (Madec, 2008) and takes a user-defined percentage of the surface k and distributes it down-depth using an exponential profile:

$$k(z) = k_0(z) + \alpha k_{\text{sur}} \exp(-z/l_k), \tag{13}$$

where k is the new turbulent kinetic energy value at depth z, k_0 is the original turbulent kinetic energy, k_{sur} is the surface turbulent kinetic energy, α is a constant for the amount (percentage) of surface turbulent kinetic energy to transfer down the column, and l_k is a length scale (m) over which this decay occurs. In this work, $\alpha = 0.05$ and $l_k = 30$.

The second equation gives the dissipation ϵ, which is described by

$$\frac{\partial \epsilon}{\partial t} + \boldsymbol{u}_i \frac{\partial \epsilon}{\partial x_i} = \frac{\partial}{\partial z}\left(\frac{\nu_M}{\sigma_\epsilon}\frac{\partial \epsilon}{\partial z}\right) + \frac{\epsilon}{k}(c_1 P + c_3 B - c_2 \epsilon), \tag{14}$$

where c_1, c_2, c_3 and σ_ϵ are constants with values given by Hill et al. (2012).

The MLD can be defined in a number of ways. Here, we use two definitions: (1) where $k < 1 \times 10^{-5}$ m^2 s^{-2} and (2) where density is 0.125 kg m^{-3} less than surface density (i.e. at $z = 0$). We use the second when comparing to observational data and the first for determining weighting of the k field when using adaptivity (see next section).

2.4 Dynamic adaptive mesh optimisation

The mesh adaptivity algorithm used in this work attempts to optimise the size as well as the shape of individual elements of the mesh in order to minimise an optimisation functional (Pain et al., 2001; Piggott et al., 2005, 2008b). We use the interpolation error which is often a reasonable indication of the error due to spatial discretisation in finite-element problems (Pichelina et al., 2000). In Fluidity, mesh adaptivity aims to increase resolution in regions of the domain with large curvatures of given fields and decrease resolution elsewhere. This approach allows good representation of the small-scale dynamics and sharp gradients without the need for high spatial resolution throughout the entire domain (Piggott et al., 2005). The mesh is adapted through a series of local topological and geometrical operations as described by Pain et al. (2001). In this work we adapt in the vertical direction only. A single column of mesh vertices is first adapted. This column is then replicated to the other three columns, which are then joined to form a quasi-1-D column of tetrahedra. The location of the vertices is constructed such that all elements in that first 1-D column have unit edge length when measured with respect to a given adaptivity metric, M.

In Fluidity a relatively simple adaptivity metric M is employed. For chosen fields f_i, adaptivity metrics M_i are defined by

$$M_i = \det|H_i|^{-\frac{1}{2p+n}} \frac{|H_i|}{\varepsilon_i}, \qquad (15)$$

where ε_i is a user-defined weight corresponding to the field under consideration and $|H_i|$ is the Hessian matrix (second-order derivatives) for that field where the absolute values of its eigenvectors have been taken, $p \in \mathbb{Z}$ and n is the dimension of the space (Loseille and Alauzet, 2011). The Hessian matrix contains information about both the magnitude and direction of the curvature of a field and hence can be used to guide generation of anisotropic elements. The final adaptivity metric used, M, is formed from a superposition of the adaptivity metrics for individual fields: $M = \bigcup_i M_i$ (Pain et al., 2001). In the work presented here we tested values of p of 2 and ∞ as both have been used in previous work, but $p = 2$ has shown superior results in resolving both weak and strong curvatures simultaneously within the same simulation (Loseille and Alauzet, 2011; Hiester et al., 2011) and is used in all simulations presented. ε may be varied spatially and temporally, but neither is utilised here. In general, for a given solution field, decreasing ε will lead to greater refinement of the mesh and increasing ε will lead to more coarsening. At this point the adaptivity metric is also modified to take into account bounds upon the maximum and minimum element size, maximum allowable aspect ratio, edge length gradation, and number of elements. For more details see Pain et al. (2001); Piggott et al. (2005, 2008b); Hiester et al. (2011) and references therein.

The mesh is adapted at run time and the frequency with which it adapts can also be specified. Here, we specify an adapt frequency of 5 h. After an adapt the solution fields must be interpolated from the pre- to post-adapt meshes. Two methods are available: "linear-interpolation" and "bounded Galerkin-projection" (Farrell et al., 2009). All prognostic fields are interpolated, along with any diagnostic fields as required. Three different interpolation methods were tested in this work: linear interpolation, which is bounded but non-conservative; Galerkin projection, which is conservative and can be made bounded at the expense of a minimal amount of diffusion (Farrell et al., 2009); and a mixture of the two, in which Galerkin projection was used for biological tracers, and linear interpolation was used for physical quantities. It is anticipated that conservation of the integral of biological quantities is crucial to obtaining a satisfactory solution, but that the physical quantities – velocity, temperature and salinity – only require linear interpolation (Hill et al., 2012). As linear interpolation is less computationally demanding than Galerkin projection, further savings in computational cost over and above those obtained through those obtained through the use of adaptivity can be gained using linear interpolation where it is adequate.

The adaptivity metric used to alter the mesh is crucial to obtaining an optimal simulation (Hiester et al., 2011). Here, we test four different adaptivity metric formulations which govern element sizes: *PAR*, *Bio*, *Bio and Phys*, and *Phys*. These use the photosynthetic active radiation only (PAR), biological fields and photosynthetic active radiation (Bio), physical fields only (Phys) or a combination (Bio and Phys). The same adaptivity metrics are used for all three test stations as we are attempting to provide an adaptivity metric formulation that works well in a variety of ocean settings and to avoid "tweaking" of the adaptivity metric for a particular location. The physical fields used are the density and velocity, and the biological fields used are the nutrients and PAR. Details of the fields used and the weighting of each field are given in Table 1. It is important to note that Fluidity allows a great deal of flexibility in choosing the adaptivity metric, unlike in the previous studies described above. Here, we investigate how the choice of which fields (physical and biological) are included in the adaptivity metric affects a simulation. We do not investigate the effects of changing the user-defined weights ε; they are chosen to give a reasonable result and may not be optimal. For the purposes of this paper ε being sub-optimal is not critical. The weights were chosen based on a preliminary parameter study and give a good compromise between good results and minimal element numbers.

The aim of choosing these fields is to enable tracking of the MLD variation on both a daily and seasonal level. This will concentrate resolution only where changes of the above quantities are large, minimising the number of elements elsewhere. The physical fields chosen are density and velocity. Density will show a steep vertical change at the base of the mixed layer, whilst velocity varies strongly in the top few metres of the water column. However, if winds strengthen, more resolution will be added as the mixing deepens. The biological tracers chosen are the nutrients and PAR. Fluxes of nutrients from depth are the main cause of plankton blooms at both India and Bermuda. However, the nutrient high, and

Table 1. Weighting of fields used for each metric used in this study. A "–" indicates that this field was not used in the metric construction. See Eq. (15).

Metric	Nutrient (ε) (mmol m^{-3})	PAR (ε) (W m−2)	Velocity (ε) (m s^{-1})	Density (ε) (kg m^{-3})
Bio	10.0	0.1	–	–
Phys	–	–	(0.1, 0.1, 10.0)	0.01
Bio and Phys	10.0	–	(0.1, 0.1, 10.0)	0.01
PAR	–	0.1	–	–

hence phytoplankton high, may not be at the surface; this is the case in Bermuda where there is a significant subsurface chlorophyll maximum. By resolving the nutrient fluxes closely we aim to also then resolve the other biological tracers as a consequence. In contrast, Station Papa shows only weakly varying surface nutrient changes. However, the upward flux of nutrients can lead to erroneous timings of the spring bloom. Therefore, the base of the MLD shows a substantial vertical nutrient change and hence adding resolution here should minimise spurious vertical numerical diffusivity. PAR is important only in the top 100 m of the domain and varies daily and hence adding this field to the adaptivity metric will add extra resolution during daylight hours down to the bottom of the photic zone.

To examine the effect of adaptive remeshing on biogeochemical models we perform a number of experiments. For each station we run fixed mesh simulations at different resolutions to examine the effects of vertical mesh resolution on the biogeochemistry. We qualitatively compare the results from the highest fixed-mesh resolution to available observations at each station. We then perform experiments varying the adaptivity metric at each station and evaluate the results compared to the highest resolution fixed-mesh simulation using the root mean square error of a number of fields to quantitatively evaluate the performance. Finally, we perform two specific examples tracking other biological tracers to demonstrate possible uses for the adaptive remeshing technology.

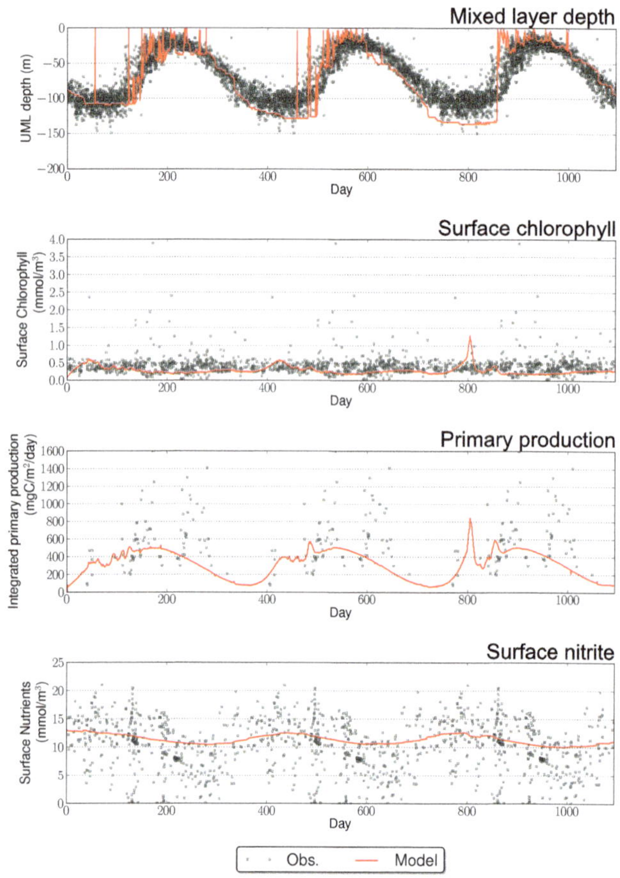

Fig. 2. Summary of simulated physical and biological behaviour at Station Papa for a uniform (2.5 m), non-adaptive simulation. From top to bottom, panels show MLD, surface chlorophyll, integrated primary productivity and surface nutrients. Where available, measured data are shown as green squares. Measured data are plotted against day of year due to lack of data for some quantities.

3 Model evaluation

We first examine a single fixed-mesh case for each station, comparing them to available observational data from Kleypas and Doney (2001) and Popova et al. (2006), before showing that the simulated response depends on the model's vertical resolution.

Station Papa in the Northwest Pacific is an ideal testing station for a 1-D simulation. There is little horizontal advection, and as such, Papa has been used to assess numerical models (Denman and Miyake, 1973; Burchard and Bolding, 2001; Hill et al., 2012). Fluidity has also been previously shown to work well at replicating the expected physics with adaptive meshes here (Hill et al., 2012). Papa's distinguishing feature is that nutrients are not limited and hence surface nutrients exhibit only a small seasonal variation. Surface values are those occurring on the uppermost element, which should be the same throughout the MLD. The results of the biogeochemical model show reasonable agreement with measured data (Fig. 2), replicating the major features of this station. The surface nutrients show the desired dip over the summer months, though this is not quite as pronounced as the observed data. The surface chlorophyll shows good

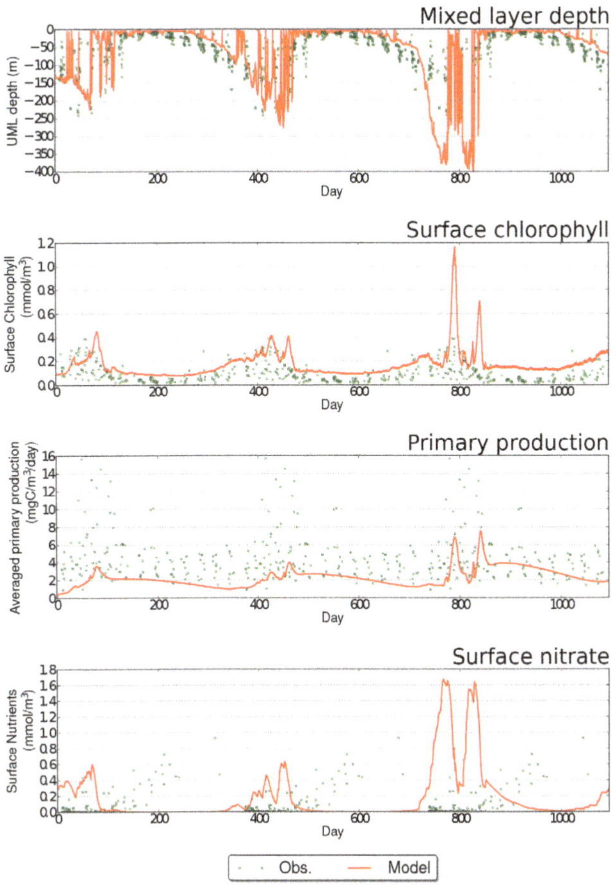

Fig. 3. Summary of simulated physical and biological behaviour at Bermuda for a uniform (2.5 m), non-adaptive simulation. From top to bottom, panels show MLD, surface chlorophyll, averaged primary productivity and surface nutrients. Where available, measured data are shown as green squares. Measured data are plotted against day of year due to lack of data for some quantities.

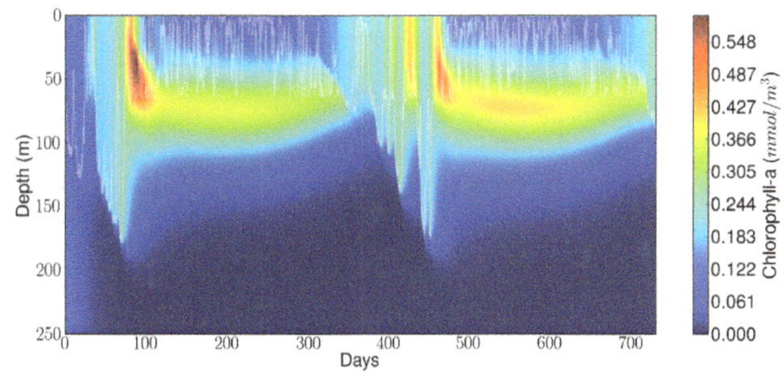

Fig. 4. Time–depth plot of chlorophyll at Bermuda, showing the clear subsurface chlorophyll maxima.

agreement to observed data, as does the integrated primary production (note: this is integrated over the mixed layer).

The model result at Bermuda, unlike Papa, shows some differences to the measured data (Fig. 3). The surface nutrient shows the observed nutrient-limited behaviour, but the limiting of nutrients occurs too early in the year. The third winter (days 700–900) shows a marked deepening of the mixed layer. This is due to surface forcing particular to that year, and longer simulations (not shown) show a return to the more normal behaviour seen in years one and two. Surface chlorophyll values lie on the upper limit of observed data, with a small peak in the spring. The primary production (note: averaged over MLD) is around a factor of two too low. However, given that we are simulating an isolated 1-D column, without any horizontal transport of quantities in or out of the domain, we believe this is a reasonable result. There is a substantial subsurface chlorophyll maximum (Fig. 4) as has been shown in measured observation and it is of similar magnitude to that obtained in previous modelling studies (e.g. Anderson and Pondaven, 2003).

Unlike for the previous stations there is a lack of MLD data for Station India. However, the model again gives a sufficiently good match to available data, although with larger discrepancies compared to other stations (Fig. 5). The spring bloom (as shown by the surface chlorophyll) happens around 30 days early, with a peak that is perhaps a factor of four too high. Similarly, the integrated primary production (note: integrated over the mixed layer) shows a peak of around 2–3 times the observed value at the same time. However, the values during the rest of the year lie around the lower limit of observed data. Surface nutrients show reasonable agreement with the timing of the spring decrease, but the level is perhaps a factor of two too high during the summer months.

3.1 Resolution dependence

For all stations we have run the simulations on a number of fixed meshes, varying the vertical resolution between 20 m and 2.5 m. We use the highest resolution (2.5 m) fixed-mesh simulation as "truth" when assessing the performance of the adaptive mesh simulations. In addition, we use qualitative comparisons to observational data at each station to ensure that the model performs as expected, given the lack of horizontal dynamics.

A standard root mean square (rms) error was used to assess model performance. The rms error is calculated as

$$\epsilon_{\text{rms}} = \sqrt{\frac{\sum_{i=1}^{n}(x_i - y_i)^2}{n}}, \qquad (16)$$

where x_i is the quantity being assessed in the high-resolution simulation and y_i is the value of the quantity produced by the simulation being considered. All simulations were run for three years. Output for all runs was produced every 12 h, giving $n = 2190$. For each field of interest a separate ϵ_{rms} is calculated, giving ϵ_{MLD} for the MLD. For biological quantities the L2_{norm} of the water column was used, giving ϵ_{N} for

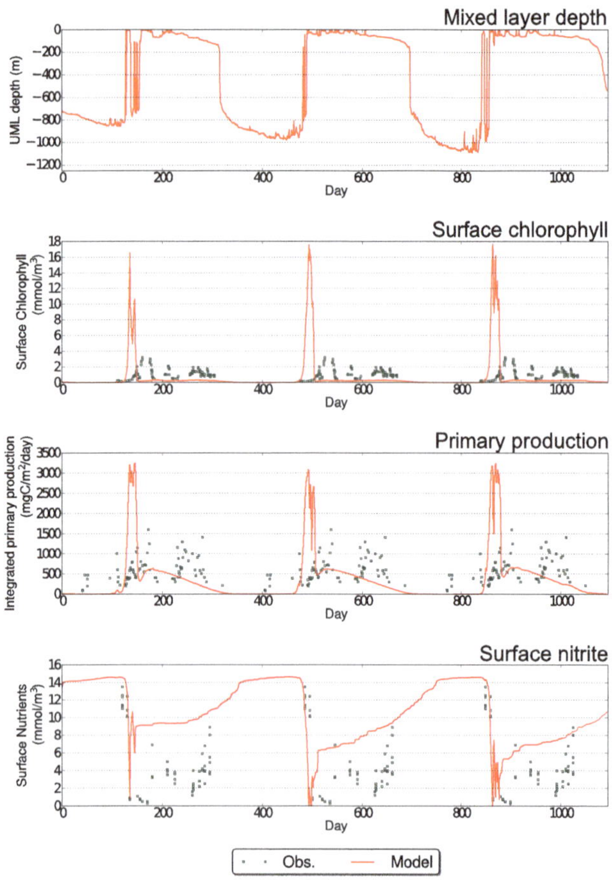

Fig. 5. Summary of simulated physical and biological behaviour at Station India for a uniform (2.5 m), non-adaptive simulation. From top to bottom, panels show MLD, surface chlorophyll, integrated primary productivity and surface nutrients. Where available, measured data are shown as green squares. Measured data are plotted against day of year due to lack of data for some quantities.

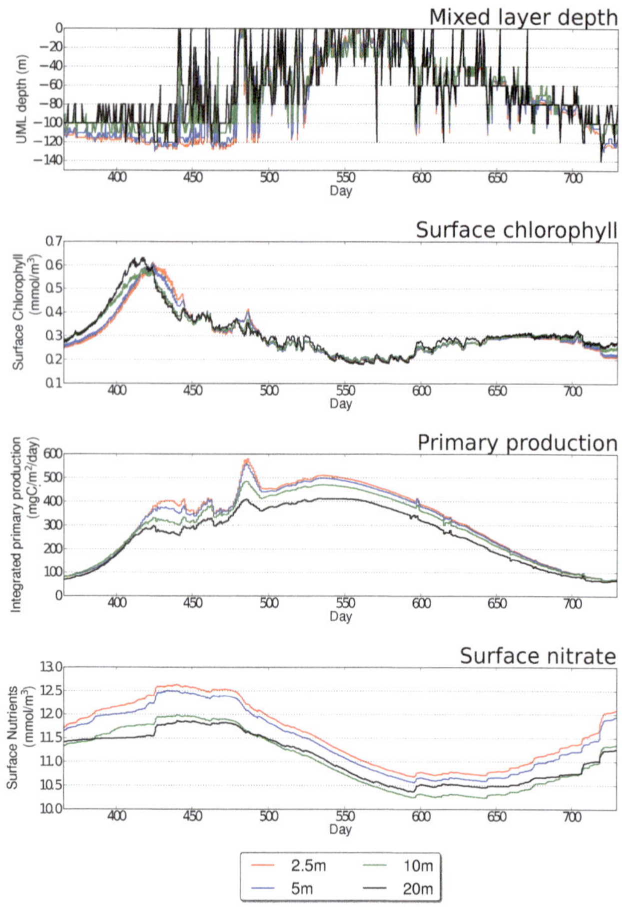

Fig. 6. Summary of simulated physical and biological behaviour at Station Papa using uniform meshes at a number of resolutions. From top to bottom, panels show MLD, surface chlorophyll, integrated primary productivity and surface nutrients.

nutrients, ϵ_P for primary productivity, ϵ_C for chlorophyll and ϵ_Z for zooplankton. Here, the $L2_{norm}$ is defined as

$$L2_{norm} = \sqrt{\int_0^n |S|^2 dV}, \tag{17}$$

where V is the volume of the domain and S is the scalar field in question.

Figures 6–8 show a single year (year 2 of the 3-year simulation to allow for model spin-up) for each station. For Station Papa (Fig. 6) there is a noticeable difference in MLD behaviour with higher resolutions showing deeper winter mixing. This in turn affects the upward mixing of nutrients, which show a marked jump when resolution is refined from 10 to 5 m. Both 5 and 2.5 m resolutions show broadly similar patterns. The difference in upward mixing of nutrients then affects the primary productivity, and surface chlorophyll shows a difference in peak surface chlorophyll of around 20 days. Bermuda shows a similar pattern (Fig. 7) with increased resolution producing higher peak nutrients due to increased upmixing, which in turn leads to increased surface chlorophyll. Finally, Station India (Fig. 8) shows a marked increase in primary productivity within the MLD, which is doubled when resolution is refined from 20 to 2.5 m. The surface nutrient data show little difference with resolution, so a sensible interpretation is that this is due to the slight increase in summer MLD depths with increased resolution. It is therefore clear that all stations show a response to vertical resolution which is the result of a complex interaction between the mixing caused by the vertical turbulence model and the biological sources and sinks.

Resolution dependence can also be seen in vertical profiles. Chlorophyll vertical profiles were taken at day 547 (mid-summer of the second year) and day 730 (mid-winter at the end of the second year) (Fig. 9). Each station shows a change in the simulated value with increasing resolution. This is perhaps most pronounced at Station Papa during the winter, where the MLD increases from around 100

Table 2. The rms error, ϵ, of fixed-mesh simulations compared to the simulation with 2.5 m vertical resolution at Station Papa. ϵ is shown for MLD, and the $L2_{norm}$ of nutrient, primary productivity, chlorophyll and zooplankton. See Fig. 6 also.

Res (m)	ϵ_{MLD} (m)	ϵ_N (mmol m^{-3})	ϵ_P (mmol m^{-3})	ϵ_C (mmol m^{-3})	ϵ_Z (mmol m^{-3})	No. elements
5 m	5.244	0.119	0.098	0.0095	0.0045	1200
10 m	10.640	0.506	0.342	0.0445	0.0169	600
20 m	16.521	0.421	0.628	0.0486	0.0283	300

Table 3. The rms error, ϵ, of fixed-mesh simulations compared to the simulation with 2.5 m vertical resolution at Bermuda. ϵ is shown for MLD, and the $L2_{norm}$ of nutrient, primary productivity, chlorophyll and zooplankton. See Fig. 7 also.

Res (m)	ϵ_{MLD} (m)	ϵ_N (mmol m^{-3})	ϵ_P (mmol m^{-3})	ϵ_C (mmol m^{-3})	ϵ_Z (mmol m^{-3})	No. elements
5 m	7.079	0.408	0.140	0.0129	0.0113	1200
10 m	12.630	0.402	0.149	0.0160	0.0236	600
20 m	18.710	0.397	0.226	0.0285	0.0410	300

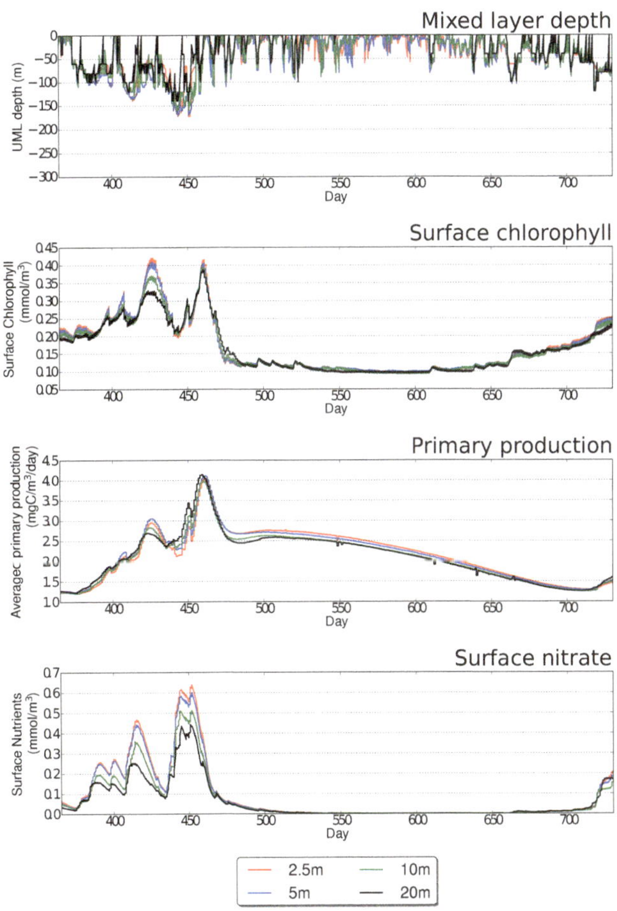

Fig. 7. Summary of simulated physical and biological behaviour at Bermuda using uniform meshes at a number of resolutions. From top to bottom, panels show MLD, surface chlorophyll, MLD averaged primary productivity and surface nutrients.

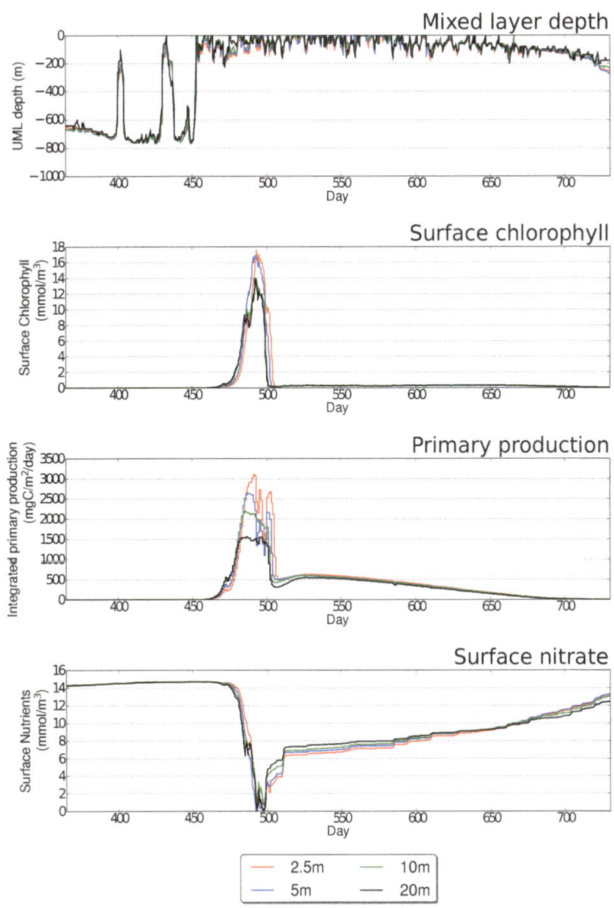

Fig. 8. Summary of simulated physical and biological behaviour at Station India using uniform meshes at a number of resolutions. From top to bottom, panels show MLD, surface chlorophyll, integrated primary productivity and surface nutrients.

Table 4. The rms error, ϵ, of fixed-mesh simulations compared to the simulation with 2.5 m vertical resolution at Station India. ϵ is shown for MLD, and the $L2_{norm}$ of nutrient, primary productivity, chlorophyll and zooplankton. See Fig. 8 also.

Res (m)	ϵ_{MLD} (m)	ϵ_N (mmol m^{-3})	ϵ_P (mmol m^{-3})	ϵ_C (mmol m^{-3})	ϵ_Z (mmol m^{-3})	No. elements
5 m	137.03	1.04	303.258	1.538	0.222	2400
10 m	138.18	1.21	357.834	1.770	0.247	1200
20 m	142.71	1.50	455.565	1.992	0.264	600

to nearly 120 m when resolution is increased from 20 to 2.5 m. Bermuda shows a decrease of both winter and summer subsurface chlorophyll maxima with increasing resolution. These vertical profiles show that the model is numerically stable, producing adequate results at even low resolution, but that vertical resolution does affect the profile simulated.

The response to resolution can be examined more quantitatively using a simple convergence test. Although convergence is non-trivial for nonlinear dynamics (Hill et al., 2012), a decrease in error should be seen with increasing vertical resolution. For all stations there is clear convergence (a decrease in error) for the MLD (Tables 2–4). Ideally, for the set-up described previously, this should be at least first-order convergence. Both Bermuda and Station Papa show this behaviour but Station India does not (though there is still a decrease in error with increasing resolution). However, for most variables there is a decrease in the error measure compared to the highest resolution simulations at each station.

The surface nutrients error stays approximately constant at both Bermuda and Station India (Tables 3 and 4). Despite these exceptions there is a clear dependence on resolution, with higher resolutions generally matching the highest resolution simulation with higher accuracy. At Papa, all biological quantities bar nutrients show a general convergence in error as resolution is increased (Table 2). The error at 10 m vertical resolution appears to be double that expected, but there is a convergence in error from 10 to 5 m. Bermuda shows clear first-order convergence of MLDs and zooplankton; and less certain convergence of chlorophyll (Table 3). Surface nutrient error appears to be constant, as does primary productivity (average over the MLD). Finally, Station India shows a general convergence with increasing resolution for all biological quantities, though not at first order (Table 4).

From these results we can see that there is a general decrease in error to the highest resolution run with increasing resolution. Therefore, using vertical adaptivity should allow a minimisation in the number of elements within the computational mesh whilst ensuring that error does not increase to an unreasonable level.

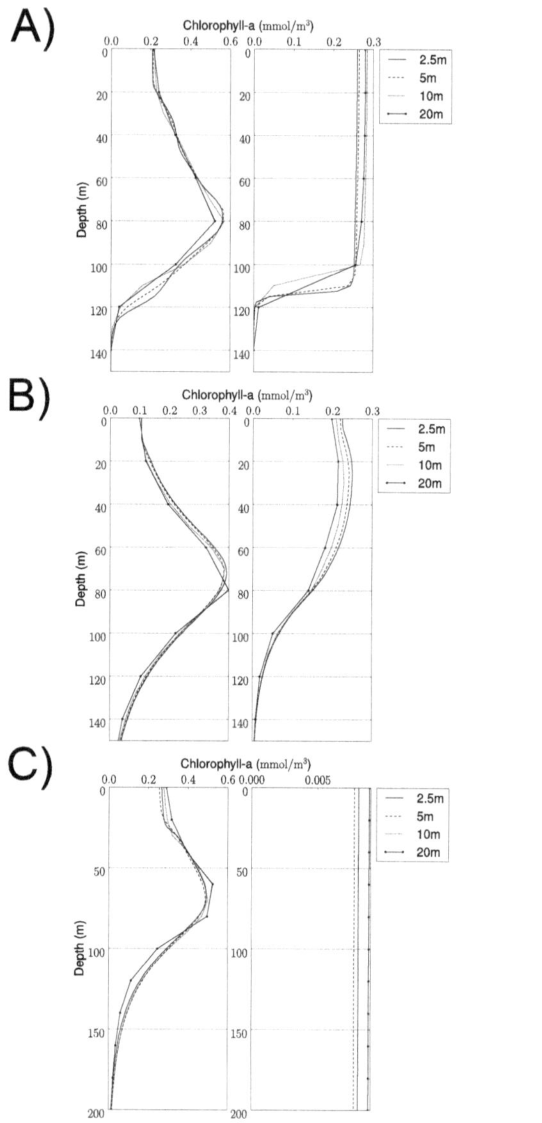

Fig. 9. Vertical profiles of chlorophyll at **(A)** Station Papa, **(B)** Bermuda and **(C)** Station India over the top 150, 150 and 200 m respectively using uniform meshes at a number of resolutions. The left-hand column shows the profile in summer (day 182) and the right-hand column shows the profile in winter (day 365). There are changes in values at key depths of up to 15 % depending on the vertical resolution used.

4 Adaptivity

We have carried out the same simulations as above using an adaptive mesh guided by a variety of different adaptivity metrics and, in addition, we have tested different interpolation

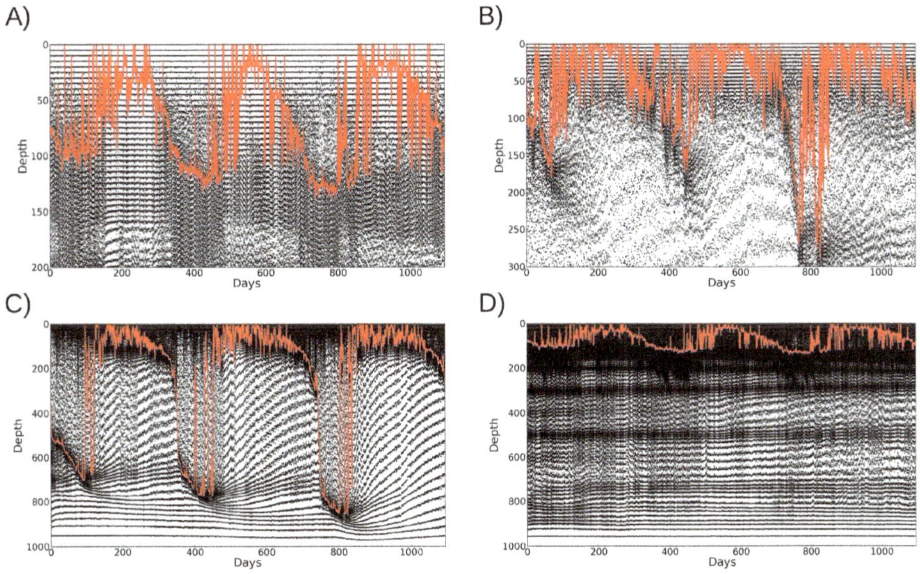

Fig. 10. Representation of the meshes obtained via adaptivity for all stations using the *bio and phys* metric. A dot is placed on each vertex in the mesh and this is repeated for each output time. Clustering of vertices therefore indicate higher resolution. Station Papa **(A)** shows reduced resolution under the mixed layer during summer, but high resolution persists for some distance below the mixed layer. Bermuda **(B)** shows substantial reduction of resolution below the mixed layer, but with the minimum resolution being maintained during the summer in the upper layers. Similarly, India **(C)** tracks the MLD, with decreased resolution within the mixed layer, whilst maintaining good resolution in the upper portion of the water column. Gradients in density mean that high-resolution zones are maintained at Station Papa **(D)** at up to 800 m depth.

methods at Station Papa. For simplicity, simulations at the three test station used the same adaptivity settings. Adaptivity was performed every 5 h. This allows changes in ocean surface forcing (which has a temporal frequency of 6 h) to be captured, along with diurnal fluctuations. Over a 3-year simulation a total of 5256 adapts are thus performed. This is a large number and therefore any additional numerical diffusivity or noise derived from adapting the mesh will be evident in the final simulation results when compared to the fixed-mesh simulations. The minimum and maximum edge lengths permitted are set to 5 and 50 m respectively. We therefore hope to find that the adaptive simulations are equivalent to the 5 m fixed resolution simulations, but use substantially fewer elements.

The adaptive algorithm was performed on a single vertical column of mesh vertices and the position of these were replicated to the other three columns. In this way we obtained a layered mesh, with vertical resolution of the layers varying according to the chosen adaptivity metric and the simulated state at the time of the adapt. Apart from the adaptive mesh, the simulations were completely identical to the fixed-mesh simulations.

The meshes produced by the adaptive algorithm showed broadly similar features between different adaptivity metrics for each particular site (Fig. 10). Comparing those produced by the adaptivity metric using both the biological and physical fields shows the mesh tracking the behaviour of the MLD. In addition, high resolution is maintained in the photic layer, but reduces with the mixed layer when the MLD increases substantially.

4.1 Effect of interpolation method

The tests at Station Papa (Fig. 11) show that Galerkin projection for all fields gives the best result for both physical behaviour (MLD) and biological quantities. In particular, there is a significant surface nutrient drift when using linear interpolation, and less so when using a mixed formulation. The effect is clear by day 1000, where surface nutrients are 25 % higher when using a mixed formulation and over 50 % higher than the fixed-mesh simulation when using linear only. The cause of this drift is additional numerical diffusion due to the linear interpolation in the physical fields (mixed and linear only) and lack of conservation (linear only). Using Galerkin projection on both physical and biological fields gives a result that is almost identical to the fixed-mesh simulation. The MLD behaviour changes in response to the interpolation method used, with a marked shoaling of the MLD around day 800 when using linear or mixed interpolation. The mixed interpolation formulation performed only slightly better in this regard and using Galerkin projection on the turbulent parameterisation quantities only improved the solution slightly (results not shown). These results contrast to those presented by Hill et al. (2012), where linear interpolation was used and adaptivity performed well at Station Papa.

Table 5. The rms error, ϵ, of adaptive mesh simulations compared to the simulation with 2.5 m vertical resolution at Station Papa. ϵ is shown for MLD, and the $L2_{norm}$ of nutrient, primary productivity, chlorophyll and zooplankton. See Fig. 6 also.

Res (m)	ϵ_{MLD} (m)	ϵ_N (mmol m^{-3})	ϵ_P (mmol m^{-3})	ϵ_C (mmol m^{-3})	ϵ_Z (mmol m^{-3})	No. elements (Mean, min, max)
5 m	5.244	0.119	0.0986	0.0095	0.0045	1200
Bio and Phys	4.769	0.053	0.385	0.0038	0.0020	458.5, 372, 558
Bio Only	14.211	0.589	0.529	0.0426	0.0300	386.1, 336, 450
Phys Only	4.837	0.070	0.398	0.0034	0.0021	310.1, 240, 408
PAR	23.324	2.664	0.592	0.0790	0.0330	612.2, 532, 660

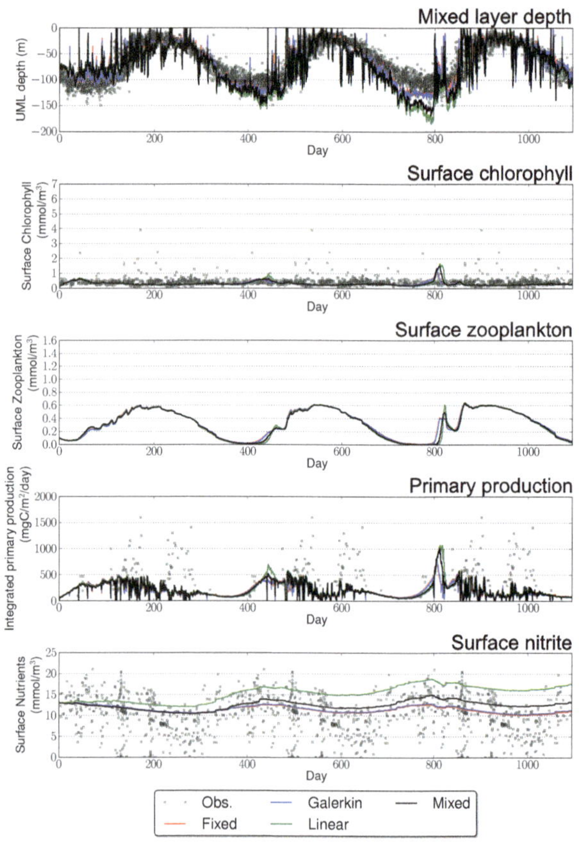

Fig. 11. Summary of simulation results from Station Papa comparing the fixed high-resolution (2.5 m) simulation with adaptive simulation using different interpolation methods between meshes. Linear and mixed perform poorly, inducing extra vertical diffusivity, compared to Galerkin projection. Panels and data are the same as in Fig. 6.

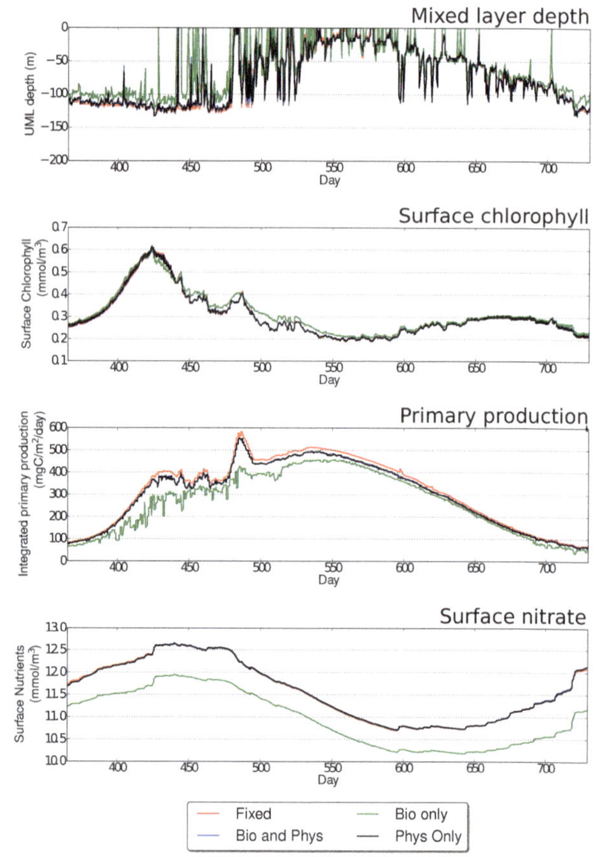

Fig. 12. Summary of results comparing different adaptivity metrics against measured data and the high-resolution (2.5 m) fixed-mesh simulation for Station Papa.

However, here we have added the additional term to simulate internal wave breaking (Eq. 13) and the adaptivity metric in Hill et al. (2012) was tuned to Station Papa only, with a lower minimum edge length. It is also worth noting that the temperature and salinity fields showed little or no difference between the fixed and adaptive simulations as was shown in Hill et al. (2012); it is the biological tracers that highlight undesired behaviour, i.e. upward fluxing of tracers, of the adaptive runs. Galerkin projection is therefore used for all subsequent adaptive simulations.

4.2 Station Papa

All adaptive simulations at Station Papa completed successfully and produced results that are a reasonable fit to measured data (Fig. 12). Using an adaptivity metric formed of physical or physical and biological quantities produces results that are not significantly different from that of the uniform 5 m resolution simulation (Table 5). All tracers give similar or better results than the 5 m fixed simulation, though given the lack of convergence shown in some of these quantities some caution must be used in interpreting these values. It is clear that using only biology or just PAR to form

Table 6. The rms error, ϵ, of adaptive mesh simulations compared to the simulation with 2.5 m vertical resolution at Bermuda. ϵ is shown for MLD, and the $L2_{norm}$ of nutrient, primary productivity, chlorophyll and zooplankton. Note that the PAR and Bio only simulations failed and recorded no result. See Fig. 7 also.

Res (m)	ϵ_{MLD} (m)	ϵ_N (mmol m^{-3})	ϵ_P (mmol m^{-3})	ϵ_C (mmol m^{-3})	ϵ_Z (mmol m^{-3})	No. elements (Mean, min, max)
5 m	7.079	0.408	0.140	0.0129	0.0113	1200
Bio and Phys	7.309	0.407	0.598	0.0084	0.0096	310.3, 264, 444
Phys Only	8.432	0.408	0.615	0.0076	0.0082	228.3, 174, 436

Table 7. The rms error, ϵ, of adaptive mesh simulations compared to the simulation with 2.5 m vertical resolution at Station India. ϵ is shown for MLD, and the $L2_{norm}$ of nutrient, primary productivity, chlorophyll and zooplankton. See Fig. 8 also.

Res (m)	ϵ_{MLD} (m)	ϵ_N (mmol m^{-3})	ϵ_P (mmol m^{-3})	ϵ_C (mmol m^{-3})	ϵ_Z (mmol m^{-3})	No. elements (Mean, min, max)
5 m	137.03	1.04	303.258	1.538	0.222	2400
Bio and Phys	141.98	1.12	356.798	1.668	0.236	287.74, 150.0, 456.0
Bio Only	112.72	1.12	324.777	1.052	0.097	220.63, 120.0, 360.0
Phys Only	141.30	1.08	344.102	1.683	0.237	249.53, 150.0, 414.0
PAR	131.84	1.17	677.312	1.920	0.245	120.00, 120.0, 121.0

the adaptivity metric is not adequate at this location as the values of the rms errors, ϵ, for all tracers are substantially larger, apart from primary productivity, which has already been identified as potentially problematic in use as an assessment of performance.

4.3 Bermuda

Not all adaptive simulations were effective at Bermuda. Using either PAR or biology only to form the adaptivity metric results in simulations failing with a solver error soon after the first or second adapt. This is attributed to insufficient mesh resolution to ensure stability for the GLS turbulence parameterisations. Unlike Station Papa, the MLD is well below the photic zone in the initial stages of the simulation. However, both simulations using either physics only or biology and physics performed well. Both gave similar results, quantitatively (Table 6) and qualitatively (Fig. 13). The two adaptivity metrics also gave lower values of ϵ_C and ϵ_Z, but, as with Station Papa, these values should interpreted with some caution.

4.4 Station India

All adaptive simulations produced good results at Station India when compared to the high-resolution fixed-mesh resolution regardless of adaptivity metric used (Table 7, Fig. 14). Minor differences in the timing of the spring bloom occurred with the biology-only adaptivity metric occurring some 25 days later than the fixed-mesh simulation. The biology-only adaptivity metric also showed an increase in the number of shoaling excursions in the spring. These did not occur when using other adaptivity metrics or in the fixed-mesh simulations. There are also minor differences in the magnitude of the integrated primary productivity, but these variations are much lower than those observed when changing resolution in the fixed-mesh simulations (Figs. 8).

4.5 Summary of adaptive results

Adaptive remeshing can clearly be successfully applied at the three ocean stations successfully using a variety of adaptivity metrics. Some adaptivity metric/station combinations perform better than others. As well as reproducing the surface values (i.e. at depth $z = 0$) and the MLD the adaptive simulations also reproduce the vertical profiles of biological parameters (see Fig. 15 for chlorophyll and compare to Fig. 9).

The effect of adaptivity is clearly seen in the meshes produced by the simulations (Fig. 10). All stations show much higher resolution around the MLD, as expected, with decreased resolution when the MLD is deep (for example at Station India). The meshes contain far fewer elements than the high-resolution fixed-mesh simulations and are therefore more computationally efficient.

For all stations it is important to include velocity and density in the adaptivity metric (labelled as "physics" in all figures) and the inclusion of nutrients appears to be only of minor importance. However, using nutrients only or PAR can lead to a numerically unstable simulations and moreover, when the simulation is numerically stable, it gives much larger errors than the adaptivity metrics that include the physics. The simulated physics drives the biology and it is therefore crucial to include measures of the physical system in the adaptivity metric to correctly simulate the physical behaviour of the system. The physical measures used here are the vertical changes in velocity and density. However, turbulent kinetic energy, temperature or salinity may also be

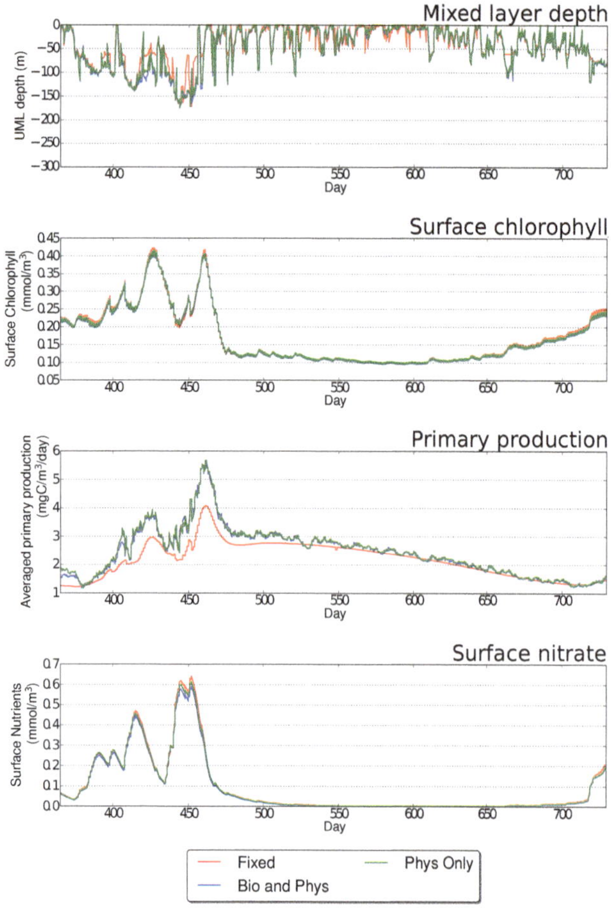

Fig. 13. Summary of results comparing different adaptivity metrics against measured data and the high-resolution (2.5 m) fixed-mesh simulation at Bermuda. Note that the simulations using PAR and biology only failed after only a few adapts.

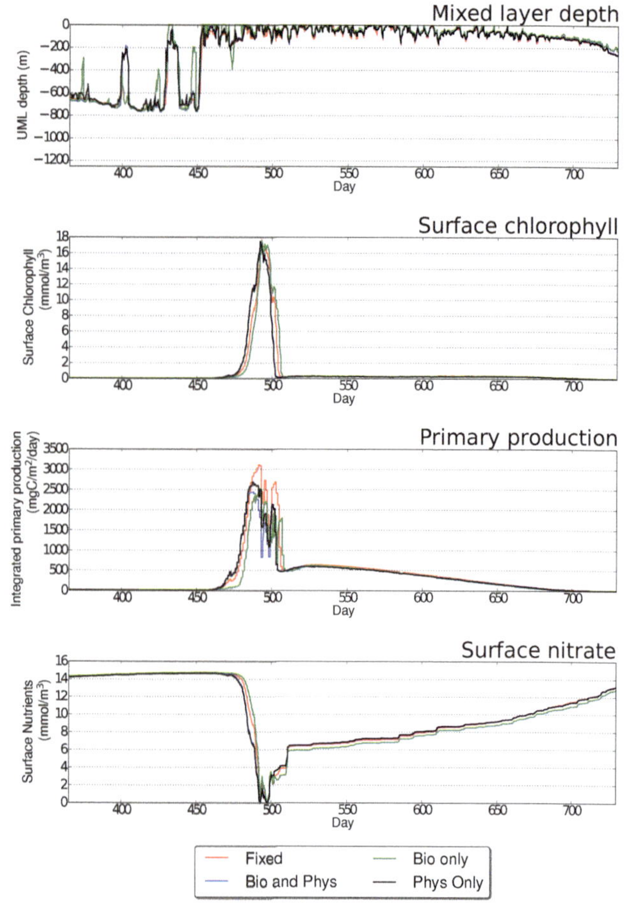

Fig. 14. Summary of results comparing different adaptivity metrics against measured data and the high-resolution (2.5 m) fixed-mesh simulation for Station India.

sensible choices (Hill et al., 2012). Accounting for physical behaviour in the adaptivity metric appears to be sufficient for a successful simulation of biological behaviour. However, if the physical properties are well simulated then the biological processes do not necessarily also need including in the adaptivity metric in order to achieve a reasonable output.

5 Specific adaptive examples

One of the primary advantages of the approach outlined above is that the adaptivity metric used to calculate the mesh edge length can be composed of any simulated or diagnosed fields. We show the potential of that method here by simulating Bermuda with a adaptivity metric focusing on chlorophyll, and Station Papa concentrating on sinking detritus. These simulations show how the mesh is able to adapt to the particulars of the simulation, tracking transient behaviour but with a lower computational overhead than a high-resolution fixed-mesh simulation.

The first experiment attempts to track falling detritus in the Station Papa case. The detritus field is added to the adaptivity metric. The result, compared to the previous adaptive simulation using only nutrients, velocity and density in the adaptivity metric, shows a considerable change in the detritus concentration at depth (Fig. 16). In the original adaptive simulation, the detritus field is smoothed out at depths of over 300 m as the resolution here is relatively coarse. The effect of adding detritus to the adaptivity metric is as expected – the field maintains the sharp boundaries as it sinks, replicating the highest resolution fixed-mesh simulation. This is clearly seen in the resulting computational mesh (Fig. 17). In addition, we have carried out the same simulation based on a fixed mesh with decreasing resolution with increasing depth, as is typical in current ocean models. In this case we used the resolution change specified by the NEMO model (Madec, 2008) where resolution varies from 3 m at the surface to 100 by 1000 m depth. The results show increased diffusion of detritus by 100 m depth and by 700 m depth all detail in the detrital pulse is lost.

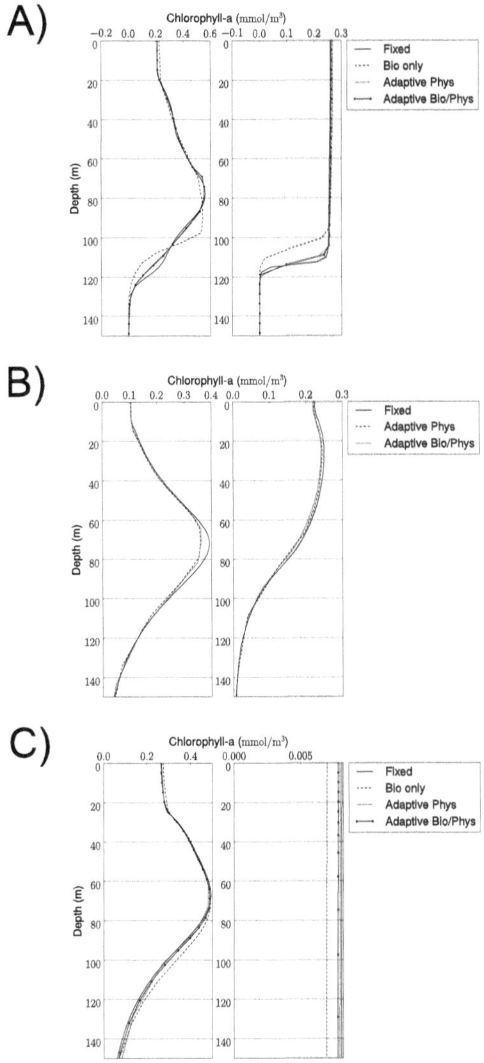

Fig. 15. Vertical profiles of chlorophyll at **(A)** Station Papa, **(B)** Bermuda and **(C)** Station India over the top 150 m, 150 m and 200 m respectively comparing the performance of different adaptivity metrics. The left-hand column shows the profile in summer (day 182) and the right-hand column shows the profile in winter (day 365). Note that all metrics, bar biology only, show similar results.

A similar result is seen at Bermuda where chlorophyll is added to the adaptivity metric. Here, we see that the subsurface chlorophyll maximum is maintained correctly (Fig. 18), whereas in the previous adaptive simulation the simulated value is lower around day 130 than for the fixed-mesh simulation. This is not the case when chlorophyll is added to the adaptivity metric. The simulation using a typical ocean model vertical resolution shows a lower chlorophyll peak (day 100) and a slightly reduced subsurface chlorophyll maximum throughout the summer compared to the high-resolution simulation. The effect of adding chlorophyll to the adaptivity metric can be seen in the resulting mesh (Fig. 17).

For both simulations there is, of course, an increase in the number of elements used compared to the original adaptive simulations, but the average number of elements is still much lower than the high-resolution fixed-mesh simulation, and accordingly, the run times are much lower. The Bermuda simulation used an average of 437 elements (576 maximum, 301 minimum). Compared to a fixed mesh of uniform resolution 2.5 m (2400 elements) this is a fivefold reduction in elements on average. The simulation based on a typical ocean model vertical resolution uses 216 elements at the coast of reduced resolution of the chlorophyll maximum. Similarly, simulating detritus at Station Papa used an average of 726 elements (507 minimum, 1120 maximum), compared to 2400 elements used in the 2.5 m fixed mesh simulation – a threefold reduction. The simulation using typical ocean model vertical resolution uses 216 elements (as at Bermuda) and is clearly incapable of accurately resolving sinking detritus.

6 Conclusions

This work has shown that the Fluidity model can successfully replicate expected biogeochemical behaviour at three key ocean stations. Both fixed and adaptive mesh simulations show very similar behaviour, but adaptive remeshing requires careful consideration of the adaptivity metric used. The physical quantities must be included in this adaptivity metric, and if the physical properties are well simulated then biological tracers do not necessarily need including in the adaptivity metric for reasonable output. This is consistent with a number of other studies where physical processes are the dominate control of ocean biogeochemical models (Berline et al., 2007). It is important to note that we have not shown an increase in model skill with either increasing resolution of when using adaptive meshes. We have, however, attempted to demonstrate numerical convergence of the biological model with increasing resolution.

It has been shown that a key component of successful adaptive simulation is to conservatively interpolate fields between the previous and new mesh following an adapt when using mesh hr-adaptivity. This is likely due to a change in mesh topology inducing spurious numerical diffusion and non-conservative methods are not suitable for biological tracers as conservation is important. The requirement of using conservative interpolation is likely down to topological changes to the mesh. Other adaptive mesh techniques such as mesh movement or *p* adaptivity (where the order of the function space is altered) may not have such requirements. There is already work exploring the effectiveness of higher-order discretisations with biogeochemical models (Ueckermann and Lermusiaux, 2010). The work presented here will be a useful guide to future efforts in this area. In addition, we also found that for this quasi-1-D domain each column of vertices in the domain must be identical, such that there are no horizontal gradients. Fluidity currently uses a columnar mesh

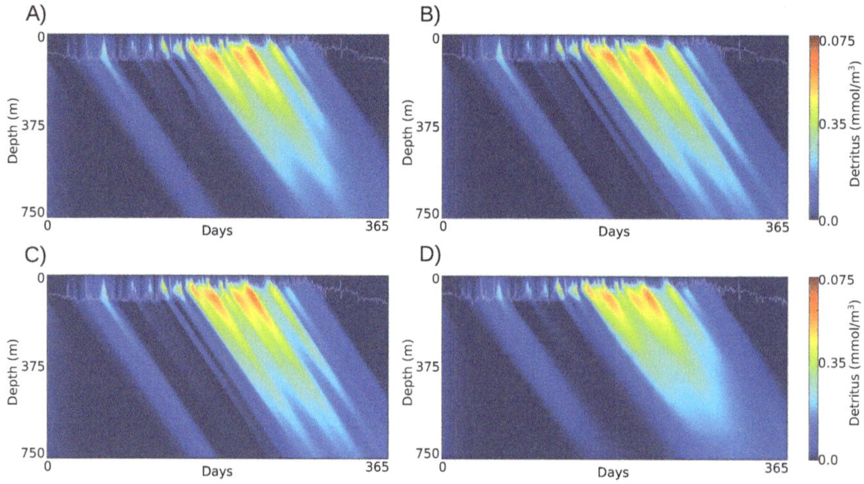

Fig. 16. Time–depth plot of detritus at Station Papa from the original adaptive mesh simulation **(A)**, the adaptive run with detritus included in the metric **(B)**, the high-resolution fixed-mesh simulation **(C)**, and **(D)** the same simulation using decreasing resolution with increasing depth. Including detritus in the adaptive metric increases the performance of the model at tracking detritus to depth, resolving detail in the timing and level of detrital pulses. Note that **(D)** smooths those events into a single unresolved event by day 300.

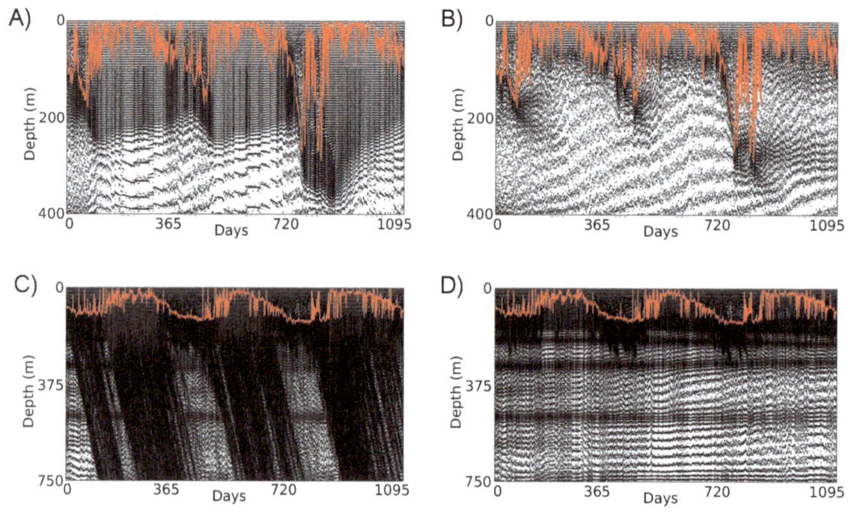

Fig. 17. Comparison of meshes produced by adaptive simulations. **(A)** Bermuda with the addition of chlorophyll to the adaptivity metric, **(B)** the original adaptive simulation at Bermuda using velocity, density and nutrients. Similarly, for Station Papa: **(C)** adaptive simulation with detritus included in the metric and **(D)** original adaptive simulation

in the vertical for large-scale ocean models (Piggott et al., 2008a) and is similar to the method used by Hofmeister et al. (2010). The work here is therefore compatible with how Fluidity will simulate global- or regional-scale models, using an unstructured horizontal mesh, with each vertical column then being extruded to the required depth. However, each column contains a different number of nodes in these cases which may induce additional vertical diffusivity due to horizontal gradients in the mesh. More work is required to ascertain if this effect is present in such simulations.

Ocean biogeochemical numerical models may require high numerical resolution in order to simulate measured data accurately (Oschlies, 2002). In addition, ocean biogeochemical models show sensitivity to a number of factors, such as light spectra (Kettle and Merchant, 2008) and surface fluxes (Friedrichs et al., 2006), but the main sensitivity is to the representation of the physics controlling the biology (e.g. Berline et al., 2007). A key part of this physical representation is the vertical advection and MLD behaviour. Regardless of whether a high horizontal resolution or adaptive horizontal meshes are used, the vertical resolution does not

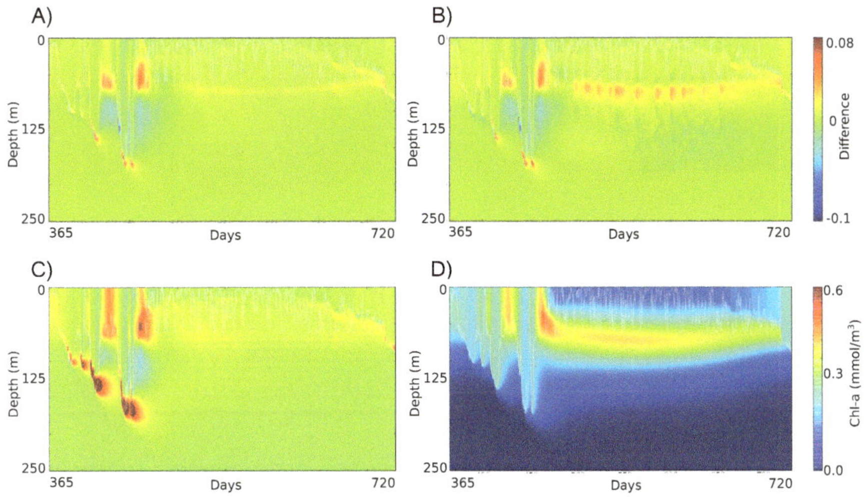

Fig. 18. Difference plots of chlorophyll at Bermuda comparing adaptive simulations (**A** and **B**) and a fixed, but varying resolution mesh (**C**) to the highest resolution (2.5 m) uniform mesh simulation (**D**). A clear improvement can be made in the subsurface chlorophyll maximum representation by adding chlorophyll to the adaptivity metric. (**A**) The original adaptive mesh simulation shows larger discrepancies around the subsurface maximum (**B**). Using fixed variable resolution as in other ocean models shows clear discrepancies in the spring and throughout the summer (**C**).

need to be constant for each spatial location. Instead, as is shown here, the resolution only needs to be placed at key locations in the vertical. Enabling higher resolution around the MLD and within the upper layers of the ocean is sufficient for replicating high-resolution fixed-mesh simulations, hence adaptive remeshing may be advantageous. However, note that we have kept other parameters that may depend on resolution constant, including the biogeochemical model parameters. The turbulence parameterisation is not resolution dependant provided the model does not attempt to resolve the turbulent scales included in the parameterisation. This is the case here. For the stations simulated here , using the normal ocean model vertical resolution where resolution decreases with depth is sufficient to obtain a reasonable solution. However, mesh adaptivity also has another powerful feature: additions to the adaptivity metric of other fields allows for the tracking of transient features and of other fields, such as detritus or chlorophyll. This is a powerful tool in tracking features of interest whilst minimising the number of elements in the computational mesh.

In summary, adaptive remeshing shows good potential to reduce the number of required elements in the computational mesh whilst maintaining, or even increasing, the accuracy of vertical processes, but only where and when it is required.

Appendix A

Biology model

As stated in the main body of the text, the biogeochemical model used in this report was based on that of Popova et al. (2006). Note, however, that the equations here are continuous for all depths, which was not true in Popova et al. (2006) where different source terms where used below the photic zone. The model parameters have also been fitted to match the data at all three test sites. The six components of the model are: nutrients (ammonium and nitrate), phytoplankton, chlorophyll, zooplankton and detritus.

A1 Biological source terms

The source terms for phytoplankton (P), chlorophyll (Chl), zooplankton (Z), nitrate (N), ammonium (A) and detritus (D) respectively are given by the following expressions:

$$S_{\mathrm{P}} = \mathrm{P}J(Q_{\mathrm{N}}+Q_{\mathrm{A}}) - G_{\mathrm{P}} - \mathrm{De}_{\mathrm{P}}, \tag{A1}$$

$$S_{\mathrm{Chl}} = (R_{\mathrm{P}}\cdot J\cdot(Q_{\mathrm{N}}+\mathrm{Q}_{A})\cdot \mathrm{P} + (-G_{\mathrm{P}}-\mathrm{De}_{\mathrm{P}}))\cdot\theta/\zeta), \tag{A2}$$

$$S_{\mathrm{Z}} = \delta\cdot(\beta_{\mathrm{P}}\cdot G_{\mathrm{P}} + \beta_{\mathrm{D}}\cdot G_{\mathrm{D}}) - \mathrm{De}_{\mathrm{Z}}, \tag{A3}$$

$$S_{\mathrm{N}} = -J\cdot\mathrm{P}\cdot Q_{\mathrm{N}} + \mathrm{De}_{\mathrm{A}}, \tag{A4}$$

$$S_{\mathrm{A}} = -J\cdot\mathrm{P}\cdot Q_{\mathrm{A}} + \mathrm{De}_{\mathrm{D}} + (1-\delta)\cdot(\beta_{\mathrm{P}}\cdot G_{\mathrm{P}} + \beta_{\mathrm{D}}\cdot G_{\mathrm{D}}) + (1-\gamma)\cdot\mathrm{De}_{\mathrm{Z}} - \mathrm{De}_{\mathrm{A}}, \tag{A5}$$

$$S_{\mathrm{D}} = -\mathrm{De}_{\mathrm{D}} + \mathrm{De}_{\mathrm{P}} + \gamma\cdot\mathrm{De}_{\mathrm{Z}} + (1-\beta_{\mathrm{P}})\cdot G_{\mathrm{P}} - \beta_{\mathrm{D}}\cdot G_{\mathrm{D}}. \tag{A6}$$

The terms in these equations are given in Table A1.

Note that unlike the model of Popova et al. (2006) we use a continuous model, with no change of equations (bar one exception) above or below the photic zone. For our purposes, the photic zone is defined as 100 m water depth. First we calculate θ:

$$\theta = \frac{\mathrm{Chl}}{\mathrm{P}\zeta}. \tag{A7}$$

However, at low light levels, Chl might be zero, therefore we take the limit that $\theta \to \zeta$ at low levels (1×10^{-7}) of chlorophyll or phytoplankton.

We then calculate α:

$$\alpha = \alpha_c\theta \tag{A8}$$

using the PAR available at each vertex of the mesh the light-limited phytoplankton growth rate, J, is then calculated:

$$J = \frac{v\alpha I_n}{\sqrt{v^2+\alpha^2+I_n^2}}. \tag{A9}$$

This can be used to calculate the limiting factors on nitrate and ammonium:

$$Q_{\mathrm{N}} = \frac{\mathrm{N}\exp^{-\Psi\mathrm{A}}}{K_{\mathrm{N}}+\mathrm{N}}, \tag{A10}$$

$$Q_{\mathrm{A}} = \frac{\mathrm{A}}{K_{\mathrm{A}}+\mathrm{A}} \tag{A11}$$

From these the diagnostic field, primary production (X_{P}), can be calculated:

$$X_{\mathrm{P}} = J(Q_{\mathrm{N}}+Q_{\mathrm{A}})\mathrm{P}. \tag{A12}$$

The chlorophyll growth scaling factor is given by

$$R_{\mathrm{P}} = Q_{\mathrm{N}}Q_{\mathrm{A}}\left(\frac{\theta_m}{\theta}\right)\left(\frac{v}{\sqrt{v^2+\alpha^2+I_n^2}}\right). \tag{A13}$$

The zooplankton grazing terms are now calculated:

$$G_{\mathrm{P}} = \frac{g p_{\mathrm{P}}\mathrm{P}^2\mathrm{Z}}{\epsilon+(p_{\mathrm{P}}\mathrm{P}^2+p_{\mathrm{D}}\mathrm{D}^2)}, \tag{A14}$$

$$G_{\mathrm{D}} = \frac{g p_{\mathrm{D}}\mathrm{D}^2\cdot\mathrm{Z}}{\epsilon+(p_{\mathrm{P}}\mathrm{P}^2+p_{\mathrm{D}}\mathrm{D}^2)}. \tag{A15}$$

Finally, the four death rates and re-mineralisation rates are calculated:

$$\mathrm{De}_{\mathrm{P}} = \frac{\mu_p\mathrm{P}^2}{\mathrm{P}+k_p} + \lambda_{\mathrm{bio}}\cdot\mathrm{P}, \tag{A16}$$

$$\mathrm{De}_{\mathrm{Z}} = \frac{\mu_z\mathrm{Z}^3}{\mathrm{Z}+k_z} + \lambda_{\mathrm{bio}}\cdot\mathrm{Z}, \tag{A17}$$

$$\mathrm{De}_{\mathrm{D}} = \mu_{\mathrm{D}}\mathrm{D} + \lambda_{\mathrm{bio}}\cdot\mathrm{P} + \lambda_{\mathrm{bio}}\cdot\mathrm{Z}, \tag{A18}$$

$$\mathrm{De}_{\mathrm{A}} = \lambda_{\mathrm{A}}\mathrm{A} \text{ where } z < 100. \tag{A19}$$

A2 Photosynthetically active radiation (PAR)

The photosynthetically active radiation, or PAR is calculated by

$$\mathrm{PAR} = (A_{\mathrm{water}} + A_{\mathrm{P}}\mathrm{P})I, \tag{A20}$$

where A_{water} and A_{P} are the absorption rates of photosynthetically active radiation by water and phytoplankton respectively.

A3 Detritus sinking velocity

Detritus is assumed to be denser than water and so sinks slowly through the water column. This is modelled by modifying the advecting velocity in the advection–diffusion equation for detritus by subtracting a sinking velocity u_{sink} from the vertical component of the advecting velocity.

Table A1. Symbols used to describe the six-component NPZD model. Typical values are provided for externally set parameters.

Symbol	Meaning	Typical value	Equation
α	initial slope of P–I curve in $(\mathrm{W\,m^{-2}})^{-1}\,\mathrm{day}^{-1}$		Eq. (A8)
α_c	Chl-a specific initial slope of $P-I$ curve	$2\,\mathrm{gC\,gChl^{-1}(W\,m^{-2})^{-1}\,day^{-1}}$	
$\beta_\mathrm{P}, \beta_\mathrm{D}$	assimilation coefficients of zooplankton	0.75	
De_D	rate of breakdown of detritus to ammonium		Eq. (A18)
De_P	rate of phytoplankton natural mortality		Eq. (A16)
De_Z	rate of zooplankton natural mortality		Eq. (A17)
De_A	ammonium nitrification rate		Eq. (A19)
δ	excretion parameter	0.7	
ϵ	grazing parameter relating capture of prey items to prey density	0.4	
G_P	rate of zooplankton grazing on phytoplankton		Eq. (A14)
G_D	rate of zooplankton grazing on detritus		Eq. (A15)
g	zooplankton maximum growth rate	$1.3\,\mathrm{day}^{-1}$	
γ	fraction of zooplankton mortality going to detritus	0.5	
I_0	photosynthetically active radiation (PAR) immediately below surface of water. Assumed to be 0.43 of the surface radiation		
J	light-limited phytoplankton growth rate in day^{-1}		Eq. (A9)
k_A	half-saturation constant for ammonium uptake	$0.5\,\mathrm{mmol\,m^{-3}}$	
k_N	half-saturation constant for nitrate uptake	$0.5\,\mathrm{mmol\,m^{-3}}$	
k_P	half-saturation constant for phytoplankton mortality	$1\,\mathrm{mmol\,m^{-3}}$	
k_Z	half-saturation constant for zooplankton mortality	$3\,\mathrm{mmol\,m^{-3}}$	
k_w	light attenuation due to water	$0.04\,\mathrm{m^{-1}}$	
k_c	light attenuation due to phytoplankton	$0.03\,\mathrm{m^2\,mmol^{-1}}$	
λ_bio	rate of the phytoplankton and zooplankton transfer into detritus	$0.05\,\mathrm{day}^{-1}$	
λ_A	nitrification rate	$0.03\,\mathrm{day}^{-1}$	
μ_P	phytoplankton mortality rate	$0.05\,\mathrm{day}^{-1}$	
μ_Z	zooplankton mortality rate	$0.2\,\mathrm{day}^{-1}$	
μ_D	detritus reference mineralisation rate	$0.05\,\mathrm{day}^{-1}$	
Ψ	strength of ammonium inhibition of nitrate uptake	$2.9\,(\mathrm{mmol\,m^{-3}})^{-1}$	
p_P	relative grazing preference for phytoplankton	0.75	
p_D	relative grazing preference for detritus	0.25	
Q_N	non-dimensional nitrate limiting factor		Eq. (A10)
Q_A	non-dimensional ammonium limiting factor		Eq. (A11)
R_P	Chl growth scaling factor		Eq. (A13)
v	Maximum phytoplankton growth rate	$1\,\mathrm{day}^{-1}$	
w_g	detritus sinking velocity	$10\,\mathrm{m\,day}^{-1}$	
z	depth		
θ	Chl to carbon ratio in $\mathrm{mgChl\,mgC^{-1}}$		
θ_m	maximum Chl to carbon ratio	$0.05\,\mathrm{mg\,Chl\,mgC^{-1}}$	
ζ	conversion factor from gC to mmolN based on C : N ratio of 6.5	$0.0128\,\mathrm{mmolN\,ngC^{-1}}$	

Acknowledgements. All authors were funded by NERC under grants NE/F00270X/1 and NE/F004184/1. The authors would like to acknowledge the use of the Imperial College London HPC service to perform some of the simulations presented here. The authors would like to thank the two anonymous reviewers for their helpful comments and suggestions.

Edited by: E. J. M. Delhez

References

Anderson, T. R. and Pondaven, P.: Non-redfield carbon and nitrogen cycling in the Sargasso Sea: pelagic imbalances and export flux, Deep Sea Res.-Pt. I, 50, 573–591, doi:10.1016/S0967-0637(03)00034-7, 2003.

Berline, L., Brankart, J.-M., Brasseur, P., Ourmières, Y., and Verron, J.: Improving the physics of a coupled physical–biogeochemical model of the North Atlantic through data assimilation: Impact on the ecosystem, J. Mar. Syst., 64, 153–172, doi:10.1016/j.jmarsys.2006.03.007, 2007.

Burchard, H. and Beckers, J.-M.: Non-uniform adaptive vertical grids in one-dimensional numerical ocean models, Ocean Model., 6, 51–81, doi:10.1016/S1463-5003(02)00060-4, 2004.

Burchard, H. and Bolding, K.: Comparative Analysis of Four Second-Moment Turbulence Closure Models for the Oceanic Mixed Layer, J. Phys. Oceanogr., 31, 1943–1968, doi:10.1175/1520-0485(2001)031<1943:CAOFSM>2.0.CO;2, 2001.

Denman, K. and Miyake, M.: Upper Layer Modification at Ocean Station Papa: Observations and Simulation, J. Phys. Oceanogr., 3, 185–196, 1973.

Farrell, P., Piggott, M., Pain, C., Gorman, G., and Wilson, C.: Conservative interpolation between unstructured meshes via supermesh construction, Comput. Method. Appl. M., 198, 2632–2642, doi:10.1016/j.cma.2009.03.004, 2009.

Ford, R., Pain, C. C., Piggott, M. D., Goddard, A. J. H., de Oliveira, C. R. E., and Umpleby, A. P.: A Nonhydrostatic Finite-Element Model for Three-Dimensional Stratified Oceanic Flows. Part I: Model Formulation, Mon. Weather Rev., 132, 2816–2831, doi:10.1175/MWR2824.1, 2004.

Friedrichs, M. A., Hood, R. R., and Wiggert, J. D.: Ecosystem model complexity versus physical forcing: Quantification of their relative impact with assimilated Arabian Sea data, Deep Sea Res.-Pt. II, 53, 576–600, doi:10.1016/j.dsr2.2006.01.026, 2006.

Gaspar, P., Grégoris, Y., Lefevre, J.-M., and Goris, Y. G. R. I.: A Simple Eddy Kinetic Energy Model for Simulations of the Oceanic Vertical Mixing: Tests at Station Papa and Long-Term Upper Ocean Study Site, J. Geophys. Res., 95, 16179–16193, doi:10.1029/JC095iC09p16179, 1990.

Hanert, E., Deleersnijder, E., and Legat, V.: An adaptive finite element water column model using the Mellor-Yamada level 2.5 turbulence closure scheme, Ocean Model., 12, 205–223, doi:10.1016/j.ocemod.2005.05.003, 2006.

Hiester, H., Piggott, M., and Allison, P.: The impact of mesh adaptivity on the gravity current front speed in a two-dimensional lock-exchange, Ocean Model., 38, 1–21, doi:10.1016/j.ocemod.2011.01.003, 2011.

Hill, J., Piggott, M., Ham, D., Popova, E., and Srokosz, M.: On the performance of a generic length scale turbulence model within an adaptive finite element ocean model, Ocean Model., 56, 1–15, doi:10.1016/j.ocemod.2012.07.003, 2012.

Hofmeister, R., Burchard, H., and Beckers, J.-M.: Non-uniform adaptive vertical grids for 3D numerical ocean models, Ocean Model., 33, 70–86, doi:10.1016/j.ocemod.2009.12.003, 2010.

Ji, R., Davis, C., Chen, C., and Beardsley, R.: Influence of local and external processes on the annual nitrogen cycle and primary productivity on Georges Bank: A 3-D biological–physical modeling study, J. Mar. Syst., 73, 31–47, doi:10.1016/j.jmarsys.2007.08.002, 2008.

Kettle, H. and Merchant, C. J.: Modeling ocean primary production: Sensitivity to spectral resolution of attenuation and absorption of light, Prog. Oceanogr., 78, 135–146, doi:10.1016/j.pocean.2008.04.002, 2008.

Khangaonkar, T., Sackmann, B., Long, W., Mohamedali, T., and Roberts, M.: Simulation of annual biogeochemical cycles of nutrient balance, phytoplankton bloom(s), and DO in Puget Sound using an unstructured grid model, Ocean Dynam., 62, 1353–1379, doi:10.1007/s10236-012-0562-4, 2012.

Kleypas, J. A. and Doney, S. C.: Nutrients, Chlorophyll, Primary Production and related Biogeochemical Properties in the Ocean Mixed Layer, Tech. Rep. TN-447+STR, NCAR, available at: http://rda.ucar.edu/datasets/ds259.0/ (last access: 8 May 2014), 2001.

Large, W. G. and Yeager, S. G.: Diurnal to decadal global forcing for ocean and sea-ice models: The data sets and flux climatologies, Tech. Rep. Technical Report TN-460+STR, NCAR, 2004.

Lévy, M., Iovino, D., Resplandy, L., Klein, P., Madec, G., Tréguier, A.-M., Masson, S., and Takahashi, K.: Large-scale impacts of submesoscale dynamics on phytoplankton: Local and remote effects, Ocean Model., 43-44, 77–93, doi:10.1016/j.ocemod.2011.12.003, 2012.

Loseille, A. and Alauzet, F.: Continuous Mesh Framework Part II: Validations and Applications, SIAM J. Numer. Anal., 49, 61–86, doi:10.1137/10078654X, 2011.

Luo, L., Wang, J., Schwab, D. J., Vanderploeg, H., Leshkevich, G., Bai, X., Hu, H., and Wang, D.: Simulating the 1998 spring bloom in Lake Michigan using a coupled physical-biological model, J. Geophys. Res.-Oceans, 117, C10011, doi:10.1029/2012JC008216, 2012.

Madec, G.: NEMO ocean engine, Note du Pole de modélisation, Institut Pierre-Simon Laplace (IPSL), France, No. 27 ISSN No 1288-1619, 2008.

McGillicuddy, D. J., Anderson, L. A., Doney, S. C., and Maltrud, M. E.: Eddy-driven sources and sinks of nutrients in the upper ocean: Results from a 0.1° resolution model of the North Atlantic, Global Biogeochem. Cy., 17, 1035, doi:10.1029/2002GB001987, 2003.

Menendez, A. N., Badano, N. D., Lopolito, M. F., and Re, M.: Water quality assessment for a coastal zone through numerical modeling, J. Appl. Water Eng. Res., 1, 8–16, doi:10.1080/23249676.2013.827892, 2013.

Oschlies, A.: Can eddies make ocean deserts bloom?, Global Biogeochem. Cy., 16, 1106, doi:10.1029/2001GB001830, 2002.

Oschlies, A. and Garçon, V.: An eddy-permitting coupled physical-biological model of the North Atlantic: 1. Sensitivity to advection numerics and mixed layer physics, Global Biogeochem. Cy., 13, 135–160, doi:10.1029/98GB02811, 1999.

Pain, C. C., Umpleby, A. P., de Oliveira, C. R. E., and Goddard, A. J. H.: Tetrahedral mesh optimisation and adaptivity for steady-state and transient finite element calculations, Comput. Method Appl. M., 190, 3771–3796, doi:10.1016/S0045-7825(00)00294-2, 2001.

Pain, C., Piggott, M., Goddard, A., Fang, F., Gorman, G., Marshall, D., Eaton, M., Power, P., and de Oliveira, C.: Three-dimensional unstructured mesh ocean modelling, Ocean Model., 10, 5–33, doi:10.1016/j.ocemod.2004.07.005, the Second International Workshop on Unstructured Mesh Numerical Modelling of Coastal, Shelf and Ocean Flows, 2005.

Pichelina, É., Fortina, M., and Boivinb, S.: Étude numérique d'estimations d'erreur a posteriori, Revue Européenne des Éléments, 9, 467–486, 2000.

Piggott, M. D., Pain, C. C., Gorman, G. J., Power, P. W., and Goddard, A. J. H.: h, r, and hr adaptivity with applications in numerical ocean modelling, Ocean Model., 10, 95–113, doi:10.1016/j.ocemod.2004.07.007, 2005.

Piggott, M., Pain, C., Gorman, G., Marshall, D., Marshall, D. P., and Killworth, P. D.: Unstructured adaptive meshes for ocean modeling, in: Ocean Modeling in an Eddying Regime. Geophysical Monograph Series, 177, doi:10.1029/177GM22, 2008a.

Piggott, M. D., Gorman, G. J., Pain, C. C., Allison, P. A., Candy, A. S., Martin, B. T., and Wells, M. R.: A new computational framework for multi-scale ocean modelling based on adapting unstructured meshes, Int. J. Numer. Meth. Fl., 56, 1003–1015, doi:10.1002/fld.1663, 2008b.

Piggott, M. D., Farrell, P. E., Wilson, C. R., Gorman, G. J., and Pain, C. C.: Anisotropic mesh adaptivity for multi-scale ocean modelling, Philos. T. Roy. Soc. A, 367, 4591–4611, doi:10.1098/rsta.2009.0155, 2009.

Popova, E. E., Coward, A. C., Nurser, G. A., de Cuevas, B., Fasham, M. J. R., and Anderson, T. R.: Mechanisms controlling primary and new production in a global ecosystem model – Part I: Validation of the biological simulation, Ocean Sci., 2, 249–266, doi:10.5194/os-2-249-2006, 2006.

Ueckermann, M. P. and Lermusiaux, P. F. J.: High-order schemes for 2D unsteady biogeochemical ocean models, Ocean Dynam., 60, 1415–1445, doi:10.1007/s10236-010-0351-x, 2010.

Umlauf, L. and Burchard, H.: A generic length-scale equation for geophysical turbulence models, J. Mar. Res., 61, 235–265, doi:10.1357/002224003322005087, 2003.

Uppala, S. M., Kållberg, P. W., Simmons, A. J., Andrae, U., Bechtold, V. D. C., Fiorino, M., Gibson, J. K., Haseler, J., Hernandez, A., Kelly, G. A., Li, X., Onogi, K., Saarinen, S., Sokka, N., Allan, R. P., Andersson, E., Arpe, K., Balmaseda, M. A., Beljaars, A. C. M., Berg, L. V. D., Bidlot, J., Bormann, N., Caires, S., Chevallier, F., Dethof, A., Dragosavac, M., Fisher, M., Fuentes, M., Hagemann, S., Hólm, E., Hoskins, B. J., Isaksen, L., Janssen, P. A. E. M., Jenne, R., Mcnally, A. P., Mahfouf, J.-F., Morcrette, J.-J., Rayner, N. A., Saunders, R. W., Simon, P., Sterl, A., Trenberth, K. E., Untch, A., Vasiljevic, D., Viterbo, P., and Woollen, J.: The ERA-40 re-analysis, Q. J. Roy. Meteorol. Soc., 131, 2961–3012, doi:10.1256/qj.04.176, 2005.

Weber, L., Völker, C., Oschlies, A., and Burchard, H.: Iron profiles and speciation of the upper water column at the Bermuda Atlantic Time-series Study site: a model based sensitivity study, Biogeosciences, 4, 689–706, doi:10.5194/bg-4-689-2007, 2007.

15

The role of subpolar deep water formation and Nordic Seas overflows in simulated multidecadal variability of the Atlantic meridional overturning circulation

K. Lohmann[1], J. H. Jungclaus[1], D. Matei[1], J. Mignot[2,3], M. Menary[4], H. R. Langehaug[5,6], J. Ba[7], Y. Gao[5,6], O. H. Otterå[6,8], W. Park[7], and S. Lorenz[1]

[1]Max Planck Institute for Meteorology, Hamburg, Germany
[2]LOCEAN, Institute Pierre Simon Laplace, University Pierre and Marie Curie, Paris, France
[3]Climate and Environmental Physics, and Oeschger Centre of Climate Change Research, University of Bern, Bern, Switzerland
[4]Met Office Hadley Center, Exeter, UK
[5]Nansen Environmental and Remote Sensing Center, Bergen, Norway
[6]Bjerknes Centre for Climate Research, Bergen, Norway
[7]Helmholtz Centre for Ocean Research Kiel, Kiel, Germany
[8]Uni Research, Bergen, Norway

Correspondence to: K. Lohmann (katja.lohmann@mpimet.mpg.de)

Abstract. We investigate the respective role of variations in subpolar deep water formation and Nordic Seas overflows for the decadal to multidecadal variability of the Atlantic meridional overturning circulation (AMOC). This is partly done by analysing long (order of 1000 years) control simulations with five coupled climate models. For all models, the maximum influence of variations in subpolar deep water formation is found at about 45° N, while the maximum influence of variations in Nordic Seas overflows is rather found at 55 to 60° N. Regarding the two overflow branches, the influence of variations in the Denmark Strait overflow is, for all models, substantially larger than that of variations in the overflow across the Iceland–Scotland Ridge. The latter might, however, be underestimated, as the models in general do not realistically simulate the flow path of the Iceland–Scotland overflow water south of the Iceland–Scotland Ridge. The influence of variations in subpolar deep water formation is, on multimodel average, larger than that of variations in the Denmark Strait overflow. This is true both at 45° N, where the maximum standard deviation of decadal to multidecadal AMOC variability is located for all but one model, and at the more classical latitude of 30° N. At 30° N, variations in subpolar deep water formation and Denmark Strait overflow explain, on multimodel average, about half and one-third respectively of the decadal to multidecadal AMOC variance. Apart from analysing multimodel control simulations, we have performed sensitivity experiments with one of the models, in which we suppress the variability of either subpolar deep water formation or Nordic Seas overflows. The sensitivity experiments indicate that variations in subpolar deep water formation and Nordic Seas overflows are not completely independent. We further conclude from these experiments that the decadal to multidecadal AMOC variability north of about 50° N is mainly related to variations in Nordic Seas overflows. At 45° N and south of this latitude, variations in both subpolar deep water formation and Nordic Seas overflows contribute to the AMOC variability, with neither of the processes being very dominant compared to the other.

1 Introduction

The strength of the Atlantic meridional overturning circulation (AMOC) significantly influences the climate in northwestern Europe through the associated northward heat transport (e.g. Hall and Bryden, 1982). A detailed understanding of the processes that govern the AMOC variability is therefore a topic of much interest. In particular, the respective role of deep water formation variability north and south of the Greenland-Scotland Ridge (GSR) in modulating the AMOC strength is still not fully understood.

Several studies based on multicentury-long control simulations with coupled climate models (e.g. Bentsen et al., 2004; Jungclaus et al., 2005; Msadek and Frankignoul, 2009; Medhaug et al., 2012; Escudier et al., 2013) suggest that AMOC variability is closely linked to convective mixing in the subpolar North Atlantic, but not strongly related to convective mixing in the Nordic Seas. Using multicentury-long control simulations with 10 coupled climate models, Ba et al. (2014), however, find in some of the models significant correlation between AMOC variability and convective mixing in the Nordic Seas. Since the source waters of the overflow water masses at the GSR are relatively complex (e.g. Tanhua et al., 2008; Våge et al., 2013), variations in Nordic Seas convection do not necessarily match with variations in the overflows across the GSR. In the CCSM4 (Community Climate System Model) preindustrial control simulation (Danabasoglu et al., 2012), AMOC variability is primarily driven by deep convection changes in the Labrador Sea. Increased overflow transports from the Nordic Seas do not lead to a stronger AMOC, but rather appear to be a precursor to a weaker AMOC through enhanced stratification in the Labrador Sea. In contrast, Köhl and Stammer (2008), using an ocean reanalysis for the last 50 years, find a high correlation of the AMOC variability at 48° N with the Denmark Strait overflow as well as with the density at intermediate depth south of the Denmark Strait, but a low correlation with the convective mixing in the Labrador Sea. Based on a 50 year-long simulation with a coupled model of the Arctic–North Atlantic Ocean, Bailey et al. (2005) find that the AMOC variability is well correlated with the integrated surface heat flux over the Labrador Sea. Their model simulation, however, has a shortage of buoyant surface water, leading to an overly strong convection. Using a regional model for the Atlantic Ocean forced with atmospheric reanalysis fields, Böning et al. (2006) show that convection events in the Labrador Sea are followed by a strengthening of the AMOC. Their model configuration, however, excludes changes in the Nordic Seas overflows by imposing climatological conditions at the northern boundary. Based on observations, Latif et al. (2006) suggest that the Atlantic dipole sea surface temperature anomaly index over the last century, which is a fingerprint for AMOC variability (Latif et al., 2004), is a response to the North Atlantic Oscillation and associated changes in Labrador Sea convection. Based on relations derived from ocean-only simulations, they estimate, however, that the observed changes in the density of the Denmark Strait overflow over the last decades translate into AMOC changes that are of comparable order as the AMOC variability associated with Labrador Sea convection during the last century. We note that the respective role of subpolar deep convection and Nordic Seas overflows has also been discussed regarding the intensity of the subpolar gyre (e.g. Levermann and Born, 2007, using a coupled climate model). This question remains nevertheless also widely open.

Here we investigate the respective role of deep water formation variability north and south of the GSR for the low-frequency AMOC variability at different latitudes in the North Atlantic in a multimodel framework by analysing long (order of 1000 years) control simulations with five coupled climate models. Different from some of the previous studies mentioned above (e.g. Bentsen et al., 2004; Jungclaus et al., 2005; Msadek and Frankignoul, 2009; Escudier et al., 2013; Ba et al., 2014), we use the thickness of Labrador Sea Water and the strength of the Nordic Seas overflows rather than convective mixing as indicators for deep water formation south and north of the GSR respectively. The Labrador Sea Water thickness and Nordic Seas overflow strength have a more direct effect on the AMOC. Regarding the Nordic Seas overflows, we consider the transport both through the Denmark Strait and across the Iceland–Scotland Ridge. The influence of the latter on the AMOC variability has so far received little attention in the literature. Based on control simulations, the influence of subpolar deep water formation and Nordic Seas overflow variations on the AMOC variability can only be estimated by means of statistical analysis. In order to more directly estimate the influence of these two processes, we perform and analyse two sensitivity experiments with one of the models, in which we artificially suppress the variations in either subpolar deep water formation or Nordic Seas overflows.

In this study, we focus on the decadal to multidecadal AMOC variability. Variability on other timescales, such as centennial or multicentennial, can be found in previous studies: Fischer (2011) discusses centennial variability of AMOC, subpolar deep water formation and Nordic Seas overflows in a 6000 year-long transient Holocene simulation with the MPI coupled climate model applying the same grid configuration as MPI-ESM-CR (Max Planck Institute for Meteorology Earth System Model). Park and Latif (2008) suggest that multidecadal AMOC variability in the KCM (Kiel Climate Model) control simulation originates in the North Atlantic, whereas multicentennial variability is driven by the Southern Ocean. Menary et al. (2012) link centennial AMOC variability in the HadCM3 (Hadley Center Climate Model), KCM and MPI-ESM-CR control simulations to a meridional shift in the intertropical convergence zone and the advection of the associated salinity anomalies into the deep water formation regions. Regarding the seasonal to interannual AMOC variability, according to the literature, it appears to be mainly wind-driven through local Ekman transport (e.g. Köhl, 2005; Biastoch et al., 2008).

The paper is organised as follows: the coupled climate models and the simulated state in the North Atlantic are described in Sect. 2. In Sect. 3, we present the analysis of the multimodel control simulations. In Sect. 4, the sensitivity experiments with one of the models are presented. We discuss the results and summarise the main conclusions in Sect. 5.

2 Model description and simulated state in the North Atlantic Ocean

We analyse millennial-scale control simulations with five global, coupled climate models; namely the MPI-ESM-CR (Jungclaus et al., 2010), the KCM (Park and Latif, 2008; Park et al., 2009), the coupled climate model developed at the Institute Pierre-Simon Laplace (IPSLCM4 v2, hereafter IPSLCM4; Marti et al., 2010; Servonnat et al., 2010), the HadCM3 (Gordon et al., 2000; Pope et al., 2000) and the Bergen Climate Model (BCM; Furevik et al., 2003; Otterå et al., 2009). These simulations were available within the EU project THOR (ThermoHaline Overturning – at Risk?). We limit our study to simulations from the project partners, as non-standard model output, such as the overflow transports across the GSR, is needed. The models use relatively coarse-resolution grid configurations and are sufficiently computationally efficient to perform long simulations (1000 years, BCM 700 years), allowing robust results on decadal to multidecadal timescales. The horizontal resolution in the North Atlantic is of the order of 100 km. We also include a control simulation with a higher-resolution grid configuration in this study: a 500 year-long control simulation with a previous version of the MPI coupled climate model (MPI-AO-LR; Jungclaus et al., 2006) with an approximately double-resolution grid configuration compared to MPI-ESM-CR. An overview of the model configurations used in this study is given in Table 1.

Regarding the simulated AMOC and North Atlantic climate variability, recent multimodel control simulation studies (including many of the simulations used here) discuss differences among the various coupled climate models in both the representation of the low-frequency North Atlantic climate variability as well as in the mechanisms and feedbacks involved (e.g. Menary et al., 2012; Langehaug et al., 2012b; Gastineau and Frankignoul, 2012; Ba et al., 2014).

The overall maximum AMOC strength in the North Atlantic in the different control simulations (Fig. 1) varies between about 10 and 21 Sv (1 Sverdrup = $10^6\,m^3\,s^{-1}$); and is located at a depth of about 1000 m between 30 and 35° N, with the exception of IPSLCM4, where it is located at 45° N. Compared to the observational estimate at 26.5° N (Cunningham et al., 2007; Kanzow et al., 2010), suggesting a mean AMOC strength of about 18.5 Sv with an interannual variability of about ±2 Sv, the simulated AMOC strength lies

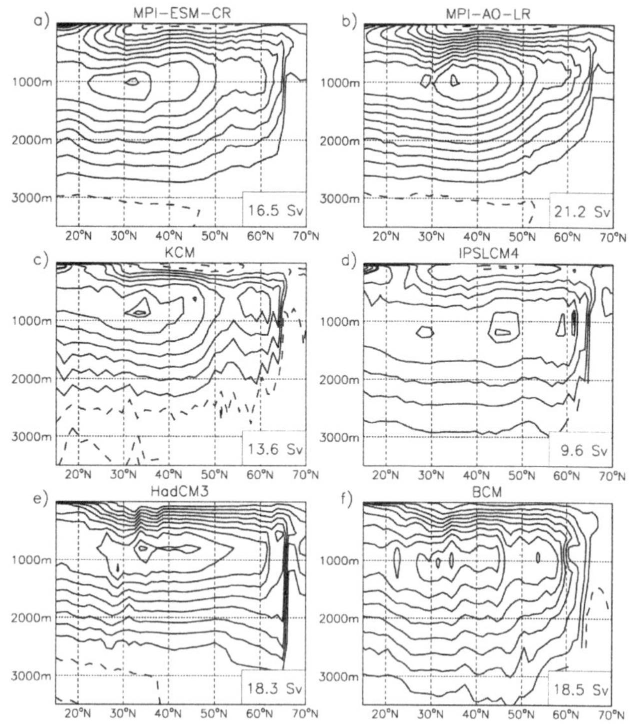

Fig. 1. Mean Atlantic meridional overturning streamfunction (in Sv) in the control simulation with **(a)** MPI-ESM-CR, **(b)** MPI-AO-LR, **(c)** KCM, **(d)** IPSLCM4, **(e)** HadCM3 and **(f)** BCM. Solid (dashed) lines indicate positive (negative) values. Contour interval is 2 Sv, starting from ±1 Sv. An additional contour line is added to show the respective location of the overall maximum strength. The respective value of the overall maximum strength is given in the lower right corner.

within the observed range at this latitude in all models except for KCM (13 Sv) and IPSLCM4 (9 Sv). For IPSLCM4, we note that the AMOC strength is drifting throughout the simulation due to a relatively short spin-up phase. Most of the models show a too shallow meridional overturning cell. This is indeed a common problem in z coordinate models due to unrealistically large diapycnal mixing associated with the Nordic Seas overflows (e.g. Willebrand et al., 2001; Danabasoglu et al., 2010; Yeager and Danabasoglu, 2012). The excessive mixing results in a density of the simulated overflow water masses which is too low compared to observations. The simulated lower branch of the AMOC extends below 3000 m only in BCM – the only model using an isopycnic vertical coordinate.

We have determined the dominant periodicity of decadal to multidecadal AMOC variability at different latitudes in the North Atlantic in the various control simulations based on peaks in the power spectra exceeding the 95 % confidence level when compared to a red-noise process. The AMOC strength at a particular latitude, we define as the maximum value of the Atlantic meridional overturning streamfunction over all depth levels at that latitude. The dominant

Table 1. Overview of the coupled climate models used in this study.

Model	Ocean component	Atmosphere component	Ocean grid	Resolution in North Atlantic	Atmosphere grid	Ocean levels	Atmosphere levels	Length of control simulation
MPI-ESM-CR	MPI-OM	ECHAM5	conformal mapping, 3°, northern pole over Greenland	50–200 km	spectral, T31[a]	40 z-levels	19	1000 years
MPI-AO-LR	MPI-OM	ECHAM5	conformal mapping, 1.5°, northern pole over Greenland	25–100 km	spectral, T63*	40 z-levels	31	500 years
KCM	NEMO[b]	ECHAM5	conformal mapping, 2°, tripolar (ORCA2 grid)	100–200 km	spectral, T31[a]	31 z-levels	19	1000 years
IPSLCM4	OPA8.2	LMDz4	conformal mapping, 2°, tripolar (ORCA2 grid)	100–200 km	3.75° (long) × 2.535° (lat)	31 z-levels	19	1000 years
HadCM3	No specific name	HadAM3	1.25° × 1.25°	~ 100 km	3.75° (long) × 2.5° (lat)	20 z-levels	19	1000 years
BCM	MICOM	ARPEGE	conformal mapping, 2.4°, northern pole over Siberia	150–200 km	spectral, T63[a]	34 isopycnic layers, non-isopycnic surface mixed layer	31	700 years

[a] Corresponding to about 3.75° (1.875°) on a Gaussian grid for T31 (T63). [b] NEMO (Nucleus for European Modelling of the Ocean) is based on the OPA model.

periodicity varies among models and latitudes and provides a rather wide range of variability from decadal to multidecadal timescales (20–75 years). Furthermore, Zanchettin et al. (2010, their Fig. 2), using the MPI-ESM-CR control simulation, and Menary et al. (2012, their Fig. 1), using the HadCM3, KCM and MPI-ESM-CR control simulations, show that the dominant periodicity of AMOC variability at 30° N is non-stationary in time. To focus on the decadal to multidecadal timescales, we apply a 10th-order Butterworth bandpass filter for periods ranging from 15 to 80 years to all model data. In the following, we refer to this variability as multidecadal variability.

The mean mixed layer depth (MLD) in March, which indicates the location of deep water formation, differs among the various control simulations (Fig. 2). Deep convection in or off the Labrador Sea is found in MPI-ESM-CR, MPI-AO-LR and BCM. Compared to the observed winter convection depth in the Labrador Sea, which varies from about 800 to 2300 m, MPI-AO-LR and KCM show a relatively large convection depth exceeding 2500 m in the mean. Apart from deep convection in the Labrador Sea, observations show relatively shallow convection in the Irminger Sea, with a winter convection depth generally not exceeding 500 m. Shallow convection in the Irminger Sea is also found in all models except for KCM and IPSLCM4. In the latter two models, deep convection is shifted south of Iceland. Regarding BCM, where an isopycnic vertical coordinate is used, MLD is not the best indicator for deep water formation (Medhaug et al., 2012). We note, however, that in our definition of subpolar deep water formation variability in BCM, the MLD is not taken into account.

For the definition of a subpolar deep water formation index, we follow Zhu and Jungclaus (2008), using the thickness of Labrador Sea Water. Nevertheless, due to the varying location of subpolar deep water formation in the different models (Fig. 2), we use the expression subpolar deep water, rather than Labrador Sea Water, in the present study. A description of how we determine the thickness of subpolar deep water in the different models is given in Appendix A. As an index for subpolar deep water formation (SDWI), we average the annual thickness of subpolar deep water over the subpolar region (60–15° W and 45–65° N). The entire subpolar region is chosen to account for the different location of the subpolar deep water formation in the various models. Based on the MPI-ESM-CR control simulation, we have also tested defining the SDWI by using the subpolar deep water thickness in March or by averaging the subpolar deep water thickness over the convection region (defined by the grid cells where the MLD in March exceeds 1000 m in at least 100 integration years). The multidecadal variability of the different indices is, however, rather similar, with correlation coefficients exceeding 0.9 (not shown). We are therefore confident that our results do not critically depend on the definition of the SDWI.

Apart from the subpolar deep convection, deep convection is found in the Nordic Seas (Fig. 2). Also in the Nordic Seas, mean location and depth of the deep water formation vary among the different control simulations. As mentioned

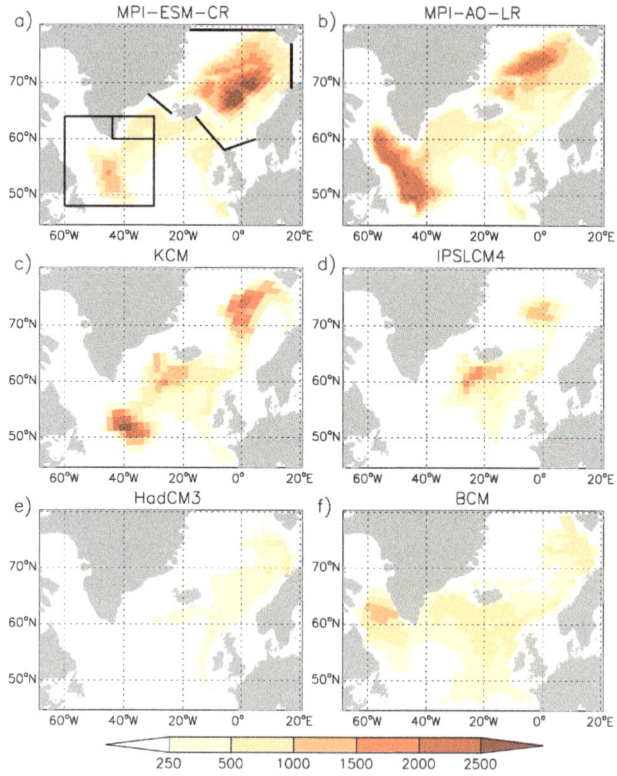

Fig. 2. Mean mixed layer depth in March (in metres) in the control simulation with **(a)** MPI-ESM-CR, **(b)** MPI-AO-LR, **(c)** KCM, **(d)** IPSLCM4, **(e)** HadCM3 and **(f)** BCM. The boxes in **(a)** indicate the area where in the sensitivity experiments temperature and salinity are relaxed towards climatological values.

earlier, we do not directly consider deep water formation in the Nordic Seas, but rather use the overflow transports across the GSR. The overflow transports through the Denmark Strait (DSO) and across the Iceland–Scotland Ridge (ISRO) are defined based on density (and depth) criteria according to the respective water mass properties at the GSR in the different models. In general, the total transport out of the Nordic Seas with a density threshold of $\sigma > 27.8\,\mathrm{kg\,m^{-3}}$ is used. A description of the criteria used in the different models is given in Appendix B.

Mean simulated values of DSO and ISRO lie between about 2 and 4 Sv (Table 2), in reasonable agreement with observational estimates (e.g. Hansen et al., 2008). An exception is the DSO in KCM, IPSLCM4 and HadCM3, where mean values of the order of 10 Sv are found. The reason for the unrealistically large DSO in the three models is not clear, but we note that all models have realistic sill depths at the GSR. In contrast to observations, where an overflow transport of about 1 Sv is found between Iceland and the Faroe Islands (e.g. Østerhus et al., 2008), the simulated ISRO is (mostly) restricted to the Faroe–Shetland Channel in all models. South of the Iceland–Scotland Ridge, one major bias in all models except for KCM is that (a large part of) the Iceland–Scotland overflow water spreads southward in the eastern part of the North Atlantic Basin (e.g. Langehaug et al., 2012a; their Fig. 5), rather than flowing around the Reykjanes Ridge (through fracture zones in the Mid-Atlantic Ridge) and joining the Denmark Strait overflow water and the deep western boundary current.

Table 2. Mean value of the overflow transport through the Denmark Strait (DSO) and across the Iceland–Scotland Ridge (ISRO) in the different control simulations.

Model	Mean DSO	Mean ISRO
MPI-ESM-CR	2.8 Sv	2.3 Sv
MPI-AO-LR	2.8 Sv	3.5 Sv
KCM	12.7 Sv	2.3 Sv
IPSLCM4	10.8 Sv	2.4 Sv
HadCM3	8.0 Sv	3.4 Sv
BCM	1.9 Sv	3.8 Sv

3 Analysis of multimodel control simulations

In this section, we present the results from a correlation analysis between the maximum AMOC strength at different latitudes in the North Atlantic and respectively SDWI, DSO and ISRO in the multimodel control simulations.

Maximum correlation coefficients between SDWI and AMOC strength on multidecadal timescales (Fig. 3) are found at 45° N in all models, with the exception of MPI-ESM-CR, where they are located at about 40° N. In IPSLCM4, similarly high correlation coefficients are found between about 45 and 55° N. Maximum correlation coefficients amount to 0.85–0.9, with the exception of HadCM3, where they reach a value of only 0.55. We note that HadCM3 also exhibits substantially weaker regression coefficients (per standard deviation of SDWI) compared to the other models (not shown). Maximum regression coefficients amount to 0.35 Sv (per standard deviation of SDWI) in HadCM3 compared to 0.5–0.8 Sv (per standard deviation of SDWI) in the other models. The negative correlation coefficients alternating with the positive ones (e.g. Fig. 3b) indicate the dominant periodicity of AMOC variability and are not further discussed here.

Significant correlation coefficients between SDWI and AMOC strength are found, except for MPI-ESM-CR, at all latitudes in the North Atlantic (latitude range 0 to 15° N not shown). In MPI-ESM-CR, correlation coefficients at zero lag or with the AMOC strength lagging are significant only between about 25 and 50° N. The reason for this difference is not clear, but horizontal resolution might play a role in the MPI model simulations. In MPI-AO-LR, where a higher-resolution grid configuration is used compared to MPI-ESM-CR, significant correlation coefficients extend to higher mid-latitudes as well as into the subtropics (Fig. 3a,

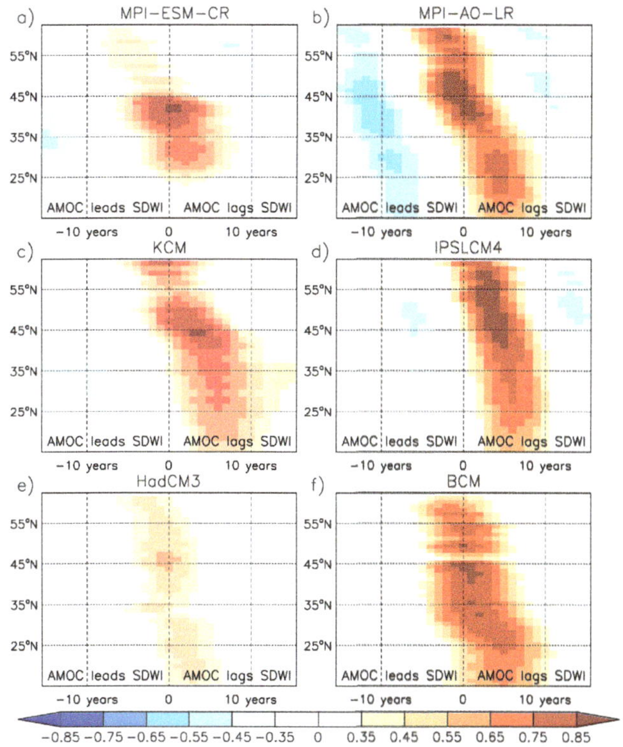

Fig. 3. Lag correlation coefficients between the bandpass-filtered (15–80 years) subpolar deep water formation index (SDWI) and maximum Atlantic meridional overturning streamfunction at different latitudes in the North Atlantic in the control simulation with **(a)** MPI-ESM-CR (depth level 1020 m), **(b)** MPI-AO-LR (depth level 1020 m), **(c)** KCM (depth level 870 m), **(d)** IPSLCM4 (depth level 1210 m), **(e)** HadCM3 (depth level 800 m) and **(f)** BCM (depth level 1000 m). Only correlation coefficients significant at the 95 % level for a two-tailed t test (taking into account both length and decorrelation timescale of the SDWI and the overall maximum strength of the Atlantic meridional overturning streamfunction) are plotted (significance level: 0.31 for MPI-AO-LR, 0.27 for BCM and 0.23 for the other models). Positive (negative) lags indicate that the SDWI is leading (lagging).

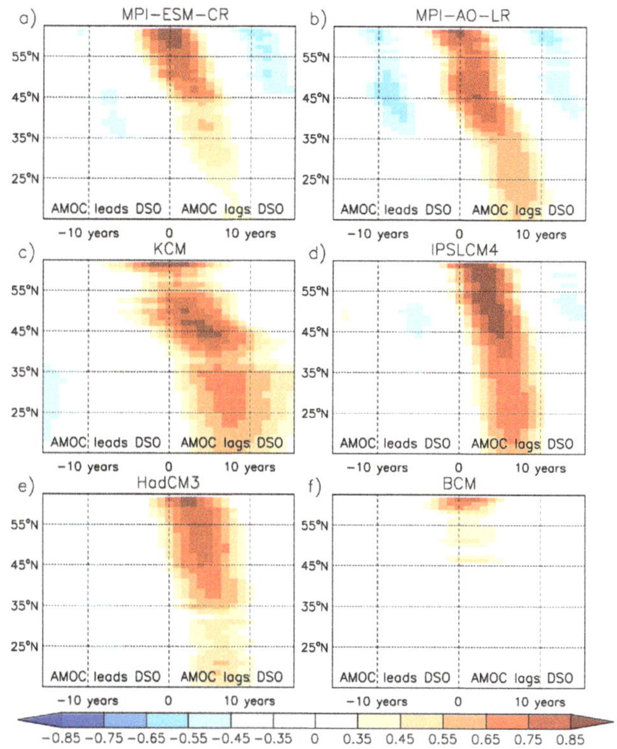

Fig. 4. Same as Fig. 3, but for the correlation between the Denmark Strait overflow transport (DSO) and Atlantic meridional overturning streamfunction.

b). In HadCM3, correlation coefficients are relatively low compared to the other models (partly except for MPI-ESM-CR) at all latitudes. The same is true, if regression coefficients (per standard deviation of SDWI) rather than correlation coefficients are considered (not shown). We speculate that this is due to the fact that HadCM3 exhibits a rather shallow subpolar deep water formation (Fig. 2e).

Maximum correlation coefficients between DSO and AMOC strength on multidecadal timescales (Fig. 4) are found at about 60° N in all models. They amount to 0.85–0.9, with the exception of BCM, where they reach a value of 0.75. In addition, strong correlation coefficients are found at about 45° N in MPI-AO-LR and KCM as well as for all subpolar latitudes in IPSLCM4. We note that KCM and IPSLCM4 exhibit a much larger variability of the DSO than the other models (standard deviation of the order of 1.5 Sv compared to less than 0.5 Sv in the other models).

Significant correlation coefficients between DSO and AMOC strength are found at all latitudes in the North Atlantic (latitude range 0–15° N not shown), with the exception of BCM. In the latter, significant correlation coefficients are only found close to the GSR. We note that BCM is the only model with an isopycnic vertical coordinate. In another multimodel study using partly the same climate models as here, Swingedouw et al. (2013) underline the limited grid resolution of this model in the Denmark Strait region. In MPI-ESM-CR, correlation coefficients south of about 30° N are barely significant. As mentioned above, horizontal resolution might play a role in the MPI model simulations, as MPI-AO-LR, where a higher-resolution grid configuration is used compared to MPI-ESM-CR, shows higher correlation coefficients south of about 45° N (Fig. 4a, b). In KCM and IPSLCM4, correlation coefficients between AMOC strength and respectively SDWI and DSO are rather similar (Fig. 3c, d compared to Fig. 4c, d). In both models, however, SDWI and DSO are highly correlated with each other (correlation coefficient of about 0.8). We note that this could be due to a specific feature of the oceanic component as these two coupled climate models use the same one.

Regarding the correlation between ISRO and AMOC strength on multidecadal timescales (Fig. 5), correlation coefficients are generally rather low compared to the correlation between DSO and AMOC strength. The same is true if regression coefficients (per standard deviation of ISRO and DSO respectively) rather than correlation coefficients are considered (not shown). In IPSLCM4, HadCM3 and BCM, an out-of-phase relation of DSO and ISRO is found on multidecadal timescales. Such an out-of-phase relation on timescales of 3–10 years is described by Serra et al. (2010), based on an ocean-only simulation with the MITgcm (Massachusetts Institute of Technology general circulation model) model forced with atmospheric reanalysis fields. Consistent with the out-of-phase relation of the two overflow branches, positive correlation coefficients are found when the AMOC strength leads the ISRO by a few years in IPSLCM4 and HadCM3. BCM shows almost no significant correlation coefficients between ISRO and AMOC strength.

In MPI-ESM-CR and MPI-AO-LR, significant correlation coefficients between ISRO and AMOC strength are found for the subpolar latitudes. In MPI-ESM-CR, correlation coefficients lie between 0.5 and 0.7, while in MPI-AO-LR they are just above the significance level. KCM is the only model, where significant correlation coefficients are found at all latitudes between 10 and 60° N (latitude range 10–15° N not shown), with correlation coefficients of 0.4–0.6. We note that the simulated influence of the ISRO on the multidecadal AMOC variability might, however, be underestimated, as the models (except for KCM) do not realistically simulate the flow path of the Iceland–Scotland overflow water south of the Iceland–Scotland Ridge.

The latitude where the maximum correlation coefficients between SDWI and AMOC strength are found (about 45° N), coincides, with the exception of HadCM3, with the latitude where the maximum standard deviation of multidecadal AMOC variability is located. The maximum standard deviation is of the order of 0.5–1 Sv and is, with the exception of HadCM3, located at about 45° N at a depth of 1500 m (Fig. 6), below the depth at which the overall maximum AMOC strength in the North Atlantic is found. In HadCM3, the maximum standard deviation of multidecadal AMOC variability is found at 60° N, coinciding with the latitude where the maximum correlation coefficients between DSO and AMOC strength are found. We note that HadCM3 shows relatively low correlation coefficients between SDWI and AMOC strength compared to the other models (except for partly MPI-ESM-CR, Fig. 3). Whether the multidecadal AMOC variability at 45 and 60° N is, to a large extent, indeed driven by variations in subpolar deep water formation and Denmark Strait overflow respectively, can, however, not definitely be determined from the statistical analysis presented in this section.

Based on the correlation analysis for the individual models, we have deduced the multimodel average of the multidecadal AMOC variance explained by SDWI and DSO respectively (Table 3). The ISRO explains substantially less variance than the DSO. Furthermore, in some models (IPSLCM4, HadCM3 and BCM), it is out of phase with the AMOC variability. Thus we do not comment further its influence here. At 45° N, where the maximum standard deviation of multidecadal AMOC variability is found, the variance explained by SDWI and DSO amounts, on multimodel average, to 68 and 57 % respectively. The sum of the explained variances by SDWI and DSO amounts to more than 100 %, indicating that these two processes are not completely independent from each other. At the more classical latitude of 30° N, the variance explained by SDWI and DSO amounts, on multimodel average, to 50 and 30 % respectively, suggesting that the influence of both subpolar deep water formation and Nordic Seas overflow on the multidecadal AMOC variability is weaker at this latitude.

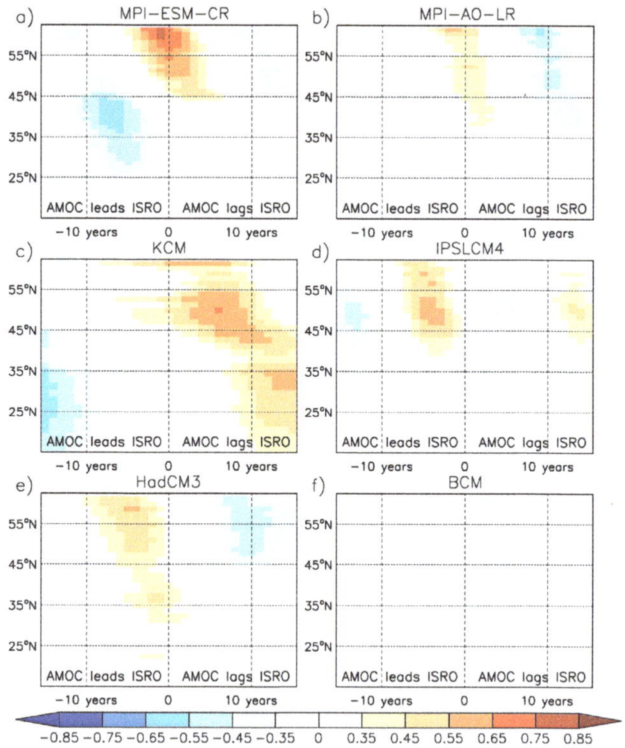

Fig. 5. Same as Fig. 3, but for the correlation between overflow transport across the Iceland–Scotland Ridge (ISRO) and Atlantic meridional overturning streamfunction.

One possible explanation for the lower variance explained by SDWI and DSO at 30 compared to 45° N is the local influence of varying Ekman transport. Time series of the bandpass-filtered (15–80 years) Ekman transport in MPI-ESM-CR, where the variance explained by SDWI and DSO is similar to the multimodel average (Table 3), show a standard deviation of 0.18 Sv at 45° N, compared to a standard deviation exceeding 0.6 Sv for multidecadal AMOC time series at the same latitude (Fig. 6a). At 30° N, however,

Table 3. Summary of the maximum variance of the bandpass-filtered (15–80 years) Atlantic meridional overturning streamfunction at 45 and 30° N explained SDWI and DSO in the different control simulations, deduced from Figs. 3 and 4 using (correlation coefficient)^2.

Model	SDWI – AMOC 45° N	DSO – AMOC 45° N	SDWI – AMOC 30° N	DSO – AMOC 30° N
MPI-ESM-CR	72 %	49 %	49 %	25 %
MPI-AO-LR	81 %	72 %	64 %	36 %
KCM	72 %	72 %	42 %	49 %
IPSLCM4	81 %	72 %	56 %	49 %
HadCM3	30 %	56 %	25 %	20 %
BCM	72 %	20 %	64 %	not significant
Multimodel average	68 %	57 %	50 %	30 %

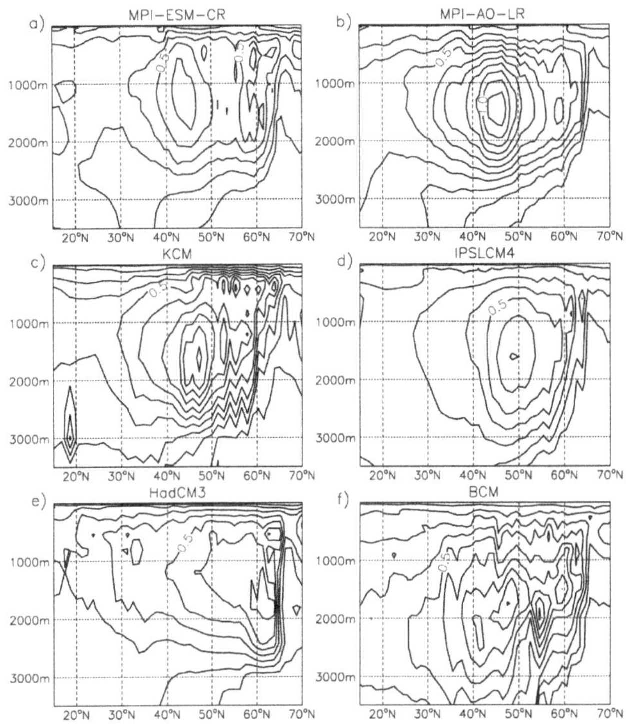

Fig. 6. Standard deviation of the bandpass-filtered (15–80 years) Atlantic meridional overturning streamfunction (in Sv) in the control simulation with **(a)** MPI-ESM-CR, **(b)** MPI-AO-LR, **(c)** KCM, **(d)** IPSLCM4, **(e)** HadCM3 and **(f)** BCM. Contour interval is 0.1 Sv, starting from 0.1 Sv.

a standard deviation of 0.25 Sv is found for the bandpass-filtered Ekman transport, compared to a standard deviation of less than 0.4 Sv for the multidecadal AMOC.

4 Sensitivity experiments with MPI-ESM-CR

4.1 Design of sensitivity experiments

The statistical analysis of the multimodel control simulations presented in the previous section could be biased by the fact that subpolar deep water formation and Nordic Seas overflows are somehow coupled. In order to more directly estimate the respective role of these two processes, we have performed two 1000 year-long sensitivity experiments with the MPI-ESM-CR model set-up (Table 1) by suppressing the variations in either subpolar deep water formation (in the following referred to as SDW experiment) or Nordic Seas overflows (in the following referred to as overflow experiment). Initial conditions are the same as for the control simulation.

In the SDW experiment, we relax temperature and salinity in the western subpolar region (60–30° W, 48–64° N; see box in Fig. 2a) towards monthly climatological values obtained from the control simulation. The relaxation timescale is one day. The region where relaxation is applied corresponds to the subpolar deep water formation site. We also include the Irminger Sea in this area, as previous modelling studies suggest a significant correlation between convective mixing in the Irminger Sea and decadal AMOC variability (e.g. Bentsen et al., 2004; Ba et al., 2013). Ba et al. (2014) show that this is true also for the MPI-ESM-CR control simulation. In the Irminger Sea region (45–30° W, 60–64° N; see small box in Fig. 2a), where the convection is relatively shallow, relaxation is applied down to a depth of 500 m, otherwise down to a depth of 900 m. The deeper levels as well as the upper four levels (approximately the upper 40 m of the water column) are excluded from relaxation to keep the variability of the overflow water masses as well as the coupling between ocean and atmosphere unaffected.

In the overflow experiment, we relax temperature and salinity in the Nordic Seas and along the GSR (boundaries indicated by the black lines in Fig. 2a) towards monthly climatological values obtained from the control simulation. The relaxation timescale is 1 day. The relaxation is applied in the entire water column, excluding the upper four levels to keep the coupling between ocean and atmosphere unaffected. The relaxation of Nordic Seas temperature and salinity towards climatological values also leads to a density of the overflow water masses at the GSR, which is almost constant with time.

Since temperature and salinity are relaxed towards climatological mean values taken from the control simulation, the simulated mean state in the North Atlantic in the two sensitivity experiments remains approximately unchanged compared to the simulated mean state in the control simulation (as described in Sect. 2). SDWI, DSO and ISRO in the

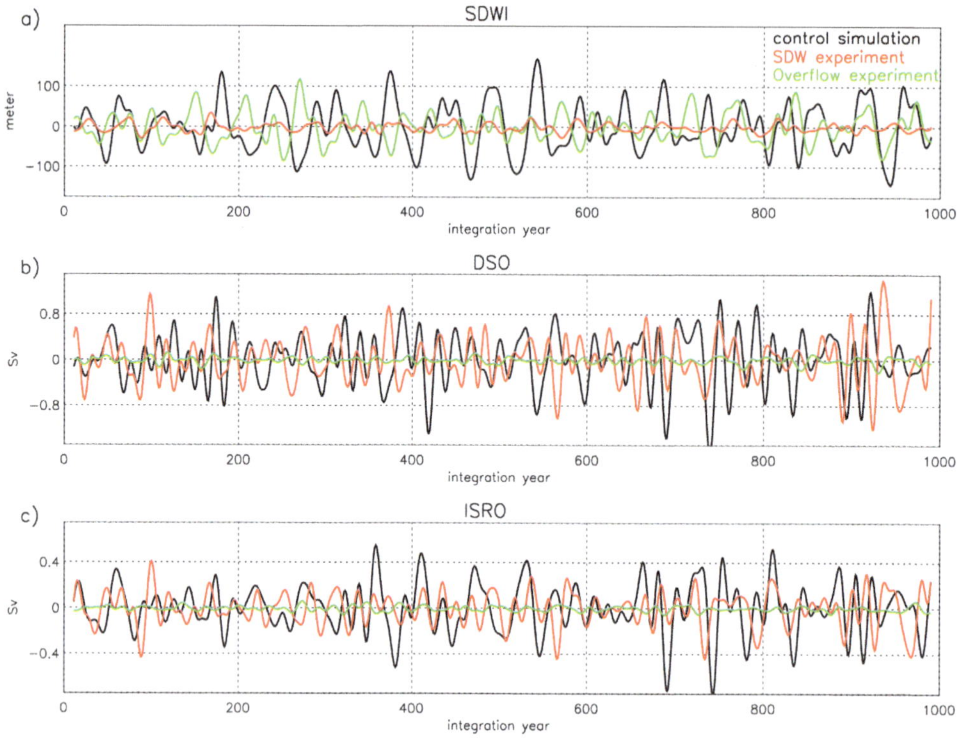

Fig. 7. Anomalous bandpass-filtered (15–80 years) **(a)** SDWI (in m), **(b)** DSO (in Sv) and **(c)** ISRO (in Sv) in the control simulation (black lines), the SDW experiment (red lines) and the overflow experiment (green lines) with MPI-ESM-CR. The respective standard deviations are given in the table. Numbers in brackets in the table give the reduction (in %) in the standard deviation in the respective sensitivity experiment compared to the control simulation.

sensitivity experiments are defined in the same way as described in Sect. 2 for the control simulation.

4.2 Results of sensitivity experiments

In the SDW experiment, the multidecadal variability of the SDWI is reduced by 82 % compared to the control simulation (Fig. 7a and table within). A reduction of 29 % is, however, also found in the multidecadal variability of the ISRO (Fig. 7c and table). One possible explanation of this phenomenon is as follows: reduced AMOC variability (due to reduced subpolar deep water formation variability) leads to reduced SST (sea surface temperature) variability in the eastern part of the Nordic Seas, as indicated by the correlation pattern between the maximum AMOC strength and North Atlantic SST in the control simulation (not shown). The variability of the overflow across the Iceland–Scotland Ridge in ocean-only and coupled simulations using the ocean component of MPI-ESM-CR is significantly correlated with the SST variability in the eastern part of the Nordic Seas and related changes in surface density, convective mixing and sea surface height (e.g. Olsen et al., 2008).

In the overflow experiment, the multidecadal variability of both DSO and ISRO is reduced by about 90 % compared to the control simulation (Fig. 7b, c and table). A reduction of 38 % is, however, also found in the multidecadal variability of the SDWI (Fig. 7a and table). This might be due to reduced AMOC variability (due to reduced overflow variability) inducing reduced subpolar deep water formation variability caused by ocean internal modes such as the one described in Zhu and Jungclaus (2008). Variations in Nordic Seas overflow and subpolar deep water formation might thus not be completely independent, at least in this model. We note, however, that in the control simulation, SDWI and DSO/ISRO are not significantly correlated with each other. Another possible explanation is that the relaxation of Nordic Seas temperature and salinity towards climatological values reduces, via advection along the East Greenland Current, upper-ocean temperature and salinity variability in the subpolar region, especially in the Irminger Sea, and consequently probably affects variability of subpolar deep water formation.

Regarding the multidecadal AMOC variability, the amplitude of variability is reduced in the sensitivity experiments compared to the control simulation (Fig. 8). We quantify the reduction in the standard deviation of the AMOC strength in the sensitivity experiments compared to the control simulation as follows: $100 \cdot [1 - (\sigma(\text{AMOC}_{\text{sensitivity experiment}})/\sigma(\text{AMOC}_{\text{control simulation}}))]$. Note that positive values correspond to a reduction in the

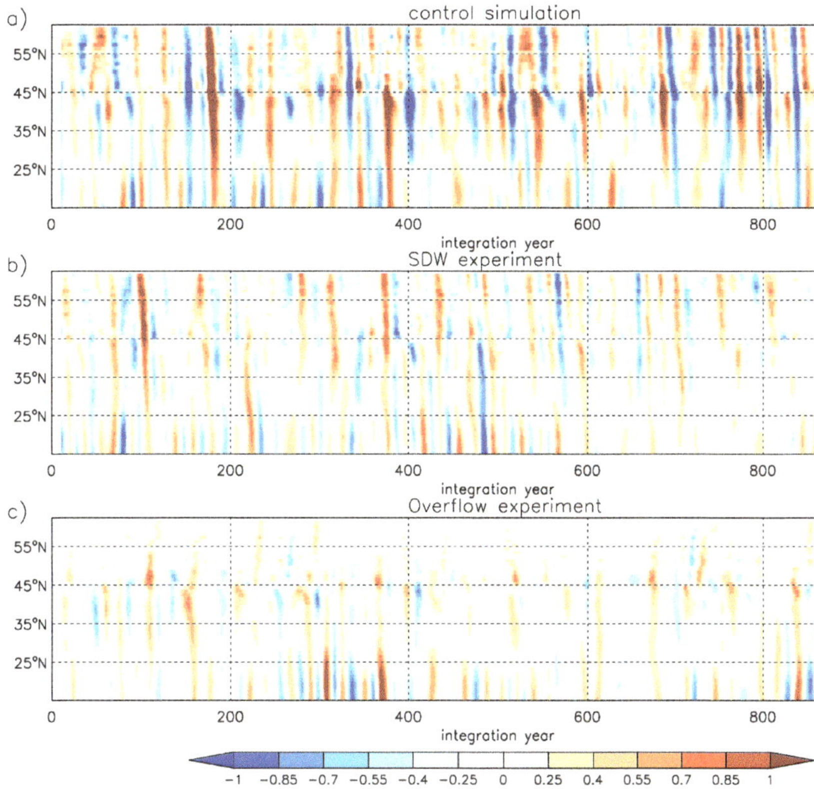

Fig. 8. Anomalous bandpass-filtered (15–80 years) maximum Atlantic meridional overturning streamfunction (in Sv) at different latitudes in the North Atlantic (depth level 1020 m) in **(a)** the control simulation, **(b)** the SDW experiment and **(c)** the overflow experiment with MPI-ESM-CR.

standard deviation of the AMOC strength in the sensitivity experiments.

In the SDW experiment (Fig. 9a), the maximum reduction in the standard deviation of multidecadal AMOC variability amounts to about 60 % and is found at about 45° N. This latitude is similar to the latitude where the maximum correlation coefficients between SDWI and AMOC strength are found in the control simulation (Fig. 3a). At 30° N, the reduction in the standard deviation of multidecadal AMOC variability in the SDW experiment amounts to about 50 %.

In the overflow experiment (Fig. 9b), the maximum reduction in the standard deviation of multidecadal AMOC variability amounts to about 80 % and is found at 55 to 60° N, again similarly to the latitude of the maximum correlation coefficients between both DSO and AMOC strength and ISRO and AMOC strength in the control simulation (Figs. 4a, 5a). At 45° N and south of this latitude, the reduction in the standard deviation of multidecadal AMOC variability is similar to what is found in the SDW experiment. We remind the reader, however, that in the overflow experiment also the variability of the SDWI is reduced by almost 40 % (Fig. 7a and table). It is interesting to note that the maximum reduction in the standard deviation in the overflow experiment is found below the depth at which the overall maximum AMOC strength in the North Atlantic is located (solid line in Fig. 9b). This indicates that the overflow water masses, as in the real ocean, contribute to the lower North Atlantic Deep Water.

In summary, we conclude from the sensitivity experiments that the multidecadal AMOC variability north of about 50° N is mainly related to variations in Nordic Seas overflows. At 45° N and south of this latitude, variations in both subpolar deep water formation and Nordic Seas overflows contribute to the multidecadal AMOC variability, with neither of the processes being very dominant compared to the other.

5 Discussion and conclusions

In this study, we have analysed long control simulations with five coupled climate models as well as sensitivity experiments performed with one of the models in order to add to the discussion in the literature about the respective role of variations in subpolar deep water formation and Nordic Seas overflows for the multidecadal AMOC variability.

Regarding the multimodel control simulations (Table 3), the multidecadal AMOC variance explained by subpolar deep water formation is higher than or of similar order as the one explained by the Denmark Strait overflow. This is true for all models at both 45° N, where the maximum standard

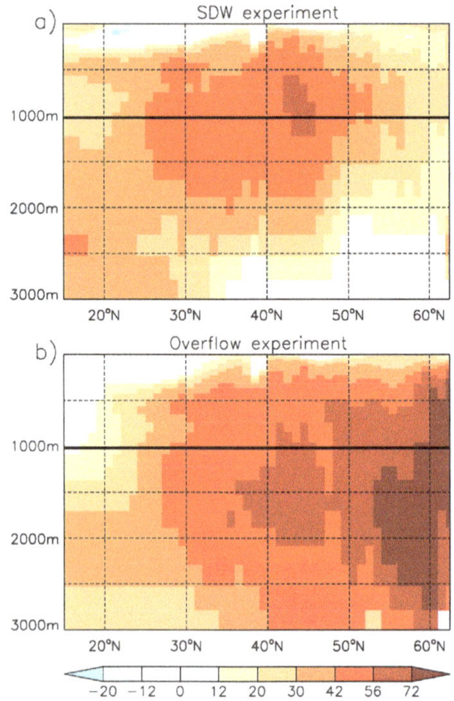

Fig. 9. Change (in %, for definition see text) in the standard deviation of the bandpass-filtered (15–80 years) Atlantic meridional overturning streamfunction in **(a)** the SDW experiment and **(b)** the overflow experiment compared to the control simulation with MPI-ESM-CR. Note that positive (negative) values correspond to a reduction (increase) in the standard deviation in the sensitivity experiments. The colour scale is chosen to match the explained variance, i.e. (correlation coefficient)^2, in Figs. 3–5. The depth level presented in Figs. 3–5 is indicated by the solid line.

deviation of multidecadal AMOC variability is found, and at the more classical latitude of 30° N, with the exception of 45° N in HadCM3. In the latter, the Denmark Strait overflow seems to have a relatively strong influence on the AMOC variability. We note that HadCM3 indeed shows a rather shallow subpolar deep water formation compared to the other models (Fig. 2). We also note that HadCM3 is the only model, in which the maximum standard deviation of multidecadal AMOC variability is found at 60° N, rather than at 45° N. The multidecadal AMOC variance explained by the Denmark Strait overflow is, for all models, substantially larger than the one explained by the overflow across the Iceland–Scotland Ridge. The latter might, however, be underestimated, as the models in general do not realistically simulate the flow path of the Iceland–Scotland overflow water south of the Iceland–Scotland Ridge.

The respective role of subpolar deep water formation and Nordic Seas overflows on the multidecadal AMOC variability suggested from the control simulations is supported by the sensitivity experiments. The latter indicate that the maximum influence of Nordic Seas overflow variations is found below the depth at which the overall maximum AMOC strength in the North Atlantic is located. However, the conclusions of our study do not change if the correlation analysis based on the multimodel control simulations is repeated for depth levels located in the lower branch of the AMOC.

Our results are in contrast with previous studies, which find the major influence from the Denmark Strait overflow (e.g. Köhl and Stammer, 2008), and rather in line with the ones suggesting a larger influence of subpolar deep water formation on the AMOC variability (e.g. Bailey et al., 2005; Latif et al., 2006; Böning et al., 2006). We note, however, that in only one of the models (BCM) the multidecadal AMOC variance explained by subpolar deep water formation is substantially larger than the one explained by the Denmark Strait overflow.

In some previous modelling studies (e.g. Jungclaus et al., 2005; Msadek and Frankignoul, 2009; Escudier et al., 2013), the AMOC variability was found not to be strongly related to convective mixing in the Nordic Seas. Variations in Nordic Seas convection, however, do not necessarily match with variations in the overflows across the GSR, since the source waters of the overflow water masses at the GSR are relatively complex (e.g. Tanhua et al., 2008; Våge et al., 2013). Our study suggests that the Denmark Strait overflow, though its influence is, on multimodel average, smaller than that of subpolar deep water formation, still explains a substantial part of multidecadal AMOC variance. At the classical latitude of 30° N, variations in Denmark Strait overflow and subpolar deep water formation explain, on multimodel average, about one-third and half of multidecadal AMOC variance respectively.

However, AMOC variations related to either subpolar deep water formation or Nordic Seas overflow variability will, through varying northward heat and fresh water transport, influence the surface density in the northern North Atlantic and might consequently affect the other of these two processes. The conclusions discussed above about the respective role of variations in subpolar deep water formation and Nordic Seas overflows for multidecadal AMOC variability might therefore be biased by the fact that these two processes are somehow coupled. If the respective role of variations in subpolar deep water formation is overestimated (due to influence by Nordic Seas overflows), then the multidecadal AMOC variance explained by subpolar deep water formation and Denmark Strait overflow will be of comparable order. If, on the other hand, the respective role of variations in Denmark Strait overflow is overestimated (due to influence by subpolar deep water formation), then only a minor part of multidecadal AMOC variance will be explained by the Denmark Strait overflow. We note, however, that in MPI-ESM-CR, which is the only model (apart from HadCM3) where subpolar deep water formation index and Denmark Strait overflow are not significantly correlated in the control integration, the respective role of these two processes for multidecadal AMOC variations is similar to the conclusions drawn from the presented analysis (Table 3).

One caveat of our study is the coarse horizontal and vertical resolution of the model grid configurations used. The resolution is non-eddy permitting, with, to some degree, the exception of MPI-AO-LR. Such coarse-resolution grid configurations allow the performance of sufficiently long simulations to ensure robust results on multidecadal timescales but, however, lead to model biases, which might affect the realism of the discussed influence of subpolar deep water formation and Nordic Seas overflows on multidecadal AMOC variations. Apart from the unrealistic flow path of the Iceland–Scotland overflow water mentioned above, one major model bias is the location of the subpolar deep water formation site, which also differs among models (Fig. 2). Only in MPI-AO-LR, where a higher-resolution grid configuration is used compared to the other models, and in BCM, deep water formation is, as in the real ocean, found in the central Labrador Sea. Another caveat of the model simulations used here is the generally biased representation of the overflow water masses in z coordinate models (e.g. Willebrand et al., 2001; Danabasoglu et al., 2010; Yeager and Danabasoglu, 2012), resulting in a too shallow meridional overturning cell. Furthermore, the structure of the meridional overturning cell differs among the models, both with respect to mean state and multidecadal variability (Figs. 1, 6). Regarding the latter, also the dominant periodicity varies among models. The largest value of the maximum AMOC strength as well as the largest variability are found in MPI-AO-LR, where a higher-resolution grid configuration is used compared to the other models.

Currently, millennial-scale simulations with eddy-permitting or even eddy-resolving grid configurations are not available. The differences among models as well as their biases, however, underline the importance of understanding multidecadal AMOC variability based on simulations with high-resolution grid configurations in the future.

Acknowledgements. This work was supported by the European Community's 7th framework programme (FP7/2007-2013) under grant agreement no. GA212643 (THOR: Thermohaline circulation – at risk?, 2008–2012). K. Lohmann also received funding through the Cluster of Excellence "CLISAP", funded by the German Science Foundation (DFG). D. Matei was supported by the Federal Ministry for Education and Research (BMBF) NORTH ATLANTIC and RACE projects. The authors wish to thank Michael Botzet and Helmuth Haak for their help with setting up the sensitivity experiments, Manfred Bersch for his help with the definition of the subpolar deep water formation index and Xiuhua Zhu for a critical review of the manuscript prior to submission. Helpful comments, which further improved the manuscript, by Anders Levermann and two anonymous reviewers are greatly acknowledged. The MPI-ESM-CR and MPI-AO-LR simulations were conducted at the German Climate Computing Center (DKRZ). This study is a contribution to the Centre for Climate Dynamics at the Bjerknes Centre.

The service charges for this open access publication
have been covered by the Max Planck Society.

Edited by: S. Josey

References

Ba, J., Keenlyside, N., Park, W., Latif, M., Hawkins, E., and Ding, H.: A mechanism for Atlantic multidecadal variability in the Kiel Climate Model, Clim. Dynam., 41, 2133–2144, doi:10.1007/s00382-012-1633-4, 2013.

Ba, J., Keenlyside, N., Latif, M., Park, W., Ding, H., Lohmann, K., Mignot, J., Menary, M., Otterå, O. H., Wouters, B., Salas, D., Oka, A., Belluci A., and Volodin, E.: A multi-model comparison for Atlantic multidecadal variability, Clim. Dynam., online first, doi:10.1007/s00382-014-2056-1, 2014.

Bailey, D., Rhines P., and Häkkinen, S.: Formation and pathways of North Atlantic Deep Water in a coupled ice–ocean model of the Arctic–North Atlantic Oceans, Clim. Dynam., 25, 497–516, 2005.

Bentsen, M., Drange, H., Furevik, T., and Zhou, T.: Simulated variability of the Atlantic meridional overturning circulation, Clim. Dynam., 22, 701–720, 2004.

Biastoch, A., Böning, C., Getzlaff, J., Molines, J.-M., and Madec, G.: Causes of interannual–decadal variability in the meridional overturning circulation of the midlatitude North Atlantic Ocean, J. Climate, 21, 6599–6615, 2008.

Böning, C., Scheinert, M., Dengg, J., Biastoch, A., and Funk, A.: Decadal variability of subpolar gyre transport and its reverberation in the North Atlantic overturning, Geophys. Res. Lett., 33, L21S01, doi:10.1029/2006GL026906, 2006.

Cunningham, S., Kanzow, T., Rayner, D., Baringer, M. O., Johns, W. E., Marotzke, J., Longworth, H. R., Grant, E. M., Hirschi, J. J.-M., Beal, L. M., Meinen, C. S., and Bryden, H. L.: Temporal Variability of the Atlantic Meridional Overturning Circulation at 26.5° N, Science, 317, 935–938, 2007.

Danabasoglu, G., Large, W. G., and Briegleb, B. P.: Climate impacts of parameterized Nordic Sea overflows, J. Geophys. Res., 115, C11005, doi:10.1029/2010JC006243, 2010.

Danabasoglu, G., Yeager, S. G., Kwon, Y.-O., Tribbia, J. J., Phillips, A. S., and Hurrell, J. W.: Variability of the Atlantic meridional overturning circulation in CCSM4, J. Climate, 25, 5153–5172, 2012.

Escudier, R., Mignot, J., and Swingedouw, D.: A 20-yrs coupled ocean-sea ice-atmosphere variability mode in the North Atlantic in an AOGCM, Clim. Dynam., 40, 619–636, doi:10.1007/s00382-012-1402-4, 2013.

Fischer, N.: Holocene and Eemian climate variability, PhD thesis, Berichte zur Erdsystemforschung No. 91, Max Planck Institute for Meteorology, Hamburg, Germany, 106 pp., 2011.

Furevik, T., Bentsen, M., Drange, H., Kindem, I., Kvamstø, N., and Sorteberg, A.: Description and evaluation of the Bergen climate model: ARPEGE coupled with MICOM, Clim. Dynam., 21, 27–51, 2003.

Gastineau, G. and Frankignoul, C.: Cold-season atmospheric response to the natural variability of the Atlantic meridional overturning circulation, Clim. Dynam., 39, 37–57, 2012.

Gordon, C., Cooper, C., Senior, C., Banks, H., Gregory, J., Johns, T., Mitchell, J., and Wood, R.: The simulation of SST, sea ice extents and ocean heat transports in a version of the Hadley Centre coupled model without flux adjustments, Clim. Dynam., 16, 147–168, 2000.

Hall, M. and Bryden, H.: Direct estimates and mechanisms of ocean heat transport, Deep Sea Res.-Pt. A, 29, 339–359, 1982.

Hansen, B., Østerhus, S., Turrell, B., Jónsson, S., Valdimarsson, H., Hátún, H., and Olsen, S.: The inflow of Atlantic water, heat, and salt to the Nordic Seas across the Greenland-Scotland Ridge, in: Arctic–Subarctic Ocean Fluxes, edited by: Dickson, R., Meincke, J., and Rhines, P., Springer Verlag, 15–43, 2008.

Jungclaus, J. H., Haak, H., Latif, M., and Mikolajewicz, U.: Arctic–North Atlantic interactions and multidecadal variability of the meridional overturning circulation, J. Climate, 18, 4013–4031, 2005.

Jungclaus, J. H., Keenlyside, N., Botzet, M., Haak, H., Luo, J.-J., Latif, M., Marotzke, J., Mikolajewicz, U., and Roeckner, E.: Ocean circulation and tropical variability in the coupled model ECHAM5/MPI-OM, J. Climate, 19, 3952–3972, 2006.

Jungclaus, J. H., Lorenz, S. J., Timmreck, C., Reick, C. H., Brovkin, V., Six, K., Segschneider, J., Giorgetta, M. A., Crowley, T. J., Pongratz, J., Krivova, N. A., Vieira, L. E., Solanki, S. K., Klocke, D., Botzet, M., Esch, M., Gayler, V., Haak, H., Raddatz, T. J., Roeckner, E., Schnur, R., Widmann, H., Claussen, M., Stevens, B., and Marotzke, J.: Climate and carbon-cycle variability over the last millennium, Clim. Past, 6, 723–737, doi:10.5194/cp-6-723-2010, 2010.

Kanzow, T., Cunningham, S. A., Johns, W. E., Hirschi, J. J.-M., Marotzke, J., Baringer, M. O., Meinen, C. S., Chidichimo, M. P., Atkinson, C., Beal, L. M., Bryden, H. L., and Collins, J.: Seasonal variability of the Atlantic meridional overturning circulation at 26.5 degrees N, J. Climate, 23, 5678–5698, 2010.

Köhl, A.: Anomalies of meridional overturning: Mechanisms in the North Atlantic, J. Phys. Oceanogr., 35, 1455–1472, 2005.

Köhl, A. and Stammer, D.: Variability of the meridional overturning in the North Atlantic from the 50-year GECCO state estimate, J. Phys. Oceanogr., 38, 1913–1930, 2008.

Langehaug, H. R., Medhaug, I., Eldevik, T., and Ottera, O. H.: Arctic/Atlantic exchanges via the subpolar gyre, J. Climate, 25, 2421–2439, 2012a.

Langehaug, H. R., Rhines, P. B., Eldevik, T., Mignot, J., and Lohmann, K.: Water mass transformation and the North Atlantic Current in three multicentury climate model simulations, J. Geophys. Res., 117, C11001, doi:10.1029/2012JC008021, 2012b.

Latif, M., Roeckner, E., Botzet, M., Esch, M., Haak, H., Hagemann, S., Jungclaus, J., Legutke, S., Marsland, S., Mikolajewicz, U., and Mitchell, J.: Reconstructing, monitoring and predicting decadal-scale changes in the North Atlantic thermohaline circulation with sea surface temperature, J. Climate, 17, 1605–1614, 2004.

Latif, M., Böning, C., Willebrand, J., Biastoch, A., Dengg, J., Keenlyside, N., Schweckendiek, U., and Madec, G.: Is the thermohaline circulation changing?, J. Climate, 19, 4631–4637, 2006.

Levermann, A. and Born, A.: Bistability of the Atlantic subpolar gyre in a coarse-resolution climate model, Geophys. Res. Lett., 34, L24605, doi:10.1029/2007GL031732, 2007.

Marti, O., Braconnot, P., Dufresne, J.-L., Bellier, J., Benshila, R., Bony, S., Brockmann, P., Cadule, P., Caubel, A., Codron, F., de Noblet, N., Denvil, S., Fairhead, L., Fichefet, T., Foujols, M.-A., Friedlingstein, P., Goosse, H., Grandpeix, J.-Y., Guilyardi, E., Hourdin, F., Idelkadi, A., Kageyama, M., Krinner, G., Lévy, C., Madec, G., Mignot, J., Musat, I., Swingedouw, D., and Talandier, C.: Key features of the IPSL ocean atmosphere model and its sensitivity to atmospheric resolution, Clim. Dynam., 34, 1–26, 2010.

Medhaug, I., Langehaug, H. R., Eldevik, T., Furevik, T., and Bentsen, M.: Mechanisms for decadal scale variability in a simulated Atlantic meridional overturning circulation, Clim. Dynam., 39, 77–93, 2012.

Menary, M., Park, W., Lohmann, K., Vellinga, M., Palmer, M., Latif, M., and Jungclaus, J.: A multimodel comparison of centennial Atlantic meridional overturning circulation variability, Clim. Dynam., 38, 2377–2388, 2012.

Msadek, R. and Frankignoul, C.: Atlantic multidecadal oceanic variability and its influence on the atmosphere in a climate model, Clim. Dynam., 33, 45–62, 2009.

Olsen, S. M., Hansen, B., Quadfasel, D., and Østerhus, S.: Observed and modeled stability of overflow across the Greenland-Scotland ridge, Nature, 455, 519–523, 2008.

Østerhus, S., Sherwin, T., Quadfasel, D., and Hansen, B.: The overflow transport east of Iceland, in: Arctic–Subarctic Ocean Fluxes, edited by: Dickson, R., Meincke, J., and Rhines, P., Springer Verlag, 427–441, 2008:

Otterå, O. H., Bentsen, M., Bethke, I., and Kvamstø, N. G.: Simulated pre-industrial climate in Bergen Climate Model (version 2): model description and large-scale circulation features, Geosci. Model Dev., 2, 197–212, doi:10.5194/gmd-2-197-2009, 2009.

Park, W. and Latif, M.: Multidecadal and multicentennial variability of the meridional overturning circulation, Geophys. Res. Lett., 35, L22703, doi:10.1029/2008GL035779, 2008.

Park, W., Keenlyside, N., Latif, M., Ströh, A., Redler, R., Röckner, E., and Madec, G.: Tropical Pacific climate and its response to global warming in the Kiel Climate Model, J. Climate, 22, 71–92, 2009.

Pope, V., Gallani, M., Rowntree, P., and Stratton, R.: The impact of new physical parametrizations in the Hadley Centre climate model – HadCM3, Clim. Dynam., 16, 123–146, 2000.

Serra, N., Käse, R., Köhl, A., Stammer, D., and Quadfasel, D.: On the low-frequency phase relation between the Denmark Strait and the Faroe-Bank Channel overflows, Tellus, 62A, 530–550, 2010.

Servonnat, J., Yiou, P., Khodri, M., Swingedouw, D., and Denvil, S.: Influence of solar variability, CO_2 and orbital forcing between 1000 and 1850 AD in the IPSLCM4 model, Clim. Past, 6, 445–460, doi:10.5194/cp-6-445-2010, 2010.

Swingedouw, D., Rodehacke, C., Behrens, E., Menary, M., Olsen, S., Gao, Y., Mikolajewicz, U., Mignot J., and Biastoch, A.: Decadal fingerprints of fresh water discharge around Greenland in a multi-model ensemble, Clim. Dynam., doi:10.1007/s00382-012-1479-9, 2013.

Tanhua, T., Olsson, K., and Jeansson, E.: Tracer evidence of the origin and variability of Denmark Strait Overflow Water, in: Arctic–Subarctic Ocean Fluxes, edited by: Dickson, R., Meincke, J., and Rhines, P., Springer Verlag, 475–503, 2008.

Våge, K., Pickart, R., Spall, M., Moore, G., Valdimarsson, H., Torres, D., Erofeeva, S., and Nilsen, J. E.: Revised circulation scheme north of the Denmark Strait, Deep Sea Res.-Pt. I, 79, 20–39, 2013.

Willebrand, J., Barnier, B., Böning, C., Dieterich, C., Killworth, P. D., Le Provost, C., Jia, Y., Molines, J.-M., and New, A. L.: Circulation characteristics in three eddy-permitting models of the North Atlantic, Prog. Oceanogr., 48, 123–161, 2001.

Yeager, S. and Danabasoglu, G.: Sensitivity of Atlantic meridional overturning circulation variability to parameterized Nordic Sea overflows in CCSM4, J. Climate, 25, 2077–2103, 2012.

Zanchettin, D., Rubino, A., and Jungclaus, J. H.: Intermittent multidecadal-to-centennial fluctuations dominate global temperature evolution over the last millennium, Geophys. Res. Lett., 37, L14702, doi:10.1029/2010GL043717, 2010.

Zhu, X. and Jungclaus, J. H.: Interdecadal variability of the meridional overturning circulation as an ocean internal mode, Clim. Dynam., 31, 731–741, 2008.

Appendix A

Definition of subpolar deep water thickness

In all models except for BCM, the density range of subpolar deep water is defined based on mean profiles of temperature, salinity and σ_2 (potential density with reference depth 2000 m). σ_2 rather than σ values are chosen, as Labrador Sea Water in the North Atlantic spreads at a depth of about 2000 m. In MPI-ESM-CR, MPI-AO-LR, KCM and IPSLCM4, profiles are averaged over all integration years and the subpolar (60–10° W, 45–65° N) grid cells where the MLD in March exceeds 1000 m in at least 10 % of the integration years. These grid cells correspond approximately to the grid cells where the mean MLD in March exceeds 500 m (Fig. 2).

In HadCM3, no subpolar grid cells are found, where the MLD in March exceeds 1000 m in at least 10 % of the integration years. The same is true, if other winter months (December, January, February) rather than March are considered. If all grid cells, where the MLD in March exceeds 1000 m in at least one integration year, are taken into account, subpolar deep convection is found in the Labrador and Irminger seas. In HadCM3, profiles of temperature, salinity and σ_2 are therefore averaged only over those integration years and subpolar grid cells, when and where the MLD in March exceeds 1000 m.

The upper isopycnal defining subpolar deep water is chosen below the temperature and salinity maximum representing water from the subtropical Atlantic, at a depth of about 800 m. The lower isopycnal is chosen at the beginning of the deep pycnocline, at a depth of about 2500–3000 m. The resulting density range of subpolar deep water for the various models is as follows: *MPI-ESM-CR*: $\sigma_2 = 36.62\text{–}36.76\,\mathrm{kg\,m^{-3}}$, *MPI-AO-LR*: $\sigma_2 = 36.65\text{–}36.77\,\mathrm{kg\,m^{-3}}$, *KCM*: $\sigma_2 = 36.90\text{–}37.01\,\mathrm{kg\,m^{-3}}$, *IPSLCM4*: $\sigma_2 = 36.76\text{–}36.87\,\mathrm{kg\,m^{-3}}$, *HadCM3*: $\sigma_2 = 36.74\text{–}37.03\,\mathrm{kg\,m^{-3}}$. The thickness of subpolar deep water we define as the depth interval between the upper and lower isopycnal specifying the density range of subpolar deep water.

For BCM, where an isopycnic vertical coordinate is used, we follow Langehaug et al. (2012a, analysing the same simulation as used in our study), defining the layers with $\sigma_2 = 36.846\,\mathrm{kg\,m^{-3}}$ and $\sigma_2 = 36.946\,\mathrm{kg\,m^{-3}}$ as (upper) Labrador Sea Water. These density layers occupy approximately the depth range between 600 and 1800 m in the Labrador Sea (Langehaug et al., 2012a; their Fig. 9). The thickness of subpolar deep water we define as the sum of the layer thickness of these two density layers.

Appendix B

Definition of Nordic Seas overflow transports

The following criteria are used to define the overflow transport through the Denmark Strait and across the Iceland–Scotland Ridge in the various models: *MPI-ESM-CR*: transport out of Nordic Seas for $\sigma > 27.76\,\mathrm{kg\,m^{-3}}$ and depth > 300 m (through Denmark Strait) and for $\sigma > 27.8\,\mathrm{kg\,m^{-3}}$ (across the Iceland–Scotland Ridge), *MPI-AO-LR*: transport out of Nordic Seas for $\sigma > 27.8\,\mathrm{kg\,m^{-3}}$, *KCM*: transport out of Nordic Seas for $\sigma > 27.8\,\mathrm{kg\,m^{-3}}$, *IPSLCM4*: transport out of Nordic Seas for $\sigma > 27.75\,\mathrm{kg\,m^{-3}}$, *HadCM3*: transport out of Nordic Seas for depth > 300 m, and *BCM*: net transport for $\sigma_2 > 36.946\,\mathrm{kg\,m^{-3}}$ (corresponding to about $\sigma > 27.83\,\mathrm{kg\,m^{-3}}$). The transport is restricted to depth levels greater than 300 m in some of the models to ensure a separation between the Denmark Strait overflow and the lower East Greenland Current. In *BCM*, σ_2 isopycnal layers are used as the vertical coordinate in the ocean component. Therefore, the density threshold is specified as the σ_2 value. In this model, also the net transport rather than the transport out of the Nordic Seas is used. We note, however, that this difference is negligible, as a transport into the Nordic Seas with the given density threshold generally does not exist.

16

Interannual correlations between sea surface temperature and concentration of chlorophyll pigment off Punta Eugenia, Baja California, during different remote forcing conditions

H. Herrera-Cervantes[1]**, S. E. Lluch-Cota**[2]**, D. B. Lluch-Cota**[2]**, and G. Gutiérrez-de-Velasco**[3]

[1]Unidad La Paz, Centro de Investigación Científica y de Educación Superior de Ensenada (CICESE), La Paz, Baja California Sur 23050, México
[2]Centro de Investigaciones Biológicas del Noroeste (CIBNOR), Instituto Politécnico Nacional 195, La Paz, Baja California Sur 23096, México
[3]Departamento de Física, Universidad de Guadalajara, México

Correspondence to: H. Herrera-Cervantes (hherrera@cicese.mx)

Abstract. Interannual correlation between satellite-derived sea surface temperature (SST) and surface chlorophyll *a* (Chl *a*) are examined in the coastal upwelling zone off Punta Eugenia on the west coast of the Baja California Peninsula, an area than has been identified as having intense biological productivity and oceanographic transition between midlatitude and tropical ocean conditions. We used empirical orthogonal functions (EOF) analysis separately and jointly on the two fields from 1997 through 2007, a time period dominated by different remote forcing: ENSO (El Niño–Southern Oscillation) conditions (weak, moderate and strong) and the largest intrusion of subarctic water reported in the last 50 years. Coastal upwelling index anomalies (CUI) and the multivariate ENSO index (MEI) were used to identify the influence of local (wind stress) and remote (ENSO) forcing over the interannual variability of both variables. The spatial pattern of the individual EOF_1 analysis showed the greater variability of SST and Chl *a* offshore, their corresponding amplitude time series presented the highest peaks during the strong 1997–2000 El Niño–La Niña cycles and during the 2002–2004 period associated to the intrusion of subarctic water. The MEI is well correlated with the individual SST principal component ($R \approx 0.67$, $P < 0.05$) and poorly with the individual Chl *a* principal component ($R = -0.13$). The joint EOF_1 and the SST–Chl *a* correlation patterns show the area where both variables covary tightly; a band near the coast where the largest correlations occurred ($|R| > 0.4$) mainly regulated by ENSO cycles. This was spatially revealed when we calculated the homogeneous correlations for the 1997–1999 El Niño–La Niña period and during the 2002–2004 period, the intrusion of subarctic water period. Both, SST and Chl *a* showed higher coupling and two distinct physical–biological responses: on average ENSO influence was observed clearly along the coast mostly in SST, while the subarctic water influence, observed offshore and in Bahía Vizcaíno, mostly in Chl *a*. We found coastal chlorophyll blooms off Punta Eugenia during the 2002–2003 period, an enrichment pattern similar to that observed off the coast of Oregon. These chlorophyll blooms are likely linked to high wind stress anomalies during 2002, mainly at high latitudes. This observation may provide an explanation of why Punta Eugenia is one of the most important biological action centers on the Pacific coast.

1 Introduction

Continuous oceanographic observations carried out on the west coast of Baja California by the CalCOFI (California Cooperative Oceanic Fisheries Investigations) and IMECOCAL (Mexican Research of the California Current) programs have helped define the region of Punta Eugenia (Fig. 1) as an oceanographic transitional zone (Durazo and Baumgartner, 2002), characterized by the interaction of subarctic and tropical waters (Almazán-Becerril et al., 2012) as a result of

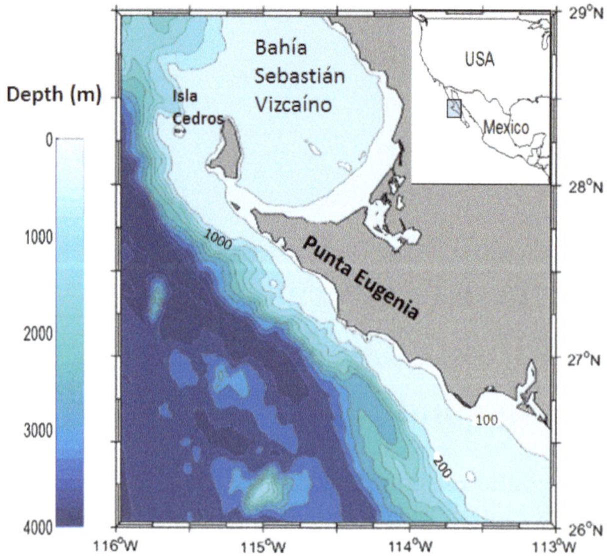

Fig. 1. Location and bathymetry characteristics of the study area.

the convergence of the southern part of the California Current (CC) and the North Equatorial Current (NEC). Its shelf is about 50–90 km wide, including Bahía Vizcaíno, this becomes narrow and almost disappears off Punta Eugenia (González-Rodriguez et al., 2012). The area is also influenced by warm and dense water originating in the Gulf of California (Parés-Sierra et al., 1997), creating a complex mixing zone between coastal and oceanic flows and intense mesoscale variability characterized by a complex pattern of filaments, meanders, and semipermanent eddy structures (Gallaudet and Simpson, 1994). These structures carry nutrient-rich coastal waters to deep areas, causing important seasonal variability and interannual and long-term changes in the mean field of variables as SST and Chl *a* (Espinosa-Carreón et al., 2004).

Seasonal wind forcing over the Punta Eugenia area is controlled regionally by the position and intensity of the North Pacific high pressure and the California semipermanent low thermal (Parés-Sierra et al., 1997). This wind pattern generates an intense coastal upwelling process that, together with the local contribution of the coastal lagoons (Guerrero Negro, Ojo de Liebre, and San Ignacio), produces one of the most important biological action centers (BAC) of the western coast of North America, which is characterized by above-average pigment concentrations (Lluch-Belda et al., 2000). The coastal ecosystem of this region is a natural refuge and a feeding and breeding area for many ecological and commercially important species (gray whale, sea turtles, spiny lobster, abalone, and clams). Maintenance of this ecosystem is based on three factors: rich coastal waters associated with an intense upwelling regime, successful implementation of cooperatives to safeguard existing resources, and relative isolation. These factors help support the fishery and natural resources in this area, which includes red lobster (value of USD 65 million annually; Vega et al., 2010; SAGARPA, 2011), abalone (value of USD 26 million annually; SAGARPA, 2001), and extraction of salt by solar evaporation (7000 t annually; SEMARNAT, 1997) contributing to the economy of the entire peninsula.

Oceanographic features off the west coast of the Baja California Peninsula are dramatically affected by global-scale ENSO (El Niño–Southern Oscillation) and interdecadal variability. El Niño events have a negative effect on fisheries: an increase in SST, sea level height, change in composition of the zooplankton community, and microbial pollution (Strub and James, 2002; Lavaniegos et al., 2002; Hereu et al., 2003; Boehm et al., 2004). Larval reproduction, embryonic development and changes in abundance of spiny lobster (*Panulirus interruptus*) are heavily impacted by El Niño–La Niña events (Vega et al., 2010) because the onset and duration of breading is accelerated or delayed (Vega, 2003), which dramatically reduces the captures in this region. Carreón-Palau et al. (2003) and Muciño-Días et al. (2004), describe serious decline of abalone because survival, growth, and larval recruitment are heavily dependent on cooler environmental conditions. These relationships between biological and environmental factors demonstrate strong physical–biological coupling in this region.

Climate effects on SST and Chl *a* in this coastal environment are well documented here due to more than a decade of satellite measurements (1997–2007). This period includes different ENSO conditions (weak, moderate and strong) well illustrated by the 1999–2004 period of increasing MEI (multivariate ENSO index; Behrenfeld et al., 2006), beginning with strong El Niño/La Niña cycles between 1997 and 1999, followed by a weak El Niño between 2002 and 2004, and finally a moderate El Niño during the 2006–2007 period. Additionally, during the 2002–2004 period the California Current System (CCS), remained in the cold phase (Goericke et al., 2005), a state it has maintained since the 1999 La Niña phase with the presence of the largest intrusion of subarctic water reported in the last 50 years, which is characterized as a cold and fresh anomaly in the upper halocline (Huyer, 2003).

In this study, we explored the interannual covariance between SST and Chl *a* off Punta Eugenia, an adequate area as a reproductive habitat due to high levels of biological production and its responses to two different large-scale processes; the ENSO cycles and the intrusion of subarctic water. Individual and joint empirical orthogonal functions (EOF) analyses were used to extract the principal modes of interannual variability and correlation patterns to analyze statistically the temporal and spatial coupled modes of variability between SST and Chl *a* and their response to warming and cooling climate processes. Additionally, we examine time series of coastal anomalies of SST, Chl *a* and wind stress to observe the propagation of the ENSO signals and subarctic water intrusion during two different time periods: 1997–2007 and 2002–2004.

2 Data and methods

The data used in this analysis are monthly composites of SST and Chl *a* satellite images covering the period from September 1997 through to December 2007 for an area 26–29° N and 113–116° W, centered at Punta Eugenia (see Fig. 1). These data are derived from an advanced very high resolution radiometer (AVHRR-Pathfinder) sensor for SST and sea-viewing wide field-of-view sensor (SeaWiFS) for Chl *a*, available at their respective internet addresses (http://podaac.jpl.nasa.gov/ and http://oceancolor.gsfc.nasa.gov/SeaWiFS). O'Really et al. (2000), Kahru and Mitchell (2001) as well as Mitchell (2004) have shown that the CalCOFI and IMECOCAL regions are ideal locations to evaluate the accuracy, precision and suitability of ocean color chlorophyll algorithms. Comparison of in situ and SeaWiFS Chl *a* estimates using these algorithms tends to produce a higher coefficient of determination ($r^2 \approx 0.9$) and a smaller root-mean-square error (RMSE ≈ 0.2); nevertheless, Kahru and Mitchell (2001) and Mitchell (2004) found that discrepancies between satellite-derived Chl *a* data and in situ values at near-shore CalCOFI stations may be caused by small-scale spatial variability.

After rotating and orienting the images along the coast, the 9 km × 9 km Chl *a* initial grid was interpolated using a MATLAB function into a 4 km × 4 km pixel array to produce the same spatial resolution used in the SST. These processes resulted in matrices of monthly composite images of SST (x,t) or Chl *a* (x,t) values where x stands for the cells of each image and t stands for each month between September 1997 and December 2007. To remove the annual and semiannual signals, we fitted the monthly time series of each pixel to the periodic function as follows:

$$\boldsymbol{F}(t) = A_0 + A_1\cos(w_1 t - \varphi_1) + A_2\cos(w_2 t - \varphi_2), \tag{1}$$

where A_0 is the annual mean, A_1, w_1, and φ_1 are the amplitude, frequency, and phase of the annual signal and A_2, w_2, and φ_2 are the equivalent of the semiannual signal. Next, we obtained the matrices of SST and Chl *a* anomalies (interannual variability), subtracting from each time series of each cell its corresponding fitted periodic function (Eq. 1) as

$$\mathbf{A}(x,t) = \mathbf{M}(x,t) - \mathbf{F}(x,t). \tag{2}$$

Each non-seasonal anomalies matrix was transformed into normalized anomalies NA(x,t) by dividing $A(x,t)$ series of each pixel by its standard deviation. The resulting NA(x,t) matrices of both variables were used in the individual EOF analysis to identify the dominant modes of SST and Chl *a* interannual variability and its evolution over time (principal components). Joint EOF analysis is calculated from the covariance matrix constructed from both variables to highlight how they covary with each other (Wilson and Adamec, 2001). Additionally we calculated homogeneous correlations (Bretherton et al., 1992), the correlation at each point between the SST and Chl *a* data and its corresponding EOF_1 temporal components for two subsets of the time series, between September 1997 and December 1999 and between January 2002 and December 2003, these time periods are chosen using the temporal evolution of the individual SST and Chl *a* principal component alongside the MEI. The integral timescale of these homogeneous correlations and the effective number of degrees of freedom would lead to an expected artificial correlation (Davis, 1976; Sciremammano, 1979; Chelton, 1982; and Tremberth, 1983), however its spatial distribution could delineate the regions that contribute the most to the mode 1 during different time periods (Wilson and Ademec, 2001).

Indices of the intensity of large-scale, wind-induced coastal upwelling (CUI) are generated by the NOAA/NMFS (National Marine Fisheries Service) Pacific Fisheries Environmental Laboratory (PFEL) at 15 standard locations along the west coast of North America (Schwing and Mendelssohn, 1996; http://www.pfeg.noaa.gov/). We used the CUI centered on 27° N, 116° W, as representative of the Punta Eugenia coastal region, filtering out the seasonal cycle by subtracting the corresponding climatological monthly average (CUI interannual anomalies time series). ENSO activity was represented by the MEI, a measure of the variability of the Pacific (Wolter and Timlin, 1993; Behrenfeld et al., 2006). We note that, though these indices summarize variability over different regions and the correlation between them are not large, wind data used to build monthly CUI anomalies in this region are strongly affected by ENSO events and are not completely independent (Storch and Zwiers, 1999). CUI and MEI values were normalized by dividing the time series by its standard deviation prior to calculating correlation values to make a more meaningful indicator of the true degree of linear relation with local SST and Chl *a* changes (Sciremammano, 1979).

Finally, to observe the evolution of the intrusion of subarctic water within the ~ 50 km closest to the coast of North America (22–45° N) including the coast of the study area and its relationship with the Chl *a* and wind forcing, we built time plots (Hovmöller diagrams) based on time series of coastal Chl *a* and wind stress anomalies data between January 2002 and December 2003. For this last analysis we used Chl *a* weekly 18 km × 18 km pixel array composites filtering out the seasonal cycle subtracting the Chl *a* weekly climatology provided by the SeaWiFS (Sea-viewing Wide Field-of-view Sensor) project. The wind stress anomalies ($\mathrm{N\,m^{-2}}$) was calculated in the same way that the matrices of SST and Chl *a* used in the individual EOF analysis, using monthly mean ocean surface wind in a 0.25° × 0.25° pixel array composite provided by the Cross-Calibrated Multi-Platform (CCMP) project website at http://podaac.jpl.nasa.gov/dataset/ and rebinned into 18 km × 18 km to have the same spatial resolution used in Chl *a* weekly composites.

3 Results

To determine the regions off Punta Eugenia where SST and Chl *a* have the largest seasonal fluctuations, the standard deviation (SD) variability is calculated for each monthly time series of the satellite-derived SST$(x,\tilde{t})$ and Chl $a(x,\tilde{t})$ observations (Fig. 2). Both patterns had large gradients along the coast and the largest seasonal fluctuations in SST ($>3\,°\mathrm{C}$) and Chl *a* ($>1\,\mathrm{mg\,m^3}$) near to the coast, mainly south of Punta Eugenia and north of Isla de Cedros. The deep zone presented the lowest values, this included Bahía Vizcaíno where the barotropic circulation associated to a semipermanent anticyclonic gyre (Palacios et al., 1996) could mask the amount of biological productivity. This pattern is consistent with near shore eutrophic conditions and/or oligotrophic conditions farther offshore (Kahru and Mitchell, 2000). The bottom panels represent the time plot of SST, Chl *a*, and CUI (space-average SST$(x,\tilde{t})$ and Chl $a(x,\tilde{t})$ and CUI monthly time series), all dominated by strong annual cycle (right panel), although large interannual changes were still evident. High peaks of Chl *a* and CUI (spring–summer) seem to show a trend to increasing annual maxima mostly (higher maxima) during the 2002–2003 period, Kahru and Mitchel (2001), using monthly composites of a four ocean color sensors-derived Chl *a* data set, showed a similar behavior in coastal areas off Baja California. The three parameters reveal coincidences between them.

Figure 3 shows the interannual variability of SST and Chl *a* off Punta Eugenia conformed by a single dominant mode. The spatial patterns of the individual EOF analyses (surface plot of the EOF_1 loadings; panels a, b), account for 78 and 45 % of the total variance for SST and Chl *a* (with the sign-reversed) respectively. Both variability patterns were not spatially homogeneous, strong gradients and high loadings were observed in a broad band along the coast but tended to be low near to the coast as distinct from the standard deviation maps (Fig. 2). The amplitude time series (time plot of the EOF_1 score; lower panel c) showed that the SST interannual variability was forced by ENSO-related events (e.g., the signals of the 1997–1998 El Niño, 1999–2000 La Niña and 2006–2007 El Niño) showing a major correlation with the MEI while the principal components of Chl *a* presented the higher amplitude peaks during the 2002–2004 period. The correlations values between the principal components of individual EOF analyses (SST and Chl *a*) and the normalized time series of MEI and CUI anomalies are presented in Table 1. The MEI is well correlated with the individual SST principal component ($R = 0.67$, $P < 0.05$) and poorly with the individual Chl *a* principal component ($R = -0.13$), while the CUI correlation with these same principal components is low in both cases ($R = -0.13$ and $R = 0.20$), suggesting that the SST interannual variability was forced principally by ENSO conditions observed during the entire period (weak, moderate, and strong) while the main forcing in the Chl *a* interannual variability could be associated to the

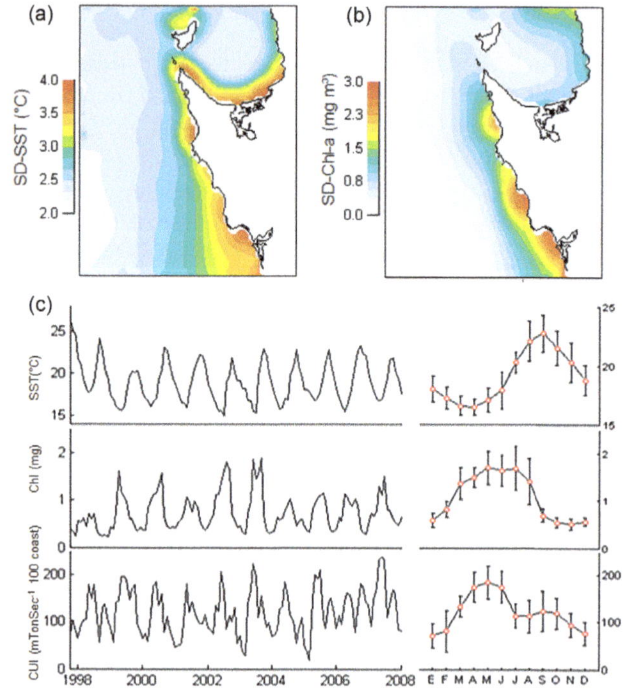

Fig. 2. Surface plots of standard deviation variability (time averaged) and monthly time series (spatial averaged) of the satellite-derived SST and chlorophyll *a* compared with the monthly CUI time series calculated from the Pacific Fisheries Environmental Group (PFEG) website (http://www.pfeg.noaa.gov/) for the Punta Eugenia region (27° N, 116° W). The curves on the right side correspond to the average seasonal cycle for each variable. Observed averages (red circles) and vertical bars indicate standard deviation.

intrusion of subarctic water represented by the high peaks in the amplitude time series of the Chl *a* during the 2002–2004 period.

Figure 4 shows the spatial pattern of the first mode of the joint EOF_1 analysis for SST and Chl *a* (panels a, b), accounting for 80 % of the total variance. The amplitude time series corresponding to the joint EOF_1 is not shown since it is virtually identical to those of individual SST components shown in Fig. 3c, having a correlation of 0.98 (see Table 1). The joint spatial modes (SST and Chl *a*), differ to those of the individual EOF_1 (Fig. 3a, b). The primary difference is that the coastal and northern region dominates more in the joint EOF, mostly in the chlorophyll mode (Fig. 4b) where the northern area of Punta Eugenia and Bahía Vizcaíno shows the largest spatial loadings. In the joint mode, low variability corresponds to the depth and southern region as a continuous feature. The spatial distribution of the correlation coefficient between monthly anomalies of SST and Chl *a* (Fig. 4c), present the area where both parameters strongly co-vary. This area (a band parallel to the coast) is enclosed with the highest correlations values ($|R| > 0.25$) and is similar with the area where the joint EOF_1 scores of the SST (Fig. 4a) are > 2.0. Correlations above R_{crit} ($P < 0.05$) $= 0.211$, i.e., the strongest

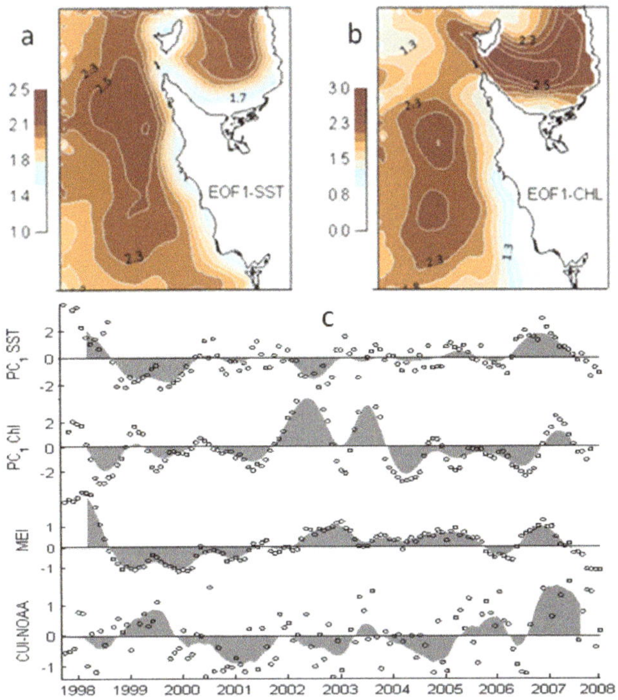

Fig. 3. Spatial patterns for mode 1 of the individual EOF analyses. **(a)** SST, **(b)** chlorophyll *a*. This mode accounts for 78 and 45 % of the total variance for SST and chlorophyll a (with the sign reversed), respectively. **(c)** Time-series data indicate the temporal evolution of the EOF1 alongside the MEI and monthly CUI anomalies (dots), all smoothed by a five-term running mean (shaded area).

negative correlation in the blue areas, are statistically significant ($P < 0.05$). Correlations that are not statistically significant at the 95 % level are shaded light blue.

Figure 5 shows contours of mean homogeneous correlation calculated for two subsets of the time series, between September 1997 and December 1999 and between January 2002 and December 2003, in both SST and Chl *a* respectively. Correlations between the shorter SST and Chl *a* time series of the data and its corresponding temporal components are given in Table 2 along with N the number of values of each subset, τ the integral timescale, N_{eff} the effective degree of freedom, the Student t value and P_{val} the statistical significance. The integral timescale of these homogeneous correlations for the ENSO cycle ($\tau_{SST} = 3.7$ and $\tau_{Chl\ a} = 2.0$) and for the subarctic period ($\tau_{SST} = 1.5$, $\tau_{Chl\ a} = 1.9$) could not be regarded as significant but can be used to assess what regions dominate during these periods. There are significant similarities between the homogeneous correlations distribution during the El Niño–La Niña period (solid and dotted contours in Fig. 5a); a narrow area confined along the coast dominates during this period, this distribution is virtually identical to the spatial correlation pattern showed in Fig. 4c (gray regions delineate areas with absolute correlations greater than 0.25), meaning that the physical–biological

Table 1. Correlation coefficients between principal components from different SST and chlorophyll *a* EOF_1 analyses (individual (ind.) and joint) and MEI and monthly CUI anomalies using the original time series ($N = 124$).

	EOF_1 ind. SST	EOF_1 ind. Chl	EOF_1 joint	MEI	CUI
EOF_1 ind. SST	1.0				
EOF_1 ind. Chl	0.21	1.0			
EOF_1 joint	0.98	0.97	1.0		
MEI	0.67	−0.13	0.63	1.0	
CUI	−0.13	0.20	0.17	−0.13	1.0

coupling shown in the coastal region is driven by both increase and decrease of SST and chlorophyll during ENSO cycles. In contrast, during the period of subarctic water intrusion the highest mean homogeneous correlations distribution (solid contours in Fig. 5b) are located in the central basin along the coast and over Bahía Vizcaíno, a region where individual EOF patterns show the high variability mainly in the chlorophyll mode (dark gray regions delineate areas with absolute Chl *a* spatial loadings greater than 2.3). These results delineate the regions that contribute the most to the individual and coupled modes during different remote forcing periods, indicating that the individual variability pattern of both parameters could be driven more by the intrusion of subarctic water than by ENSO cycles.

Figure 6 shows the time series of (a) the MEI index (bars) and CUI anomalies (curves), both compared with the temporal evolution of (b) SST and (c) Chl *a* coastal anomalies signals plotted in two Hovmöller diagrams (i.e., time series of coastal pixels plotted as contours) from 26° N (at the bottom of the diagrams) to the north of Bahía Vizcaíno (at the top, the black line in the middle indicates the position of Punta Eugenia on the coast). Monthly CUI anomalies time series are partially impacted by the different ENSO cycles (weak, moderate and strong), in phase but negatively correlated with the MEI in most of the time period. Panel b shows that coastal anomalies of SST are associated with ENSO events mainly at the strong 1997–1998 El Niño event, where the maximum SST anomalies (from ~3 to −2 °C) in the south of Punta Eugenia are transmitted to Bahía Vizcaino with a slight attenuation mainly during the 1998–2000 La Niña event whose influence abruptly finalizes prior to that shown by MEI. Previous to the weak 2002–2004 El Niño event, the SST anomalies showed a short warm period. A strong cold event was observed at the beginning of 2002, with high concentration of negative anomalies (~ −3 °C) observed mainly in Bahía Vizcaíno, which remained stationary along the entire coast through the end of 2005 with a slight attenuation after 2003 south of Punta Eugenia. The 2006–2007 El Niño event, caused the end of the previous cold event, SST anomalies of ~2 °C were observed mainly on the southern

Table 2. Correlations between N monthly SST and Chl a time series of the data and its corresponding principal components of individual EOF analyses for two subsets: ENSO period (September 1997–December 1999) and Subarctic period (January 2002 and December 2003) along with τ the integral timescale, N_{eff} the effective degree of freedom, Student's t value and P_{val} the statistical significance. Correlation significance levels were obtained consulting classical statistical tables and using the methodology proposed by Davis (1976) and Trenberth (1983).

		R	N	τ	N_{eff}	Student's t value	P_{val}
ENSO	SST vs. PC_1	0.99	28	3.7	7	96.0	5.60×10^{-10}
period	CHL vs. PC_1	−0.83	28	2.0	13	5.1	2.50×10^{-4}
Subarctic	SST vs. PC_1	0.99	24	1.5	15	48.0	3.13×10^{-16}
period	CHL vs. PC_1	−0.91	24	1.9	12	7.2	2.48×10^{-5}

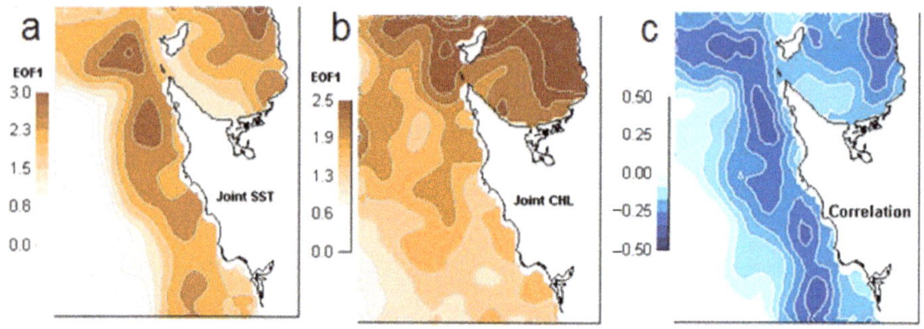

Fig. 4. Spatial patterns for mode 1 of the joint EOF analyses of **(a)** SST, **(b)** chlorophyll a (accounting for 80 % of the total variance) and **(c)** the correlation map between monthly anomalies of SST and chlorophyll a. The largest negative correlations (< -0.2, the blue areas) are seen along the coast and coincide with the high joint EOF_1 scores (> 2.0) of SST **(a)**. Absolute correlations above R_{crit} ($P < 0.05$) = 0.211, (i.e., the strongest negative correlation) are significant at the 95 % confidence level.

coast of Punta Eugenia with a slight attenuation in Bahía Vizcaíno.

Panel c shows that coastal anomalies of Chl a are associated with ENSO events only at the beginning of the period; the strong 1997–1999 ENSO events where the Chl a anomalies showed high negative anomalies (from ~ -2 to $1.5\,mg\,m^{-3}$). Moderated concentrations occurred during the peaks of the 1998–2000 La Niña conditions, mainly south of Punta Eugenia with a strong attenuation when the signal are transmitted to Bahía Vizcaino. During the weak 2002–2004 El Niño event, Chl a showed unusual behavior, higher concentrations (1.8–$2.0\,mg\,m^{-3}$) occurred during this period along the entire coast. Positive anomalies remained stationary mainly in the coast of Bahía Vizcaíno through the end of 2006. The moderate 2006–2007 El Niño did not appear to have a noticeable effect on the coastal anomalies of Chl a, though it continued to have positive anomalies, maybe influenced by the positive phase showed by the CUI during the 2005–2007 period. These conditions (positives anomalies) were maintained until the end of the study period.

Figure 7 shows the relationship between the wind stress and Chl a anomalies during the period of subarctic water intrusion (January 2002–December 2003) in an area close to the coast of North America (22–45° N). Latitude–time plots of (a) anomalies of monthly wind stress (magnitude and direction; $N\,m^{-2} \times 10^{-2}$) showed weak wind stress anomalies in early 2002 but increasing in intensity during spring–summer of 2002 (onshore direction). Anomalous wind stress along the northeastern Pacific coast may drive enhanced coastal upwelling and anomalous Ekman transport along the immediate coast (Taylor et al., 2008). Coastal Chl a anomalies (b) showed in detail that the southern coast, which includes from the study area (shaded in green) to the tip of the peninsula, presented high positive anomalies in Chl a ($> 2.5\,mg\,m^{-3}$). It should be noted that the Chl a signals off Punta Eugenia, reach bloom conditions compared to that found north of 42° N (off the Oregon coast), where the presence of the cold and fresh subarctic water masses in the CCS was first noticed (Kosro, 2003; Freeland et al., 2003; Goericke et al., 2005). High positive Chl a coastal anomalies, continued off Baja California during spring and summer 2003, without any significant delay or attenuation by the co-occurring 2002–2003 El Niño event (Bograd and Lynn, 2003; Durazo et al., 2005), while wind stress anomalies showed a pronounced decline.

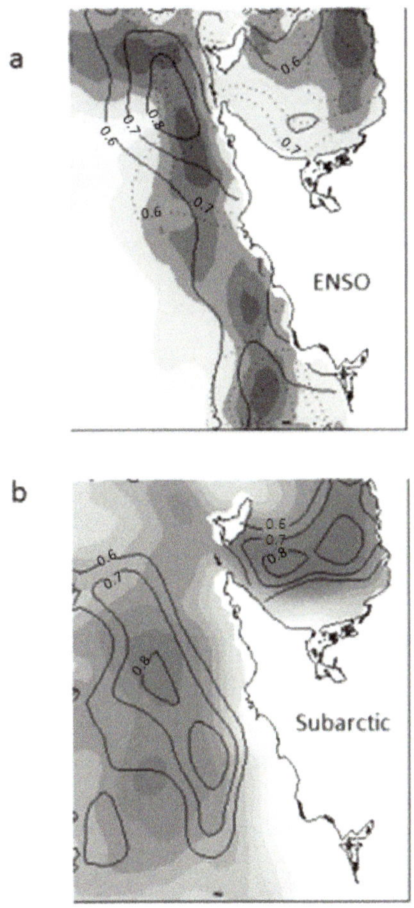

Fig. 5. Contours of mean homogeneous correlation between SST and Chl *a* for the El Niño and La Niña period (solid and dotted contours) overlaid on **(a)** the correlation map. Mean homogeneous correlation for the intrusion of subarctic water period (solid contours) overlaid on **(b)** mode 1 of the individual EOF analyses of chlorophyll *a*. The El Niño correlation is done using data from September 1997 through to December 1998; the La Niña correlation is done using data from September 1998 through to December 2000; and the subarctic water correlation is done using data from January 2002 through to December 2003. The contour interval is 0.1 and the minimum contour shown is 0.6.

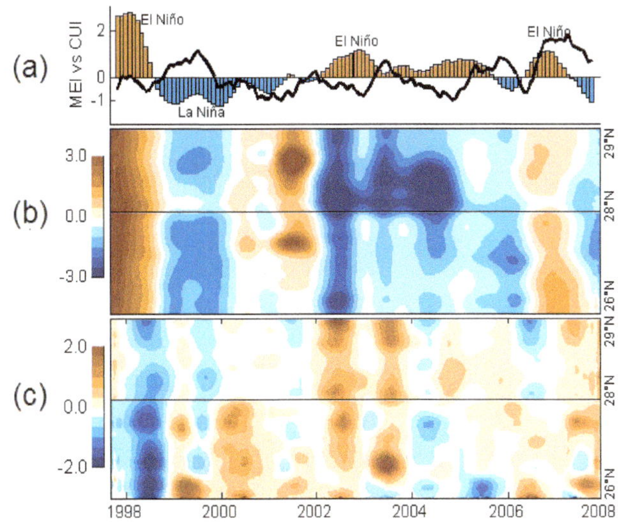

Fig. 6. Temporal evolution of **(a)** the MEI and monthly CUI anomalies. El Niño (La Niña) episodes are indicated by orange (blue) bars respectively as reported by the Climate Prediction Center of the National Center of Environmental Prediction at the National Oceanic and Atmospheric Administration (CPC-NCEP-NOAA). Hovmöller diagram of monthly averaged coastal SST anomalies **(b)** and chlorophyll *a* anomalies **(c)**, from September 1997 to December 2007, covering approximately 300 km of coastline, from 26 (bottom graph) to 29° N (upper graph). Solid line indicates the position of Punta Eugenia on the coast (28° N).

4 Discussion

In this study, we used EOF analysis both separately and jointly over satellite-derived SST and Chl *a* anomaly data off Punta Eugenia, a region with high levels of biological productivity that is associated with wind-driven upwelling processes (González-Rodriguez et al., 2012; Lluch-Belda et al., 2000), which has allowed us a wide-ranging view of the biophysical coupling during ENSO events and the largest intrusion of subarctic water reported in the last 50 years (Huyer, 2003; Venrick et al., 2003). Moreover, by relating each SST and Chl *a* series within ∼ 30 km of the coast of Punta Eugenia to two independent variables, MEI and CUI, and by following the coastal signal of the Chl *a* related to the intrusion of subarctic water from 45 to 22° N; to our knowledge, this is the first time that physical–biological covariability driven by remote forcing (tropical and subarctic origin) off Punta Eugenia has been examined at this level of detail.

The spatial patterns of individual EOF_1 (Fig. 3a, b) showed that the interannual variability of SST and Chl *a* are not homogeneous and have a strong gradient along the coast. Although both patterns are similar, its principal components are markedly different ($R \approx 0.21$; see Table 1). The statistical relationship between the principal components of the SST mode and the MEI showed similar trends and were well correlated ($R = 0.67$, $P < 0.05$), except with the CUI anomalies time series (see Table 1). This may indicate that they were forced at the beginning of the time period by ENSO-related events but during the 2002–2004 period both SST and Chl *a* were forced by the large intrusion of subarctic water reported in the last 50 years (Venrick et al., 2003; Bograd and Lynn, 2003; and Durazo et al., 2005; Goericke et al., 2005), which masked the presence of the El Niño event in this period. Chl *a* reached the highest peak of variability during this period (Fig. 3c), suggesting that Chl *a*, unlike SST, was mostly affected by the intrusion of subarctic water. In this case the individual variability pattern of Chl *a* could be defined as the subarctic water mode. For the 1999–2006 period, a period with a sustained increase in MEI values (Berenhfeld et al., 2006), the time series show a similar behavior. At the regional scale both warming and cooling

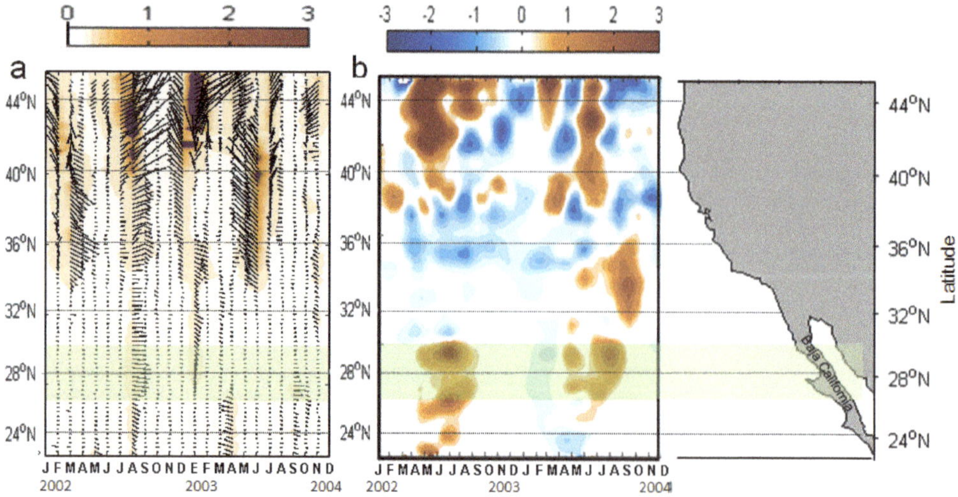

Fig. 7. Hovmöller diagrams of interannual anomalies of **(a)** monthly wind stress (magnitude and direction; $\mathrm{N\,m^{-2}} \times 10^{-2}$) and **(b)** weekly chlorophyll *a* ($\mathrm{mg\,m^{-3}}$) from January 2002 to December 2003 along the northeastern Pacific coast from 22 to 45° N (values within 50 km of the coast), including Punta Eugenia area (shaded in green). The wind data are provided by the CCMP project website at http://podaac-opendap.jpl.nasa.gov/opendap/allData/ccmp/L3.0/flk/, and weekly chlorophyll *a* by the SeaWiFS project of NASA Goddard Space Flight Center website at http://oceancolor.gsfc.nasa.gov/.

events are registered in SST and Chl *a*; these observations are consistent with the overall increasing (decreasing) trend of Chl *a* in the coastal (deep) zone of Punta Eugenia as indicative of weaker El Niño events (Kahru et al., 2012).

In contrast, the spatial patterns of joint $\mathrm{EOF_1}$ showed that the SST–Chl *a* interannual covariation off Punta Eugenia is the ENSO mode. The spatial pattern of SST (Fig. 4a) showed an intense variability within the near-shore band, while the spatial pattern for Chl *a* (Fig. 4b) differs little by having high loadings more evenly centered north of Punta Eugenia, including Bahía Vizcaíno. The spatial pattern of joint SST mode coincides to the correlation map between SST and Chl *a* (Fig. 4c), having a spatial correlation of $|R| = 0.89$. During El Niño events, coastal-trapped waves (Durazo and Baumgartner, 2002; Dever and Winant, 2002; Jacobs et al., 1994) reach the Baja California Peninsula coast accompanied by meandering poleward flow, positive SST anomalies increase the sea surface height that deepens the nutricline and reduces the availability of nutrients to the euphotic zone. While the signals associated with La Niña events are accompanied by positive anomalies of Chl *a* associated with an intensification of the northern winds, from which the upwelling events raise the nutricline and thus increase the availability of nutrients in the coastal region of Baja California (Espinosa-Carreón et al., 2012; Chavez et al., 1999; Wooster and Hollowed, 1995). The correlation pattern, suggests that the different physical–biological responses to events such as the 1997–1999 ENSO oscillations arise from a combination of ecological and physical dynamics (Wilson and Adamec, 2001), suggesting a high biophysical coupling off Punta Eugenia during ENSO cycles (Espinosa et al., 2004). Beyond the obvious considerations, any signal of ENSO on SST and Chl *a* would be masked by local processes (oceanic and atmospheric forcing, dissipative phenomena, upwelling) thus obscuring their statistical relationships. Therefore we believed that the similarity of the trends between the series shown in Fig. 3c and the spatial patterns shown in Fig. 4 indicate an important role of ENSO in forcing interannual variations off Punta Eugenia.

In a similar fashion, we overlaid the mean homogeneous correlation calculated for the periods El Niño–La Niña and the intrusion of subarctic water on the spatial correlation pattern (Fig. 5a) and the chlorophyll mode of the individual $\mathrm{EOF_1}$ analysis (Fig. 5b). The region dominated for the El Niño–La Niña cycle coincide with those of the spatial distribution of intense co-variability and the high correlations between SST and Chl *a* (Fig. 4c) coincident with the region where intense coastal wind-driven upwelling and complex mesoscale variability are found (Espinosa-Carreón et al., 2012; Gonzalez-Rodrìguez et al., 2012; Soto-Mardones et al., 2004) and where the propagation of coastal waves has been observed during the El Niño event (Parés-Sierra and O'Brien, 1989; Durazo and Baumgartner, 2002). The mean homogeneous correlation, calculated during the intrusion of subarctic water (solid contours in Fig. 5b), coincides significantly to that of individual $\mathrm{EOF_1}$ (Fig. 3a, b) when analyzed over the entire time series. Although the integral timescale and the small number of effective number of degrees off freedom of these homogeneous correlations (see Table 2) leads to expected artificial correlations and cannot be regarded as significant (Davis, 1976; Chelton and Davis, 1982; and Tremberth, 1983), the spatial patterns showed that the spatial trends vary during these different remote forcing conditions (Wilson and Ademec, 2001).

Time plots of SST and Chl *a* coastal anomalies (Fig. 6b, c) showed that, MEI and CUI anomalies are not related with the intrusion of subarctic water. Espinosa-Carreón et al. (2004) suggested that on an interannual timescale, changes in the monthly CUI anomalies do not appear to be the primary source of variability in the oceanic parameters such as SST and Chl *a*, since removing the seasonal component results in low correlation values (see Table 1). The chlorophyll bloom showed during the 2002–2003 period was stronger than during the strong 1999–2000 La Niña event, evidencing that the distribution of biological groups in areas as far south as 28° N were influenced mostly by the intrusion of water of northern origin, than by events of both local and equatorial origin (positive CUI anomalies and ENSO event). These conditions, driven by the southern part of the California Current (Bograd and Lynn, 2003; Durazo et al., 2005), caused a reversal in environmental conditions opposite to the Pacific Ocean's variability shown by the MEI in this period. Durazo et al. (2005) observed at the surface the presence of zooplankton groups (salps) associated with an intrusion of subarctic water but also at depths of 100 m zooplankton groups (chaetognaths) associated with water of tropical origin that accompanies El Niño events. These observations are consistent with the overall increasing trend of Chl *a* in the upwelling coastal zone off Punta Eugenia showed by Kahru et al. (2012), either due to increased upwelling or increased nitrate concentration in the upwelled water.

The evolution of exceptionally high Chl *a* concentration along the western coast of North America during the period of subarctic water intrusion (2002–2003), can be observed in Fig. 7. This was extended toward the Equator more than 3000 km (from 45 to 22° N), a coastal area where the extension of the continental shelf, the geometry of the coastline and the local processes (wind-driven coastal upwelling and advection) play an important role in modulating the productivity. Wind stress anomalies developed north of 35° N and strengthened during spring and summer 2002, coinciding with positive Chl *a* anomalies observed along the North America coast. This could be associated with onshore large-scale advection induced by winds in the central Pacific (Chelton and Davis, 1982) or anomalous Ekman transport that may drive enhanced coastal upwelling and/or more vigorous along-coast currents (Taylor et al., 2008). It is worth noting that positive Chl *a* anomalies observed off Punta Eugenia during spring and summer 2002, could be compared to those observed north of 42° N, both could be result of the strong west wind stress anomalies and enhanced coastal upwelling developed in spring–summer 2002, increasing levels of pigment concentration above the average for the coastal zone. This bloom of positive Chl *a* anomalies occurs again during summer 2003 and is coincident with the peaks of amplitude time series of the Chl *a* individual EOF_1 scores (Fig. 3c).

Our study is consistent with the notion that changes in Chl *a* concentration could be due to other factors such as decadal variability and, thus, changes in grazing pressures on the phytoplankton (Lavaniegos and Jiménez-Pérez, 2006; Wilson and Adamec, 2001) that may cause an impact on biological communities located along the coast. Gaxiola-Castro et al. (2008), using data obtained during IMECOCAL cruises (2001–2007), observed that the intrusion of subarctic water represented by an anomalous low salinity condition in the southern sector of the California Current appears to be coupled with the 2002–2006 warm phase of the Pacific Decadal Oscillation index (PDO). Neither the local upwelling (represented by the CUI anomalies) nor the zonal Ekman drift velocity used as a proxy for coastal upwelling showed signals associated with the presence of the intrusion of subarctic water. In this case, the positive signals of wind stress developed along the coast were in agreement with the bloom of the Chl *a*, mainly north of 35° N (Fig. 7).

5 Conclusions

Unlike the data gathered during oceanographic cruises, the temporal and spatial resolution of satellite-derived SST and Chl *a* data allowed for observation (both separately and jointly on the two fields) of the physical–biological coupling of very near-shore environments during large-scale processes that affected the region off Punta Eugenia. This region of intense biological productivity and oceanographic transition is highly influenced by the intrusion of subarctic water and by subtropical signatures triggered by poleward flow (e.g., ENSO cycles). The spatial patterns of individual EOF_1 could be forced mainly by an unusual enhanced onshore transport of subarctic water observed mainly during the period 2002–2004, defined by Goericke et al. (2005) as a cold phase of the CCS. Whereas the physical–biological coupling was forcing mainly by interannual variability of local and equatorial origin (CUI and ENSO events) tightly observed in an alongshore band of ~ 40 km wide. The Hovmöller diagram of coastal SST and Chl *a* anomalies calculated from September 1997 to December 2007 and the homogeneous correlation calculated during the strong 1997–1999 El Niño–La Niña time period, delineated the regions that contribute the most to the ENSO mode.

Although, some ENSO events dominated the area for a long time, the presence of the intrusion of subarctic water off Punta Eugenia might have had a greater influence on the individual interannual variability of both SST and Chl *a* variables. This remote forcing results in a large-scale chlorophyll bloom that extends for more than 3000 km of the northeastern Pacific coast (22–45° N) showing that Punta Eugenia is one of the most important biological action centers of the western coast of North America with levels of pigment concentration comparable to those found at higher latitudes (Oregon coast). These results successfully support the data from previous hydrographic surveys (CaLCOFI and IMECOCAL programs) off Baja California.

Acknowledgements. Ira Fogel of CIBNOR provided editorial improvements. We thank E. Beier (CICESE-BCS) for helpful comments. We thank the Pathfinder and SeaWiFS projects and the NASA Physical Oceanography Distributed Active Archive Center for the production and distribution of AVHRR–SST and SeaWiFS data. This research was funded by SAGARPACONACYT grants 2009-113C. Additional thanks go to the anonymous reviewers who gave valuable comments on an earlier draft of the manuscript.

Edited by: M. Kosro

References

Almazán-Becerril, A., Rivas, D., and García-Mendoza, E.: The influence of mesoscale physical structures in the phytoplankton taxonomic composition of the subsurface chlorophyll maximum off Baja California, Deep-Sea Res.-Pt. I, 70, 91–102, 2012.

Behrenfeld, M. J., O'Malley, R. T., Siegel, D. A., McClain, C. R., Sarmiento, J. L., Feldman, G. C., Milligan, A. J., Falkowski, P. G., Letelier, R. M., and Boss, E. S.: Climatic-driven trends in contemporary ocean productivity, Nature, 444, 752–755, 2006.

Bretherton, C. S., Smith, C., and Wallace, J. M.: An intercomparison of methods for finding coupled patterns in climate data, J. Climate, 5, 541–560, 1992.

Boehm, A. B., Lluch-Cota, D. B., Davis, K. A., Winnan C. D., and Monismith, S. G.: Covariation of coastal water temperature and microbial pollution at interannual to tidal periods, Geophys. Res. Lett., 31, L06309, doi:10.1029/2003GL019122, 2004.

Bograd, S. J. and Lynn, R. J.: Anomalous subarctic influence in the southern California Current during 2002, Geophys. Res. Lett., 30, 8020, doi:10.1029/2003GL017446, 2003.

Carreón-Palau, L., Guzman Del, P. S. A., Belmar, J. P., Carrillo, J. L., and Herrera, R. F.: Microhábitat y biota asociada a juveniles de abulón Haliotis fulgens y Haliotis corrugata en bahía Tortugas, Baja California Sur, México, Ciencias Marinas, 29, 325–341, 2003.

Chavez, F. P., Strutton, P. G., Friederich, G. E., Feely, R. A., Feldman, G. C., Foley, D. G., and McPhaden, M. J.: Biological and chemical response of the equatorial Pacific to the 1997–1998 El Niño, Science, 286, 2126–2131, 1999.

Chelton, D. B.: Statistical reliability and the seasonal cycle: comments on "Bottom pressure measrements across the Antartic Circumpolar Current and their relation to the wind", Deep-Sea Res., 29, 1381–1388, 1982.

Chelton, D. B. and Davis, R. E.: Monthly mean sea level variability along the west coast of North America, J. Phys. Oceanogr., 12, 757–784, 1982.

Davis, R. E.: Predictability of sea surface temperature and sea level pressure anomalies over the North Pacific Ocean, J. Phys. Oceanogr., 6, 249–266, 1976.

Dever, E. P. and Winant, C. D.: The evolution and depth structure of shelf and slope temperatures and velocities during the 1997–1998 El Niño near Point Conception, California, Prog. Oceanogr., 54, 77–103, 2002.

Durazo, R. and Baumgartner, T. R.: Evolution of oceanographic condition off Baja California: 1997–1998, Prog. Oceanogr. 54, 7–31, 2002.

Durazo, R., Gaxiola-Castro, G., Lavaniegos, B., Castro-Valdez, R., Gómez-Valdez, J., and Mascarenhas Jr., A. S.: Oceanographic conditions off the western Baja California coast, 2002–2003: A weak El Niño and subarctic water enhancement, Ciencias Marinas, 31, 537–552, 2005.

Espinosa-Carreón, L., Strub, P. T., Beier, E., Ocampo-Torres, F., and Gaxiola-Castro, G.: Seasonal and interannual variability of satellite-derived chlorophyll pigment surface height, and temperature off Baja California, J. Geophys. Res., 109, C03039, doi:10.1029/2003JC002105, 2004.

Espinosa-Carreón, L., Gaxiola-Castro, G., Beier, E., Strub, P. T., and Kurczyn, J. A.: Effect of mesoscale processes on phytoplankton chlorophyll off Baja California, J. Geophys. Res., 117, C04005, doi:10.1029/2011JC007604, 2012.

Freeland, H. J., Gatien, G., Huyer, A., and Smith, R. L.: A cold halocline in the northern California current: an invasion of subarctic water, Geophys. Res. Lett., 30, 1141, doi:10.1029/2002GL016663, 2003.

Gallaudet, T. C. and Simpson, J. J.: An empirical orthogonal function analysis of remotely sensed sea surfaced temperature variability and its relations to interior oceanic processes off Baja California, Remote Sens. Environ., 47, 374–389, 1994.

Gaxiola-Castro, G., Durazo, R., Lavaniegos, B., De-la-Cruz-Orozco, M. E., Millán-Nuñez, E., Soto-Mardones, L., and Cepeda-Morales, J.: Pelagic ecosystem response to interannual variability off Baja California, Ciencias Marinas, 34, 263–270, 2008.

Goericke, R., Venrick, E., Mantyla, A., Bograd, S. J., Schwing, F. B., Huyer, A., Smith, R. L., Wheeler, P. A., Hoff, R., Peterson, W. T., Chavez, F., Collins, C., Marinovic, B., Lo, N., Gaxiola-Castro, G., Durazo, R., Hyrenbach, K. D., and Sydeman, W. J.: The state of the California Current, 2004–2005: Still cool?, CalCOFI Rep., 46, 32–71, 2005.

González-Rodriguez, E., Trasviña-Castro, A., Gaxiola-Castro, G., Zamudio, L., and Cervantes-Duarte, R.: Net primary productivity, upwelling and coastal currents in the Gulf of Ulloa, Baja California, México: Ocean Science, 8, 703–711, 2012.

Hereu, C. M., Jiménez-Pérez, L. C., and Lavaniegos, B. E.: Effects of the El Niño 1997–1998 and the rapid transition to cool conditions in the winter 1998–1999 on the copepods and salps of the California Current. 3rd International Zooplancton Production Symp., 20–23 May 2003, Gijón Spain, Program and Abstracts, p. 53, 2003.

Huyer, A.: Preface to special section on enhanced subarctic influence in the California Current, 2002, Geophys. Res. Lett., 30, 8019, doi:10.1029/2003GL017724, 2003.

Jacobs, G. A., Hulburt, H. E., Kindle, J. C., Metzger, E. J., Mitchell, J. L., Teague, W. J., and Wallcraft, A. J.: Decade-scale trans-Pacific propagation and warming effect of an El Niño anomaly, Nature, 370, 360–363, 1994.

Kahru, M. and Mitchell, B. G.: Influence of the 1997–98 El Niño on the surface chlorophyll in the California Current, Geophys. Res. Lett., 27, 2937–2940, 2000.

Kahru, M. and Mitchell, B. G.: Seasonal and nonseasonal vaiability of satellite-derived chlorophyll and colored dissolved organic matter concentration in the California Current, J. Geophys. Res., 106, 2517–2529, 2001.

Kahru, M., Kudela, R. M., Manzano-Sarabia, M., and Mitchell, B. G.: Trends in the surface Chlorophyll of the California Current: Merging data from multiple ocean color satellites, Deep-Sea Res.-Pt. II, 77–80, 89–98, 2012.

Kosro, P. M.: Enhanced southward flow over the Oregon shelf in 2002: A conduit for subarctic water, Geophys. Res. Lett., 30, 8023, doi:10.1029/2003GL017436, 2003.

Lavaniegos, B. E. and Jiménez-Pérez, L. C.: Biogeographic inferences of shifting copepod distribution during 1997–1999 El Niño and La Niña in the California Current, Contributions to the Study of East Pacific Crustaceans, 4, 113–158, 2006.

Lavaniegos, B. E., Jiménez-Pérez, L. C., and Gaxiola-Castro, G.: Plankton response to El Niño 1997–1998 and La Niña 1999 in the southern region of the California Current, Prog. Oceanogr., 54, 33–58, 2002.

Lluch-Belda, D., Elorduy-Garay, J., Lluch-Cota, S. E., and Ponce-Díaz, G.: BAC Centros de Actividad Biológica del Pacifico Mexicano. Centro de Investigaciones Biológicas del Noroeste, Centro Interdisciplinario de Ciencias Marinas, Consejo Nacional de Ciencia y Tecnología, La Paz, B.C.S., Mexico, 2000.

Mitchell, B. G.: Bio-optical measurements and modeling of the California Current and Southern Oceans, Final Report, Scrips Institution of Oceanography, University of California San Diego, La Jolla, CA 92093-0218, 27 pp., 2004.

Muciño-Días, M., Sierra-Rodríguez, O. P., Castro, J. J., Talavera, J. J., Turrubiates, J. R., Caballero, F., and Rivera, J. L.: Estado de las poblaciones de Abulon azul Haliotis fulgens y amarillo H. corrugata por zona reglamentada en la costa occidental de la península de Baja California, Temporada 2003/2004, Dictamen Técnico, Instituto Nacional de la Pesca. Dirección general de Investigación pesquera en el Pacífico Norte, 12 pp., 2004.

O'Really, J. E., Maritorena, S., O'Brien, M. C., Siegel, D. A., Toole, D., Menzies, D., Smith, R. C., Mueller, J. L., Mitchell, B. G., Kahru, M., Chavez, F. P., Strutton, P., Cota, G. F., Hooker, S. B., McClain, C. R., Carder, K. L., Müller-Karger, F., Harding, L., Magnuson, A., Phinney, D., Moore, G. F., Aiken, J., Arrigo, K. R., Letelier, R., and Culver, M.: SeaWiFS Postlaunch Calibration and Validation Analyses, Part 3, edited by: Hookerm, S. B. and Firestone, E. R., NASA Tech. Memo. 2000-206892. NASA Goddard Space Flight Center, Vol. 11, 49 pp., 2000.

Palacios, H. E., Argote, M. L., Amador A., and Mancilla, M.: Simulación de la circulación barotrópica inducida por viento en Bahía Sebastián Vizcaíno, B.C., Atmósfera, 9, 171–188, 1996.

Parés-Sierra, A. and O'Brien, J. J. : The seasonal and interannual variability of the California Current System: A numerical model, J. Geophys. Res., 93, 3159–3180, 1989.

Parés-Sierra, A., Lopez, M., and Pavía, E.: Oceanografía Física del Océano Pacífico Nororiental, in Contribuciones a la Oceanografía Física en México, Monograf. Ser. Vol. 3, Unión Geofísica Mexicana, Puerto Vallarta, México, 1–24, 1997.

SAGARPA: Anuario estadístico de pesca, 2001, Comisión Nacional de Pesca y Acuacultura, México, 2001.

SAGARPA: Anuario estadístico de pesca, 2011, Comisión Nacional de pesca y Acuacultura, México, 2011.

SEMARNAT: Secretaria de Medio Ambiente Recursos Naturales: Salitrales de San Ignacio, sal y ballenas en Baja California, México, 1997.

Sciremammano, F: A suggestion for the presentation of correlation and their significance level, J. Phys. Oceanogr., 9, 1273–1276, 1979.

Schwing, F. B. and Mendelssohn, R.: Increased coastal upwelling in the California Current System, J. Geophys. Res., 102, 3421–3438, doi:10.1029/96JC03591, 1996.

Soto-Mardones, L., Parés-Sierra, A., García, J., Durazo, R., and Hormazabal, S.: Analysis of the mesoscale structure in the IMECOCAL region (off Baja California) from hydrographic, ADCP and altimetry data, Deep-Sea Res.-Pt. II, 51, 785–798, 2004.

Storch, H. V. and Zwiers, F. W. (Eds.): Statistical analysis in climate research. Cambridge University Press, Cambridge, 1999.

Strub, P. T. and James, C.: Altimeter-derived surface circulation in the large-scale Pacific gyres. Part 2: 1997–1998 El Niño anomalies, Prog. Oceanogr., 53, 185–214, 2002.

Taylor, S. V., Cayan, D. R., Graham, N. E., and Georgakakos, K. P.: Northerly surface winds over the eastern North Pacific ocean spring and summer, J. Geophys. Res., 113 D02110, doi:10.1029/2006JD008053, 2008.

Trenberth, K. E.: Signal Versus Noise in the Southern Oscillation, Mon. Weather Rev., 112, 326–332, 1983.

Vega, A. V.: Reproductive strategies of the spiny lobster Panulirus interruptus related to the marine environmental variability off central Baja California, México: management implications, Fish. Res., 65, 123–135, 2003.

Vega, A., Treviño, E., Gracia, G., Espinoza-Castro, L., and Zuñiga-Pacheco, C.: Evaluación de la langosta roja (Panulirus Interruptus) en la región centro occidental de la península de Baja California, mediante modelos dinámicos de biomasa: puntos de referencia y recomendaciones de manejo. Informe de Investigación, Instituto nacional de la Pesca, 21 pp., 2010.

Venrick, E., Bograd, S., Checkley, D., Cummings, S., Durazo, R., Gaxiola-Castro, G., Hunter, J., Huyer, A., Hyrenbach, K., Lavaniegos, B., Mantyla, E. A., Schwing, F. B., Smith, R. L., Syderman, W. J., and Wheeler, P. A.: The state of the California Current, 2002–2003: Tropical and subarctic influences vie for dominance, CalCOFI Report, 44, 28–60, 2003.

Wilson, C. and Adamec, D.: Correlation between chlorophyll and sea surface height in the tropical Pacific during 1997–1999 El Niño Southern Oscillation event, J. Geophys. Res., 106, 31175–31188, 2001.

Wolter, K. and Timlin, M. S.: Monitoring ENSO in COADS with a seasonally adjusted principal component index. Proc. 17th Climate Dynamics Workshop; Norman, OK, NOAA/NMC/CAC, NSSL, Oklahoma Clim. Survey, CIMMS and School of Meteorology, Univ. Oklahoma, Norman, OK, 52–57, 1993.

Wooster, W. S. and Hollowed, A. B.: Decadal scale variations in the eastern subarctic Pacific. I. Winter ocean conditions, Climate change and northern fish populations, edited by: Beamish, R. J., Canadian Spec. Publ. Fish. Aquatic Sci., 121, 81–85, 1995.

Sources of 21st century regional sea-level rise along the coast of northwest Europe

T. Howard[1], A. K. Pardaens[1], J. L. Bamber[2], J. Ridley[1], G. Spada[3], R. T. W. L. Hurkmans[2], J. A. Lowe[1], and D. Vaughan[4]

[1]Met Office Hadley Centre, FitzRoy Road, Exeter, EX1 3PB, UK
[2]Bristol Glaciology Centre, School of Geographical Sciences, University of Bristol, University Road, Bristol, BS8 1SS, UK
[3]Dipartimento di Scienze di Base e Fondamenti, Urbino University "Carlo Bo", Via Santa Chiara, 27, 61029 Urbino (PU), Italy
[4]British Antarctic Survey, Madingley Road, High Cross, Cambridge, CB3 0ET, UK

Correspondence to: T. Howard (tom.howard@metoffice.gov.uk)

Abstract. Changes in both global and regional mean sea level, and changes in the magnitude of extreme flood heights, are the result of a combination of several distinct contributions most, but not all, of which are associated with climate change. These contributions include effects in the solid earth, gravity field, changes in ocean mass due to ice loss from ice sheets and glaciers, thermal expansion, alterations in ocean circulation driven by climate change and changing freshwater fluxes, and the intensity of storm surges. Due to the diverse range of models required to simulate these systems, the contributions to sea-level change have usually been discussed in isolation rather than in one self-consistent assessment. Focusing on the coastline of northwest Europe, we consider all the processes mentioned above and their relative impact on 21st century regional mean sea levels and the 50-year return flood height. As far as possible our projections of change are derived from process-based models forced by the A1B emissions scenario to provide a self-consistent comparison of the contributions. We address uncertainty by considering both a mid-range and an illustrative high-end combination of the different components.

For our mid-range ice loss scenario we find that thermal expansion of seawater is the dominant contributor to change in northwest European sea level by 2100. However, the projected contribution to extreme sea level, due to changes in storminess alone, is in some places significant and comparable to the global mean contribution of thermal expansion. For example, under the A1B emissions scenario, by 2100, change in storminess contributes around 15 cm to the increase in projected height of the 50-year storm surge on the west coast of the Jutland Peninsula, compared with a contribution of around 22 cm due to thermal expansion and a total of 58 cm from all of the contributions we consider. An illustrative combination of our high-end projections suggests increases in the 50-year return level of 86 cm at Sheerness, 95 cm at Roscoff, 106 cm at Esbjerg, and 67cm at Bergen. The notable regional differences between these locations arise primarily from differences in the rates of vertical land movement and changes in storminess.

1 Introduction

In recent years, several scientific communities have put considerable effort into establishing projections of the various individual contributions to global mean sea-level rise. These include: projections of thermal expansion of the oceans (e.g. Kuhlbrot and Gregory, 2012); ice loss from Antarctica (e.g. Vizcaíno et al., 2010, Ligtenberg et al., 2013), Greenland (e.g. Goelzer et al., 2013; Nick et al., 2013; Shannon et al., 2013) and mountain glaciers (e.g. Giesen and Oerlemans, 2013); and from changes in the terrestrial storage of water as ground water and in reservoirs (e.g. Lempérière, 2006).

At intervals, these efforts have been usefully consolidated into "consensus" projections of global mean sea-level rise by the Intergovernmental Panel on Climate Change (IPCC;

e.g. Church et al., 2013), which have guided the global debate on options for adaptation to, and mitigation of, the impacts of climate change. Rather less effort has gone into understanding the complex processes that modify global mean sea level to give regional sea-level changes, which include vertical land movements, changes in the Earth's gravity field, regional variations in thermal expansion, and changes in ocean circulation and changing freshwater fluxes driven by climate change. Very few studies indeed have attempted to produce projections of changes in extreme sea level (i.e. that experienced during occasional damaging storm events), although it is well known that the most acute impacts of sea-level change on coastal communities, assets and environments can be best understood through changes in the severity and/or frequency of such extreme events (Lowe and Gregory, 2005). On a global scale these changes are dominated by changes in mean sea level (e.g. Woodworth and Blackman, 2004), but on a local scale changes in atmospheric circulation, and the consequent impact on the generation of storm surges, may also be significant.

A valuable metric in understanding these potential impacts is the change in height of the highest sea level experienced during a storm over a particular return period. For example, a change in the height of the most extreme storm surge expected once in every 50-year period, the so-called 50-year storm surge, gives a good indication of the increase in flood risk for a particular portion of coastline.

One key study, by Katsman et al. (2011), examined this issue for the Netherlands coastline, and further included the potential that river discharge will exacerbate problems in coastal flood defence. Their study used various sources for the contributions to sea level and storminess. Some contributions were based on process-based models tied to specific emissions scenarios, but others, specifically the contributions from ice loss, were based on extrapolation of short time series of observations. In addition, that study focused on the Dutch coastline, without a consideration of how the various contributions varied across a wider geographical area. A more recent study made a global assessment of regional variations in sea level (Slangen et al., 2012) but excluded a number of the processes mentioned above. In particular, they did not incorporate freshwater forcing, storm surge changes, or process-based model estimates of ice sheet contributions.

Other European studies, based on limited spatial areas, have found that the high natural variability in water level makes detection of trends in the extremes problematic. Haigh et al. (2010), for example, considered data from 18 sites around the English Channel and found changes in extreme water level similar to rates of observed mean sea-level change over the 20th century. Araújo and Pugh (2008) reached a similar conclusion for Newlyn in the UK, based on data from 1915 to 2005. For a recent review of past changes in factors contributing to extreme sea levels in the North Sea, see Weisse et al. (2012).

In this study, we use projections of regional sea-level change arising from 21st century ice loss from ice sheets and glaciers (Spada et al., 2013; Howard et al., 2014), including the effects of changes to the gravity field and Earth rotation. We also include modelled changes in the climate of storm surges produced in support of the UK Climate Projections, Marine and Coastal Projections 2009 (UKCP09; Lowe et al., 2009) – a project to investigate the projected major contributions to 21st century change in coastal flood risk and shelf-sea hydrology around the coast of the United Kingdom. The UKCP09 project included both a (primarily process-based) ensemble projection of sea-level change and a proxy-based assessment of plausible upper limits (the so-called "H++" scenario). Their ensemble analysis gave projections of UK coastal geocentric sea-level rise of 13 to 60 cm over the 21st century (5th to 95th percentile, not including vertical land movement or changes in the climate of storm surges) under the A1B emissions scenario. The corresponding figure from the "upper end" of their H++ scenario range was 190 cm. Here we extend the UKCP09 approach to include more of the northwest (NW) European coast, and we augment the UKCP09 results with new process-based ice-melt projections.

In addition to the sources of sea-level variability listed above, we include projected contributions from thermal expansion, changes in dynamic sea level associated with the freshwater flux from ice melt, changes in dynamic sea level associated with projected global warming under the A1B emissions scenario, and projected changes in relative sea level due to vertical land movements arising from ice loss since the Last Glacial Maximum, termed glacial isostatic adjustment (GIA).

Our aim is to provide a comparison of geographic variations and uncertainties in the different contributions to NW European sea level. To do this we consider two scenarios for 21st century change. The first, which we describe as a "mid-range projection", is taken from close to the middle of the uncertainty range for each process considered. The second, which we describe as a "high-end projection", is taken from the upper end of the distribution. For contributions *TE*, *ADSL* and *SRG* (*TE*: Global mean thermal expansion of the ocean, *ADSL*: Regional changes in dynamic sea level for A1B emissions scenario, excluding the effects of ice loss determined in offline models; *SRG*: Local 21st century change in 50-year return level of storm (skew) surge associated with changes in atmospheric storminess under the A1B scenario; see Table 1 for summary of all such abbreviations), these two scenarios represent the 50th and 95th percentiles respectively of an ensemble of projections to which we fitted a normal distribution (see Sect. 3.3). For the other contributions, we do not have access to a substantial ensemble of projections, and so we cannot specify percentiles. For these, the mid-range projection gives a representative estimate, and the high-end projection gives an illustrative measure of the upper end of the range of possible outcomes, based on a combination of

Table 1. Summary of the contributions to 21st century sea-level change included in this study.

Abbreviation	Description	Source
TE	Global mean thermal expansion of the ocean	Developed herein following Lowe et al. (2009)
TIM	Global mean terrestrial ice melt from offline ice models	Howard et al. (2014)
GCFF	Gravitationally consistent fingerprint of projected future ice melt from offline ice models (fast elastic lithosphere response) expressed as departures from the global mean	SELEN, Spada and Stocchi (2007)
GIA	Ongoing vertical land movement, due to glacial isostatic adjustment (slow viscous lithosphere response)	ICE-5G(VM2) Peltier (2004)
IDSL (DSL due to ice)	Regional changes in dynamic sea level associated with the freshwater forcing from ice loss determined in offline models	HadCM3, Howard et al. (2014) and refs therein
ADSL (DSL directly due to A1B)	Regional changes in dynamic sea level for A1B emissions scenario, excluding the effects of ice loss determined in offline models	Developed herein following Lowe et al. (2009)
SRG	Local 21st century change in 50-year return level of storm (skew) surge associated with changes in atmospheric storminess under the A1B scenario	HadCM3 PPE downscaled with CS3, Lowe et al. (2009) and references therein

our expert judgement (see Sect. 3.2) and analysis of process-based models. The one exception to this is the contribution from GIA, for which we present only a single modelled estimate.

Some of the factors *not* included in this study, but which may contribute to changes in extreme sea level, include changes in terrestrial water storage (e.g. Lettenmaier and Milly, 2009), changes in the climate of waves and swell (e.g. Lowe et al., 2009), changes in the tidal constituents or tidal range (e.g. Mudersbach et al., 2013; Woodworth, 2010; Müller et al., 2011; Pickering et al., 2012) and changes in the seasonal cycle (e.g. Wahl et al., 2014).

2 Nomenclature

We use the abbreviations MR and HE to refer to the representative mid-range and illustrative high-end model projections throughout. "MME" (Multi-Model Ensemble) refers to the ensemble based on some, or all, of the models that contributed data to the Climate Model Intercomparison Project Phase 3 or Phase 5 (henceforth CMIP3, CMIP5 respectively). "PPE" (see Sect. 3.3) refers to the Perturbed Physics Ensemble of HadCM3 variants as described by Murphy et al. (2007).

In order to be unambiguous and concise regarding each of the contributions to 21st century sea-level change considered in this study, they are tabulated (Table 1) together with the abbreviations used for each.

Whilst discussing the impacts on the NW European coast we consider the components listed above evaluated within geographical masks formed by the ocean model grid boxes nearest to the coast. Again to avoid ambiguity we tabulate (Table 2) abbreviations which we use for each component as it is evaluated on a particular coast mask.

3 Models and methods

3.1 Contribution from *TE* and *ADSL*

For the contributions *TE* and *ADSL*, we used 11 model projections from the CMIP3 project (see, http://www-pcmdi.llnl.gov/ipcc/about_ipcc.php). This project produced model data that were used in the IPCC Fourth Assessment Report (Meehl et al., 2007; henceforth AR4).

We used 11 of the CMIP3 models. These were selected from the full set of CMIP3 models on the basis that DSL projections under the A1B scenario and the accompanying parallel sections of simulations with fixed greenhouse gas concentration were both available to us (together with global thermal expansion). Using this subset of models therefore allowed a common approach to be taken in removing model drift in the DSL pattern changes.

Table 2. Definitions and abbreviations used for those components which are evaluated on a local coastal mask.

Abbreviation	Description
Mn_ADSL	*ADSL* averaged over the NW Europe coastal mask shown in Fig. 2
Mn_IDSL	*IDSL* averaged over the NW Europe coastal mask shown in Fig. 2
Mn_GCFF	*GCFF* averaged over the NW Europe coastal mask shown in Fig. 2
Max_GCFF	Maximum of *GCFF* over the NW Europe coastal mask shown in Fig. 2
Mn_SRG	*SRG* averaged over the high-resolution coastal mask shown in Fig. 3
Max_SRG	Maximum of *SRG* over the high-resolution coastal mask shown in Fig. 3

A common land mask was applied to each of the model *(TE + ADSL)* projections and a NW European coastally averaged *(TE + Mn_ADSL)* then calculated for each model (the masked region is shown in Fig. 2; the values are very similar when using a masked region of half this width), where *Mn_ADSL* is *ADSL* averaged over the masked region. We also used *TE* data for each of the 11 models to partition the sum *(TE + Mn_ADSL)* into its components *TE* and *Mn_ADSL*. This also enabled us to check our distribution of 11 *TE* values against the two percentiles reported in AR4, and to make a basic comparison with the corresponding values in AR5 (see Sect. 4.1).

3.2 Contribution from *TIM and GCFF*

We derived projections of two contributions, *TIM* and *GCFF*, from two time series which are the same as those used in Spada et al. (2013). We used expert judgement in the selection of these two time series which we regard as a representative mid-range (MR) and an illustrative high-end (HE) time series taken from a plausible perturbed ice parameter ensemble. Some of the simulations from the full ensemble were deemed unrealistic based on expert judgement. We partitioned each time series into a global mean (*TIM*[1]) and spatially varying component (*GCFF*). We selected data for the 2080–2099 period, calculating changes with respect to 1992 (see Howard et al., 2014 for further details). This model output was provided on a grid (described by Spada and Stocchi, 2007) with lower resolution than our surge model grid (shown, for example, in Fig. 3). However, this lower-resolution grid does not overlap all the specific coastal sites chosen for this study, and so for each of these sites, we interpolated the low-resolution grid using values from grid points to the east and west (for example around the British Isles). For grid points on the high-resolution grid with no low-resolution point to the east, we used the value for the nearest low-resolution value to the west. This is justified by the fact that, in this region, there is only a small east–west gradient, compared to the dominant north–south gradient in the *GCFF* component.

[1]This is the global mean contribution due to a change in mass of the oceans, sometimes referred to as the eustatic contribution but more correctly termed the barystatic contribution.

3.3 Contribution from *SRG*

There is good evidence (e.g. Howard et al., 2010; Sterl et al., 2009) that changes in storm surge contribution to sea level can be added linearly to mean sea-level change and so, following Lowe et al. (2009), that is the approach we took in this study.

Our approach for deriving projections of the contribution of changes in storm surge to extreme sea level also follows that of Lowe et al. (2009). The approach is based on a perturbed atmospheric physics ensemble (PPE) of the HadCM3 global climate model, downscaled by a matching ensemble of perturbed HadCM3 variants in atmospheric regional climate model (RCM) format.

We acknowledge that the uncertainty range of our PPE will not in general encompass the structural uncertainty of the MME that we were able to use for the *TE* and *ADSL* contributions. Nevertheless, the range of climate sensitivity exhibited by the PPE is comparable to that of the MME (Collins et al., 2011).

The surface wind and atmospheric pressure fields output by the RCMs were used to drive a barotropic surge model, CS3 (Flather et al., 1998), in combined surge-and-tide mode. A parallel simulation in tide-only mode allows extraction of the non-tidal surge residual or the skew surge. Changing bathymetry due to increasing mean sea level is not included in the surge model but is added linearly in a separate step (this does not include changing coastline due, for example, to newly inundated areas; the land/sea mask does not change). The model integration ran from 1850 to 2100. The domains of CS3 and the RCM are shown in Fig. 1. CS3 has been extensively validated for the present-day climate (see for example, Lowe et al., 2009).

As in Lowe et al. (2009), we present projections of the "skew surge", which is a good measure of the impact of a surge. It is defined as the difference between the height of the highest water level predicted by the model and the nearest-in-time astronomical high tide at the same point. A statistical generalised extreme-value model (Coles, 2001), with linear time trend in the location and scale parameter, was fitted to the five largest independent skew surges in each year (see for example Tawn, 1992; Coles, 2001). From this, we diagnosed a maximum likelihood estimator of the trend in the 50-year return level. Thus our approach allows for the possibility of a linear time trend in the extremes (forced by the changing atmospheric storminess). The magnitude of this trend compared with the corresponding uncertainty gives a measure of the statistical significance.

Figure 1. Domain of the storm surge model, CS3 (shaded). The outer frame indicates the domain of the regional atmospheric climate model.

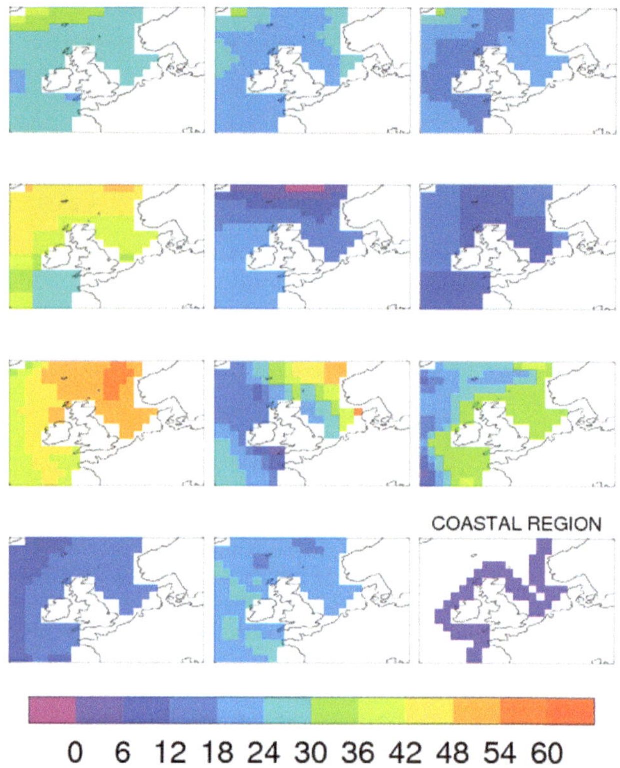

Figure 2. *(TE + ADSL)* in centimetres for the NW European region derived from 11 CMIP3 models. Projections are for the SRES A1B scenario and for the period 2080–2099 relative to 1980–1999. The lower right frame shows the mask used to extract *Mn_ADSL*.

Again, following the approach taken by Lowe et al. (2009), we treated the ensemble member projections as alternative, but equally valid, realisations of a naturally variable future storm surge climate. Under this assumption, it is reasonable to take the mean trend in the ensemble as a representative central projection, and the standard deviation of the ensemble as a measure of spread. Assuming that the 11 trends in the extremes are normally distributed, we could then estimate the 95th percentile of the trend, independently at each location. This assumption is reasonable, even though we do not expect the *extremes themselves* to be normally distributed. We acknowledge that the approach is limited by the fact that we disregard the spatial dependence in the trends, but argue that it is still reasonable because we make no claims about the joint probability of the spatial patterns.

4 Results

4.1 Contribution from *TE* and *ADSL*

Patterns of combined *TE* + *ADSL* for 11 members of the CMIP3 MME are shown in Fig. 2, along with the coast mask used to evaluate the local spatial mean (i.e. a spatial mean representative of the NW European coast). The 11 members that we used were: ccma_cgcm3_1_t47, gfdl_cm2_0, gfdl_cm2_1, giss_aom, giss_model_e_h, giss_model_e_r, miroc3_2medres, miub_echo_g, mpi_echam5, mri_cgcm2_3_2a and ukmo_hadcm3. The range of projections of *TE* reported in AR4 for A1B and for 2090–2099 relative to 1980–1999, were 13 cm (5th percentile) and 32 cm (95th percentile), giving a mean of 22.5 ± 5.8 cm, where the uncertainty is expressed as a standard deviation assuming a normal distribution. The corresponding value from our 11-member ensemble is 21.4 ± 5.5 cm. For this contribution, the agreement is very good, and so it makes little difference whether we adjust our estimates to take account of the range quoted in AR4 or not; however, we have made this adjustment in order to take advantage of the additional information used in AR4 (the *TE* range quoted in AR4 comes from more than 11 models). This ensured that our approach, applied to the NW European coastline, was consistent with the methodology used previously (Lowe et al., 2009). The adjustment consists of replacing the 11-member mean and variance of *TE* (21.4 cm and 30.25 cm^2) with the corresponding values from AR4 (22.5 cm and 33.64 cm^2). We also made a consistent adjustment to the mean and variance of the sum *(TE + Mn_ADSL)*, whilst preserving the covariance as given by the 11 ensemble members. Although the agreement between the 11-member *TE* and the range of projections of *TE* reported in AR4 is very good, we acknowledge that with the available data we cannot know how well our 11-member assessment of the *Mn_ADSL* contribution would match that of the full MME, if suitable data were available to calculate it. For each of the 11 members, the partitioning of the sum *(TE + Mn_ADSL)* into its components *TE* and *Mn_ADSL* is shown by the dotted points in Fig. 4.

A direct comparison with CMIP5 results is not possible due to the different approach to emissions in AR5 compared to AR4, however we note that both mean and the spread of

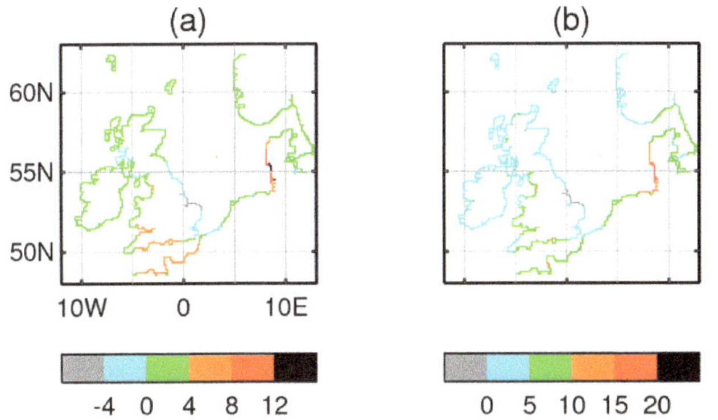

Figure 3. Projected contribution in centimetres of the change in storm surge climate to 21st century change in the 50-year return storm surge for **(a)** central estimate, **(b)** 95th percentile. Both are given in terms of the skew surge above astronomical high tide, and represent the contribution due to atmospheric storminess changes only.

(*TE* + *Mn_ADSL*) are comparable to the values for the NW European region given in AR5 under RCP4.5 (Church et al., 2014; their Fig. 13.16).

4.2 Contribution from *TIM and GCFF*

Patterns of these contributions are not shown here because they are similar to those presented previously (Spada et al., 2013). The *TIM* contributions are 20 cm (mid-range) and 50 cm (high-end).

4.3 Contribution from *SRG*

The projected component of extreme water level changes due only to changes in storm surge climate is shown in Fig. 3. We note that the pattern is broadly consistent with the expected increase in the frequency and/or intensity of westerly or southwesterly winds. This is addressed further in the discussion section below.

5 Comparison and addition of contributions

The distribution of values, shown in Fig. 4, allows us to explore the relative significance of the various contributions to 21st century extreme sea-level change around NW Europe.

The contributions shown in Fig. 4 come from very different types of model and in general their error covariances are not known (the exception is *TE* and *Mn_ADSL*, see below). Furthermore, the HE TIM member is selected as an illustrative high-end realisation with an unquantified probability of exceedance. Nevertheless, previous studies (e.g. Sterl et al., 2009; Howard et al., 2010) have shown that it is reasonable to combine surge changes additively with mean sea-level change. Similarly Howard et al. (2014) show that for their model *IDSL* and *ADSL* changes appear to be additive.

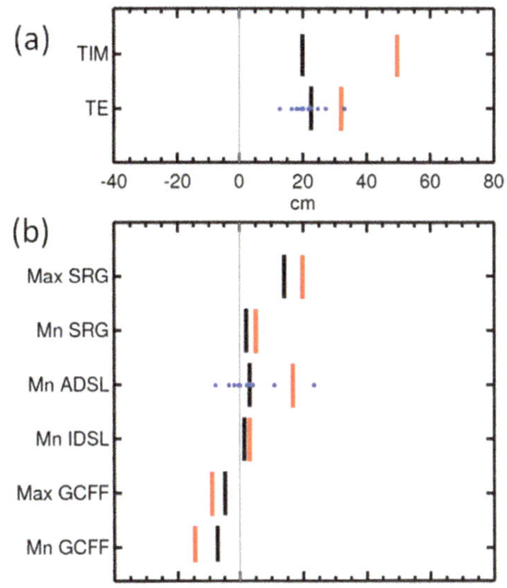

Figure 4. Comparison of contributions to the 21st century change in 50-year storm surge height around NW Europe (centimetres). **(a)** Global mean contributions, **(b)** local contributions. Black bars indicate the representative mid-range, and red, the illustrative high-end contributions. Blue dots show the individual ensemble members where these are available. The abbreviations used in the figure are those given in Tables 1 and 2.

Therefore we present an illustrative addition of HE contributions to the change in the height of the 50-year storm surge, for seven locations around NW Europe in Fig. 5. The locations are Aberdeen, Sheerness, Cork Harbour, Roscoff, The Hague, Esbjerg and Bergen. Again the abbreviations used for the contributions are those of Tables 1 and 2. Note that the HE contribution from (*TE* + *Mn_ADSL*) is *not* simply the sum of the HE *TE* and HE *Mn_ADSL* contributions shown in Fig. 4 because in the case of these two contributions only we *do* know about their covariance and we can select a HE value directly from the distribution of their sum.

6 Discussion

Under our representative mid-range projection we found that *TE* and *TIM* are the largest contributors to 21st century extreme sea-level change around NW Europe (Figs. 4 and 5). However, the largest uncertainties arise from *TIM*, and *TIM* made the largest contribution to the illustrative high-end projection. The uncertainty due to *Mn_ADSL* was also large. Within the limitations of our method (i.e. the use of only one model formulation with two different meltwater forcings), we found both contribution and uncertainty to be very small for *Mn_IDSL* (MR: 1.4 cm; HE: 3 cm). The *SRG* contribution is important in some locations, though within the limitations of our *SRG* modelling approach, it appeared to be relatively well-constrained (see Sect. 6.1). Spatial variations due to *SRG* and *GIA* are both considerable.

Table 3. High-end 21st century sea-level change figures taken from Lowe et al. (2009), Katsman et al. (2011) and this study. The Lowe et al. figures are the top of the range of their "H++" scenario, with "Local" referring to the coast of the UK. The Katsman figures are the top of the range of their high-end scenario, with "Local" referring to the coast of the Netherlands. For this study the "Local" figure refers to The Hague. Figures for Lowe et al. (2009) and for this study exclude both *GIA* and *SRG* because these were not treated in the same way in all three studies.

Lowe et al. "H++"	Global	2.5 m
	Local (UK)	1.9 m
Katsman et al. A1FI	Global	1.15 m
	Local (Netherlands)	1.05 m
This study A1B, HE	Global	0.82 m
	Local (The Hague)	0.84 m

In Table 3 we compare illustrative high-end projections from our study with two other studies addressing sea-level change around the coast of NW Europe. In order to make a like-for-like comparison we exclude contributions from *SRG* and *GIA* in Table 3 because these contributions were addressed differently in the three studies. The IPCC AR5 (Church et al., 2013) note that there is low agreement within semi-empirical model projections and no consensus in the scientific community about their reliability. Thus we have not quoted projections from such models here. However, for a summary of such projections in comparison to process-based projections the reader may see for example Nicholls et al. (2011).

Lowe et al. (2009) provide a mostly process-based projection[2] of sea-level rise for the coast of the UK (with which our *TE, GIA, ADSL* and *SRG* contributions are all consistent). However, Lowe et al. (2009) also offer a (partly proxy-based) projection for vulnerability testing, their so-called "H++" scenario range. It is informative to compare this scenario with our results (Table 3), although the H++ range was intended to look at plausible upper limits based partly on proxy data and it should be borne in mind that recent observations and model projections do not provide any evidence to suggest that such high levels of sea-level rise will occur within the 21st century. Church et al. (2013), acknowledging that the potential additional contribution to global mean sea-level rise from the collapse of marine-based sectors of the Antarctic Ice Sheet cannot be precisely quantified, judge that there is "medium confidence that it would not exceed several tenths of a metre of sea level rise during the 21st century".

Katsman et al. (2011) developed a plausible scenario of high-end 21st century sea-level rise for both the global mean and for the coast of the Netherlands under the A1FI emissions scenario. Owing to the different approach and different emissions scenario, our results are not directly comparable but comparison for the global and regional total is possible

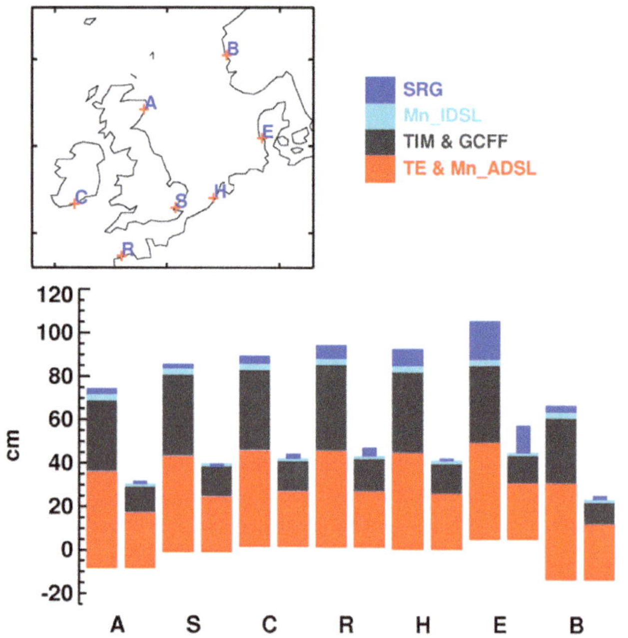

Figure 5. Addition of illustrative high-end (and representative mid-range) projections of contributions to 21st century change in the height of the 50-year storm surge for seven locations around NW Europe. The locations are Aberdeen, (A); Sheerness, (S); Cork Harbour (C); Roscoff, (R); The Hague, (H); Esbjerg (E) and Bergen (B). For each location, the larger (left-hand) bar shows the high-end estimate, and the smaller (right-hand) bar shows the mid-range estimate. The projected contribution from *GIA* is shown as an offset to the zero of each bar. The mid-range *SRG* projection at Sheerness is negative, and so that this can be seen, the mid-range *SRG* projections are shown as half-width bars. Further details are given in the main text.

(Table 3). The differences between the Katsman figures and this study are mostly due to the larger projected contribution from the Antarctic ice sheet in the "severe" scenario used in Katsman et al. (2011). However, it is also noticeable that the local sea-level rise is greater than the global mean in our study, but smaller in the other two. The main reason for this is the smaller contribution from the Greenland ice sheet in our projections (see Howard et al., 2014). (The fingerprint of Greenland melt is characterised by less-than-global-mean increases around NW Europe).

Our *TE* and *Mn_ADSL* contributions are taken from the CMIP3 models and have been discussed elsewhere (e.g. Pardaens et al., 2011; Landerer et al., 2007; Lowe and Gregory, 2006; Suzuki et al., 2005). Our *TIM* and *Mn_GCFF* are based on work described in Spada et al. (2013). Our *Mn_IDSL* contribution is discussed extensively by Howard et al. (2014). Consequently, we focus here on the *SRG* contribution for NW Europe.

[2]Their projection of ice sheet behaviour was not process based.

6.1 Changes in surges

Following Lowe et al. (2009) we assessed the local statistical significance of the *SRG* contribution by comparing the PPE mean with the PPE spread. We considered a range of geographically distributed locations (shown in Fig. 5). We found that the *SRG* contribution was positive and locally statistically significant in this sense at Aberdeen, Cork Harbour, Roscoff, Esbjerg and Bergen (p value less than 0.05, one-sided test).

The changes in *TE* and *ADSL* (Fig. 2) were based on the CMIP3 MME, whereas the surge changes were based on the more limited PPE. Despite this, even under the limited range of changes in storminess emerging from the PPE, we find that surge changes are a significant contributor in some locations. Surge changes are critically dependent on changes in atmospheric storminess and it should be borne in mind that previous work has shown that the range of storminess changes emerging from the PPE in our region of concern are smaller than those emerging from the CMIP3 MME (Lowe et al., 2009). As a consequence, the surge changes could be more important than implied by Fig. 5 under a more comprehensive multi-model assessment (see Lowe et al., 2009 chapter 4, for their discussion of the "H++" approach).

A general pattern of a projected increase in storm surge extremes in the eastern North Sea, particularly on the west-facing coast of the Jutland Peninsula (Fig. 3), in contrast to a much smaller signal on the east coast of the UK, can be seen in projections made by a number of previous studies (based on emissions scenarios including A1B, A2 and B2), e.g. Gaslikova et al. (2013), Debernard and Røed (2008), Woth (2006) and Lowe and Gregory (2005, in particular their projection for the B2 scenario). These previous studies, however, used only single realisations or small ensembles, typically combined with a comparison of two 30-year time slices representing the projection and the present day respectively. Such an approach is susceptible to aliasing information from natural multi-decadal variations in storminess over the region (see for example Jenkins et al., 2008, in particular their Fig. 1.14). Some recent studies (e.g. Sterl et al., 2009; Lowe et al., 2009) have addressed this issue by using larger model ensembles and/or considering the linear change in storm surge characteristics over the full period of the simulation, typically 150 years. We use both approaches here. Consistent with Sterl et al. (2009) we find no significant increase in storm surge characteristics on the Dutch coast. As noted by Sterl et al. (2009) this in turn is not inconsistent with an increase in westerly or southwesterly winds, which do not tend to produce severe surges on that coast.

6.2 Surge change relationship with wind changes

Sterl et al. (2009) projected a small increase in maximum wind speeds over the southern North Sea over the 21st century, noting that since this was due to winds from the southwest it may not lead to a significant increase in surge heights along the Dutch coast. Our model results support this conclusion for the Dutch coast; for example, we find a small, statistically insignificant increase in surge amplitude at The Hague. We also note that the larger (and in some areas statistically significant) trends in the ensemble-mean surge projections are mostly on west or southwest facing coasts, broadly consistent with an increase in the westerly or southwesterly component of wind. However, there is no strong signal of winter mean wind speed change in our PPE (see for example Brown et al. (2009), particularly their Fig. 10 referring to the UK or their Fig. 13a which refers to the entire RCM domain). In validating their driving climate models Sterl et al. (2009) and Howard et al. (2010) consider the annual maximum daily mean wind speed at 5° E, 55° N. We looked at our model relationship between this quantity and the changing surge levels at Esbjerg, where some of the largest changes are projected. We find a statistically significant but weak correlation between modelled annual maximum daily mean wind speed at 5° E, 55° N and the modelled annual maximum skew surge at Esbjerg (Pearson's $r = 0.2$, p value less than 0.001 over the 150 years of model simulation, using data from all 11 members of the PPE) and a weak correlation between the *century-scale change* in the 50 year return level of daily mean wind speed at 5° E, 55° N (11 values, one from each PPE ensemble member) and the *century-scale change* in the 50-year return level of skew surge at Esbjerg (11 values, one from each PPE ensemble member; $r = 0.37$, p value 0.26). Our ensemble mean projected century-scale change in the 50-year return level of daily mean wind speed at 5° E, 55° N is of the order of 10 cm per second over the 21st century (90 % confidence interval of [−16, 37] centimetres per second per century). Although this is *consistent* with the increase in surge amplitude at The Hague, in view of the weak correlation it is of limited utility in *explaining* the increase. Many other factors need to be considered: wind direction, duration, spatial extent and storm track, for example. Such an analysis is beyond the scope of the present work.

Gaslikova et al. (2013) report somewhat similar patterns of projected storm surge change to ours under the A1B emissions scenario and their changes appear to be related to their reported increase in both the annual 99th percentile of wind speed and the frequency of occurrence of westerly or southwesterly winds with speeds greater than 17.2 m s^{-1}, both change assessments being based on the 30-year time slice approach. De Winter et al. (2013) consider projections of 12 members of the CMIP5 ensemble and they also find an indication of annual extreme winds over the North Sea coming more often from western directions.

7 Conclusions

In this study, we compared representative mid-range and illustrative high-end projections of 21st century contributions to extreme sea-level change around NW Europe. These contributions arise from changes in atmospheric storminess, terrestrial ice melt from ice sheets and glaciers, thermal expansion of the oceans, changes in dynamic sea level associated with the projected ice loss, and with projected climate change, and effects of glacial isostatic adjustment associated with loss of ice since the Last Glacial Maximum. As far as possible we have used projections based on the SRES A1B emissions scenario to provide a self-consistent, process-based model-derived set of projections of mean sea-level rise and changes in extreme surge levels expressed as the change in the height of the 50 year flood. In regard to sea-level rise, the A1B scenario is broadly similar to both representative concentration pathways RCP4.5 and RCP6.0, which differ only in the latter half of the 21st century.

We find that the largest sources of uncertainty within these single-emission-scenario simulations are in the contribution from dynamic sea-level changes associated with the direct warming effects of radiative forcing, and the global mean contribution of ice loss. Our approach does have some limitations; for example, our analysis of the changing distribution of surges is based on a perturbed-physics ensemble generated by a single model, rather than a multi-model ensemble. Nevertheless, we find that the change in the statistical distribution of surges (associated with changes in their atmospheric forcing only) contributes significantly to the spatial variations in the projected changes of the 50-year flood height, and in some locations may make a significant contribution, particularly along parts of the European mainland coast not considered by Lowe et al. (2009). For example, our central estimate of this contribution to the rise in the 21st century extreme sea level on the west coast of the Jutland Peninsula is 15 cm, which can be compared with a central estimate of around 22 cm from global mean thermal expansion.

Acknowledgements. This work was supported by funding from the ice2sea programme from the European Union 7th Framework Programme, grant number 226375. Ice2sea contribution number #164. This work was also supported by the Joint UK DECC/Defra Met Office Hadley Centre Climate Programme (GA01101). We acknowledge the significant contribution made by Kevin Horsburgh and his colleagues at the National Oceanography Centre. We thank the two anonymous reviewers whose comments helped to improve this paper.

Edited by: A. Sterl

References

Araujo, I. B. and Pugh, D. T.: Sea levels at Newlyn 1915–2005: analysis of trends for future flooding risks, J. Coastal Res., 24, 203–212, doi:10.2112/06-0785.1, 2008.

Brown, S., Boorman, P., McDonald, R., and Murphy, J.: Use and interpretation of surface wind projections from the 11-member Met Office Regional Climate Model ensemble, UKCP09 Technical Note. Met Office Hadley Centre, Exeter, UK, available at: http://ukclimateprojections.defra.gov.uk (last access: 12 December 2013), 2009.

Church, J. A., Clark, P. U., Cazenave, A., Gregory, J. M., Jevrejeva, S., Levermann, A., Merrifield, M. A., Milne, G. A., Nerem, R. S., Nunn, P. D., Payne, A. J., Pfeffer, W. T., Stammer, D., and Unnikrishnan, A. S.: Sea level change, in: Climate Change 2013: The Physical Science Basis Contribution of Working Group I to the Fifth Assessment Report of the Intergovernmental Panel on Climate Change, edited by: Stocker, T. F., Qin, D., Plattner, G.-K., Tignor, M., Allen, S. K., Boschung, J., Nauels, A., Xia, Y., Bex, V., and Midgley, P. M., Cambridge University Press, Cambridge, UK and New York, NY, USA, in press, 2013.

Coles, S.: An introduction to statistical modeling of extreme values, Springer, 208 pp., 2001.

Collins, M, Booth, B. B. B., Bhaskaran, B, Harris, G. R., Murphy, J. M., Sexton, D. M. H., and Webb, M. J.: Climate model errors, feedbacks and forcings: a comparison of perturbed physics and multi-model ensembles, Clim. Dynam., 36, 1737–1766, doi:10.1007/s00382-010-0808-0, 2011.

Debernard, J. B. and Roed, L. P.: Future wind, wave and storm surge climate in the Northern Seas: a revisit, Tellus A, 60, 427–438, doi:10.1111/j.1600-0870.2008.00312.x, 2008.

de Winter, R. C., Sterl, A., and Ruessink, B. G.: Wind extremes in the North Sea Basin under climate change: An ensemble study of 12 CMIP5 GCMs, J. Geophys. Res.-Atmos., 118, 1601–1612, doi:10.1002/jgrd.50147, 2013.

Flather, R. A., Smith, J. A., Richards, J. D., Bell, C., and Blackman, D. L.: Direct estimates of extreme storm surge elevations from a 40 yr numerical model simulations and from observations, The Global Atmos. Ocean Syst., 6, 165–176, 1998.

Gaslikova, L., Grabemann, I., and Groll, N.: Changes in North Sea storm surge conditions for four transient future climate realizations, Nat. Hazards, 66, 1501–1518, doi:10.1007/s11069-012-0279-1, 2013.

Giesen, R. H. and Oerlemans, J.: Climate-model induced differences in the 21st century global and regional glacier contributions to sea-level rise, Clim. Dynam., 41, 1–18, 2013.

Goelzer, H., Huybrechts, P., Fürst, J. J., Nick, F. M., Andersen, M. L., Edwards, T. L., Fettweis, X., Payne, A. J., and Shannon, S. R.: Sensitivity of Greenland ice sheet projections to model formulations, J. Glaciol., 59, 733–749, 2013.

Haigh, I., Nicholls, R., and Wells, N.: Assessing changes in extreme sea levels: application to the English Channel, 1900–2006, Cont. Shelf. Res., 30, 1042–1055, doi:10.1016/j.csr.2010.02.002, 2010.

Howard, T., Lowe, J., and Horsburgh, K.: Interpreting century-scale changes in Southern North Sea storm surge climate derived from coupled model simulations, J. Climate, 23, 6234–6247, doi:10.1175/2010jcli3520.1, 2010.

Howard, T., Ridley, J., Pardaens, A. K., Hurkmans, R. T. W. L., Payne, A. J., Giesen, R. H., Lowe, J. A., Bamber, J. L., Edwards, T. L., and Oerlemans, J.: The land-ice contribution to 21st-century dynamic sea level rise, Ocean Sci., 10, 485–500, doi:10.5194/os-10-485-2014, 2014.

Jenkins, G. J., Perry, M. C., and Prior, M. J.: The climate of the UK and recent trends. Met Office Hadley Centre, Exeter, UK, available at: http://ukclimateprojections.defra.gov.uk (last access: 12 December 2013), 2008.

Katsman, C. A., Sterl, A., Beersma, J. J., van den Brink, H. W., Church, J. A., Hazeleger, W., Kopp, R. E., Kroon, D., Kwadijk, J., Lammersen, R., Lowe, J., Oppenheimer, M., Plag, H. P., Ridley, J., von Storch, H., Vaughan, D. G., Vellinga, P., Vermeersen, L. L. A., van de Wal, R. S. W., and Weisse, R.: Exploring high-end scenarios for local sea level rise to develop flood protection strategies for a low-lying delta-the Netherlands as an example, Climatic Change, 109, 617–645, doi:10.1007/s10584-011-0037-5, 2011.

Kuhlbrodt, T. and Gregory, J. M.: Ocean heat uptake and its consequences for the magnitude of sea level rise and climate change, Geophys. Res. Lett., 39, L18608, doi:10.1029/2012GL052952, 2012.

Landerer, F. W., Jungclaus, J. H., and Marotzke, J.: Regional Dynamic and Steric Sea Level Change in Response to the IPCC-A1B Scenario, J. Phys. Oceanogr., 37, 296–312, doi:10.1175/JPO3013.1, 2007.

Lempérière, F.: The role of dams in the XXI century: achieving a sustainable development target, International Journal on Hydropower and Dams, 13, 99–108, 2006.

Lettenmaier, D. P. and Milly, P. C. D.: Land waters and sea level, Nature Geosci., 2, 452–454, doi:10.1038/ngeo567, 2009.

Ligtenberg, S. R. M., van de Berg, W. J., van den Broeke, M. R., Rae, J. G. L., and van Meijgaard, E.: Future surface mass balance of the Antarctic ice sheet and its influence on sea level change, simulated by a regional atmospheric climate model, Clim. Dynam., 41, 867–884. doi:10.1007/s00382-013-1749-1, 2013.

Lowe, J. A. and Gregory, J. M.: The effects of climate change on storm surges around the UK, Philos. T. R. Soc., 363, 1313–1328, doi:10.1098/rsta.2005.1570, 2005.

Lowe, J. A. and Gregory, J. M.: Understanding projections of sea level rise in a Hadley Centre coupled climate model. J. Geophys. Res., 111, C11014, doi:10.1029/2005JC003421, 2006.

Lowe, J. A., Howard, T. P., Pardaens, A., Tinker, J., Holt, J., Wakelin, S., Milne, G., Leake J, Wolf, J., Horsburgh, K., Reeder, T., Jenkins, G., Ridley, J., Dye, S., and Bradley, S.: UK Climate Projections science report: Marine and coastal projections, Met Office Hadley Centre, Exeter, UK, available at: http://ukclimateprojections.defra.gov.uk (last access: 12 December 2013), 2009.

Meehl, G. A., Stocker, T. F., Collins, W. D., Friedlingstein, P., Gaye, A. T., Gregory, J. M., Kitoh, A., Knutti, R., Murphy, J. M., Noda, A., Raper, S. C. B., Watterson, I. G., Weaver, A. J., and Zhao, Z.: Global climate projections, in: Climate change 2007: The Physical Science Basis, Contribution of Working Group I to the Fourth Assessment Report of the Intergovernmental Panel on Climate Change, edited by: Solomon, S., Qin, D., Manning, M., Chen Z, Marquis, M., Averyt, K. B., Tignor, M., and Miller, H. L., Cambridge University Press, London, 2007.

Mudersbach, C., Wahl, T., Haigh, I. D., and Jensen, J.: Trends in high sea levels of German North Sea gauges compared to regional mean sea level changes, Cont. Shelf Res., 65, 111–120, doi:10.1016/j.csr.2013.06.016, 2013.

Muller, M., Arbic, B. K., and Mitrovica, J. X.: Secular trends in ocean tides: Observations and model results, J. Geophys. Res.-Oceans, 116, C05013, doi:10.1029/2010JC006387, 2011.

Murphy, J. M., Booth, B. B. B., Collins, M., Harris, G. R., Sexton, D. M. H., and Webb, M. J.: A methodology for probabilistic predictions of regional climate change from perturbed physics ensembles, Philos. T. R. Soc. A, 365, 1993–2028, doi:10.1098/rsta.2007.2077, 2007.

Nicholls, R. J., Marinova, N., Lowe, J. A., Brown, S., Vellinga, P., De Gusmao, D., Hinkel, J., and Tol, R. S. J.: Sea-level rise and its possible impacts given a "beyond 4 °C world" in the twenty-first century, Philos. T. R. Soc. A, 369, 161–181, doi:10.1098/rsta.2010.0291, 2011.

Nick, F. M., Vieli, A., Andersen, M. L., Joughin, I., Payne, A., Edwards, T. L., Pattyn, F., and van de Wal, R. S. W.: Future sea-level rise from Greenland's main outlet glaciers in a warming climate, Nature, 497, 235–238, doi:10.1038/nature12068, 2013.

Pardaens, A. K., Gregory, J. M., and Lowe, J. A.: A model study of factors influencing projected changes in regional sea level over the twenty-first century. Climate Dynamics 36, 2015–2033, doi:10.1007/s00382-009-0738-x, 2011.

Peltier, W. R.: Global glacial isostasy and the surface of the ice-age earth: The ice-5G (VM2) model and grace, Annu. Rev. Earth Pl. Sc., 32, 111–149, doi:10.1146/annurev.earth.32.082503.144359, 2004.

Pickering, M. D., Wells, N. C., Horsburgh, K. J., and Green, J. A. M.: The impact of future sea-level rise on the European Shelf tides, Cont. Shelf Res., 35, 1–15, doi:10.1016/j.csr.2011.11.011, 2012.

Shannon, S. R., Payne, A. J., Bartholomew, I. D., van den Broeke, M. R., Edwards, T. L., Fettweis, X., Gagliardini, O., Gillet-Chaulet, F., Goelzer, H., Hoffman, M. J., Huybrechts, P., Mair, D. W. F., Nienow, P. W., Perego, M., Price, S. F., Smeets, C. J. P. P., Sole, A. J., van de Wal, R. S. W., and Zwinger, T: Enhanced basal lubrication and the contribution of the Greenland ice sheet to future sea-level rise, Proc. Natl. Acad. Sci. USA, 110, 14156 14161, doi:10.1073/pnas.1212647110, 2013.

Slangen, A. B. A., Katsman, C. A., van de Wal, R. S. W., Vermeersen, L. L. A., and Riva, R. E. M.: Towards regional projections of twenty-first century sea-level change based on IPCC SRES scenarios, Clim. Dynam., 38, 1191–1209, doi:10.1007/s00382-011-1057-6, 2012.

Spada, G. and Stocchi, P.: SELEN: a Fortran 90 program for solving the "sea-level equation", Comput. Geosci., 33, 538–562, doi:10.1016/j.cageo.2006.08.006, 2007.

Spada, G., Bamber, J. L., and Huurkmans, R.: The gravitationally-consistent fingerprint of future terrestrial ice loss, Geophys. Res. Lett., 40, 482–486, doi:10.1029/2012GL053000, 2013.

Sterl, A., van den Brink, H., de Vries, H., Haarsma, R., and van Meijgaard, E.: An ensemble study of extreme storm surge related water levels in the North Sea in a changing climate, Ocean Sci., 5, 369–378, doi:10.5194/os-5-369-2009, 2009.

Suzuki, T., Hasumi, H., Sakamoto, T. T., Nishimura, T., Abe-Ouchi, A., Segawa, T., Okada, N., Oka, A., and Emori, S.: Projection of future sea level and its variability in a high-resolution climate model: Ocean processes and Greenland and Antarctic ice-melt contributions, Geophys. Res. Lett., 32, L19706, doi:10.1029/2005GL023677, 2005.

Tawn, J.: Estimating Probabilities of Extreme Sea Levels, Appl. Stat., 41, 77–93, 1992.

Vizcaíno, M., Mikolajewicz, U., Jungclaus, J., and Schurgers, G.: Climate modification by future ice sheet changes and consequences for ice sheet mass balance, Clim. Dynam., 34, 301–324, 2010.

Wahl, T. F., Calafat, M., and Luther, M. E.: Rapid changes in the seasonal sea level cycle along the US Gulf coast from the late 20th century, Geophys. Res. Lett., 41, 491–498, doi:10.1002/2013GL058777, 2014.

Weisse, R., von Storch, H., Niemeyer, H. D., and Knaack, H.: Changing North Sea storm surge climate: an increasing hazard?, Ocean Coast. Manage., 68, 58–68, doi:10.1016/j.ocecoaman.2011.09.005, 2012.

Woodworth, P. L.: A survey of recent changes in the main components of the ocean tide, Cont. Shelf Res., 30, 1680–1691, doi:10.1016/j.csr.2010.07.002, 2010.

Woodworth, P. L. and Blackman, D. L.: Evidence for systematic changes in extreme high waters since the mid-1970s, J. Climate, 17, 1190–1197, doi:10.1175/1520-0442(2004)017<1190:efscie>2.0.CO;2, 2004.

Woth, K., Weisse, R., and von Storch, H.: Climate change and North Sea storm surge extremes: an ensemble study of storm surge extremes expected in a changed climate projected by four different regional climate models, Ocean Dynam., 56, 3–15, doi:10.1007/s10236-005-0024-3, 2006.

Permissions

All chapters in this book were first published in OSD, by Copernicus Publications; hereby published with permission under the Creative Commons Attribution License or equivalent. Every chapter published in this book has been scrutinized by our experts. Their significance has been extensively debated. The topics covered herein carry significant findings which will fuel the growth of the discipline. They may even be implemented as practical applications or may be referred to as a beginning point for another development.

The contributors of this book come from diverse backgrounds, making this book a truly international effort. This book will bring forth new frontiers with its revolutionizing research information and detailed analysis of the nascent developments around the world.

We would like to thank all the contributing authors for lending their expertise to make the book truly unique. They have played a crucial role in the development of this book. Without their invaluable contributions this book wouldn't have been possible. They have made vital efforts to compile up to date information on the varied aspects of this subject to make this book a valuable addition to the collection of many professionals and students.

This book was conceptualized with the vision of imparting up-to-date information and advanced data in this field. To ensure the same, a matchless editorial board was set up. Every individual on the board went through rigorous rounds of assessment to prove their worth. After which they invested a large part of their time researching and compiling the most relevant data for our readers.

The editorial board has been involved in producing this book since its inception. They have spent rigorous hours researching and exploring the diverse topics which have resulted in the successful publishing of this book. They have passed on their knowledge of decades through this book. To expedite this challenging task, the publisher supported the team at every step. A small team of assistant editors was also appointed to further simplify the editing procedure and attain best results for the readers.

Apart from the editorial board, the designing team has also invested a significant amount of their time in understanding the subject and creating the most relevant covers. They scrutinized every image to scout for the most suitable representation of the subject and create an appropriate cover for the book.

The publishing team has been an ardent support to the editorial, designing and production team. Their endless efforts to recruit the best for this project, has resulted in the accomplishment of this book. They are a veteran in the field of academics and their pool of knowledge is as vast as their experience in printing. Their expertise and guidance has proved useful at every step. Their uncompromising quality standards have made this book an exceptional effort. Their encouragement from time to time has been an inspiration for everyone.

The publisher and the editorial board hope that this book will prove to be a valuable piece of knowledge for researchers, students, practitioners and scholars across the globe.

List of Contributors

B. Sundarabalan
Ocean Optics and Imaging Laboratory, Department of Ocean Engineering, Indian Institute of Technology Madras, Chennai-600036, India

P. Shanmugam
Ocean Optics and Imaging Laboratory, Department of Ocean Engineering, Indian Institute of Technology Madras, Chennai-600036, India

J. Kämpf
School of the Environment, Flinders University, Adelaide, Australia

V. Cardin
OGS, Istituto Nazionale di Oceanografia e di Geofisica Sperimentale, Borgo Grotta Gigante 42/c, 34018 Sgonico, Trieste, Italy

G. Civitarese
OGS, Istituto Nazionale di Oceanografia e di Geofisica Sperimentale, Borgo Grotta Gigante 42/c, 34018 Sgonico, Trieste, Italy

D. Hainbucher
CEN, Institut für Meereskunde, University of Hamburg, Bundesstraße 53, 20146 Hamburg, Germany

M. Bensi
OGS, Istituto Nazionale di Oceanografia e di Geofisica Sperimentale, Borgo Grotta Gigante 42/c, 34018 Sgonico, Trieste, Italy

A. Rubino
Università Ca' Foscari di Venezia, Dorsoduro 2137, 30121 Venice, Italy

B. Loose
Graduate School of Oceanography, University of Rhode Island, 215 South Ferry Rd., Narragansett, RI, 02882, USA

W. R. McGillis
Lamont-Doherty Earth Observatory of Columbia University, 61 Route 9W, Palisades, NY 10964, USA

D. Perovich
US Army Corps of Engineers Cold Regions Research and Engineering Laboratory, 72 Lyme Rd., Hanover, NH, 03755, USA

C. J. Zappa
Lamont-Doherty Earth Observatory of Columbia University, 61 Route 9W, Palisades, NY 10964, USA

P. Schlosser
Lamont-Doherty Earth Observatory of Columbia University, 61 Route 9W, Palisades, NY 10964, USA
Department of Earth and Environmental Sciences, Columbia University, Room 106 Geoscience Bldg., Palisades, NY 10964, USA
Department of Earth and Environmental Engineering, Columbia University, 918 Seeley Mudd Bldg., Columbia University, 500 West 120th St, New York, NY 10027, USA

M. Ablain
Collecte Localisation Satellite (CLS), Ramonville Saint-Agne, France

A. Cazenave
Laboratoire d'Etudes en Géophysique et Océanographie Spatiales (LEGOS), Toulouse, France

G. Larnicol
Collecte Localisation Satellite (CLS), Ramonville Saint-Agne, France

M. Balmaseda
European Centre for Medium-Range Weather Forecasts (ECMWF), Reading, UK

P. Cipollini
National Oceanography Centre (NOC), Southampton, UK

Y. Faugère
Collecte Localisation Satellite (CLS), Ramonville Saint-Agne, France

M. J. Fernandes
Faculdade de Ciências, Universidade do Porto, 4169-007 Porto, Portugal
Centro Interdisciplinar de Investigação Marinha e Ambiental (CIIMAR/CIMAR), Universidade do Porto, 4050-123 Porto, Portugal

O. Henry
Laboratoire d'Etudes en Géophysique et Océanographie Spatiales (LEGOS), Toulouse, France

J. A. Johannessen
Nansen Environmental and Remote Sensing Center (NERSC), Bergen, Norway

P. Knudsen
Technical University of Denmark (DTU), Lyngby, Denmark

O. Andersen
Technical University of Denmark (DTU), Lyngby, Denmark

J. Legeais
Collecte Localisation Satellite (CLS), Ramonville Saint-Agne, France

B. Meyssignac
Laboratoire d'Etudes en Géophysique et Océanographie Spatiales (LEGOS), Toulouse, France

N. Picot
Centre National d'Etudes Spatiales (CNES), Toulouse, France

M. Roca
isardSAT, Barcelona, Catalonia, Spain

S. Rudenko
Helmholtz Centre Potsdam GFZ German Research Centre for Geosciences, Telegrafenberg 14473 Potsdam, Germany

M. G. Scharffenberg
University of Hamburg, Hamburg, Germany

D. Stammer
University of Hamburg, Hamburg, Germany

G. Timms
CGI, London, UK

J. Benveniste
European Space Agency (ESA), ESRIN, Frascati, Italy

L. Biermann
Sea Mammal Research Unit, Scottish Oceans Institute, St. Andrews, UK

C. Guinet
Centre National de la Recherche Scientifique, Centre d'Etudes Biologiques de Chizé, Villiers en Bois, France

M. Bester
Mammal Research Institute, Department of Zoology and Entomology, University of Pretoria, Pretoria, South Africa

A. Brierley
Pelagic Ecology Research Group, Scottish Oceans Institute, St. Andrews, UK

L. Boehme
Sea Mammal Research Unit, Scottish Oceans Institute, St. Andrews, UK

M. Álvarez
IEO - Instituto Español de Oceanografía, Apd. 130, A Coruña, 15001, Spain

H. Sanleón-Bartolomé
IEO - Instituto Español de Oceanografía, Apd. 130, A Coruña, 15001, Spain

T. Tanhua
GEOMAR Helmholtz Centre for Ocean Research Kiel, Düsternbrooker Weg 20, 24105 Kiel, Germany

L. Mintrop
MARIANDA, Tulpenweg 28, 24145 Kiel, Germany

A. Luchetta
CNR ISMAR - Consiglio Nazionale delle Ricerche, Istituto di Scienze Marine, viale R. Gessi 2, 34123, Trieste, Italy

C. Cantoni
CNR ISMAR - Consiglio Nazionale delle Ricerche, Istituto di Scienze Marine, viale R. Gessi 2, 34123, Trieste, Italy

K. Schroeder
CNR ISMAR - Arsenale, Tesa 104, Castello 2737/F, 30122 Venice, Italy

G. Civitarese
OGS - Istituto Nazionale di Oceanografia e di Geofisica Sperimentale, Borgo Grotta Gigante 42/c, 34018 Sgonico, Trieste, Italy

R. Bozzano
National Research Council of Italy (CNR-ISSIA) - Via de Marini 6, 16149, Genova, Italy

E. Fanelli
Marine Environment Research Centre (ENEA) - Santa Teresa, Pozzuolo di Lerici 19032, Italy

S. Pensieri
National Research Council of Italy (CNR-ISSIA) - Via de Marini 6, 16149, Genova, Italy

P. Picco
Marine Environment Research Centre (ENEA) - Santa Teresa, Pozzuolo di Lerici 19032, Italy

M. E. Schiano
National Research Council of Italy (CNR-ISMAR) - Via de Marini 6, 16149, Genova, Italy

A. Oviedo
Institute of Environmental Science and Technology (ICTA), Universitat Autònoma de Barcelona (UAB), 08193 Bellaterra, Spain

P. Ziveri
Institute of Environmental Science and Technology (ICTA), Universitat Autònoma de Barcelona (UAB), 08193 Bellaterra, Spain
Earth & Climate Cluster, Department of Earth Sciences, FALW, Vrije Universiteit Amsterdam,
FALW, HV1081 Amsterdam, the Netherlands
ICREA, Institució Catalana de Recerca i Estudis Avançats, Barcelona, Spain

M. Álvarez
IEO - Instituto Español de Oceanografia, Apd. 130, A Coruña 15001, Spain

T. Tanhua
4GEOMAR Helmholtz-Zentrum für Ozeanforschung Kiel, Marine Biogeochemistry, Düsternbrooker Weg 20, 24105 Kiel, Germany

A. Bracher
Alfred Wegener Institute Helmholtz Centre for Polar and Marine Research, Bussestraße 24, 27570 Bremerhaven, Germany
Institute of Environmental Physics, University of Bremen, Otto-Hahn-Allee 1, 28359 Bremen, Germany

M. H. Taylor
Leibniz Center for Tropical Marine Ecology, Fahrenheitstraße 6, 28359 Bremen, Germany

B. Taylor
Alfred Wegener Institute Helmholtz Centre for Polar and Marine Research, Bussestraße 24, 27570 Bremerhaven, Germany

T. Dinter
Alfred Wegener Institute Helmholtz Centre for Polar and Marine Research, Bussestraße 24, 27570 Bremerhaven, Germany
Institute of Environmental Physics, University of Bremen, Otto-Hahn-Allee 1, 28359 Bremen, Germany

R. Röttgers
Helmholtz Zentrum Geesthacht Center of Materials and Coastal Research, Max-Planck-Str., 21502 Geesthacht, Germany

F. Steinmetz
HYGEOS, Euratechnologies, 165 Avenue de Bretagne, 59000 Lille, France

O. S. Pokrovsky
Georesources and Environment Toulouse GET, UMR5563, CNRS, Université Paul Sabatier, 14 Avenue Edouard Belin, 31400 Toulouse, France
BIO-GEO-CLIM Laboratory, Tomsk State University, Tomsk, Russia

L. S. Shirokova
Georesources and Environment Toulouse GET, UMR5563, CNRS, Université Paul Sabatier, 14 Avenue Edouard Belin, 31400 Toulouse, France
Institute of Ecological Problems of the Northern Regions, Ural Branch of the Russian Academy of Science, 23 Naberezhnaja Sev. Dviny, Arkhangelsk, Russia

J. Viers
Georesources and Environment Toulouse GET, UMR5563, CNRS, Université Paul Sabatier, 14 Avenue Edouard Belin, 31400 Toulouse, France

V. V. Gordeev
P.P. Shirshov Institute of Oceanology of the Russian Academy of Science, 36 Nakhimovsky Prospekt,

V. P. Shevchenko
P.P. Shirshov Institute of Oceanology of the Russian Academy of Science, 36 Nakhimovsky Prospekt,

A. V. Chupakov
Institute of Ecological Problems of the Northern Regions, Ural Branch of the Russian Academy of Science, 23 Naberezhnaja Sev. Dviny, Arkhangelsk, Russia

T. Y. Vorobieva
Institute of Ecological Problems of the Northern Regions, Ural Branch of the Russian Academy of Science, 23 Naberezhnaja Sev. Dviny, Arkhangelsk, Russia

F. Candaudap
Georesources and Environment Toulouse GET, UMR5563, CNRS, Université Paul Sabatier, 14 Avenue Edouard Belin, 31400 Toulouse, France

C. Causserand
Georesources and Environment Toulouse GET, UMR5563, CNRS, Université Paul Sabatier, 14 Avenue Edouard Belin, 31400 Toulouse, France

A. Lanzanova
Georesources and Environment Toulouse GET, UMR5563, CNRS, Université Paul Sabatier, 14 Avenue Edouard Belin, 31400 Toulouse, France

C. Zouiten
Georesources and Environment Toulouse GET, UMR5563, CNRS, Université Paul Sabatier, 14 Avenue Edouard Belin, 31400 Toulouse, France

H. Nefzi
Faculté des Sciences de Tunis, Université Tunis ElManar, Tunisia

D. Elhmaidi
Faculté des Sciences de Tunis, Université Tunis ElManar, Tunisia

X. Carton
Laboratoire de Physique des Océans-Université Occidentale de Bretagne Brest, France

E. Aas
Department of Geosciences, University of Oslo, Gaustadalleen 21, 0349 Oslo, Norway

J. Høkedal
Narvik University College, Lodve Langesgt. 2, 8508 Narvik, Norway

K. Sørensen
Norwegian Institute for Water Research, Gaustadalleen 21, 0349 Oslo, Norway

J. Hill
Applied Modelling and Computation Group, Department of Earth Science and Engineering, Imperial College London, SW7 2AZ, UK

E. E. Popova
National Oceanography Centre, Southampton, University of Southampton Waterfront Campus, European Way

D. A. Ham
Applied Mathematics and Mathematical Physics, Department of Mathematics, Imperial College London, SW7 2AZ, UK

M. D. Piggott
Applied Modelling and Computation Group, Department of Earth Science and Engineering, Imperial College London, SW7 2AZ, UK
Grantham Institute for Climate Change, Imperial College London, SW7 2AZ, UK

M. Srokosz
National Oceanography Centre, Southampton, University of Southampton Waterfront Campus, European Way Southampton, SO14 3ZH, UK

K. Lohmann
Max Planck Institute for Meteorology, Hamburg, Germany

J. H. Jungclaus
Max Planck Institute for Meteorology, Hamburg, Germany

D. Matei
Max Planck Institute for Meteorology, Hamburg, Germany

J. Mignot
LOCEAN, Institute Pierre Simon Laplace, University Pierre and Marie Curie, Paris, France
Climate and Environmental Physics, and Oeschger Centre of Climate Change Research, University of Bern, Bern, Switzerland

M. Menary
Met Office Hadley Center, Exeter, UK

H. R. Langehaug
Nansen Environmental and Remote Sensing Center, Bergen, Norway
Bjerknes Centre for Climate Research, Bergen, Norway

J. Ba7, Y. Gao
Nansen Environmental and Remote Sensing Center, Bergen, Norway
Bjerknes Centre for Climate Research, Bergen, Norway

O. H. Otterå
Bjerknes Centre for Climate Research, Bergen, Norway
Uni Research, Bergen, Norway

W. Park
Helmholtz Centre for Ocean Research Kiel, Kiel, Germany

S. Lorenz
Max Planck Institute for Meteorology, Hamburg, Germany

H. Herrera-Cervantes
Unidad La Paz, Centro de Investigación Científica y de Educación Superior de Ensenada (CICESE), La Paz, Baja California Sur 23050, México

S. E. Lluch-Cota
Centro de Investigaciones Biológicas del Noroeste (CIBNOR), Instituto Politécnico Nacional 195, La Paz, Baja California Sur 23096, México

D. B. Lluch-Cota
Centro de Investigaciones Biológicas del Noroeste (CIBNOR), Instituto Politécnico Nacional 195, La Paz, Baja California Sur 23096, México

G. Gutiérrez-de-Velasco
Departamento de Física, Universidad de Guadalajara, México

T. Howard
Met Office Hadley Centre, FitzRoy Road, Exeter, EX1 3PB, UK

A. K. Pardaens
Met Office Hadley Centre, FitzRoy Road, Exeter, EX1 3PB, UK

J. L. Bamber
Bristol Glaciology Centre, School of Geographical Sciences, University of Bristol, University Road, Bristol, BS8 1SS, UK

J. Ridley
Met Office Hadley Centre, FitzRoy Road, Exeter, EX1 3PB, UK

G. Spada
Dipartimento di Scienze di Base e Fondamenti, Urbino University "Carlo Bo", Via Santa Chiara, 27, 61029 Urbino (PU), Italy

R. T. W. L. Hurkmans
Bristol Glaciology Centre, School of Geographical Sciences, University of Bristol, University Road, Bristol, BS8 1SS, UK

J. A. Lowe
Met Office Hadley Centre, FitzRoy Road, Exeter, EX1 3PB, UK

D. Vaughan
British Antarctic Survey, Madingley Road, High Cross, Cambridge, CB3 0ET, UK